Cleaner Production

Francisco José Gomes da Silva ·
Ronny Miguel Gouveia

Cleaner Production

Toward a Better Future

Springer

Francisco José Gomes da Silva
ISEP—School of Engineering
Polytechnic of Porto
Porto, Portugal

Ronny Miguel Gouveia
ISEP—School of Engineering
Polytechnic of Porto
Porto, Portugal

ISBN 978-3-030-23164-4 ISBN 978-3-030-23165-1 (eBook)
https://doi.org/10.1007/978-3-030-23165-1

This Springer imprint is published by the registered company Springer Nature Switzerland AG
The registered company address is: Gewerbestrasse 11, 6330 Cham, Switzerland

To my daughter and wife, and to the memory of my father.
Francisco José Gomes da Silva

To my wife Elisabete, my parents and brother for all the patience and unconditional support. Also, for all of those who want to change the world into a better place.
Ronny Miguel Gouveia

Foreword

Nowadays, the need to change the worldview paradigm within society currently operates has become an extensively discussed topic.

The essential challenge of sustainable consumption and production is how to achieve economic development without environmental degradation, in order to operate within the limits of the planet's ecosystems. Meeting this challenge will require technological innovation, rethinking current business models and political determination.

This book is focused upon various subjects related with Cleaner Production, such as the early stages and evolution of Cleaner Production, effects of business activity on the environment, effects of industrial pollution on public health and human behavior, key challenges in sustainable consumption, environmental managing systems concepts' or drivers and barriers to Cleaner Production.

The main focus is the way these attitudes and developments are evolving, how they can help us to prevent or adapt to climate change and how these approaches are likely to evolve in the next decades. These approaches are addressed in ten chapters: (1) Cleaner Production Definition and Evolution, (2) Cleaner Production Main Concept and History, (3) Global Population Growth and Industrial Impact on the Environment (4) Sustainable Consumption, (5) Regulations and Environmental Management Systems, (6) Cleaner Production Tools and Environmental Management Practices, (7) Practices on Cleaner Production and Sustainability, (8) Sustainable Production Cases, (9) Drivers and Barriers to Cleaner Production and (10) Tracking Environmental Performance.

The book provides different, but complimentary approaches to help industry and society in advancing on their paths towards sustainability. Initiatives and challenges are included, which systematically address problems affecting raw material changes, technological modifications, product and policy changes.

This work was conceptualized for an audience of graduate students mainly; however, it can also be consulted by engineers and company managers who search for the state-of-the-art applications on Cleaner Production.

Maria João Viamonte, Ph.D.
ISEP's President
Porto, Portugal

Preface

We are all aware that humankind has mistreated the planet we live in, and that natural phenomena are occurring that were not common. However, since this is not a constant concern, it tends to be ignored almost always. Sometimes we are confronted with phenomena in which Nature shows "anger" in the way it has been treated. Here we reflect on the problems we induce in the environment, take measures that are easy to take and can be easily circumvented or ignored, and continue to the new phenomenon of Nature, and the whole cycle repeats itself.

Although humanity is endowed with intelligence, it has not yet been able to put it into practice, in order to establish an adequate balance between quality of life, adequate income, and sustainable use of resources.

In fact, politicians need the economy to grow, so that taxes generate more revenue, and they can develop the policies they want. For the economy to grow, consumption must also grow, either internally or externally to a given economic system or country. The growth of consumption generates the need for greater production and, consequently, greater consumption of resources. Both the production and the consumption of resources, generate problems for the environment, because there is still a balance between these factors. This creates problems for the environment, culminating in severe phenomena of Nature, which manifest themselves in the form of storms, hurricanes, earthquakes, tsunamis, etc. At the time this book was written, one more of these phenomena was recorded: the Idai storm in Mozambique (2019).

Although regulation is an important factor in preventing further environmental damage, we must be aware that improving environmental conditions just depends on us. To do so, we will have to be much less sensitive to marketing actions, curbing consumption. We will have to be much more selective in the products we consume, and we will have to have increased concerns about how they are produced. Are we prepared for this?

In fact, science plays an extremely important role in the development of new, more environmentally friendly production techniques, as well as in the development of tools for an adequate assessment of progress in improving environmental management. However, one of the main factors that need to be properly managed is education and awareness of all of humanity for these problems, inducing citizens with increased environmental concerns, creating increased pressure on producers to do so. Also, at the time of writing this book, there has been a drastic change in consumption habits related to energy used in motor vehicles. The demand by diesel-powered cars is clearly decreasing, while gasoline-powered, hybrid and electric vehicles are growing sharply. This movement will have to be applied in many other fields, in order to make our planet a pleasant and safe place to live, ensuring that future generations will be able to enjoy the planet even better than us.

Acknowledgements The authors would like to acknowledge the following people, agencies and organizations for kindly allowing the use of several images and tables throughout this book: United Nations Environment Programme Agency, theoildrum.com, Jean-Marc Jancovici, NASA—National Aeronautics and Space Administration, World Air Quality Index Project—AQICN.Org, European Commission, OECD—Organisation For Economic Co-Operation and Development, United Nations University, International Council of Chemical Associations, Finnish Environment Institute, Energy Information Administration and Elsevier.

Porto, Portugal Francisco J. G. Silva
Porto, Portugal Ronny Miguel Gouveia

Introduction

The environment has been a serious concern in the last decades, with promises still remaining unfulfilled while peculiar climatic phenomena and other changing expressions are felt year after year. In fact, the Earth has known various climate changes throughout its history, with the advance and retreat of seven glacial cycles during the last 650,000 years, with the sudden culmination of the last ice age circa 7000 year ago, giving rise to the human civilization and the modern climate era (NASA, 2017). Despite these cycles and corresponding fluctuations of the carbon dioxide level, this indicator had never exceeded the barrier of 300 parts per million before the beginning of the 20th century, having just recently surpassed the unprecedented 400 parts per million mark. The growth of this indicator is particularly severe after mid of the 20th century, being coincident with the world's Gross Domestic Product (GDP) per capita evolution, which quadrupled between 1950 and 2018. In the meantime, the world's population and corresponding needs continues to grow. However, there is not a direct relation between the GDP per capita arising and population growth, as the number of processed products related to food and other goods has been permanently increasing alongside with productivity improvements which have enabled a faster GDP per capita growth comparatively to the increase in population, mainly in developed countries. On the other hand, several paradigms have changed immensely. The movement of people based essentially on public transportation in the first half of the 20th century has changed to the use of personal vehicles powered by thermal engines in the second half of the 20th century, drastically increasing CO_2 emissions. Moreover, the use of electricity at home harshly enlarged due to the arrival of numerous housewares able to make easier and more comfortable the domestic life. This increasing need of electric power augmented the number of power generation facilities, with a consequent increase of CO_2 emissions, until the advent of the renewable power generation sources. Furthermore, in the second half of the 20th century, there was a significant enlargement of the chemical products available for the most diverse application, such as laundry detergents, hairspray, insecticides, and so on. The conservation of some food products and the processing of others in order to widen market offerings powered the chemical industry with a consequent environmental impact, leading to the production of dangerous effluents and management of processes with increasing nefarious impact on the environment. More recently, the globalization of companies and enhancing network action have been responsible for the increase in movement of semi-products or products, growing the need for transportation and consequently blowing up CO_2 emissions.

In fact, the GDP per capita growth effectively felt in the last seventy years brought an improved quality of life, mainly provided by a drastic technological development. However, even though the population's general well-being has increased, thoughtful concerns surround us, regarding pollution, global warming, glacial retreat, ocean acidification and extreme climate events, among others, preventing a full use of the improved social conditions. Furthermore, technological developments have also been made in terms of pollution control, but this technology is expensive and some countries are not exerting the sufficient pressure over companies in order to adopt the best practices in this field.

Contents

Abbreviations

3R	Reduce, Reuse, Recycle
Al_2O_3	Aluminum Oxide
ANP	Analytical Network Process
AOX	Adsorbable Organic Halogens
APRSCP	Asia-Pacific Roundtable on Sustainable Consumption and Production
AQI	Air Quality Index
ATP	Adenosine Triphosphate
BEV	Battery Electric Vehicles
BFBR	Bubbling Fluidized Bed Reactor
BLL	Blood Lead Levels
BOD	Biochemical Oxygen Demand
BREEAM	Building Research Establishment Environmental Assessment Methodology
BTA	Best Available Technique
BTYK	Supreme Council for Science and Technology
$CaCO_3$	Calcium Carbonates
CASBEE	Comprehensive Assessment System for Built Environment Efficiency
CCN	Consumer Citizen Network
CCS	Carbon Capture and Storage
CE	Circular Economy
CED	Cumulative Energy Demand
CERCLA	Comprehensive Environmental Response, Compensation and Liability Act
CERES	Coalition for Environmental Responsible Economies
CF	Carbon Footprint
CFED	Cumulative Fossil Energy Consumption
CH_4	Methane
Cl	Chloride

ClO_2	Chlorine Dioxide
CNCPC	Chinese National Cleaner Production Center
CO	Carbon Monoxide
CO_2	Carbon Dioxide
COD	Chemical Oxygen Demand
COHb	Carboxyhemoglobin
$CoO\text{-}MoO_3$	Cobalt Molybdenum Oxide
CP	Cleaner Production
CPC	Cleaner Production Centers
Cr(VI)	Hexavalent Chromium
CRED	Cumulative Renewable Energy Demand
CSIR	Council for Scientific and Industrial Research
CSPD	Checklist for Sustainable Product Development
CSR	Corporate Social Responsibility
CWF	Conscious Water Footprint
DEIE	Developing and Emerging Industrial Economies
DESIRE	Demonstrations in Small Industries for Reducing Waste
DfA	Design for Assembly
DfD	Design for Deconstruction
DfE	Design for the Environment
DfM	Design for Manufacturing
DfS	Design for Sustainability
DfX	Design for X
DOE	Design of Experiments
DPF	Date Palm Fiber
EC	European Commission
ECF	Elemental Chlorine-Free
EcoM2	Eco-Design Maturity Model
Eco-VA	Eco-Value Analysis
ECQFD	Environmental Conscious Qfd
EEA	European Environment Agency
E-FMEA	Environmental Failure Mode Effects Analysis
EI	Eco-Innovation
EIFA	Environmental Impact and Factor Analysis
EIQ	Environment Impact Quantification
EMS	Environmental Management System
EOD	Environmental Objectives Deployment
EP	Eutrophication Potential
EPA	Environmental Protection Agency
EPACI	Environmental Performance Assessment Composite Index
EPHEIS	European Information System
EPIP	Environmental Performance of Industrial Processes
EPLC	Environmental Product Life Cycle Matrix
EPR	Extended Producer Responsibility
E-QFD	Environmental-QFD

ERPA	Environmentally Responsible Product Assessment Matrix
ESGP	Earth System Governance Project
EST	Environmentally Sound Technologies
EU	European Union
FAHP	Fuzzy Analytical Hierarchy Process
FAO	Food and Agriculture Organization
FBSBR	Fixed-Bed Sequence Batch Reactor
GA	Genetic Algorithm
GEC	Global Environment Centre Foundation
GEF	Global Environmental Facility
GFRP	Glass-Fiber Reinforced Plastics
GGGI	Global Green Growth Institute
GHG	Greenhouse Gases
GLO	Ground Level Ozone
GQFD	Green-QFD
GWP	Global Warming Potential
H^+	Hydrogen
H_2O	Water
H_2SO_4	Sulfuric Acid
Hb	Hemoglobin
HCl	Hydrochloric Acid
HDPE	High Density Polyethylene
HF	Hydrofluoric Acid
HOE	House of Ecology
IATF	International Automotive Task Force
ICAST	Institute for Communication and Analysis of Science and Technology
ICCM	International Conference on Chemicals Management
IEE	Industrial Energy Efficiency
ILO	International Labor Organization
ILUC	Indirect Land Use Change
IPCC	Intergovernmental Panel on Climate Change
IPPC	Integrated Pollution Prevention and Control Directive
ISM	Interpretive Structural Modeling
ISO	International Organization for Standardization
IT	Information Systems
ITC	International Trade Centre
IUCN	International Union for Conservation of Nature
IV	Ultraviolet
KPI	Key Performance Indicator
LCA	Life-Cycle Assessment
LCC	Life Cycle Costing
LCCA	Life Cycle Costs Analysis
LCECA	Life Cycle Environmental Cost Analysis
LC-HoQ	Life Cycle House of Quality

LCM	Life Cycle Management
LEED	Leadership in Energy and Environmental Design
LPC	Life Cycle Planning
LUC	Direct Land Use Change
MCDM	Multi-Criteria Decision-Making
MDG	Millennium Development Goals
MECO	Materials, Energy, Chemicals and Others
MEIM	Minimum Environmental Impact Methodology
MILP	Mixed Integer Linear Programming
MSW	Municipal Solid Waste
MVA	Manufacturing Value Added
MVOC	Microbial Volatile Organic Compounds
N_2	Dinitrogen
N_2O	Nitrous Oxide
Na	Sodium
NACIPPE	National Cleaner Industrial Production Project of Ethiopia
NAICS	North American Industry Classification System
NASA	National Aeronautics and Space Administration
NCPC	National Cleaner Production Center
NEP	New Environmental Paradigm
NEPA	National Environment Policy Act
NETT21	New Environmental Technology Transfer in the 21st Century
NGO	Non-Governmental Organizations
NH_4	Ammonia
NO_2	Nitrogen Dioxide
NO_x	Nitrogen Oxides
NP	Potential Nutrition
NPC	National Productivity Council
NPPR	National Pollution Prevention Roundtable
O_2	Oxygen
O_3	Ozone
OCEES	Oxford Center for Environment, Ethics and Society
OCSC	Oxford Commission on Sustainable Consumption
ODL	Oxygen Delignification
ODP	Ozone Depleting Potential
OECD	Organisation for Economic Co-Operation and Development
OEM	Original Equipment Manufacturers
OGT	On-Site Green Techniques
OPM	Oil Point Method
PAO	Poly α-Olefin
Pb	Lead
$PbCO_3$	Lead Carbonate
PDCA	Plan-Do-Check-Act
PE	Person Equivalent
PEMFC	Polymeric Electrolyte Membrane Fuel Cells

PEMSEA	Partnerships in Environmental Management for The Seas of East Asia
PEPA	Process Environmental Performance Assessment
PET	Polyethylene Terephthalate
PGPR	Plant Growth Promoting Rhizobacteria
PILOT	Product Investigation, Learning and Optimization Tool
PM	Particulate Matters
PM10	Particulate Matter of Less Than 10 Microns in Diameter
PM2.5	Particulate Matter of Less Than 2.5 Microns in Diameter
POCP	Photochemical Ozone Creation Potential
PP	Polypropylene
PPA	Pollution Prevention Act
PPIN	Pollution Prevention Information Network
PS	Polystyrene
PSI	Product Sustainability Index
PSO	Particle Swarm Optimization
PVC	Polyvinyl Chloride
QFDE	Quality Function Deployment for Environment
RAILS	Readiness Assessment for Implementing DFE Strategies
RECP	UNIDO-UNEP Joint Global Resource Efficient and Cleaner Production
RH	Relative Humidity
RoHS	Restriction of Hazardous Substances Directive
RPUF	Rigid Polyurethane Foams
RR	Relative Risk
RRM	Renewable Raw Materials
SBR	Sequencing Batch Reactor
SBTool	Sustainable Building Tool
SCP	Sustainable Consumption and Production
SCR	Selective Catalytic Reactions
SDG	Sustainable Development Goals
SECO	Swiss Secretary for Economic Affairs
SEPA	State Environmental Protection Administration
S-LCA	Social Life Cycle Assessment
SLCA	Streamlined Life Cycle Assessment
SMART	Sustainable, Measurable, Responsible, Replicable and Time-Sensitive
SME	Small and Medium Enterprise
SO_2	Sulfur Dioxide
SO_3	Sulfur Trioxide
SOFC	Solid Oxide Fuel Cells
SO_x	Sulfur Oxides
TAP	Technical Assistance Program/Provider
TCCD	8-Tetrachlorodibenzo-P-Dioxin
TEST	Transfer of Environmentally Sound Technology

TiO_2	Titanium Dioxide
TRI	Toxic Release Inventory
TTGV	Technology Development Foundation of Turkey
TUBITAK	The Scientific and Technological Research Council of Turkey
TWC	Three-Way Catalyst
U.S.A.	United States of America
UNCTAD	United Nations Conference on Trade and Development
UNDO	United Nations Industrial Development Organization
UNEP	United Nations Environment Programme
UNEP-DTIE	United Nations Environmental Programme—Division of Technology, Industry and Environment
UNEP-IEO	United Nations Environmental Programme—Industry and Environment Office
UNEP-IETC	United Nations Environmental Programme—International Environmental Technology Center
UNESCO	United Nations Educational, Scientific and Cultural Organization
UNGC	United Nations Global Compact
US EPA	United States Environmental Protection Agency
USAEP	United States Asia Environment Partnership
USAID	United States Agency for International Development
USDA	United States Department of Agriculture
VAT	Value Added Tax
VOC	Volatile Organic Compounds
WBCSD	World Business Council for Sustainable Development
WCPS	World Cleaner Production Society
WEC	World Environmental Center
WEEE	Waste Electrical and Electronic Equipment
WEF	Water Emission Footprint
WF	Water Footprint
WHO	World Health Organization
WMC	Waste Minimization Circles
WTP	Water Treatment Plant
ZED	Zero Effect, Zero Defect

Chapter 1
Cleaner Production Definition and Evolution

Abstract Cleaner Production has been defined differently throughout the past decades, mostly due to the advances of technologies and the learning with past mistakes. Cleaner Production, lean manufacturing, and pollution prevention are sometimes wrongfully used as synonyms and, however, represent different approaches to a similar problem. The definition and evolution of the Cleaner Production concept are discussed, briefly framing each evolutionary step of the term and their authors. An exposure and comparison between wrongfully thought synonyms of Cleaner Production such as pollution prevention and lean manufacturing are made.

Keywords Cleaner production · Cleaner production definition · Pollution prevention · Cleaner production evolution

Cleaner Production (CP) and pollution prevention are synonyms, although the sense of the second seems broader than the first. Indeed, pollution prevention is usually used in the USA, as this term has its roots linked to the first approach done, with the Pollution Prevent Act, in 1990. The terminology CP is widely used and has taken different definitions, depending on the authors or institutions involved.

1.1 Definition of Cleaner Production

The first definition of Cleaner Production (CP) known was launched in 1990 by UNEP-IEO (United Nations Environmental Program—Industry and Environment Office) and defined CP as *the continuous application of an integrated preventive environmental strategy applied to processes, products, and services to increase overall efficiency and reduce risks to humans and the environment*. This definition constituted the first ideas around the CP concept, being perceived as broad enough to overcome the environmental concerns identified and regarding the scientific knowledge at that time.

In 1994, the 1st European Roundtable on Cleaner Production Programs (Hens et al. 2018) issued a new definition for CP as *the conceptual and procedural approach to production that demands that all phases of the life cycle of a product or of a process*

F. J. Gomes da Silva and R. M. Gouveia, *Cleaner Production*,
https://doi.org/10.1007/978-3-030-23165-1_1

should be addressed with the objective of prevention or the minimization of short and long-term risks to humans and the environment. A total societal commitment is required for making effective this comprehensive approach achieving the goal of a sustainable society.

In 1998, through the International Declaration on Cleaner Production, the UNEP promoted slight changes in the previous description issue in 1990, defining CP as *the continuous application of an integrated, preventive strategy applied to processes, products, and services in pursuit of economic, social, health, safety, and environmental benefits*. This new definition brought as main difference forsaking a policy of risk reduction that can be viewed as a passive action, by a more active one, which seeks for better conditions of economic activities in order to obtain real benefits for human beings and the environment. This increased requirement levels, not only for avoiding harmful actions for the environment, but as well as for promoting an active strategic management and a way of thinking on how to contribute to human well-being in a less polluted environment.

Also, in 1998, Fresner (1998) defined CP in his work as *a preventive strategy to minimize the impact of production and products on the environment. The principal actors are the companies, which control the production processes. They are influenced strongly by their customers and politics*. This author pointed out clearly the players responsible for acting accordingly to the necessary rules, pointing out that customers and governments must have a decisive influence on the companies' behavior by two ways: the customer must select products produced or provided by companies that respect the environment and governments must issue the necessary legislation and strictly enforce its implementation, respectively. This concept was pioneer at that time and marked the evolution of the definition in a very positive way.

Cleaner Production was also briefly defined by Kjaerheim (2005) as *an effective way of obtaining improved material utilization, reduced energy consumption, and lower emission levels. It also motivates positive preventive action and promotes a holistic view of resources, production, economy, and the environment.*

Fresner revisited the prior CP definition established by himself in 1998, now through a work developed together with Yaacoob (Yaacoob and Fresner 2006), stating that CP *is a preventive, company-specific environmental protection initiative. It is intended to minimize waste and emissions and maximize product output*. This setting connects the concept of CP to productivity but seems to represent a step backward in relation to the definition previously established by the same author, since it does not promote the same required involvement with stakeholders. Thus, this definition did not add anything interesting to the prior definitions already issued at that time.

Trying to summarize the complex definition of CP established by UNEP, Nilsson et al. (2007) refer that *Cleaner Production is a "win-win" or even "win-win-win" strategy. It protects the environment, the consumer, and the worker while improving industrial efficiency, profitability, and competitiveness*.

Also, in 2007, Glavic and Lukman (2007) provided a new definition of CP as *a systematically organized approach to production activities, which has positive effects on the environment. These activities encompass resource use minimization, improved eco-efficiency, and source reduction, in order to improve the environmental protection and to reduce risks to living organisms. It can be applied to processes used in any industrial sector and to products themselves (cleaner products).* This definition is principally focused on industrial activities, forgetting the need to extend the concept to the services sector. Moreover, like others already reviewed in this chapter, this definition does not create the atmosphere and necessary integration of all stakeholders as well as consider the pressure that markets and consumers must do in order to produce effective changes in product design and corresponding manufacturing steps. Hence, this definition can be considered as limitative regarding other definitions already issued at that time, not giving the desired impulse in order to enlarge the concept and its implementation.

The main definition was provided in 2011 by the UNEP-DTIE (United Nations Environmental Programme, Division of Technology, Industry and Environment) as follows: *Cleaner Production relies on the continuous application of an integrated preventive environmental strategy applied to processes, products, and services to increase overall efficiency and reduce risks to humans and the environment.* This definition carries us to the foundations of what CP should be: the minimization of waste generation in production processes and products, improving as well resource efficiency. However, this short definition is based on a complex previous UNEP definition that can be found in the Nilsson et al. (2007) book as follows: *Cleaner Production is the continuous application of an integrated preventive environmental strategy to processes, products, and services to increase overall efficiency and reduce risks to humans and the environment. Cleaner Production can be applied to the processes used in any industry, to products themselves, and to various services provided in society. For production processes, Cleaner Production results from one or a combination of a number of measures as conserving raw materials, water and energy; eliminating toxic and dangerous raw materials; and reducing the quantity and toxicity of all emissions and wastes at source during the production process. For products, Cleaner Production aims to reduce the environmental, health, and safety impacts of products over their entire life cycles, from raw materials extraction, through manufacturing and use, to the "ultimate" disposal of the product. For services, Cleaner Production implies incorporating environmental concerns into designing and delivering services.* This is probably the most complete definition, but is a little dispersing regarding its focus.

Also, the OECD (Organisation for Economic Co-operation and Development) has a proper CP definition, which is the following: *Technologies that extract and use natural resources as efficiently as possible in all stages of their lives; that generate products with reduced or no potentially harmful components; that minimize releases to the air, water, and soil during fabrication and use of the product; and that produce durable products which can be recovered or recycled as far as possible; output is achieved with as little energy input as is possible.* This definition can be considered as more effective, accurate, and complete, focused not only on the resources spent and emissions/effluents/waste generated in the initial production process but also in the product life cycle, highlighting as well energy concerns.

The European Commission (EC 2017) also defined Cleaner Production as *technologies that extract and use natural resources as efficiently as possible in all stages of their lives; that generate products with reduced or no potentially harmful components; that minimize releases to air, water, and soil during fabrication and use of the product; and that produce durable products which can be recovered or recycled as far as possible; output is achieved with as little energy input as is possible.* This definition is very robust, just missing the stakeholder's commitment.

Regarding the definitions previously stated, there are different approaches to the CP definition with different focus, despite performing the same purpose. When taking into account the term "technologies" regarding manufacturing processes, it can give the idea that CP intends to create the adequate means to avoid pollution anyway. However, cleaner technology is drastically different from technologies that intend pollution prevention, such as end-of-pipe engineering solutions. In this way, the European Union (EU) is peremptory in the differentiation of approaches: *End-of-pipe solutions do not usually result in efficiency or productivity gains, therefore representing a pure cost to the firms. Cleaner technology on the other hand improves process efficiency. Furthermore, cleaner technology usually reduces pollution emissions to all media instead from one to the other.* In fact, end-of-pipe engineering solutions represent an additional cost for the product while the idea of CP is to rethink products and processes on all levels, from design and raw materials up until product end-of-life, becoming more competitive as well as reducing negative impact to the environment as much as possible (EC 2017). Thus, the term "technologies" was divided into two segments: "Hard Technologies," directly related with end-of-pipe engineering solutions, and "Soft Technologies," corresponding to the methods of operation and management, able to ensure an effective reduction in resources consumption.

Some recent definitions can be also found on the Internet, such as the one issued by SD Features (2017), stating that *CP is a preventative approach to managing the environmental impacts of business processes and products. CP uses changes in technology, processes, resources, or practices to reduce waste, environmental, and health risks; minimize environmental damage; use energy and resources more efficiently; increase business profitability and competitiveness; and increase the efficiency of production processes. Cleaner production is applicable to all businesses, regardless of size or type.* This definition is wide enough to include many significant subjects related to Cleaner Production and environment, but leaves out other important aspects such as the stakeholders' engagement and their power over the manufacturers and service providers in order to always keep in mind environmental concerns. Moreover, and regarding that it is a 2017 definition, it does not include the product life cycle, which can be found in other older definitions. Thus, despite being a relatively recent definition, there are some mistakes which make this definition incomplete.

Regarding the characterization provided in 2017 by the Kaunas University of Technology Lithuania (KTU 2017), *CP can be defined as a preventive, integrated continuous strategy applied to products, processes, and services, to enhance efficiency, which improves environmental performance and reduces costs.* This definition can implicitly contain certain aspects that are of extreme importance for the overall concept, but is too generic in its form, as it does not explicitly refer to the stakeholders' importance in the whole improving process. The wide range achieved by the early part of the definition assumes the inclusion of some tools such as life cycle assessment and recyclability in the earlier stage of the product definition, which will cross the whole life of the product. However, there is another important aspect to consider in this definition: The work to be done should always be understood as incomplete, so that there is a permanent improvement in environmental performance, being this process also accomplished by a cost reduction, ensuring product competitiveness.

It can be stated that both pollution prevention and Cleaner Production underline environmental management by saving resources, rather than pollution control approaches. Similarly, for pollution prevention, CP cannot be considered as a finished process, but something that requires a constant evolution by the introduction of current technology, inventive product development, new brand materials, and innovative management concepts. The World Business Council for Sustainable Development (WBCSD) adopted the first part of the initial UNEP definition for Cleaner Production as follows: *Cleaner Production is the continuous application of an integrated preventive environmental strategy applied to processes, products, and services to increase eco-efficiency and reduce risks for humans and the environment*. This definition is short enough to produce the required effect in the mind of managers and people in general, and presents a clear robustness, pointing out that it is a never-ending process, which needs to involve the entire structure of companies in order to follow an integrated strategy in an eco-efficient way that preserves the environment and humanity.

Many authors have addressed the CP issue, providing other short definitions; however, it will be difficult to establish a better definition than the one stated by WBCSD referred above.

In the Roy M. Harrison's book *Pollution: Causes, Effects and Control* (Harrison 2001), Cleaner Production is defined as *means of providing a human benefit, which, overall, uses less resources and causes less environmental damage than alternative means which is economically competitive*. This point of view presents a broadband idea that goes from the human well-being to the business competitiveness, touching on an important issue: resource saving. Moreover, this definition also emphasizes that business competition must pay attention to environmental concerns, being a very interesting point of view.

Sometimes, Cleaner Production and lean manufacturing tend to be realized as similar concepts. However, their purposes are a little different: The first one can be understood as an approach to reduce environmental impacts, while lean is essentially focused on the reduction in costs and time with immediate repercussions on the product and market (Song and Liang, 2011). However, the concepts can have similar goals. In the industry, Lean methodology application usually leads to resource savings in terms of materials and production time, acting on the product since the design to the delivery process, including stock management and optimization. By this way, the amount of material spent can be eventually reduced, the energy consumed can be lowered, and the emission of pollutants can be decreased, which is in line with the CP principles. Nevertheless, this point of view is centered essentially on economic concerns, driving to a better competitiveness for industrial companies. The CP concept is far more comprehensive and can lead to positioning environmental concerns ahead of economical ones, through the use of environmentally friendly materials which sometimes are more expensive than the original ones, consequently reducing drastically environmental risks and presenting a broader range of action. CP considers all aspects from product design to the end-of-life phase, always based on the reduction in resources and environmental risks. Thus, lean manufacturing can be taken as a tool contributing to the implementation of CP, acting simultaneously in environmental and economical fronts, although by itself it is insufficient to fulfill all requirements set by the CP principles. However, when properly implemented, lean manufacturing leads to the saving of resources, which is one of the many areas of intervention concerning the principle of CP, acting as well in creating a more cost-effective product and helping companies become more competitive. Hence, savings achieved by lean manufacturing implementation can free resources to act in other fields, namely researching and substituting harmful materials by ones that are more inoffensive or investing in more environmentally friendly equipment. Many other approaches can be considered in this way, being performed for instance in the design phase and in the recycling process, significantly improving the compliance of products and services with the basic principles of CP. Some of those tools will be described later in this book, due to their relevance to environmental preservation and usefulness. Cleaner Production can lead to the reduction in: the quantity of material or energy used in the manufacture of a product or service (for instance, healthcare services); wasted materials removed from a component during its production process; harmful necessary material generated or used to extract mineral in metallurgical processes; emissions in the production of energy or products; water usage in industrial processes or in services; or to the improvement of heat insulation of buildings by taking advantage of the sunlight in building design, etc. Moreover, Cleaner Production can be applied on all levels of decision within a company, independently of its size, being that Cleaner Production must be the primary focus of the company's leader and the guidelines for its implementation should be conveniently described, with all milestones perfectly established in time. Regarding most of the industrial processes, the product conception should be properly thought in order to avoid or drastically reduce the emission of effluents (gases, fluids, or solid waste), minimizing as well the quantity of water and energy expended during the entire manufacturing process.

1.2 Cleaner Production Evolution Process

Since the first perception of how humanity was impacting and damaging the environment in the late 1960s to the present day, a drastic and frightening evolution can be observed. The main sources of environmental attack have been executed in ways such as emission of hazardous products to the air and water, generation of solid and toxic wastes, and abusive consumption of natural resources. This trend has forced the discussion on how to reduce pollutant emission and create Cleaner Production processes.

The world is permanently changing and the human mind is the steering factor that leads to that change. Thus, it is not abnormal to observe that concepts are usually evolving regarding the effect of previous actions or development states, creating new paradigms and solutions. As the concept of productivity has evolved in time, passing by three different stages, such as: (a) increase in production output for the same input by an improvement in the organizational effectiveness, (b) cost-effectiveness, which means increasing the profitability of products by the cost reduction, and (c) consistency, that intends to measure the quantity produced in accordance with the level of quality required by the market. New challenges to organizations were established, emerging the ISO 9000 standard, giving an answer regarding the quality management system. This evolution in the productivity concept reveals different stages of market and management maturity, being necessary to adopt adequate managing tools in order to achieve these goals.

As in the productivity case, the environment has been seen according to different points of view and following different rhythms depending on the mind-set of the company or community involved. Figure 1.1 intends to show how concepts have been changing since the mid of the twentieth century, together with the tools developed in order to improve productivity performance. The initial attitude of ignoring pollution gives rise to the next stage (which occurs when humanity perceives the impact of some wastes and emissions resulting from productive processes that were being developed), bringing new challenges, having as a first answer the dissolution and dispersion of by-products, due to the lack of technological solutions able to solve the problem. However, wastes and emissions represent a burden for the companies.

Obviously, soon after it was realized that these solutions were not sustainable, as they were only hiding pollution instead of minimizing it. Effectively, the dispersion or dissolution needed to be done in specific sites, causing lack of reputation to the companies using these places, leading to a decrease in sales. Furthermore, the constraints in terms of legislation and society standards became far more demanding. Moreover, in order to ensure these new demands, an appropriate financial structure of the companies is required, affecting their economic performance. The increased awareness about these problems gave rise to new technological developments in order to treat pollution, avoiding its emission, dispersion, or dilution. At the same time and in the middle of the turbulence caused by this awakening of consciousness, 3M developed a 3P philosophy taking advantage of an excellent interdepartmental cooperation, which would be later marked as one of the main stages achieved during this evolution phase: pollution prevention at the source and waste minimization. The programs funded by the US government take their first steps, and the American industry starts to look at the environment as a real concern. The concept of sustainable development now has as its

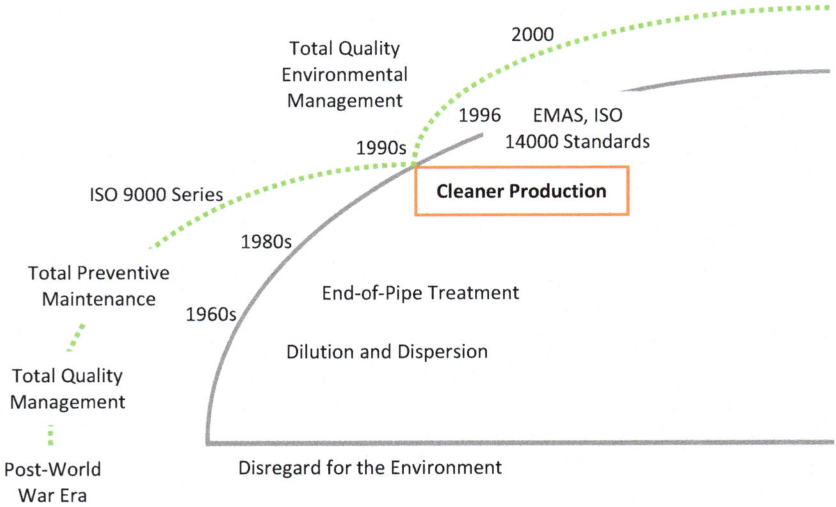

Fig. 1.1 Productivity and environmental management evolution over time. Adapted from UNIDO/UNEP Guidance Manual (2017)

main guidelines less and more efficient consumption of materials and energy, as well as the reduction or change in harmful substances by other more innocuous ones, with lower risks to human health (Hens et al. 2018). Moreover, sustainable development was defined at that time as *development that meets the needs of the present without compromising the ability of future generations to meet their own needs* by the Report of the World Commission on Environment and Development in 1987, denominated "Our common future." Regarding the report *Cleaner Production and eco-efficiency* issued by UNEP/WBSCD in 1998, governments need to promote and perform a legislation review, while stakeholders need to implement Cleaner Production effective practices, which would result in less resource consumption and waste generation by the society as a whole, better competitiveness and higher added value. Later, in 1991, the concept of Cleaner Production arises mainly in Europe and, as in the USA, the developed countries quickly adhere to the first initiatives, leading the development process and aiding developing countries to start walking in the desired direction. At that time, the main goal was reduction in waste and emissions generation, creating as well the necessary conditions to avoid the use of hazardous substances and materials. However, it was realized that companies' direction and drive depended largely on the mind-set of the senior management. Thus, the need for involving management in the conversion of processes and use of cleaner technologies was felt. Hence, the first edition of the ISO 14000 standard was launched in 1996, targeting the companies' environmental management. Indeed, the standard also intended to structure the strategic positioning of the companies facing the environment, improving the enrollment and commitment of the senior management with the principles of sustainability, and driving advantageously their own business under environmental thinking. Since that time, researchers and institutions have put a significant effort in order to improve rules and community perception of the best actions to follow in order to invert environmental degradation and catastrophic climatic phenomena. Thus, a deeper thinking has been devoted to CP, giving rise to the development of new tools able to respond to major concerns, such as how to create greener products through design; how to create the necessary logistics to promote material recyclability; how to reuse and regenerate products; or how can some of these concepts be integrated in the services sector. By this way, new tools have been developed, disseminated, and universally implemented, like the eco-design, the green economy, the circular economy, among others. The concerns regarding design have gained more and more importance. In the decade of 2010s, sustainable consumption emerged, having created serious reflections. Indeed, it is the consumption of goods and new needs and fashions that triggers and stimulates manufacturers to offer the desired products in the market. Sometimes, through aggressive marketing plans, manufacturers try to induce in the consumer the need to buy a certain product. However, it is the consumer that has the power of choice between purchasing it or not, as along as it is not an essential product. Thus, the need to create an educated consumer was felt, as he can contribute directly to the generation of less waste and consumption of resources. However, this can be a long battle as the change in social habits usually needs at least one generation (about 20 years) to occur. Effectively, the consumer needs to have enough training to quickly make choices that take into account the environmental repercussions that their purchase or option will bring to the environment. Some basic options will take many years to be truly implemented, such as the type of restaurant we can choose to have a meal in. In this simple case, one can make a decision of having a meal in a restaurant where the food is served in polymer and paper packages, constituting waste after that meal, or take a more environmentally friendly option by choosing a restaurant where the food is served in traditional porcelain dinnerware, needing just some water consumption (which can be easily treated) and a small amount of cleaning products to be ready to serve again. Such as this example, there are many consumer-dependent actions that can be taken that will pressure companies to offer products and services that are less resource consuming and have a lower impact on the environment. This is a question of consumer attitude and, together, consumers can force corporations to change their position regarding environmental action by choosing to mostly purchase products from companies associated with environmental preservation. A survey carried out in Europe in mid-2010s allowed observing that nine in ten consumers are expecting from companies something more than investment returns (Business News Daily 2017). Nowadays, consumers are aware of the behavior of companies, observing their actions in several aspects, such as how they proceed relatively to their employees or how they address environmental concerns. Following the same study, 84% of the respondents also stated that they look for sustainable products complying with the regulations and the best practices in each sector of activity whenever possible. Indeed, there is market value for the companies that, on a voluntary base, share their success with others, independently of the way used. Thus, a new concept was created called corporate social responsibility (CSR). This concept represents an evolution of the CP thinking and way of being in the market, and can be expressed regarding different ways, such as ethical labor practices, local community help, arts sponsoring, and environmental commitment. Market offerings and consumer mentality are changing, demanding an increasing engagement with social causes, e.g., climate change, poverty, regional severe accidents, seasonal and regional diseases, and many other situations which can bring notoriety to companies through a community feeling of support. It is expected that, in the future, employees will be focused on seeking out companies able to drive programs into three different strands: people, environments, and revenues. Thus, communities are watching an increasingly pronounced change in the driving factors and missions of companies, including progressively social aspects as one of the main fields of action out of their core business. CSR was studied, and in 2010, the ISO 26000:2010 standard was launched regarding this issue.

Corporate social responsibility was also included in the revision of the ISO 14001:2015 standard, becoming this standard more demanding toward companies and stakeholders, increasing the linkage between CSR and sustainability. Indeed, the ISO 26000:2010, appealing to the social responsibility, disperses its focus under seven different vectors of action, while the ISO 14001:2015 standard pulls that action mainly to sustainability, giving particular emphasis to the importance that companies should have in terms of environmental preservation. The new version of the ISO 14001:2015 standard also intended to achieve a broader action, not only focused on industrial activities, but also regarding the product conception and services sectors. Regarding the new version issued in 2015, the standard recommends the application of PDCA (Plan–Do–Check–Act) as a continuous improvement tool throughout all of the management system in a closed loop, increasing permanently the level of requirement at each cycle and pursuing the desired results in terms of sustainability. Relatively to the 2004 version, the ISO 14001:2015 standard presents the following improvements in terms of environmental management:

- Requires companies to have a continuous performance improvement regarding environmental concerns and sustainability;
- Requires compliance with environmental rules in charge;
- Defines external conditions that can be affected by the company performance. This subject was not covered in the previous version of the standard;
- Better defines the real needs and expectations of stakeholders;
- Greatly demands for leadership and commitment from top management. The environmental policy and objectives must be aligned with the strategic steering and with the organizational context;
- Requires a serious commitment of the top management with environment protection, including pollution prevention, sustainable use of resources, mitigation and adaptation to climate changes, and the protection of biodiversity and ecosystems;
- Asks for the legal requirements and other requirements subscribed by the organization be defined in accordance with the needs and expectations of the stakeholders;

- Asks for organizations to determine risks and opportunities related to: (i) environmental aspects, (ii) compliance obligations, and (iii) other issues which need to be treated to ensure that environmental management system can achieve the desired results, prevent or reduce undesirable effects, including the potential for external environmental conditions affect the organization, and achieve continuous improvement;
- Calls for organizations to consider explicitly the life cycle of products and services at different levels: (i) on the determination of the environmental aspects and (ii) operational control, particularly regarding design and development, environmental requirements on products and purchased services, communication of environmental requirements relevant to suppliers and the need to provide information concerning the significant potential environmental impacts associated with transport or distribution, the end-of-life treatments, and final destination of products and services;
- States that the organization should assess its environmental performance and effectiveness of the system of environmental management, using indicators;
- Adds the concept of continuous improvement;
- Details about how established Key Performance Indicators for environmental continuous improvement are pointed out;
- Eliminates the requirement concerning preventive actions, but corrective actions absorb the preventive ones;
- Removes procedural requirements, being replaced at various points of the standard by processes.

The ISO 14001:2015 standard on environmental management systems presents new challenges, which are also an opportunity to establish this volunteer management tool as a truly differentiating factor for organizations. Regarding the improvements performed from the initial version of the ISO 14001:2004 standard to the current one, it is possible to observe that environmental concerns are playing a more important role in several aspects, being perceived that concepts needed to change quickly and be adjusted to more ambitious targets because, even after numerous important meetings between the most industrialized countries, several political resolutions, and constraining regulations, the main focus of action needed to be the corporate top management, being the most influential entity capable of changing thought processes and consumption patterns toward sustainable development. The implementation of the ISO 14001:2015 standard is voluntary, and the period of adjustment from the previous version to the current one is between 2015 and 2018. This standard will be properly developed later in this book.

Figure 1.2 tries to illustrate the different steps taken by CP along the last 60 years. It is clear that initial regulations compelled industries to adopt certain kinds of pollution control and prevention, but without the top management commitment and the community judgment, it is seriously more difficult to reach the required level of respect by the environment.

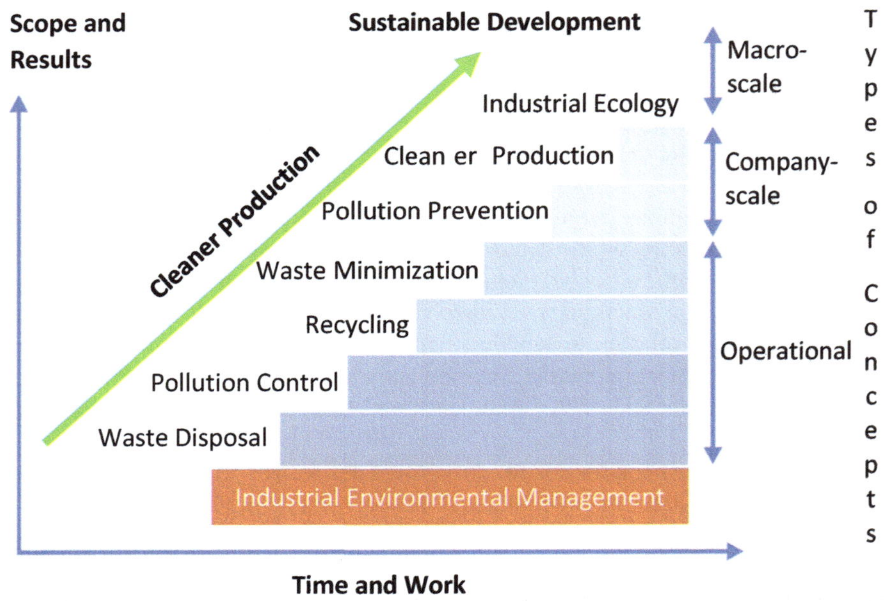

Fig. 1.2 Staircase of concepts of industrial environmental management. Adapted from Hamner (1996)

References

Business News Daily: https://www.businessnewsdaily.com/4679-corporate-social-responsibility.html, [ONLINE], Retrieved on 22 Dec 2017

J. Fresner, Cleaner production as a means for effective environmental management. J. Clean. Prod. **6**, 171–179 (1998)

P. Glavic, R. Lukman, Review of sustainability terms and their definitions. J. Clean. Prod. **15**, 1875–1885 (2007)

W.B Hamner: *What is the Relationship Among Cleaner Production, Pollution Prevention, Waste Minimization and ISO 14000?* http://www.unep.fr/shared/publications/other/WEBx0072xPA/manual_cdrom/CPlinks/pdfs/con199601.pdf [ONLINE], Retrieved on 10 Dec 2017

R.M. Harrison: *Pollution: Causes, Effects and Control*. (Royal Society of Chemistry, London, UK, 2001). ISBN: 978-0854046218

L. Hens, C. Block, J.J. Cabello-Eras, A. Sagastume-Gutierez, D. Garcia-Lorenzo, C. Chamoro, H. Mendoza, D. Haeseldonckx, C. Vandecasteele, On the evolution of "Cleaner Production" as a concept and a practice. J. Clean. Prod. **170**, 3323–3333 (2018)

G. Kjaerheim, Cleaner production and sustainability. J. Clean. Prod. **13**, 329–339 (2005)

Kaunas University of Technology Lithuania, Institute of Environmental Engineering, Definition of cleaner production. http://www.un.org/esa/sustdev/sdissues/technology/cleanerproduction.pdf [ONLINE], Retrieved on 26 Dec 2017

L. Nilsson, P.O. Persson, L. Rydén, S. Darozhka, A. Zaliauskiene, *Cleaner Production—Technologies and Tools for Resource Efficient Production* (The Baltic University Press, Uppsala, Sweden, 2007). ISBN 91-975526-1-5

Review of Cleaner Production: Annex 2 http://ec.europa.eu/environment/enveco/eco_industry/pdf/annex2.pdf [ONLINE], Retrieved on 7 Dec 2017

SD Features. https://www.gdrc.org/sustdev/concepts/02-c-prod.html [Online], Retrieved on 22 July 2017

L. Song, D. Liang, Lean construction implementation and its implication on sustainability: a contractor's case study. Can. J. Civ. Eng. **38**(3), 350–359 (2011)

UNIDO-UNEP Guidance Manual: *How to establish and operate cleaner production centres*. http://www.unep.fr/shared/publications/other/WEBx0072xPA/manual_cdrom/Guidance%20Manual/PDF%20versions/Part1.pdf [ONLINE], Retrieved on 28 Dec 2017

A. Yaacoub, J. Fresner, *Half is Enough—An Introduction to Cleaner Production* (LCPC Press, Beirut, Lebanon, 2006). ISBN 3-9501636-2-X

Chapter 2
Cleaner Production Main Concept and History

Abstract One way of defining a Cleaner Production is saying that it is a set of rules with the objective of protecting the environment and minimizing waste, ranging not only from manufacturing processes but to the entire life cycle of a product. This concept can even be applied on a personal level, addressing one's lifestyle and daily choices. The first concerns regarding pollution arouse in the 1960s in the USA and started spreading across the world. However, the rate at which this concept is diffused as well as the seriousness that it receives varies immensely from country to country. Here, the main concept of CP is debated as well as its initial stages across all continents.

Keywords Cleaner Production concept · UNIDO · UNEP · Cleaner Production evolution · Regional Cleaner Production Program Development · Country-specific Cleaner Production Evolution

2.1 Main Concept of Cleaner Production

Beginning focused on producing with less waste, Cleaner Production (CP) evolved into a concept of improving resource effectiveness in production. The Cleaner Production concept can be defined as a set of precautionary management rules that intend to protect the environment from the harmful effects of products and manufacturing processes. Moreover, the concept of CP can be viewed as something that can be applied covering the whole lifecycle of a product or service, enlarging the concept of well-being to future human generations. Indeed, the concept is broad enough to encompass several tools, which can be distributed from eco-design to lean production systems. Thus, it is possible to state that this concept is linked to resources and environmental preservation, adopting methodologies, techniques and technologies able to minimize or eliminate the aggression of some business processes and products relatively to the environment. The concept is mainly oriented toward manufacturing systems, but it can be considered wide enough to extend to the personal attitude of people regarding consumption, because everyone can prevent waste generation, increase material recycling and, above all, select products attending to their ecologi-

F. J. Gomes da Silva and R. M. Gouveia, *Cleaner Production*,
https://doi.org/10.1007/978-3-030-23165-1_2

cal footprint and knowledge about companies' practices. Obviously, these concerns have evolved significantly through time, mainly from the mid of the twentieth century, after the World War II, following the economic growth registered after that tragic episode in history.

In fact, the concept of CP can lead down different paths depending on the sector of activity. If we are talking about production processes, it is crucial to preserve energy and raw materials, avoiding as well the use of toxic materials and reducing the amount and level of toxicity of waste materials and effluents. On the other hand, if we are talking about semi-products or products, the main concern should be focused on their life cycle and corresponding nefarious effects for the environment, since the initial raw material to the end of life of the final product. In opposition, if we are thinking about services, it is mainly a question of organizational and operational practices and mindset, as in most cases, changing common practices is enough to reduce the consumption of energy, water, materials, or products.

2.2 Industrial Growth and the United Nations' Initial Concerns

The GDP growth felt since 1950 was mostly based on industrial and technological developments. Under the alerts disseminated by environmental organizations and starting to feel the extreme climate events, governmental organizations were pressed to act, preventing the increasing damage perpetrated by humanity against Nature. The main sources of pollution and emissions have been identified, allowing for studies on how to minimize or reverse these environmental attacks.

The first concerns regarding pollution were highlighted for the first time in the USA in late 1960s, by the National Environmental Policy Act, which only passed to national law in 1990 (Pollution Prevention Act), culminating in the creation of Cleaner Production Centers (CPC), denominated at that time as pollution prevention centers (UN 2017). Likewise, pollution concerns were put on top in the United Nations Conference on Human Environment, which took place in Stockholm in 1972. In this conference, some problems such as population growth, water reserves and consumption and desertification were dealt, giving rise to the Stockholm Declaration, which highlighted the need for acting on pollution prevention. Prior to 1990, the Industry and Development Seminar also debated the creation of Cleaner Production Centers, anticipating the 1992 Earth Summit. Thus, in 1991 Norwegian and Polish Non-governmental Organizations (*Norske Sivilenginores Forening* and *Naczelna Organizacja Techniczna*) created the first CPC supported by the Norwegian and Polish governments, which the main goal was to raise awareness, conduct training, develop trainers, and build a framework between Cleaner Production and environmental management system (EMS). Also, in 1992, environmental technology transfer mechanisms were faced in an effective way, through the "Rio Declaration." Indeed, in 1992 the concept of National Cleaner Production Center (NCPC) was promulgated, conducted by the principles formally drawn up in Agenda 21, as result of the United Nations Conference on Environment and Development (Earth Summit) in Rio de Janeiro (Brazil). In the same year, the Swiss government established CP as a priority field, developing assertive ways to disseminate and operationalize technology transfer for developing countries by the creation of NCPCs able to redesign manufacturing processes toward environmental sustainability. This program, carried out by Swiss Secretary for Economic Affairs (SECO), consisted in providing technological assistance to other centers installed in different countries. In 1993, under the suggestion of the forthcoming NCPC Program, Australia gave its first step in the Cleaner Production direction through the creation of the Australia Center for Cleaner Production, translating the strong commitment of Australia's main agents with Cleaner Production. The main activities of this center, as well as others, were to provide the necessary consultant services to companies, promote technological transfer and general Cleaner Production education, having as funding associated members, institutions and sponsors but, unfortunately, the center failed in 1996 and closed in 1997.

2.3 Early Stages and Evolution of the Cleaner Production

2.3.1 Cleaner Production in the Oceania

Unlike the stated by Brown and Stone (2007), Boyle (1999) had the perception that even though the increasing number of programs encouraging the practice of CP policies, the adhesion of small- and medium-sized companies were not in accordance with what was desirable, mainly due to the lack of governmental legislation and community awareness to produce less waste, suggesting at that time a closer actuation of government and market-based agents in order to promote the required mindset change through legislation and education. In that work, 11 recommendations were drawn regarding the improvement of CP policies application in New Zealand. However, the analysis carried out by Brown and Stone (2007) 8 years later shows that New Zealand faced the CP concerns in a different point of view. Indeed, the New Zealand society faced the environmental concerns in a really positive way, not needing a set of specific regulations as in Europe to act favorably regarding environment preservation and society well-being. New Zealand legislative process was based on a consultative practice, resulting in agreements with the industrial sector, namely the packaging and dairying activities, as well as new strategies for resources management (energy and waste). The activity around CP became more evident at the turn of the century with a significant growth of NGO (non-governmental organizations) working in this field, developing founded projects in a competitive business base with the industrial community and society in general. Later, the availability to share information, cooperate and act in a concerted way seems to be improved and there are initiatives showing a planned approach to funded project businesses in this field in New Zealand. Thus, New Zealand shows the power of initiative and willingness enough to pursuit its way in CP policies without a strong regulation such as in Europe. A study performed by Collins et al. (2007) through a survey taking into account more than 800 enterprises refers that big companies are more available to be engaged in CP programs and practices than the others, pointing out the need of small and medium companies to obtain more information regarding this issue. Moreover, for these kinds of companies, costs are referred to as a handicap to a more intensive application of CP practices, affecting the companies' competitiveness. However, it remains clear that excess of regulation is not needed neither is the best way to pursue the best CP practices, remaining the mindset education of senior management as the steering factor to increase the best environmental practices.

Regarding the Malaysian situation (Yusup et al. 2015), and contrary to the New Zealand willingness to adopt CP practices, the environmental protection rules and regulations are the main driver that impel companies to follow CP principles, being the ownership consciousness the key factor on the company decision to have environmental care or not. Moreover, different activity sectors show dissimilar willingness to adopt CP practices, being the electric and electronics sector one of the most significant manufacturing sectors in the Malaysian economic context (59.1% of the Malaysia' exports), also being one of the sectors that show a greater openness to the implementation of sustainable practices of CP. Furthermore, companies which present ISO certifications like ISO 14001 or ISO 9001 show a greater engagement with CP best practices. Hence, there are yet opportunities to improve logistic systems, as well as recycling activities and the use of chemicals in some industrial processes, becoming closer these activities with the real recommendations of CP practices. Despite the regulation implemented by the National Green Technology Policy of Malaysia, there is still much to do regarding the mindset of the companies' ownerships, being one of the main factors that need to be focused on in the near future.

2.3.2 *Cleaner Production in Europe*

A study regarding the Cleaner Production concept in Lithuania carried out by Kliopova and Staniskis (2006) depicts the Lithuanian situation, showing that the textile industry was the leading sector at that time, with 81% of the investment having been made in the optimization of that sector of activity, resulting in a more efficient use of electrical energy, heat energy and water, as well as a reduction of wastewater volume and contamination. On the other hand, the focus of the food industry was centered on decreasing energy consumption, reduction of the use of chemicals, and minimization of the waste generated. It was possible to observe that a strong effort was also made in the food and beverage industry in order to modernize the equipment used in the manufacturing process, helping to implement CP principles. The study also reveals that there are different needs depending on the activity sector. Thus, sectors such as the food industry, wood transformation, or stockbreeding mainly need programs able to manage the waste generated by their processes; hence, the investments around CP need to be addressed to waste processing and recycling or reuse. Furthermore, the furniture industry adopted new energy sources, burning their own waste and generating by this way the necessary heat/energy for their processes. Moreover, because many technological advances have been registered in the last decades, equipment needs to be upgraded, being an excellent opportunity to select processes and equipment complying with the main requirements of the CP concepts. Regarding the textile, furniture and electronics industries, a new chemical-related risk management system was designed regarding workplace exposure and the environment, according to the CP policies.

After some of the first initiatives previously referred, developed countries expanded their Cleaner Production centers to other developing countries, disseminating their knowledge and promoting the necessary training activities and technology transfer. By this way, they are able to provide the essential technical assistance and catalyzing investments for CP sustainable actions. These centers were usually co-funded by the countries involved through bilateral agencies for development and some international projects. However, these Cleaner Production Centers were focused mainly in technology transfer, minimizing the progress of local skills, and knowledge. This program was successfully carried out between 1994 and 1995 in Central and Eastern Europe. Also, in 1994, some Norwegian and Norwegian–Russian centers co-funded by both countries were established by initiative of the Norwegian Government, through the World Cleaner Production Society (WCPS), in cooperation with the Russian Authorities. In the meanwhile, the network of centers in Central and Eastern Europe was extended until September 1997, supported by the USAID (United States Agency for International Development) and the WEC (World Environmental Center) (UNIDO 2017).

In fact, the first eight NCPCs started their work in 1995, sparking wide-ranging interest. Since then, many other NCPCs have been established, mainly supported by UNIDO (United Nations Industrial Development Organization) and UNEP (United Nations Environmental Program), having in 2015 NCPCs around 58 countries (UNIDO-UNEP 2015), as well as other similar initiatives carried out by other developed countries. In this way, NCPCs have created local expertise and framework, sharing information, and enabling the development of international programs and activities. However, even after more than 20 years of activity, these centers remain necessary, as can be understood by the relatively small progress felt in terms of emission reduction and the difficult negotiations usually carried out among the more industrialized countries.

UNIDO also settled a procedure to associate Cleaner Production assessments with local environmental management systems, environmental management accounting and environmentally sound technology assessment, in order to boost the Transfer of Environmentally Sound Technology (TEST), increasing in this sense the environmental engagement of companies in order to reach the international standard levels. This work was mainly carried out in Croatia, Hungary, and Slovakia through the local NCPCs. The established cooperation and procedure were firstly implemented in 17 of the most polluting companies in the Danube region, encompassing five countries: Bulgaria, Croatia, Hungary, Romania, and Slovakia. The project has resulted almost in full because all industries had completed Cleaner Production and environmentally sound technology assessments, 11 of them successfully implemented an environmental management system and other six became able to proceed to environmental management accounting. The actions implemented had no cost for 212 companies, and other 109 companies implemented low-cost actions. The total investment spent in the implementation of those actions was USD1.7 million, inducing annual savings of about USD1.3 million. Moreover, an annual reduction of 4.6 million m^3 in wastewater yearly released in Danube was also attained. Other 141 high-cost actions were also identified, but just 38 had been managed to gather consensus regarding their implementation before the project completion, having been estimated costs in the order of USD47.3 million for these 33 actions (UNIDO 2017).

The situation in Turkey was also analyzed by Yüksel (2008), showing that even after running CP programs for more than 10 years and despite all the environmental concerns, the companies' managers have the perception that CP policies are a cost driver, seeing as well the cost of "hard technologies" as the main difficulty to implement end-of-pipe engineering solutions. However, regarding the survey carried out and the results obtained from 105 big companies, the managers stated that companies are complying with the legal requirements of environmental protection and the benefits obtained are according to the expectations generated by the investments performed under the pollution prevention programs. Thus, the managers seem able to adopt a more proactive position in the future. Nevertheless, the preference of these managers lies essentially on end-of-pipe technology. The concept of CP is somewhat applied to the product and processes, through design care. However, services are not usually considered regarding environmental concerns, resulting in deficient reverse logistics needs for product recycling. This problem is accentuated by the lack of programs able to support the recycling process. Furthermore, the integration of production planning with the environmental issues is referred as unruly, making more difficult the establishment of CP practices regarding materials management. Moreover, the government seems that it is not doing its best in order to improve the awareness of people and enterprises regarding the environmental concerns. Despite the excellent conditions for the utilization of renewable energy, Turkey has felt difficulties in support programs capable of inducing its use by its companies.

The post-Soviet Eastern Europe deserved the attention of The Norwegian Society for Chartered Engineers, which designed a CP training and coaching program tailored to industry technical managers. The program was firstly issued in Poland, with the collaboration of the Polish engineering society. Successful results were reported at the end of 1994, with a waste reduction in the range of 20–40% regarding the group of 150 businesses covered by the program, corresponding to 20 beneficiary sectors. Furthermore, the investment needed was almost inexistent and was identified reductions of around 30% only possible with small investments (USD67,000) in CP technologies/hardware, which predicts annual savings around USD351,000. Between 1992 and 1996, 46 companies from the Czech Republic were covered by an identical program, reaching total annual welfares of USD7 million. The ratio between the money spent and the benefits achieved was about 1:12. The successful results attained induced the Organisation for Economic Cooperation and Development (OECD) to develop special programs for transitional countries until 2001, offering professional training, focused information, in-plant demonstrations and other practical actions, in order to improve the Cleaner Production capacity of those countries (UNIDO 2017).

2.3.3 Cleaner Production in Asia

In Japan, the Global Environment Centre Foundation (GEC) worked together with the UNEP-IETC (International Environmental Technology Center) since 1995 and during three years, promoted the transfer of Environmentally Sound Technologies (ESTs) as well as information, stimulating the compilation of data and disperse information which was categorized under five electronic databases called NETT21 (New Environmental Technology Transfer in the Twenty-First Century), divided into the following issues: air pollution control technology, waste management technology, water pollution control technology, and on-site green techniques (OGT). The GEC has made this information available on the Internet, allowing access not only to Japanese institutions, but all over the world. This information has also been used as the subject of several international seminars, where GEC intends to disclose information already collected and validated. However, the technical information disseminated by GEC is essentially based on "end-of-pipe" technology, since the main purpose of this institution is pollution control, thus leaving forgotten some concepts formulated in Agenda 21. Nevertheless, the perception of the importance of ideas like zero emissions and waste reduction has been deeply ingrained in the Japanese society for a long time. Hence, the only thing missing is the effort of combining all these concepts, creating the necessary means to join the "hard technologies" used in the "end-of-pipe" techniques, with the "soft technologies" related with products and processes management. Thus, in 1998, the Cleaner Production Technology Promotion Committee and working group were created. The Promotion Committee had as main focus the formulation and evaluation of Cleaner Production technologies criteria aiming developing nations, while the working group was engaged on assessing and investigating the technologies collected, checking if they were in agreement with the criteria previously referred. In 2005, GEC carried out a study in order to realize if the first version of the database created was serving the interests of the communities assessed, realizing that the interest from the industrial community remained very high, but new focus of interest was emerging from other areas such as restaurants and other activities in the services sector. Following these conclusions, the GEC databases were updated in 2007, including the new sectors which had remarkable interest in pursuing CP policies out of the manufacturing sector. Since then, the GEC has been refining the database, not only in newly included sectors, as well as those included in the initial version (NETT21 2017).

Between 1995 and 1996, some actions were also carried out in other Asian countries such as Taiwan and Singapore through the USAEP (United States Asia Environment Partnership) and local industrial confederations, which had as main activity the promotion of seminars and workshops for industry CEO's in order to improve the sensitivity of who decides in terms of investment policies within the companies.

Other programs with less impact have been also carried out by UNIDO, such as DESIRE (DEmonstrations in Small Industries for Reducing wastE) program between 1993 and 1994 in India, with the main emphasis was in waste minimization and having as focus 12 companies distributed by sectors such as pesticides formulation, pulp and paper and textile production, giving rise to a novel methodology which attracted the attentions and followers of other Asian countries interested in waste minimization practices.

Similarly, China participated in an UNEP CP program carried out based on an Environmental Technical Assistance project assisted by the World Bank, which intended to perform Cleaner Production audits to companies, identifying and implementing new industrial practices which allow remarkable profits (2.9 Million USD) comparing with the investment (less than 100.000 USD) done in the first stage of the program. The same program was extended and in 2001, the China Cleaner Production Promotion Law was launched, imposing mandatory audit provisions, after the UNEP UNIDO Waste and Emission Audit and Reduction Manual Implementation and industrial audit training program (UNIDO-UNEP 2015). An analysis made by Peng and Liu (2016) draws a very pessimistic picture about the current situation in China, with regard to CP policies and application. In fact, these authors refer that despite all the efforts carried out to promote CP, the achieved results are clearly modest because most of the policies established by the Chinese Government are on a voluntary base, merely encouraging the adoption of those policies. Thus, Chinese companies do not feel pressured to follow the governmental recommendations by law, preventing their application because CP policies are perceived as capital consuming, not showing short-term profitability. Resuming, the motivation to adopt and follow CP policies in China is relatively weak (Chang et al. 2015). Furthermore, regarding some authors, CP does not seem to be a flag for the Chinese Government (Peng and Liu 2016). Though, due to financial limitations of many Chinese companies, it seems clear that a government financial support is being needed in order to incentive enterprises to promote CP and sustainable practices, centralizing the available information and creating a CP online service able to improve the cooperation with foreigner institutions (Zhang et al. 2016).

Also in Asia, in September 1998, the international declaration on Cleaner Production was adopted at the 5th International High-Level Seminar held in South Korea, as follows: *We understand Cleaner Production to be the continuous application of an integrated, preventive strategy applied to processes, products and services in pursuit of economic, social, health, safety, and environmental benefits*. This statement has contributed decisively so that CP should be understood as something broader than just the pollution control, enlarging the concept to the product design and life cycle.

In May 2014, the APRSCP (Asia–Pacific Roundtable on Sustainable Consumption and Production) provided a new platform able to be used by multi-stakeholders from Asia and the Pacific in order to share experiences and lessons learned in new sustainable consumption and production (SCP) technologies and strategies, through exploration of available information on how to get financing, monitoring and/or evaluating the SCP implementation (Charmondusit et al. 2016). The APRSCP assumes itself as an international institution focused on Asia concerns about SCP, remaining as a non-governmental and non-profitable institution which intends to promote the cooperation in the region, improving the skills and developing the implementation of sustainable consumption and production strategies. The APRSCP remains in action through a conference that takes place yearly in different places in that region, serving as driver force for new achievements in the CP field, impelling the industrial community to do more and more for a better environment in the region (APRSCP 2016).

The CP ideas in India were widespread through Waste Minimization Circles (WMC). Normally, each WMC comprises four to six enterprises considering the same geographical area, starting the program with a two-day intensive training about CP models related to industrial activity, covering some relevant areas such as prior environmental and economic policy experiences; Cleaner Production ideas, welfares and organization; WMC action and preparation guidelines; activity planning, and orientation. Afterward, applicants meet about twice monthly for roughly one year under the supervision of the organizer, while applicants apply CP concepts and practices in their businesses, reporting and exchanging with partners their experiences. Furthermore, one year after the formal end of WMC actions, applicants continued to meet regularly and were stimulated to establish spin-off WMCs on their own. The success of this initiative gave rise to more than 120 WMC actions between 1995 and 2003 funded by Ministry of Environment and Forestry and by World Bank, concerning more than 500 small companies from 37 industry sectors across 17 States of India in a total investment higher than USD2.5 million and estimated overall payback within 14 months.

The approach carried out by NCPC in Vietnam followed other principles, focusing essentially on technical and management training of local experts, involving Cleaner Production skilled personnel from local universities and technical institutes, as well as staff from the companies. The training program consisted of four training modules totaling 15 training days, including theoretical and practical lessons into companies. The number of plant assessments totalized 42 by the end of 2002, with 12 projects concluded in that year and generating around USD1 million of annual savings. The success of the prior actions gave rise to organization of the first national Cleaner Production roundtable, which appealed 150 participants on behalf of government, industry and academics, establishing new goals for 2005 through an action plan. Moreover, Vietnam NCPC established environmental and quality management systems, which were certified regarding ISO 14001 and ISO 9001 standards in 2002.

Also in Vietnam, some institutions such as International Trade Centre (ITC), International Labor Organization (ILO), Food and Agriculture Organization (FAO), and United Nations Conference on Trade and Development (UNCTAD) undertook a program focused on poverty reduction in villages where the main subsistence mode was handicraft activities, through the improvement of sustainable value chain for products based on natural materials such as silk, bamboo, rattan and seagrass, as well as others with associated labor such as handmade paper and lacquerware. CP assessments were performed in about 50 crafts companies, along with the introduction of sustainable design concept, preparing 65 trainers in order to later widespread their knowledge by 1430 crafts households.

2.3.4 Cleaner Production in the Middle East

Regarding the particular situation of Egypt, in the Middle East, and even after more than 20 years since the creation of CP concepts, industrial companies are yet at an early stage on their evolutionary process. The culture subjacent to the managing process has led to a reactive positioning when subject to external incitements, such as international sponsoring programs. Incentives have been given under the assumption that companies beneficiaries of these funds have carried out actions capable of being identified through environmental audits, conveying their experience to other companies. However, the conclusions of a survey performed by Sakr and Sena (2017) showed that this approach is not enough to produce the desired results, being necessary to induce a proactive mindset in managers in order to assume their responsibilities in terms of environment protection and CP action.

2.3.5 Cleaner Production in North America

The United States of America (USA) is one of the most industrialized countries, contributing heavily to the pollution generated in the world. USA adopted the name of pollution prevention instead of CP, which program is usually called 2P. The study of Miller et al. (2008) describes the main legislative steps done between 1969 and 1990, since the National Environment Policy Act (NEPA) to the Pollution Prevention Act (PPA). The initial approach was made imposing regulations which had to be met by the polluters, having programs to fund the necessary actions in order to clean waterways, reduce air pollution and decontaminate lands. But, in 1980, the USA launched the Comprehensive Environmental Response, Compensation and Liability Act (CERCLA), establishing that polluters must pay, allowing for a cleaning process of contaminated areas, situation that started to affect the competitiveness of companies' and, moreover, the Act lead the industrial community to realize that pollution treatment was not the best way for the future, being necessary to drastically cut pollution generation and energy consumption, changing products and processes, as well as looking at waste engendered and excess of spent resources in a different manner. This new sensitivity of the community was felt in mid-1980s, giving rise to the Era of Strategic Environmental Management (1985–1994), emphasizing the care in pollution prevention instead of pollution treatment. At the beginning of this Era, a greater attention was given to chemicals used in manufacturing processes which bring potential risks to human life and the environment, through the Toxic Release Inventory (TRI), impelling the industry to reduce their use and calling the community's attention to the danger that they represent. Moreover, through an

Environmental Protection Agency's (EPA) challenge which registered the participation of 1294 companies, they were invited to reduce the release or transfer of 17 chemicals (Bai et al. 2015) with recognized harmful effect on the environment by 33% from 1988 to 1992 and 50% from 1988 to 1995. The compromise established by those companies generated reductions estimated in 230 million kilograms. The program allowed to achieve a combined reduction of emissions of 64%, showing that these initiatives and programs lead to better results than pure governmental legislation. The commitment of some big companies with the environment has been commendable throughout the history of pollution prevention in the USA. Companies such as DuPont, Dow Chemicals, Xerox, or 3M launched their own programs and a competitive game among them, regarding the reduction of harmful chemicals. The positioning of 3M's manager Joe Ling deserves a vibrant applause, starting the 3P program (Pollution Prevention Pays) in 1975 and defining a new way for product development, starting by the change of materials, processes and equipment leading to reducing of waste generation since the product design phase, clearly in line with the current concept of CP. These companies reported economic benefits when establishing these programs directly linked to the reduction of pollution-related actions. From 1985 to 1990, some EPA reports were edited, trying to prepare for the environmental challenges of the 1990s. Around 1990, several initiatives boosting CP policies were carried out in the USA. EPA launched the Design for the Environment (DfE) and product stewardship initiatives, the American Chemistry Council (prior Chemical Manufacturers' Association) started with its Responsible Care program and pollution prevention was included in the ISO 14000 and ISO 9000 standards. Also, an environmental mission statement was adopted by companies as a commitment with the 10 elementary principles established by Coalition for Environmental Responsible Economies (CERES). An increasingly number of companies have been adhering to these principles, enlarging the base of companies working under the same principles, taking also advantage of economic benefits and of an improved perception seen by the community and consumers. Since 1993, the Federal Government Performance and Results Act implies that everyone receiving federal funds needs to apply metrics in order to measure their impact according to the strategic plans of the funding agencies. The report "An Ounce of Pollution Prevention is Worth Over 167 Billion lb of Cure" was edited by EPA and NPPR (National Pollution Prevention Roundtable) and intended to illustrate ten years of pollution prevention in the USA between 1990 and 2000. Besides, a succeeding report launched in 2003 revealed truly amazing achievements: 78,500,000 t of pollutants reduced, less 180,000 L of water consumed, savings for companies around 900 million Euros, and a rate of return of investment of 5:1. Nevertheless, time has showed that focus is sometime disturbed by other factors such as business competition needs, public support decrease, and high complexity of the P2 documentation. Additionally, other factors have shifted the attention of the society of the USA such as war, terrorism education, political changes and health care, relegating to the background pollution prevention and CP policies and actions. In the meantime, P2Rx were created as a national network of regional pollution prevention information centers devoted to increase the adoption of pollution prevention by improving the dissemination of relevant information, serving as the first source of information regarding P2 programs, increasing the availability of information about P2 programs and policies and enabling regional networking around these concerns. The survey carried out as a result of the evaluation process of the EPA PPIN (Environmental Protection Agency Pollution Prevention Information Network) Grant Program showed a considerable reduction of hazardous and non-hazardous wastes (29% and 27%, respectively), as well as an elimination of 7% in both wastes. However, in the same survey, 51% of the respondents did not know the amount of reduction achieved in their companies or if any reduction had been attained (EPA 2008). Regarding the same report, almost 50% of the respondents did not know if the P2Rx had promoted a positive effect on the economic performance of their business, saving time, and costs. The situation was not significantly different for companies under the TAP (Technical Assistance Program/Provider) support. Assessing to the EPA Web site, it is possible to realize that this agency is permanently working through new programs and 4 years strategic plans regarding the environmental preservation, both through information dissemination and other focalized actions. Regarding the EPA-190-K-17-002 report (EPA 2017), the main goals achieved in 2016 were the following ones:

- Providing resilience training and tools to nearly 4200 operators of small water utilities to address aging infrastructure, significant weather events, and other issues;
- Cleaning up more than 9600 additional sites and making them ready for anticipated use;
- Completing more than 1000 chemical assessments related to pesticides and commercially available chemicals;
- Working with state and tribal partners through E-Enterprise for the environment to create Web-based mechanisms and mobile phone applications which increase access to information and reduce regulatory compliance and reporting burden;
- Completing 136 confirmatory tests on emissions and fuel consumption for light-duty vehicles.

Thus, it is possible to realize that a difficult path is being traveled by EPA in order to improve environmental preservation mechanisms, trying to involve businesses, communities and the general public, making available relevant information, coordinating funding programs, investigating and identifying chemical risks, and carrying out actions leading to the best environmental practices.

2.3.6 Cleaner Production in Latin America

There was always a big difference between the CPCs in developed countries and countries under development. The CPCs in developed countries present the advantage of having a well-established funding process, allowing for better Cleaner Production policy integration and development, taking advantage of the rules already established in terms of preventing pollution and the framework implemented between institutions and companies. On the other hand, the CPCs installed in countries under development have been more oriented to provide services, information, and education about Cleaner Production, i.e., tending to act more in a passive way than in an active manner. Therefore, it is not strange for these centers to be characterized by a miscellaneous management between national and foreigner people, making harder to keep coherent and active policies that are perfectly oriented with the interests of these countries. Thus, these centers have had as main activity sharing information and promoting technology transfer than acting proactively in the development of national capabilities and influencing the establishment of appropriate legislation to comply with more restrictive rules about pollution.

In fact, the UNIDO/UNEP NCPC Program has had an important role in terms of providing stakeholders the opportunity to share information, improve their ability to increase Cleaner Production policies and implement sustainable Cleaner Production practices through international cooperation. This program was converted into another one called "UNIDO-UNEP Joint Global Resource Efficient and Cleaner Production (RECP) Program for developing and transition countries", widening the idea related to the name according to the evolution felt from the original concept to the current one. Moreover, taking advantage of the experience already acquired by UNIDO and UNEP, CECPnet was created in Latin America, aiming *to foster and contribute to the effective and efficient development, application, adaptation, and replication of RECP concepts, methods, policies, practices, and technologies in developing countries and countries with economies in transition, and to facilitate effective North–South and South–South collaboration* (UNIDO 2017). At the end of 2014, the success of the UNIDO-UNEP RECP Program was clear, with 58 centers spread by 56 different countries in Europe, Asia, Africa, and Latin America. In September of 2015, the city of Davos, in Switzerland, held a celebration.

Despite sometimes seen as a group of developing countries, many efforts have been done in order to improve manufacturing processes, trying to diminish the environmental impact of its industry. Mainly due to the proximity to the USA and economic reasons, Mexico has grown at very interesting rates in terms of GDP directly related to the industrial sector in the last three decades. In 1997, a program based on demonstration of the CP advantages was created in Mexico. For this purpose, seven foundries were selected in the city of San Louis Potosi, having been identified 103 improvement opportunities, mostly related with energy consumption reductions and materials exchange and/or usage decrease, which will lead to savings about 80,000 USD. From the opportunities identified, 42% of the recommendations were actually employed and other 36% realized as viable for implementation in the near future. Moreover, another important outcome was extracted from the program: a Cleaner Production best practice manual was created regarding the sector of castings.

Between 1998 and 2002, UNIDO supported the first regional program of NCPCs installation including three countries (Costa Rica, El Salvador, and Guatemala) under the supervision of the NCPC already existing in Nicaragua since 1997. Simultaneously, an initiative carried out by the Canadian Development Agency in Honduras about CP was also reported. Regarding the program implemented in Guatemala, El Salvador, and Costa Rica, the results at the end of the program relatively to the number of companies included were satisfactory: 70 companies in Guatemala, 48 companies in El Salvador, and 41 companies in Costa Rica. These companies act mainly in the food industry sector, as well as in agro-industry. Moreover, twelve environmental management system assessments were performed in El Salvador and Costa Rica. Furthermore, synergies among these countries were attained, allowing for the exploration of opportunities in areas such as standardization, coordination and training, claiming as one voice for new regulations in an effort to catch up with developed countries in these matters.

In order to overcome some problems related to the need of investments and lack of guarantees able to be given by companies in Colombia, the Swiss Government supported an established program leading to better assessment of companies to loans for Cleaner Production preapproved projects. This project, called Green Credit Line, was later extended to Peru and Vietnam. Furthermore, the program presented some other benefits to the adherent companies because, upon an independent evaluation of the environmental gains accomplished after the investment concluded, the company could receive a compensation of up to 15% or 25% of the loan contracted, through the NCPC local organization, depending on the level of compliance of the project and gains effectively achieved. The success of the program can be analyzed by the following numbers: between 2003 and 2015, loans contracted under the Green Credit Line program reached the amount of USD12.4 million in Colombia, and a total of USD3.4 million was reimbursed by the companies through the support of the NCPCs. The steel company Aceros Industriales was one of the beneficiaries from this program, having made a total investment of 640,000 USD replacing the chemical surface treatment by a mechanical one. The benefits were clear: less 400 ton of greenhouse gases produced and lessened downtime lead to profits of about 500,000 USD per year to the company directly related to the process exchange and investment done. Regarding the previously refereed, the company also received a reimbursement of 200,000 USD from Green Credit Line due to the successful results.

2.3.7 Cleaner Production in Africa

As is well-known, Africa is a less industrialized continent, where pollution is a concern not due to massive industrialization, but, in most cases, essentially because of more permissive regulations. Indeed, there is a lot of contradictions in this region as it is optimal for polluting companies to settle in, due to low labor costs and lack of significant restrictions in terms of environmental care requirements. A downside, however, is the fact that industries are faced with some lack of training of potential workers and other important obstacles, such as political concerns. Still, industrial activity is growing in Africa, namely in the countries in Northern Africa and in South Africa.

In 2002, Austria and Switzerland established a supporting program for South Africa in order to establish NCPCs in that country, which was kept until the end of 2008. Over these six years, the NCPC implemented Cleaner Production assessments in 52 industrial companies focused essentially on three of the most polluting sectors:

chemical industry (26 companies), agro-food (21 companies), and textile industry (five plants). The Department of Environment and Tourism was the local institution responsible for establishing the best strategy, coordinate and implement the CP action plan together with the NCPC. The program actions consisted on making available information and awareness, financial support, local installation of technology, cooperation in training people on operating the new technology, and support on how to create the necessary regulation. After the completion of the six-year program, the NCPC responsibilities were transferred to the Council for Scientific and Industrial Research (CSIR), an independent unit managed by an executive committee with mixed participation in the business sector and the government. In the meantime, the basis for a sustainable funding of the above-mentioned council was ensured under a competitiveness program developed by the South Africa Department of Trade and Industry.

Kenya was also one of the first African countries to welcome NCPCs. However, later, a specific program around Lake Victoria was developed in terms of environmental management, integrating that country, as well as Burundi, Uganda, Tanzania and Rwanda, which was carried out between 2009 and 2015 under the funding support of the Swedish Government through a World Bank loan. This program, essentially coordinated by the NCPC in Kenya, is started by the identification of the main sources of pollution providing from Kenya, Uganda, and Tanzania, targeting 88 companies as responsible for the discharge of about 7000 ton of pollutants in the lake. Several improvements were proposed and implemented, improving as well personnel training about environmental management. It can be refereed as an example the case of the Kenyan Kitumbe Tea Factory where several improvements were undertaken, namely rainwater harvesting, LED lighting and solar drying, among others, leading to a 60% reduction in specific water use, as well as a 20% reduction in energy consumption. The total investment done in the first 30 companies' adherent to the program was USD81 million, leading to annual savings estimated in USD15.5 million. From the USD81 million previously refereed, about USD65 million were spent in end-of-pipe technology regarding wastewater treatment in just three companies, conducting to savings of USD2.3 million. Thus, the investments cannot be considered in a proportional ratio with the gains achieved, as can be seen in this case, because the other 27 companies spent smaller amounts of money, which payback was surely lesser than 18 months. Later, the program was extended to Burundi and Rwanda following the same methodology.

Between 2009 and 2012, UNIDO implemented TEST in Egypt, Morocco, and Tunisia, in partnership with the NCPCs already existent in those countries, as well as with some related technical centers. Under this program, a total of 43 companies were studied and received support regarding the implementation of Cleaner Production assessment, employment of environmental management system and environmental management accounting. At the end, 76% of the actions were undertaken, among the 765 ones previously identified as necessary, in a total investment of approximately USD20 million (average of USD465,000 per company) in order to attain annual savings of approximately USD17 million (average USD400,000 per company). It can be also highlighted that this project led to annual savings of 9.7 million m^3 of freshwater and 263 GWhr, which represent important natural resources savings (UNIDO 2017).

References

Asia Pacific Roundtable for Sustainable Consumption and Production, *Asia Pacific Roundtable for Sustainable Consumption and Production* (2016), www.aprscp.net/APRSCP-Oct16.pdf. Retrieved 18 Dec 2017

Y. Bai, J. Yin, Y. Yuan, Y. Guo, D. Song, An innovative system for promoting cleaner production: mandatory cleaner production audits in China. J. Clean. Prod. **108**, 883–890 (2015)

C. Boyle, Cleaner production in New Zealand. J. Clean. Prod. **7**, 59–67 (1999)

G. Brown, L. Stone, Cleaner production in New Zealand: taking stock. J. Clean. Prod. **15**, 716–728 (2007)

I.S. Chang, J. Wu, H.M. Qiao, Z.L. Zhang, The spatio-temporal approach to regional analysis on cleaner production in China. Renew. Sustain. Energy Rev. **52**, 1491–1503 (2015)

K. Charmondusit, S. H. Gheewala, T. Mungcharoen, Green and sustainable innovation for cleaner production in the Asia-Pacific region. J. Cleaner Prod. **134**(Part B), 443–446 (2016)

E. Collins, S. Lawrence, K. Pavlovich, C. Ryan, Business networks and the uptake of sustainability practices: the case of New Zealand. J. Clean. Prod. **15**, 729–740 (2007)

EPA, *Evaluating the Effectiveness of the EPA Pollution Prevention Information Network Grant Program, Report* (2008), https://www.epa.gov/sites/production/files/2015-09/documents/eval-effectiveness-epa-p2-information-network.pdf. Retrieved 20 Dec 2017

EPA, United States Environmental Protection Agency, *Fiscal Year 2018, Justification of Appropriation Estimates for the Committee on Appropriations, Tab 14: Program Performance* (2017), EPA-190-K-17-002, https://www.epa.gov/sites/production/files/2017-06/documents/fy18-cj-14-program-performance.pdf. Retrieved 20 Dec 2017

I. Kliopova, J.K. Staniskis, The evaluation of cleaner production performance in Lithuanian industries. J. Clean. Prod. **14**, 1561–1575 (2006)

G. Miller, J. Burke, C. McComas, K. Dick, Advancing pollution prevention and cleaner production e USA's contribution. J. Clean. Prod. **16**, 665–672 (2008)

NETT21, Global Environmental Centre Foundation (2017), http://nett21.gec.jp/CP_DATA/English/background.html. Retrieved 11 Dec 2017

H. Peng, Y. Liu, A comprehensive analysis of cleaner production policies in China. J. Clean. Prod. **135**, 1138–1149 (2016)

D. Sakr, A.A. Sena, Cleaner production status in the Middle East and North Africa region with special focus on Egypt. J. Clean. Prod. **141**, 1074–1086 (2017)

UN, UNIDO/UNEP Guidance Manual (Part 2): *How to establish and operate Cleaner Production Centres* (2017), http://www.unep.fr/shared/publications/other/WEBx0072xPA/manual_cdrom/Guidance%20Manual/PDF%20versions/Part2.pdf. Retrieved 4 Dec 2017

UNIDO, NCPC 20 years (2017), https://www.unido.org/our-focus/cross-cutting-services/partnerships-prosperity/networks-centres-forums-and-platforms/national-cleaner-production-centres-ncpcs-networks/ncpc-20-years. Retrieved 12 Dec 2017

UNIDO-UNEP, *National Cleaner Production Centres: 20 Years of Achievement, Towards Decoupling Resource Use and Environmental Impact from Manufacturing Growth* (2015), http://www.recpnet.org/wp-content/uploads/2016/08/NCPC-20-yrs-final.pdf. Retrieved 9 Dec 2017

H. Yüksel, An empirical evaluation of cleaner production practices in Turkey. J. Clean. Prod. **16S1**, S50–S57 (2008)

M.Z. Yusup, W.H.W. Mahmood, M.R. Salleh, M.N.A. Rahman, The implementation of cleaner production practices from Malaysian manufacturers' perspectives. J. Clean. Prod. **108**, 659–672 (2015)

J.Y. Zhang, J.H. Jiang, G.M. Tian, The potential of fertilizer management for reducing nitrous oxide emissions in the cleaner production of bamboo in China. J. Clean. Prod. **112**, 2536–2544 (2016)

Chapter 3
Global Population Growth and Industrial Impact on the Environment

Abstract Pollution is generated from numerous sources however, since the industrial revolution, the industrial sector has become the main contributor. This sector has been responsible for some of the most tragic attacks on Nature seen throughout the history of humanity, from nuclear disasters, to large oil spills and heavy metal poisoning of people and wild life. Nevertheless, population growth coupled with damaging consumption patterns and low recycling efforts also contribute significantly to an already bad scenario. Pollution can originate from many sources, assuming different forms and states of matter, contaminating and damaging fundamental resources for human survival such as the air, water, land, ozone layer, food chain, among many others. Furthermore, economic development and environmental damage seem to go hand in hand as important aspects are sometimes overlooked in favor of rapid economic gains or as an effort to attract external investments. In this chapter, the impact of the industrial sector on the environment is discussed, recalling some of the most disastrous accidents it has generated. Also, the different forms that pollution can assume is discussed as well as how they affect the environment and human health.

Keywords Industrial environmental impact · Industrial environmental disasters · Industrial footprint · Population growth and environment · Pollution · Air pollution · Water pollution · Human health

As referred in the Introduction, industrial activities remain the principal driving force for the world's economic growth, mainly due to its strong capacity of innovation and gains in terms of productivity. In the initial decades of the twentieth century, industrial growth increased rapidly as well as the Manufacturing Value Added (MVA), contributing by this way to poverty reduction in developing countries by the creation of employment and generation of incomes. Believing in the statistics, developing countries still remain with strong capacity to improve their industrial performance and incomes due to their ability and range of progress they present in terms of technological development.

Over the twentieth century, there was a significant movement of the industrialization from Western countries to the Eastern ones, as well as from the North to the South (using the world map as reference). Also, in the second-half of the twentieth century, there was a fast growth of the world's population. In 2020 it is expected that

F. J. Gomes da Silva and R. M. Gouveia, *Cleaner Production*,
https://doi.org/10.1007/978-3-030-23165-1_3

the world's population will have tripled when compared with 1950. In parallel, the world's energy consumption continually rises, being almost 5.5 times greater relatively to 1950. These different growth rates mainly illustrate how industrial activity has increased inducing an energy growth rate clearly higher than the world's population growth. Figure 3.1 illustrates in two graphs this comparison. Indeed, the energy consumption is currently taken as a Key Performance Indicator (KPI) for industrialization evolution. In Fig. 3.2 is shown the global CO_2 emissions and Earth temperature in a similar period, allowing for a correlation among factors.

As can be also seen in Fig. 3.1, 1950 represents the starting year of an escalation in terms of the amount of produced CO_2 emissions, coincident with the commence of population growth and with the significant increase of process energy consumption, with later repercussions in terms of global temperature rise. Thus, the correlation among these factors is perfectly clear, being emissions and the Earth's heating process accelerated by the technological development occurred after the end of the Second World War.

As referred in the UNIDO (2016) report, regarding the year of 2014, industrialized countries were responsible for 64.1% of the MVA but are losing influence compared with Developing and Emerging Industrial Economies (DEIEs), as these last ones were less affected by the 2008–2010 crisis and are responding better to customer demand challenges, through lower labor costs and the incorporation of new technologies.

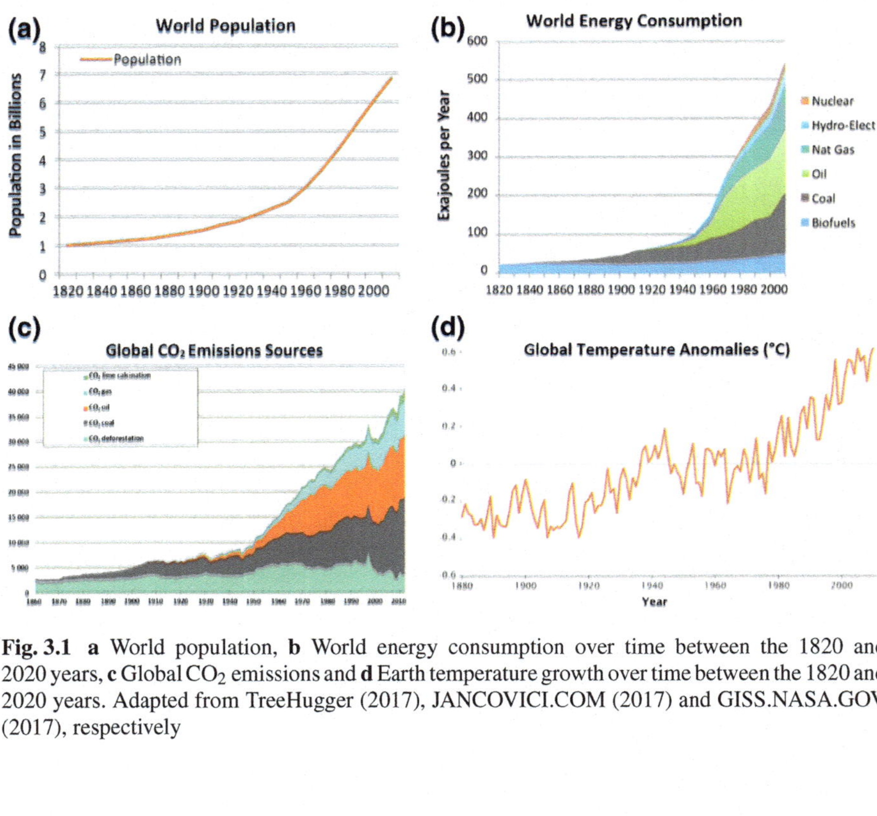

Fig. 3.1 **a** World population, **b** World energy consumption over time between the 1820 and 2020 years, **c** Global CO_2 emissions and **d** Earth temperature growth over time between the 1820 and 2020 years. Adapted from TreeHugger (2017), JANCOVICI.COM (2017) and GISS.NASA.GOV (2017), respectively

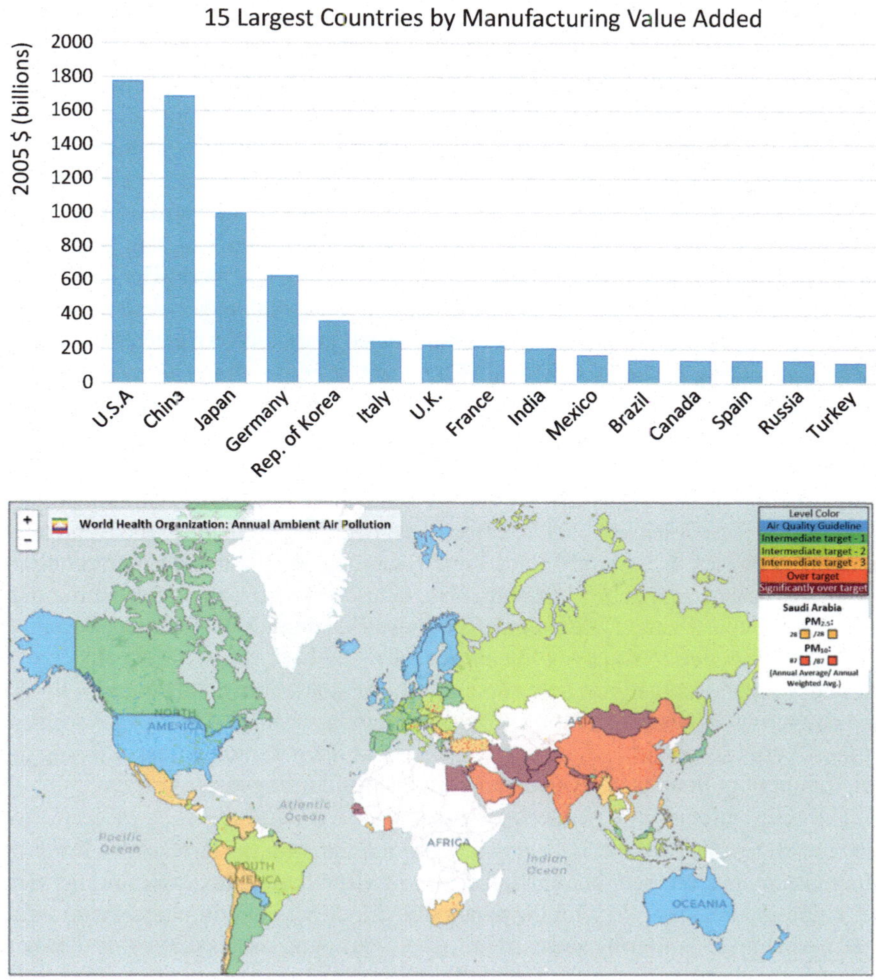

Fig. 3.2 Manufactured value added by country for the most 15 industrialized countries in the world (UNIDO 2016) and Ambient Air Pollution in the world (AQICN.ORG 2015)

China is playing an important role in the world MVA as it is responsible for 18.4% of the world's MVA in 2014, being the second largest manufacturing country in the world, after the USA. Moreover, the Chinese contribution to the DEIE between 1990 and 2014 registered a drastic increase from 15.8 to 51.3%, respectively. Furthermore, manufacturing activities assumed a relevant position in China's economy, being the largest sector of activity in that country and contributing one third to China's GDP. On the other hand, the contribution of the USA for the MVA is even larger than that presented by all non-China DEIEs countries combined.

As stated previously, industrial activity has influence on various types of pollution, namely solid wastes, water consumption and pollution, and air pollution.

Moreover, industrial activity usually needs energy, which is generated by different means, but only a minor part is renewable energy. Thus, the production of energy delivered to industrial companies can also generate pollution, of which one of the most used sources is based on hydrocarbons which is also one of the most pollutant. Regarding just air pollution, it is curious to note that the USA, despite being the most industrialized country, in general does not present concerning air pollution levels comparatively to some developing countries. Figure 3.2 shows the 15 most developed countries regarding a key parameter: MVA. Comparing this graph with the global map presented in the same figure provided by the World Health Organization (WHO) through the AQICN.ORG (2015) website, one can observe that the countries with higher air pollution rates are located in the Eastern side of the world. Indeed, according to the World Health Organization, the average air concentration of PM 2.5 particles in Delhi city (India) is 150 μg/cm^3, these extremely fine particles, with 2.5 μm diameter or smaller, are produced by vehicles engines and other types of combustion, some from industrial processes and power plants. On the other hand, London presents an average concentration of 16 μg/cm^3 of PM 2.5 particles of in the air (The Telegraph 2017). Effectively, regarding the top charts of the most polluted countries in the world issued in 2014 by the WHO, one can see countries such as Pakistan, Qatar, Afghanistan, Bangladesh, Egypt, United Arab Emirates, Mongolia, India, Bahrain and Nepal. The industrialization effect is more severe in these countries due to some significant reasons: (a) these countries usually live in difficult economic situations, being receptive to welcome some of the most pollutant and dangerous industrial facilities, (b) the regulations promoted by the government of these countries are not strong enough to delimit the pollution grade, allowing the generation of pollutants in levels not acceptable for developed countries.

The economic growth of developing countries (or DEIEs) can be faced in two antagonistic ways: without industry the economic growth is significantly lower or non-existent and, on the other hand, industry proliferation without rules brings serious environmental problems for the population. There are opinion-makers who argue that to end hunger, suffering and extreme poverty in countries it is preferable to install industrial facilities, even though outside the existing environmental safety regulations typical in most developed countries, then to prevent their installation and condemn the population to continue in misery. However, everyone can easily realize that is only a question of time as: people will live economically better but severe, health problems will be expectable in the medium term. Moreover, these companies are installing pollutant facilities in these countries taking advantage of their economic needs and lack of culture regarding personal health, wellbeing and corresponding future risks. Unfortunately, it is not expectable that developing countries share the same environmental concerns as developed countries, creating a greater gap in terms of general quality of life between ones and the others, not only mortgaging locally the health of these people, but also creating problems on a global scale. Evidently, the world would be better off if all countries abide with the established environmental rules but, it is clear those rules affect severely the day-to-day operation of some huge enterprises, being economy pillars in these countries, ensuring jobs for thousands of people and providing products with the required quality in terms of functionality, at the cost of ignoring some environmental rules and harming the environment. Thus, a blind implementation of the environmental policies cannot be employed in developing countries, otherwise it would seriously affect their development as many enterprises are not properly prepared to adjust their processes to the current demanding requirements already established in some developed countries, even entailing their closure, which negatively impacts a developing country's progress. Moreover, developing countries can learn based on important disasters over the last decade's and try to use modern technologies in order to prevent pollution and serious population harm.

3.1 Some of the Biggest Environmental Disaster Related with Industrial or Energetic Activities

The biggest environmental disasters with important impact to society are not exclusively from undeveloped or developing countries. Indeed, several industrialized and developed countries have suffered relevant setbacks, unfortunately setting examples leading to further improvements and establishment of new procedures and rules. Following enumerated are some important environmental disasters that had as origin industry or industry-related problems.

- **Itai-Itai disease** (Toyama Perfecture, Japan, 1912): Severe pain due to bone weakness and brittleness resulting from cadmium poisoning originating from Mitsui Mining and Smelting Co, Ltd. released on to the Jinzu River. The water was used for rice irrigation, entering by this way the food chain;
- **London great smog** (London, UK, 1952): through a complex combination of windless and cold weather, an intense smog resulting from burning coal, airborne pollutants engulfed the city of London, leading to the premature death of about 12,000 people.
- **Minamata disease** (Minamata, Japan, 1956): Neurological syndrome produced by severe mercury poisoning, resulting from methylmercury release in industrial wastewater providing from the chemical industry Chisso Corporation, which contaminated the fish used as food for people and animals. The harmful effect of the mercury poisoning was felt by the community over the next 36 years;
- **Ontario Minamata disease** (Ontario, Canada, 1970): The same neurological problems caused equally by mercury poisoning resulting from the Dryden Chemical Company mercury direct discharge in the Wabigoon-English River (approximately 9000 kg from 1962 to 1970). The harmful effect was felt until 1965, date on which the mercury cell was put out of service. The company closed down in the next year;
- **Seveso disaster** (Seveso, Italy, 1976): Release of 6000 kg of toxic chemicals which resulted in the highest known exposure to 2, 3, 7, 8-tetrachlorodibenzo-*p*-dioxin (TCDD) in residential populations. Around 18 km^2 were covered by the chemical contamination, affecting soils, water courses and the atmosphere. The

locals suffered some diseases such as chloracne, the only one with direct effect established with certainty. Other pathologies were felt some years after, namely cancer of the gastrointestinal sites and of the lymphatic and hematopoietic tissue.

- **Times Beach Dioxin contamination** (Missouri, EUA, 1971): Chemical waste heavily contaminated with dioxin providing from NEPACCO company, due to a concentration of dioxin resulting from the dioxin purification of the 2,4,5-trichlorophenol used in the production of hexachlorophene, was improperly cared by Russel Martin Bliss company who, unaware about the contamination potential of the product, sprayed a thick layer of waste oil in a horse farm and arena in order to control dust problems. The success of the dust control let to the use of the same product in other arenas, magnifying the negative dioxin effect. Subsequently, some months after, sixty-two horses died and the people directly in contact with the contaminated soil started having illnesses of different order, such as chloracne, diarrhea, abdominal pain and so on. Moreover, in 1972 Russel Bliss was contracted by Times Beach Authorities to spray 37 km of road with the same waste oil, magnifying even more the contamination. In 1972, a tank was discovered in the Verona facility of the NEPACCO company with more than 16,000 L of dioxin provided with a concentration higher than 340 ppm. In 1982, Times Beach suffered its worst flood in its history, disseminating the contamination. Thus, it was recommended the reallocation of its population, avoiding greater problems. The dioxin half-life, based on current knowledge, is estimated to be between seven to eleven years.
- **The Three Mile Island explosion** (Three Mile Island, Pennsylvania, USA, 1979): a partial nuclear meltdown of the reactor number 2, resulted in the release of huge amounts of radioactive gases and radioactive iodine into the environment. The accident had origin in a nuclear coolant leakage not valorized by workers and technicians due to a lack of expertise in this area to fix previous working defects felt in the nuclear reactor. The area cleanup process lasted about 14 years. No direct deaths were reported but the consequences in the populations' health in the surroundings was not properly evaluated.
- **Bhopal disaster** (Bhopal, India, 1984): perhaps, it was the worst disaster directly related to industrial activity. A methyl isocyanate leakage, as well as other chemical gases, providing from Union Carbide India Limited pesticide plant, let to the exposure of more than 500,000 people to that toxic cloud. The results vary from source to source, but following the local Authorities data, more than 3500 deaths were confirmed to be directly related with that gas release, but the number of injuries was significantly higher. The real causes of the disaster remain unclear, being attributed to sabotage or maintenance slouch. This accident was preceded by many other minor incidents in the same plant, mainly between 1981 and 1984.
- **Chernobyl disaster** (Chernobyl/Pripyat, Ukraine (old Soviet Union), 1986): this accident was strictly related with the production of energy by nuclear means. A mistake during a station blackout simulation resulted in an uncontrolled reaction condition. Water rushed into steam generated a destructive steam explosion and a consequent open-air graphite fire, which gave rise to updrafts, spreading radioactive substances in the air, with subsequent precipitation in the Western Soviet Union and Europe. The number of deaths directly related with the accident was

not considerable, but the further effects on workers and population health cannot be properly dissected.

- **Sandoz chemical spill** (Basel-Landschaft, Switzerland, 1986): the incident resulted from a fire and corresponding extinguishing process, leading to the release of toxic agrochemicals into the air and also tons of pollutants inflowing to the Rhine river, coloring it red. The pollution killed several wildlife downstream, situation that lasted more than two years to be restored. The origin of this accident was never discovered.
- **Phillips disaster** (Pasadena, Texas, USA, 1989): a release of about 40 ton of highly flammable gases took place at the Houston Chemical Complex of the Phillips 66 Company, devoted to the production of High-Density Polyethylene (HDPE) usually used for milk bottles and other plastic containers. The accident had as cause regular maintenance operations, leading to a reactor explosion. The blast felt reached 3.5 on the Richter scale and 23 workers died and more than 300 were injured, among the usual 1500 workers affected to this plant. The first explosion triggered further chain explosions.
- **AZF explosion** (Toulouse, France, 2001): the explosion in the Azote Fertilisant chemical plant in Toulouse measured 3.4 on the Richter scale and was equivalent to a detonation of 20–40 ton of TNT. It was heard in a radius of about 80 km. The accident gave rise to 29 direct deaths and caused wounds in about 2500 people.
- **Jilin explosion** (Jilin, China, 2005): the accident occurred in the Jilin Petrochemical plant in a nitration unit for aniline equipment. Six people were killed and about 70 people suffered injuries, forcing more than 10,000 people to be evacuated from the area. The blast broke windows about 200 m away. Moreover, the explosion contributed significantly to the contamination of the Songhua River, being estimated that about 100 ton of substances containing benzene and nitrobenzene made its way into the river. Human exposure to benzene is directly linked to diseases like Leukemia. The levels of benzene recorded about 80 km away, in the Amur River, were 108 times greater than the local safety limit.
- **Kingston coal fly ash slurry spill** (Roane County, Tennessee, USA, 2008): a spill of coal fly ash sludge from TVA Kingston Fossil Plant occurred in 2008, giving rise to the release of about 4 million m^3 of slurry providing from fly ash treatment. The spill reached the Emory River, expanding its influence on an area of about 1.2 km^2 and covering the surroundings with a thickness of coal fly ash slurry of about 1.8 m, damaging Nature, homes and waterways in that area. Tests performed to the river water close to the spill showed concerning levels of Arsenic, Mercury, Lead and Thallium. In the next year, an independent test detected preoccupant levels of toxic elements such as arsenic, copper, barium, cadmium, chromium, lead, mercury, nickel, and thallium in the river water.
- **Al-Mishraq fires** (Mishraq, Iraq, 2003 and 2016): regarding the political problems felt in that area, fire was deliberately set to the Mishraq Sulfur State Company, leading to the release of 21,000 ton of sulfur dioxide in 2003, giving rise to a white smoke cloud visible during 48 h and reaching Mosul, 45 km away, having had catastrophic effects in the local vegetation and originating the need of medical assistance for many local citizens. In 2016, the plant burned during seven days due

to a new terrorist attack, releasing about 161,000 ton of sulfur dioxide and killing two citizens and conducting about 1000 people to the hospital due to toxic gas inhalation. Definitively, these incidents had not origin in operation deficiencies, but just in deliberated attacks.

- **Fukushima Nuclear accident** (Fukushima, japan, 2011): a strong earthquake caused a tsunami which hit the nuclear power plant of Fukushima Daiichi, disabling the power supply and cooling system of three nuclear reactors, which partially melted in the first three days of the accident, lasting two weeks to be considered stable. However, the complete cooling process ended just 9 months later. High radioactive releases were felt in the next 6 days after the accident, however, over 100,000 people were evacuated from the area, avoiding directly-related deaths caused by radioactive radiation. The contaminated water leakage was particularly cared, preventing the release of radioactive materials. Regarding the huge dimension of the accident, the direct consequences to the environment were not so significant as other ones previously occurred.

These are just some of the most critical environmental accidents caused by industrial failures. However, the mining sector has also gave rise to severe environmental incidents, such as the Montana asbestos contamination (Libby, Montana, USA, 1919), the large crater with 70 m diameter induced by geologists exposing a huge reservoir of methane gas that does not stop burning (Derwezr, Turkmenistan, 1971), the Baia Mare cyanate spill in 2000 (Romania), amongst others. Moreover, oil tankers and oil refineries have been protagonists of numerous important accidents with catastrophic causes for the environment, such as the Amoco Cadiz oil spill in 1975, the Exxon Valdez oil spill in 1989, or more recently the BP Gulf of Mexico oil spill in 2010, this one lasting 87 days to be solved and causing around 8.8 Billion Dollars in natural resources damages.

The previously mentioned accidents are just some examples on how industrial and energy sectors are contributing on a large scale with accidents responsible for severe aggressions to the environment.

3.2 Effects of the Business Activity on the Environment

As already well-known, industrial pollution can have harmful effects on the environment, such as global warming, climate changes, air, soil or water pollution, human diseases, species extinction, amongst others. Following are presented some causes of the principal means affected by business activity.

- Air pollution: industrial activities, energy production and transportation are one of the most significant factors affecting air quality, due to the emission of Ozone (at ground-level), CO_2, NO_x and SO_2 gases, as well as particulate matters (PM), organic solvents, amongst others. CO gas is mainly emitted by ground transportation means, such as cars, bus, trucks, bulldozers, excavators (EPA 2018a). Moreover, the disintegration of some products containing asbestos fibers can release to the environment asbestos or other harmful dusts known to can cause human sickness such as lung diseases, when inhaled in high concentrations.

- Wastewater: there are several effluents discharged into waterways (treated or untreated), namely harmful products such as heavy metals and salts. The industrial effluents are usually in fewer amounts than the municipal ones, but their effect is much more harmful and intense than others. Untreated water may cause severe environmental consequences downstream, such as: land irrigation with contaminated water, groundwater reservoirs contamination, treated water degradation, water transportation means and so on. Moreover, this kind of pollution not only causes damage to the aquatic life and biodiversity on the river and its banks, but also flows into the sea, inducing further coastal pollution.
- Land pollution: hazardous products providing from industrial activities or energy production plants are the principal sources of land pollution. Soil contamination is normally originated by direct pollutants leakage, direct exposure to hazardous products or groundwater pollution. There are several activities contributing to this, namely chemical plants, oil refineries, metal surface treatment and coating plants, printing and textile industries, hazardous materials storage, oil and gas stations, amongst others. Usually, the decontamination of soils is not easy to perform, being necessary many years and a lot of money to reestablish the soil's original condition.
- Solid waste: this kind of pollution is generated by any activity in general, either be it from an industrial or from an everyday human life origin, different types of solid waste are produced: dry waste, organic waste or industrial waste. There has been a significant change regarding the policies related with this waste, from treatment and landfill deposition, to the more recent orientations related to the first steps of the product development, selecting materials and manufacturing techniques in order to allow for material recycling, reuse or conversion into another product, minimizing the need of landfills.
- Hazardous materials: this pollution has origin mainly in agriculture and industrial activities (primary and secondary sectors of business activity). In the primary sector, the use of chemical products with the intent of increasing productivity and prevention against pests and phenomena of nature are a common practice, e.g. pesticides, herbicides and fungicides. These products, if ingested in small portions by animals, can lead to poisoning or severe diseases which then can be transmitted to humans in different ways. Moreover, incorrect distribution of these products in agriculture can convey chemicals to animals through pasture, introducing in the food chain harmful products to human health. The storage of these products and possible leakage are a real concern. Regarding the secondary sector, the threats were already referred previously, but chemical and energy industries are some of the business sectors where this kind of pollution is more frequent and intense.
- Radiation: is a non-visible, non-perceptible type of pollution, that can generate serious health issues to humans. Regarding recent technologies, people and the environment are exposed to both ionizing radiation and non-ionizing radiation. Ionizing radiation provides essentially from sources such as X-ray machines, accelerators and radioactive materials. Non-ionizing radiation is coming mainly from mobile broadcasting centers, electrical installations and lasers. This kind of devices

and techniques are subjected to license from the national authorities in order to regulate the total amount of radiation emitted but, population aggregates are subjected to higher levels of radiation. Furthermore, the licensing regulation must respect the maximum levels considered as safe for the human health, following the international standards in this field.

- Noise: this is a kind of pollution is usually linked to the traffic in the big cities. However, this phenomenon is much more usual than it might seem. From amusement areas to some industrial atmospheres, noise is present in many places, affecting human health and causing many diseases. Indeed, frequent or intense exposure to loud environments may cause mental disturbance, physical fatigue, among others.

As described above, there are many implications of business activities on the environment, all of them with more or less impact on our lives, public health, weather and the future of humanity.

3.3 Effects of Industrial Pollution on Public Health and Human Behavior

The most immediate and continuous effects of pollution are usually related to human well-being and health. As previously referred, some of the environmental disasters described gave rise to perfectly identified illnesses, being possible to surely establish a relationship between the harmful product and the disease it caused. Furthermore, many studies have been carried out in order to correlate certain types of pollution with various kinds of health disorders. The vast majority of the studies have been carried out in the field of air pollution, where a cumulative effect of two different sources of pollution is acting: ground transportation and industrial. Moreover, although some studies have been carried out in other countries, China is the main target of these studies, essentially due to two circumstances: it is the most populated country in the world, with huge population concentrations in its main cities and the fast growing rate of its industry in the last decades, passing rapidly from the primary economic sector of activity (agriculture) to the secondary one (industry) with the corresponding inexperience regarding the best practices for managing environmental protection. In fact, concerning the period between 2001 and 2010, the number of premature deaths attributed to air pollution in China arose from 418,000 to 514,000, an increase of around 20%. However, these numbers are conservative when comparing with another study done by the WHO (Global Burden of Disease Study), which predicted a number of premature deaths about 1.2 million in 2010 (Wu et al. 2018).

In a wider view, the environment affects public health directly by different manners , such as through harmful exposure, poor working conditions, scarce and inefficient infrastructures and dishonored ecosystems. In addition, there are other environmental risks, such as household air pollution and climate changes. Some of these causes can be attributed to industrial pollution, but others are related to economic reasons as seen in impoverished countries. However, as those causes are inducing health problems, the working capacity of a population is also affected, decreasing their ability to generate added-value, contributing by this way to the increase of poverty, creating a vicious cycle that leads to slower development rates and less well-being of their population. Effectively, a study performed by Zheng et al. (2016) reported that when under severe air pollution, people are drastically less prone to dine out frequently, taking as well less satisfaction from dinning out. On the other hand, Graff and Neidell (2012) settled that air pollution had a noteworthy adverse impact on productivity.

3.3.1 Effects of Air Pollution

Air pollution has been responsible for a huge number of deaths. Regarding the information launched by the World Health Organization, approximately 6.5 million deaths in 2012 were related to air pollution. The same organization estimates that 3 million deaths per year are result of exposure to outdoor air pollution and about 4 million can result of household exposure to smoke from dirty cookstoves and fuels. Within this group of deaths, 36% result from lung cancer, 34% provide from strokes and 27% are consequence of heart disease (WHO 2018). Moreover, perception of air pollution induces people to avoid doing outdoor activities, which also has damaging effects to their health. Indeed, a study carried out by Sexton (2012) refers that the time spent by people in outdoor activities decreased by 18% under the consciousness that air quality in their area presents alert signs. Some cities and industrial regions assume alarming views under air pollution, as can be seen in Fig. 3.3.

A study carried out some years ago considering all air pollutants, refers that 17% of the total deaths in China were caused by air pollution, which means that about 4400 people are dying daily (Rohde and Muller 2015). This fact generates repercussions on the Chinese economy, with an average global cost of 6.5% of the Chinese GDP since 2010, with clear tendency to rise due to the increasingly growth of the population, urbanization and business volume (Crane and Mao 2015). The situation in India is similar, with 620,000 estimated deaths in 2010 (Swaniti Initiatives 2015) and with an average value of 3% in the Indian GDP corresponding to expenses directly or indirectly related to air pollution (Burney and Ramanathan 2014).

The pollutants identified as with the most damaging influence on the public health include particulate matter (PM2.5 and PM10), carbon monoxide (CO), nitrogen dioxide (NO_2), ozone (O_3), sulfur dioxide (SO_2) and Lead (Pb) (Xiao et al. 2018). The effects of these pollutants are diverse and are usually expressed by different ways:

- Particulate Matter: PM is capable of penetrating deep into lung passageways and entering the bloodstream causing cardiovascular, cerebrovascular and respiratory impacts, being classified as a cause of lung cancer by WHO's International Agency for Research on Cancer (IARC) in 2013. Furthermore, it is used as well as one of the most common indicators to assess the health impact resulting from human exposure to ambient air pollution. The health risks associated with PM10 and PM2.5 (particulate matter of less than 10 and 2.5 μm in diameter) is particularly well

Fig. 3.3 Appearance of an air polluted industrial region. *Credits* AndreyGuryanov/Shutterstock.com

documented (WHO 2018). They can be generated by combustion, atmospheric reaction or mechanical processes;

- Ozone (O_3): it is also a greenhouse gas and strong oxidant that damages cells lining in the respiratory system, causing a variety of adverse health outcomes, including lung function decrease, asthma attacks, and premature death. Cells in the deep lung are most vulnerable because they are not protected by a mucus layer. O_3 usually results from reactions in the atmosphere from primary pollutants (NOx and VOCs) in the presence of sunlight and its development is favored by high temperatures. The WHO guidelines point out a maximum concentration of 100 $\mu g/m^3$ during 8 h;
- Nitrogen dioxide (NO_2): affects lung function in people with asthma, contributes to acid rain and secondary particle formation and it is a precursor of ground-level ozone. The NO_2 usually results from high-temperature combustion;
- Carbon monoxide (CO): it inhibits the capacity of blood to carry oxygen to organs and tissues, people with chronic heart disease may experience chest pain when CO levels are high and, at very high levels, CO impairs vision, manual dexterity and learning ability, and can be fatal. CO usually results from incomplete combustion;
- Sulfur dioxide (SO_2): causes reversible declines in lung function of people with asthma, exacerbates respiratory symptoms in sensitive individuals, contributing as well to acid rain and formation of PM2.5 through atmospheric reactions. The SO_2 is usually generated from combustion of sulfur-containing coal and oil, as well as from metal smelting operations;

- Lead (Pb): delays the intellectual development of children. The main source of Lead in the air was from gasoline, but the new generation of combustion engines is prepared to use unleaded gasoline;
- Carbon Dioxide (CO_2): carbon dioxide perseveres in the atmosphere for 50–200 years. Thus, it can be expected that past and current releases continue contributing to global warming now and in the future. Global warming is generating new demands for water, while water sources tend to decrease in amount and quality. Water is essential to human health, as well as to manufacturing processes, food and energy production. Climate change is expected to promote drastic weather events in some areas, thereby causing an increase in sediment and pollutants washed into drinking water supplies. Moreover, rising sea levels will cause saltwater to infiltrate some freshwater systems, increasing the need for desalination and drinking water treatments.

Sustainable Development Goals (SDGs) reflect the concerns about air pollution: the levels of air pollution in cities are taken as an indicator for urban sustainable development (SDG 11). Access to clean energy—particularly clean household fuels and technologies—is highlighted as an indicator for sustainable energy (SDG 7) and mortality due to air pollution (ambient and household) is used an indicator for the SDG health goal (SDG 3). Consistent estimations regarding the exposure to air pollution and its impacts should constitute the work base of who decide in terms of environmental policies, improving the quality of air through pollution reduction.

3.3.1.1 Particulate Matter (PM)

It must be referred that exposure to particulate matter under 2.5 μm (PM2.5) in the European Union has decreased the human average lifespan in around 8.6 months (Orru et al. 2011). These particles present small diameter, passing easily through the respiratory filtering system, carrying numerous toxic elements able to penetrate airways and lungs, which thereafter will be passed to other parts of the human body by diffusion, causing damage and various diseases. The influence of exposure to these kinds of particles is even more severe in children under the age of 5 and elderly people (Xing et al. 2016). Moreover, Pope et al. (2002) collected data related to 500 thousand adults living in large cities, concluding that both overall mortality and mortality related to cardiopulmonary diseases, as well as lung cancer, augmented by 4, 6 and 8%, respectively, for every 10 $\mu g/m^3$ $PM_{2.5}$ increase. Furthermore, another study carried out by the American Cancer Society collected and sorted data of about 1.2 million American adults for 26 years [1982–2008], finding that lung cancer mortality enlarged by 15–27% when PM2.5 air concentrations increased by 10 $\mu g/m^3$ (Turner et al. 2011).

PM2.5 has also increase dramatically in several Chinese cities, but mainly in Beijing. Indeed, less than 1% of the 500 largest Chinese cities are in line with the air quality guidelines suggested by the WHO. Moreover, seven of these cities are usually rated among the ten most polluted cities in the world (Xing et al. 2016). Despite many actions regarding air pollution control in Beijing, according to the Environmental Statement of China in 2014, concentrations around 80 $\mu g/m^3$ of PM2.5 were usual in Beijing during 2014. However, this concentration was clearly exceeded on the 26th of December 2015, varying between 314 and 340 $\mu g/m^3$ in Tiananmen Square (IBT 2015). The World Health Organization recommends as limit an exposure to 25 $\mu g/m^3$ of PM2.5 over 24 h or 10 $\mu g/m^3$ of PM2.5 as annual average exposition. PM2.5 are considered more harmful that PM10 due to its effect of penetration in the alveolar and tracheobronchial regions, bringing systemic effects. Effectively, epidemiological researches have been carried out, showing a direct correlation between PM2.5 exposure and enlarged risks of mortality and morbidity connected to respiratory and cardiopulmonary diseases (Salomon et al. 2013) and toxicological confirmation advocates that platelet activation and inflammation in the lungs also can be directly related to PM2.5 exposure (Lippmann 2014; Van Winkle et al. 2015). In Fig. 3.4 one can find a satellite map attending the worldwide average situation of PM2.5 concentrations over 2001–2006, where it is possible to perceive high concentrations throughout Asia as well as particular cases such as New York and Africa presenting dangerous concentration of PM2.5, according to the limits above described as recommended by the WHO.

Regarding exposure to PM2.5, Yin et al. (2017) and Morelli et al. (2016) refer several studies reporting its effects on human health, such as respiratory and cardiovascular morbidity and mortality (Pope and Dockery 2006; Künzli et al. 2010; Peters 2011), other respiratory diseases (Hao et al. 2015), cardiopulmonary mortality (Samoli et al. 2005; Ostro et al. 2006; Boldo et al. 2006), cardiovascular disease and lung cancer (Pope et al. 2011; Lepeule et al. 2012; Raaschou-Nielsen et al. 2013), brain damages (Holgate et al. 2016), as well as a direct correlation between that exposure and other types of diseases, such as type II diabetes (Holgate et al. 2016; Frampton et al. 2012), obesity and hypertension (Dubowsky et al. 2006; Holgate et al.

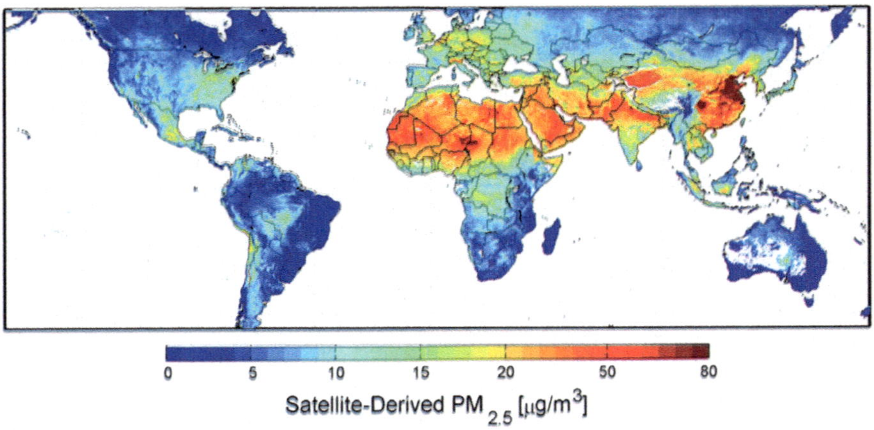

Fig. 3.4 Global satellite-derived map of PM2.5 averaged over 2001–2006. *Credit* Dalhousie University, Aaron van Donkelaar (NASA 2018)

2016). According to the referred above and other studies developed in the last years (Kioumourtzoglou et al. 2016), PM2.5 exposure is also the cause of some well-known identified mental illnesses such as dementia, Alzheimer's and Parkinson's sicknesses, in the field of neurodegenerative diseases, which usually implies hospitalization. Moreover, the diseases refereed previously can arise under either short or long-term exposure to PM2.5, being also classified underneath two categories of impact: acute or chronic (Katsouyanni et al. 1997; Analitis et al. 2006; Hoek et al. 2013). Figure 3.5 intends to give an overview of all potential health diseases caused by PM2.5 exposure, being possible to observe that the same disease can assume different levels of severity.

Estimations made through the study developed by Yin et al. (2017) leads to perceive that 0.03% of Beijing's population was victim of premature death due to PM2.5 pollution exposure, being chronic bronchitis the disease with higher impact in that number, regarding the year of 2012. Furthermore, this result can be compared with another achieved by Fattore et al. (2011) taking into account two industrialized towns in Italy, where there was a 0.03% excess of mortality per year caused by PM2.5 pollution exposure with 10 $\mu g/m^3$. These results can be compared with a wider European study performed by Boldo et al. (2006) regarding Air Pollution and Health: a European Information System (EPHEIS) considering twenty-three European cities, reported an average value of 0.03% for all-cause mortality rate caused by PM2.5 exposure. On the other hand, the study conducted by Fann et al. (2012) regarding the U.S.A. overall population showed that the percentage of premature deaths due to PM2.5 exposure in 2005 was estimated in 0.04%. However, the study carried out by King et al. (2016) regarding just New York disclosed a higher value, around 0.07% of the resident population in this state. This result matches with the situation depicted in Fig. 3.3, showing a particular situation in New York City and surroundings, comparatively to the remaining American territory.

The other side of these tragic results is the cost that Governments are facing in terms of population support and loss of lives. Returning to the studies related to Beijing, Yin et al. (2017) states that the total external costs are around 0.3–0.9% of the regional GDP, based on the Beijing Statistical Yearbook of 2013, corresponding to the situation of the 2012 year.

However, there are other studies focused on the same region reporting different values. Taking as reference the year of 2013, Wang et al. (2016a) reports a loss of 0.06% in the GDP related to the Beijing region due to PM2.5 exposure and premature deaths. Another study carried out by Wang et al. (2015) regarding the year of 2010, but focused on the Yangtze Delta River estimates an economic burden of 0.02% in the regional GDP due to the same causes. Studies carried out predicting cost impact on the GDP related to PM2.5 exposure in 2030 in China without pollution control policies (Xie et al. 2016) are pointing out losses of 2%, whereas studies focused in Italy and having as scenario a target of 20 $\mu g/m^3$ as pollution limitation (Ciucci et al. 2016) show an impact on Italian GDP of just 0.01%. Despite the huge difference in population density between these countries, the investment in pollution control presents an excellent financial return and, more than that, a considerable amount of lives can be saved from premature death, improving the well-being of the population.

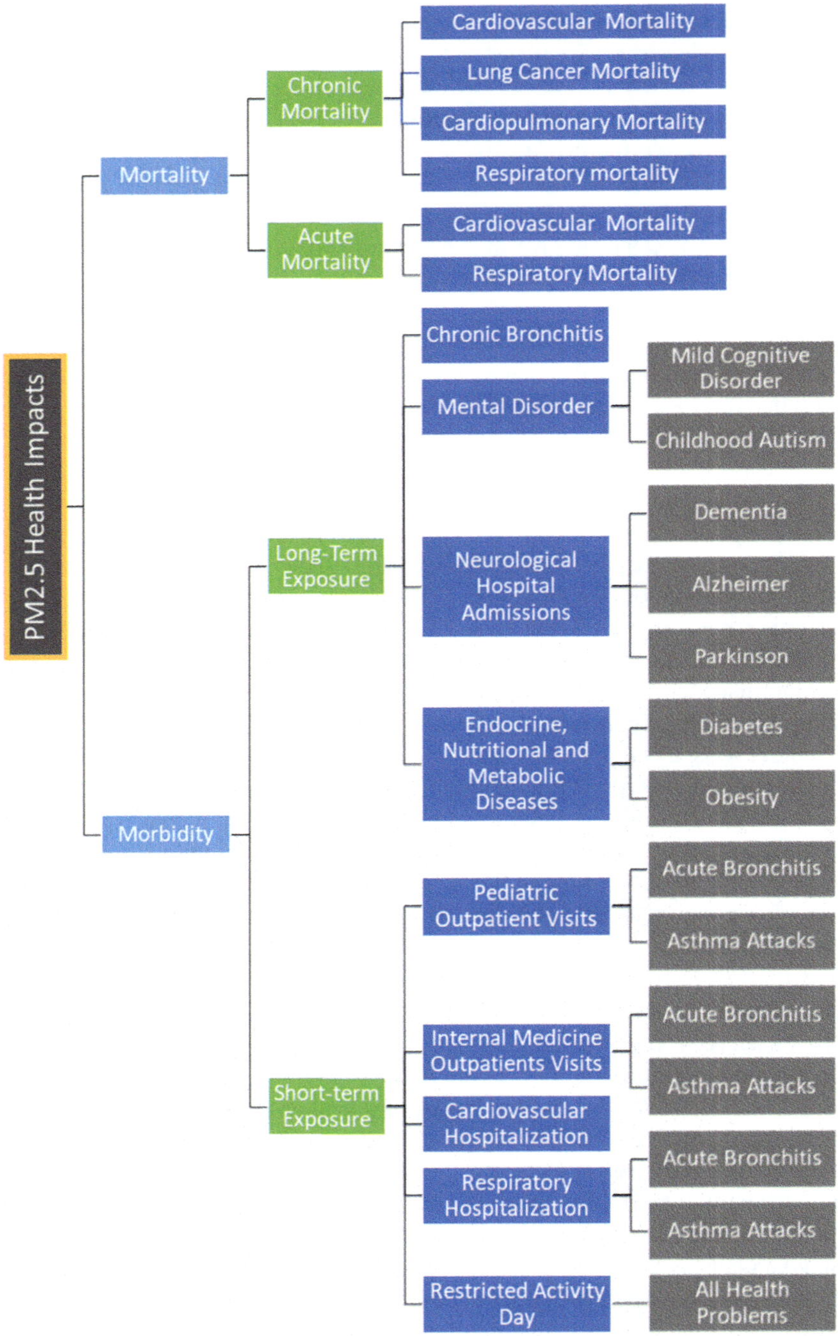

Fig. 3.5 Potential health impacts related to PM2.5 human exposure. Adapted from Yin et al. (2017)

Due to the abundance of studies concerning public health and corresponding costs in China, and because different approaches and techniques have been used, the numbers often differ. Studies carried out by the World Bank report that public health injuries in China between 1995 and 2003 can be estimated in 4–5% of the national GDP, concerning air pollution. However, other study performed by Matus et al. (2012) estimated that the public health had an impact on the Chinese GDP of about 6–9% between 1995 and 2005. This study was focused mainly in PM_{10} concentrations, which are usually considered less detrimental to public health than finer particles like $PM_{2.5}$ due to more effective nose filtration on the formers relatively to the last ones. However, in 2005, the mean concentration of PM10 in China was about 109 $\mu g/m^3$, exceeding clearly the limit recommended by WHO of 20 $\mu g/m^3$.

3.3.1.2 Ground Level Ozone (GLO)

Ozone (O_3) assumes two different facets: good Ozone and bad Ozone. The good Ozone is formed naturally in the upper atmosphere, constituting a shielding layer which avoids harmful ultraviolet rays from reaching the Earth's surface. This Ozone layer is sensible to some pollutant gases, which as is well known have been destroying it. However, a recent inversion on the harm effects caused by man is being felt and the hole created under these actions is now shrinking. The other kind of O_3 is usually called 'Ground level' Ozone (GLO), or even 'bad' Ozone, and is a harmful pollutant, which is formed through the chemical reaction between Nitrogen Oxides (NO_X) and Volatile Organic Compounds (VOC) under the action of sunlight. A sketch with the photochemical GLO formation can be seen in the work of Wang et al. (2016b). Up to date, and amongst the air pollutants previously referred, GLO is considered as the most harmful one in Europe, inducing negative effects on human health, agriculture, flora and some materials (Sicard et al. 2017). Nevertheless, this problem is not only serious in Europe. Measurements performed in three of the major cities in China (Beijing–Tianjin–Hebei, Yangtze River Delta and Pearl River Delta) from June 2016 to August 2016 allowed to realize that the emission of PM was exceeded by that of GLO (Liu et al. 2017). Regarding the year of 2015, the daily maximum 8-h Ozone concentration reached a record of 202.3 $\mu g/m^3$ on average, surpassing by 27% the values recommended by the Grade II Standard (160 $\mu g/m^3$). As stated before, these values depend on the VOCs/NO_x ratio and weather circumstances, which may result or not in favorable conditions for the creation of GLO (Simon et al. 2015). Moreover, the chemical reaction in the origin of GLO closely depends on the atmospheric pressure, relative humidity and incoming solar radiation (Pu et al. 2017). The sources of the gases which give rise to GLO are diverse, such as: industrial plants, industrial boilers, chemical plants power plants, refineries, fumes from car engines, gasoline hazes or chemical solvents. This GLO is particularly abundant and dangerous in sunny days and polluted areas but, since it can be easily conveyed by the wind, rural areas can also be affected, being possible to detect high concentrations of Ozone in these areas (EPA 2018b). The exposure to GLO is particularly harmful to people suffering of asthma, elderly people, children and people with intense outdoor activity, along with certain kind of people suffering from predisposition to malabsorption of some nutrients such as vitamin C and E. The inhalation of GLO can cause many adverse health reactions, such as pneumonia, worsening of bronchitis, allergic rhinitis, emphysema, asthma by muscles airways constriction, holding the air in the alveoli and giving rise to external symptoms such as coughing or wheezing, shortness of breath and chest tightness or pain, which may conduct people to hospital admissions. There are other studies that suggest the existence of a synergistic relationship between GLO and other pollutant which may be directly related with asthma occurrences (Alexis and Carlsten 2014). Moreover, researchers are looking for confirmation to a direct relationship between children short-term exposure to GLO and the appearance of asthma episodes (Sousa et al. 2013). The human health risk regarding pollutant exposure can be assessed by the probability of developing a related disease, regarding a certain previous health state, for each increase of 10 $\mu g/m^3$ of pollutants concentration in the air (Sicard et al. 2012). The risk values are considered on two different bases: daily 8-h maximum and daily 1-h maximum ozone concentrations. This risk is calculated and published by the WHO regarding a specific area (WHO 2001, 2004), in order to standardize the values, making them comparable and allowing for establishing the maximum exposure concentrations. The maximum level conducting to risk of death directly related to GLO exposure have been discussed and the limit previously established by the WHO was later discussed by Environmental Protection Agency (EPA) agents (EPA 2007). The first robust results with regard to health impact of long-term ozone exposure, considering maximum 1-h concentrations, were found by Jerrett et al. (2009), who's concept was later used by other researchers (Anenberg et al. 2010; Lelieveld et al. 2013), being known as the Relative Risk (RR) of death due to respiratory disease under long-term GLO exposure. Following the study carried out by Liu et al. (2017), the number of deaths directly related with GLO exposure in China during the year of 2015 was roughly estimated in the range from 50,000 to 80,000, disseminated throughout the cities of Beijing, Henan, Hubei, Shandong and in the Province of Sichuan, as well as the specific areas of Yangtze River Delta and Pearl River Delta, areas that are usually targeted in several studies due to their particularly polluted atmosphere. Regarding the effects of GLO on public health and increasing mortality related to GLO exposure, WHO (2005) suggested the reduction from 120 to 100 $\mu g/m^3$ of 8-h maximum concentration, recommendations that were observed by the Chinese authorities. Thus, new guidelines were established in China for GLO air quality standards, recommending for class 1 areas (remote) daily 8- and 1-h maximum of 100 and 160 $\mu g/m^3$ respectively, while for class 2 areas (urban/industrial and rural surroundings), the new established values were 160 and 200 $\mu g/m^3$, respectively. Some studies carried out in the Pearl River Delta established a direct relationship between the enlarged GLO concentration and the number of deaths, concluding that an increase of 10 $\mu g/m^3$ in the prior two days was directly related to an increased mortality in 0.81% (Tao et al. 2012). The Chinese values can be compared with others presented by US EPA (EPA 2013) by monitoring stations spread throughout the U.S.A., between 2007 and 2009, which presented values from 29 to 44 ppb for daily average 8-h maximum GLO concentration and 1-h maximum GLO concentration, respectively.

In addition to the public health risks, there are also effects on agricultural crops, which are mainly affected to growth problems than with problems related to photosynthesis and biomass accumulation (Wang et al. 2017). However, mainly in China, GLO concentrations in large cities may be easily swept away by the winds for several kilometers, reaching crops in the surroundings. Furthermore, studies carried out by Chameides et al. (1999), Feng et al. (2003, 2015) and Wang et al. (2012) suggested that wind-blown GLO concentrations from the largest cities may affect both winter wheat harvest and rice production. Effectively, Wang's study (2012) showed that 62 ppb concentrations of GLO negatively impacts roughly 14–20% of the China's rice production, whereas wheat can be impacted even further due to its higher level of sensitivity. In this case, the negative effect is felt differently throughout the seasons, being summer the most favorable season to generate GLO and spring the windiest. Thus, Feng et al. (2015) are estimating for 2020 spring wheat harvest losses between 2 and 29% for all of China, along with losses between 3.7 and 10% in rice crop, about 64% of the summer corn crop and around 45% in soybean crop. However, all these values may be affected drastically, depending on weather conditions and eventual diminishing of the pollution levels. In terms of public health effect, and regarding China's medical expenses, wage and leisure loss, it is possible to observe that PM causes an impact seven times higher than exposure to GLO, being the last one the second most severe pollutant in the Chinese economy (Matus et al. 2012).

3.3.1.3 Nitrogen Dioxide

Nitrogen dioxide (NO_2) is well-known as a combustion product of fossil fuels such as coal, oil or gas, having as source industrial or road traffic. NO_2 is odorless and nonflammable, being usually linked to congenital malformations. Thus, the exposure to this pollutant during the pregnancy period is particularly dangerous for the fetus. Moreover, NO_2 is also commonly referred as cause of childhood asthma and Perez et al. (2013) estimate in their study that 14% of all asthma occurrences are induced by high exposure to traffic. Studies carried out in primary schools in London, U.K. (Guerriero et al. 2016) lead to concluding that decreasing the indoor exposure to NO_2 allows a significant decrease in asthma episodes among the children in school. Furthermore, the exposure to NO_2 also decrease lungs function, leading to airways inflammation and other respiratory diseases, normally translated by difficulty breathing, coughing or wheezing. However, the studies carried out concerning the impact of NO_2 on mortality are not able to establish a direct relation between long-term NO_2 exposure and non-natural deaths. Thus, this pollutant has a negative repercussion on the human well-being, although not lethal by itself regarding the concentrations usually reported in the most polluted places. However, NO_2 as well as other NO_x products can be in the origin of particulate matter (PM) or Ozone, extremely harmful for the human health, by reaction with other chemical pollutants contained in the air.

3.3.1.4 Carbon Monoxide (CO)

Carbon monoxide (CO) is a dangerous pollutant as it is an odorless, colorless and tasteless gas, being usually a result of the incomplete combustion of carbon-containing combustibles or other compounds. The concentration of CO might be comprised between 0.1 ppm for clean atmosphere to 5000 ppm near domestic wood fire chimneys (Fawcett et al. 1992) being also present in noteworthy amounts in tobacco smoke (Hoffmann et al. 2001). In addition to tobacco, the main sources of CO are: fumes providing from vehicles with internal combustion engines, malfunctioning gas engines of heating systems, smoke related to fire events, charcoal burning, biomass burning or, less frequent, inhalation of methylene chloride or similar solvents providing from chemical products used as paint removers. CO concentrations vary pointedly from East to West or from North to South, being as well different from region to region in the same country, result of uneven levels of industrialization (Varma et al. 2015). The toxicity mechanism of CO is based on inhibition of cellular respiration and hypoxia, since the affinity of hemoglobin for CO is 210 times greater than the affinity for oxygen (Barret et al. 2015). CO can also injure the human body via CO-induced tissue hypoxia followed by re-oxygenation, damaging to the central nervous system (Prockop and Chichkova 2007). The danger presented by large concentrations of CO in the air is a consequence of CO being eagerly absorbed from the lungs into the bloodstream, forming a slowly and reversible complex with hemoglobin (Hb), known as carboxyhemoglobin (COHb) (Raub 1999; Janík et al. 2017). The longer the period of exposure to CO or the higher the concentration of CO in the inhaled air and alveolar airing, the greater the total amount of COHb is created in the blood stream. Older studies also refer other ways of CO induced hypoxia based on the decrease of oxygen release (Douglas et al. 1912; Roughton and Darling 1944), which leads to headaches or even death, depending on its intensity. CO causes as well inflammatory reactions by increasing levels of cytosolic heme and the heme oxygenase-1 protein, resulting in intracellular oxidative stress (Weaver 2009). The medical signs of CO poisoning are known being non-specific and differ with CO saturation levels, exposure duration or pre-existing health restrictions, as well as individual susceptibility (Janík et al. 2017). It is well-known that the exposure to low levels of CO (25–50 ppm) unfavorably disturbs the nervous and cardiopulmonary systems. The effects felt by humans due to CO poisoning can range from simple dizziness to death and its symptoms initiate at approximately 20% COHb, being necessary 50 and 80% COHb to be fatal (Ryter and Otterbein 2004). Nowadays, it is believed that air pollution may be roughly related to neurocognitive abnormalities as well as behavioral sicknesses like the autism spectrum (Vrijheid et al. 2012; Block et al. 2012). Regarding the industrial processes, CO poisoning only occurs under accidental leakage of concentrated chemical substances, which can cause severe damages to the people involved in the process, mainly in closed facilities with restrict air renovation. Despite this, CO poisoning is the leading cause of unnatural death in the U.S.A. (Iqbal et al. 2012).

3.3.1.5 Sulfur Dioxide (SO_2)

Sulfur Dioxide is a colorless gas characterized by a penetrant odor, which may assume the liquid state when under pressure. Moreover, it dissolves in water very easily and presents non-inflammable properties. As other pollutants, SO_2 provides from fossil fuel burning at power plants using coal or oil as energy source, along with from copper smelting. However, SO_2 may also result from volcanic eruptions in a natural scenario. When SO_2 is released into the air, it can be converted to sulfuric acid, sulfur trioxide or sulfates. However, SO_2 also dissolves in water, giving rise to sulfurous acid. Soil also can absorb SO_2 but its diffusion in the soil is not well-known. Furthermore, SO_2 presents complex behaviors, acting directly as pollutant or stimulating the formation of other pollutants, such as particulates. As it is well-known, SO_2 acts as principal precursor of sulfate aerosols through the heterogeneous pathways (Yang et al. 2016). Other well-known pollutants such as NO_2, O_3 or OH radicals can powerfully induce SO_2 oxidation in aerosol media, releasing high levels of sulfate as result of the reaction (Liu et al. 2012), which may give rise to particulate sulfate under severe haze conditions in China (Wang et al. 2016b). Moreover, sulfate is also a key haze promotor, inducing losses of visibility (Guo et al. 2014). The mixture between SO_2 and GLO with soot particles has been also studied by He et al. (2017), remaining clear that an increasing content of water in soot, taking advantage of its hygroscopic properties, induces a greater sulfate formation rate on soot, which, in its turn, changes the Relative Humidity (RH). Thus, under elevated levels of RH, SO_2 soot particles toxicity is more intense. Hence, under high RH conditions and with the co-existence of GLO and SO_2, sulfate formation is ensured, increasing the toxic effects for the environment and public health (He et al. 2017).

People may contact with SO_2 by breathing or through the skin, when it assumes the form of sulfuric acid. SO_2 assumes higher concentrations in industrial processes where it arises as a by-product in the industries abovementioned, as well as in other fields of industry with predominant use of chemicals in a continuous way, such as food preservatives, fertilizers, paper manufacture and, obviously, in the production of sulfuric acid. It is well-known that SO_2 induces cardiovascular diseases (Mar et al. 2000; Hong et al. 2002), causing cardiac myocyte damage by the increase of intracellular calcium (Zhang et al. 2013). Indeed, SO_2 clearly presents vaso-relaxant effects, which are usually felt on the coronary artery. Moreover, Zhang et al. (2013) also conclude that the SO_2 negative effect on the heart is significantly higher than SO_2 derivatives, based on experiences carried out with mice. In fact, as early as in 1987, Neuberger et al. (1987). found a temporal correlation between prominent levels of SO_2 and serious intensification of respiratory and cardiopulmonary mortality in Vienna between the years of 1972 and 1983, mainly regarding elderly people. Moreover, a study performed by Touloumi (1997) taking as target 12 European cities, allowed to conclude that daily mortality rate arose 3% at each increase of 50 $\mu g/m^3$ of SO_2 levels. Regarding a recent study carried out by Khaniabadi et al. (2018), about cardiovascular and respiratory mortality related to PM_{10} and SO_2 air pollutants in Khorramabad region of Iraq, cardiovascular mortality played key role in the overall mortality caused by air pollution.

It is curious to note that exposure to SO_2 is worst in the winter, due to mechanisms referred previously. However, it is in the summer that more severe cases are usually reported, mainly due to an increase of outdoor activities, as during the winter the major trend is for people to remain at home or in their working place.

It seems obvious that, through regulation or by self-initiative, different countries should act in order to avoid larger health problems. In the work presented by Li et al. (2016) regarding some regions in Hong Kong, it is clear that regulations imposed by the Government produced real positive effects for public health. Referring the years of 1989, 1990 and 1991, period on which restrictions to SO_2 emissions were imposed, the peak values reached in 1989 and 1990 of 88–101 mg/m^3 and 113–136 mg/m^3 dropped to 23–26 mg/m^3 forced by the legislation launched by the Honk Kong Government in July 1990. Thus, if the market is not able to care for the environment as needed, Governments need to enforce restrictions in order to protect public health and minimize health costs.

3.3.1.6 Lead (Pb)

Lead is a metal with very interesting properties, which use was largely widespread by several important fields, having been applied in the manufacture of pipes, paints, glazes for pottery, insecticides, hair dyes, as well as an additive for petrol, regarding its self-lubricant properties. These and other applications took advantage of other well-known lead properties, such as high corrosion resistance, elevated specific weight and low melting temperature. Due to studies that revealed the high toxicity of lead, mainly in children, restrictions to its use were established and many applications needed to find alternative solutions, in order to comply with regulations. As referred previously, many former applications implied direct contact with lead i.e. breathing of vapors, contact and absorption by hair or skin. Despite the strong reduction of lead use, some applications cannot yet find a substitute solution, remaining in use in some products, such as car batteries, as alloy element in filler metals for soldering (Pb–Sn alloys), lead crystal glass, weights for lifting, diving belts, pigments for paints, ammunitions and radiation protective effects, taking advantage of the properties abovementioned. However, the Governments in EU and the U.S.A. have imposed serious restrictions for the use of lead as well as increased its taxation. The restrictions were imposed by the following legislation launched in July 2006: European Union Waste Electrical and Electronic Equipment Directive (WEEE) and Restriction of Hazardous Substances Directive (RoHS).

People have exposure to lead in many ways, being the daily human intake usually about 0.1 mg. Contamination does not induce immediate symptoms. There is no safe blood level known, mainly for children, as referred by the Center for Disease Control and Prevention (CDCP 2017). Lead can affect multiple systems in the human body and tends to accumulate in the bones and teeth, which can store up to 120 mg. However, brain, kidney and liver also absorb lead. The main effect on human health is felt on the level of disturbance of embryo or fetus development, along with cancer problems. In fetuses and young children, lead also seriously affects brain development, as well as the nervous system. Moreover, children are usually more likely to have lead exposure do to the contact with old housing objects containing lead or soil and there after taking their hands to their mouth. Moreover, lead tends to be absorbed easier by children than adults. Under very low blood levels, some symptoms are almost inexpressive, such as reduced stature or growth difficulties, as well as diminished hearing acuity. Low blood lead levels as 10 mg/mL are usually associated to lacks of intelligence and weakened neurobehavioral development. Medium levels of lead absorption may conduct to disturbance of the central nervous system, decrease in renal function or hematopoietic system diseases, hypertension, cognitive deficiencies and reproductive scarcities (National Toxicology Program 2012). Elevated lead levels in children may cause acute neurologic problems, which can result in coma, convulsions or even death (Gwini et al. 2012; Chowdhury et al. 2014; Steenland et al. 1992, 2017).

A recent study carried out by Levin (2016) about the direct and indirect costs related with health problems caused by lead in the U.S.A. suggests that more than 392 million dollars are spent annually, just referring occupational activities in consequence of the contact with products and soil containing lead, thus, other activities are not included in these costs. In fact, the same study refers 804,000 industry workers and 838,000 construction industry workers as potential people also exposed to harmful lead effects as consequence of their jobs, following the data provided by Occupational Safety and Health Administration (OSHA 2017). Moreover, Alarcon (2016) refers that 94% of the adults with significant lead in their blood levels are facing jobs where they are exposed to lead in their workplace, showing a lack of information about this issue because not all employers are officially declaring the blood levels of their workers. Furthermore, OSHA acts in the U.S.A. in order to "assure so far as possible every working man and woman in the Nation safe and healthful working conditions" (Williams-Steiger Act 1970), using for that two different standards: 29 CFR 1910.1025 (USDL 2018a) devoted to general industry and 29 CFR 1926.62 (USDL 2018b) related to construction industry.

Regarding lead levels present in the air, regulations may need to be revised. Effectively, the current legislation recommends as limit an exposure of 50 $\mu g/m^3$ mean for 8-h, either for industry in general as well for construction industry. However, this goal was defined based on keeping blood lead levels (BLLs) less than 60 $\mu g/dl$ over a working lifetime, as define in the American standard (43 CFR 1910.1025 1978) (GPO 2018). Nevertheless, to accomplish the goal established for 2020 in which workers should keep their BLLs lower than 10 $\mu g/dl$ (HealthyPeople 2018), regulations need to be rapidly adjusted. In addition, studies are referring that the established BLL level target of 30 $\mu g/dl$ in some countries is not enough, as there are reports of lead reproductive effects at much lower BLLs (Kosnett et al. 2007). Regarding also wind dragged lead, it is important to refer people who are exposed to indoor firing ranges, due to the lead usually used in projectiles. Indeed, it is estimated that about 16,000–18,000 people are practicing target shot or working in indoor firing ranges in the U.S.A., being exposed to airborne lead dust released by lead-based projectiles (Beaucham et al. 2014).

Children represent the population group mostly susceptible to contamination by lead, because they absorb larger amounts of metals from the digestive system relatively to adults and elderly people (Lidsky and Schneider 2003). It is also reported that, in addition to children, people with nutritional insufficiencies, pregnant or lactating women present increased absorption of lead as substitute for calcium (Ettinger et al. 2009). The development of children's nervous systems is particularly vulnerable to toxicants (Grandjean and Landrigan 2006), which, even under low levels of exposure may cause long-term cerebral weakening (Centers for Disease Control and Prevention 2012), with nefarious effects on memory, intelligence and behavioral aspects (Lanphear et al. 2005). Obviously, these situations entail high costs to the society and economy of countries. Serious effects can also be felt by women working in places with prospective exposure to lead, as lead accumulated in the bones tends to cross the placenta during the pregnancy period, which may cause neurodevelopmental delays and other complications for the baby's health (Gershanik et al. 1974; Silbergeld 1991; Schnaas et al. 2006; Shaffer and Gilbert 2017). Moreover, Joo et al. (2018), after a deep study concerning the effect of lead on different genders, concluded that males are more vulnerable to prenatal exposure, whereas females are more defenseless to postnatal exposure, being the main effects felt on attention and sleep disturbance. There are several studies about the different impact of lead exposure in females and males, but the results are not consensual (Dietrich et al. 1987; Bellinger et al. 1990; Rabinowitz et al. 1991; Tong et al. 2000), as referred by Kasten-Jolly and Laurence (2017). In fact, those studies report disagreeing results in time exposure needed to produce effects, necessary concentrations and effects duration. Anderson et al. (2012) report that reversible effect on memory damage can be observed in males, but the same effect is not common in females. Moreover, males tend to lose the sense of smell, experiencing as well a more pronounced grey-matter change. On the other hand, females present higher tendency for sleep disturbance, depression and anxiety relatively to the males. These different effects in males and females are attributed to hormonal differences, affecting tissue damage in different ways. Kasten-Jolly et al. (2012) found as well that females are more prone to develop Alzheimer's disease than the males, when exposed to lead in childhood.

3.3.1.7 Carbon Dioxide (CO_2)

The effect of greenhouse gases (GHG) has been deeply studied by many authors. Usually, GHGs refer to gases such as: carbon dioxide, methane, nitrous oxide and fluoride. Thus, GHGs include the usual emissions released during manufacturing processes, transportation related services and product consumption. Hertwich and Peters (2009) carried out a broad study in order to identify the footprint of eight classes of products and services throughout 73 countries and 14 combined regions in the world. Some key numbers are revealed in this work, where the average national footprints per capita are very different depending on the regions, reporting values of around 1 tCO_2/year regarding African countries, which contrasts with the values pointed out for the U.S.A. or Luxembourg: 301 tCO_2/year. In a global view, the greenhouse gas emissions are divided into three main groups with the following distribution: 72% are generated by household consumption, 18% are related to investments and 10% correspond to government consumption. Into these groups, there are important subgroups, such as: food is responsible for 20% of the greenhouse gas emissions, houses maintenance and upgrades contribute with 19% and mobility represents around 17%. Moreover, this distribution is not similar in developed or developing countries: mobility and manufactured goods assume higher importance in the developed countries, while food and services undertake more relevance in developing countries (Hertwich and Peters 2009). Indeed, carbon footprint has taken on more significance for consumers, manufacturers and investors, because everyone is aware of its meaning and impact to the future of humanity, attending to the climate changes and other phenomena felt more and more often. Thus, some manufacturers are putting the corresponding footprint on the package of their products, trying to distinguish their products from competitors by this way, exhibiting a greater attention to the environment and to these details. On the other hand, consumers assume a higher responsibility in the global market, being aware that their purchasing choice can influence the manufacturers' behavior, searching for information about footprint data and selecting the products with most favorable values. Finally, investors need to pay attention to the behavior of companies, as risks are incomparably lower with a compliant company, increasing attention and attracting bigger profits. The same questions can be put in relation to the supply chain around businesses: a modern fleet will generate fewer emissions and will be more efficient. Obviously, these situations are better cared and driven by consumers in developed countries. Furthermore, the attention given by governments is also different in developed and developing countries, due to different points of view, unalike global education and knowledge of populations and dissimilar cultures, resulting in different degrees of requirements by the population in terms of environmental policies. Regarding the sources of the GHGs, it is obvious that consumption is a large impact factor and studies regarding consumption can help significantly to realize how one can act in order to decrease those emissions.

Also following the work presented by Hertwich and Peters (2009), the worldwide production of motorized vehicles by itself, originated 800 million MT of CO_2 in 2001, which is comparable to the aircraft production sector, and life cycle assessment (LCA) allowed to realize that half of the GHGs produced by automotive industry can be attributed to materials (Schweimer and Levin 2000). More so, it is important to note that automotive industry has a global activity, with an intricate supply chain and final distribution mesh, which also generates a large amount of emissions that are not considered in the numbers abovementioned. As in other studies, Hertwich and Peters (2009) conclude that the amount of emissions is intensely related to consumption expansion, which is proven by the facts: African countries are responsible for the emissions of about 1 tCO_2/person/year, while 28 tCO_2/person/year and 33 tCO_2/person/year are the emissions produced by countries as the U.S.A. or Luxemburg, respectively, being mobility and the consumption of manufactured goods the main contributors to GHGs emission values. However, the GHG emissions declared by each country cannot be directly related to the national production because there are differences between import and export activities, as well as the consumption profiles. Thus, some data is not accurate. Two studies carried out by Wiedmann et al. (2008, 2010) about the United Kingdom in the period between 1992 and 2004 reported an increase of 12% of the carbon footprint (CF), unlike the decrease of 5% declared for the territorial CF due to the increasing consumption supported by import activities. Thus, the CF launched by official national offices cannot be a clear expression of the consumption evolution in the country. One of the best examples comes from China, where 30% of emissions were related to manufacture for export in 2005 (Weber et al. 2008; Feng et al. 2013). This effect is increasingly more pronounced, principally in developed countries, with a larger discrepancy between their territorial emissions and their domestic CF. Thus, there are some authors (Kanemoto et al. 2014) proposing different procedures to determine the CF, based not only on the territorial emissions but considering as well the import activity, which generates emissions in other countries, since some developed countries are pushing the most pollutant manufacturing activities to other developing countries, diminishing their own emissions but not contributing by this way to the decrease of emissions in a global manner (Kanamoto et al. 2014). Moreover, it is important to note that manufactured goods are one of the key factors impelling international trade and only a revolution in the mobility sector can invert the current increasing trend of emissions, bringing GHGs emission values back to sustainable levels. Thus, the responsible utilization of ground transportation (public transportation) and the use of cleaner energy systems, could reduce carbon emissions and cut the burden of household air pollution, which origins about 4.3 million deaths per year, and ambient air pollution, which causes around 3 million deaths every year (WHO 2017).

3.3.1.8 Other Consequences of Air Pollution

Many studies have been carried out trying to establish a correlation between health degradation due to air pollution exposure and the socio-economic status (Deguen and Zmirou-Navier 2010). However, the results do not present a clear trend both in Europe and the United States of America. Indeed, it seems difficult to institute a linear relationship between the socio-economic level and the exposure to air pollution because, in some European areas it is reported that people with average socio-economic status are the most exposed to air pollution (Havard et al. 2009; Padilla et al. 2014), where the city of Lyon is cited as an example. Nevertheless, in other studies corresponding to other areas in Europe in general, as well as studies focused in the cities of Lille and Marseille, it was observed that the highest level of air pollution exposure is attributed to the highest social deprivation, which is in line with other studies carried out in the United States of America, where the same patters was also observed (Kruize et al. 2007; Namdeo and Stringer 2008; Padilla et al. 2014). Conversely, Forastiere et al. (2007) reports air pollution exposure in areas where the people with highest socio-economic status live. The same has been reported by Padilla et al. (2014) regarding Paris, France.

Other curious studies tried to correlate air pollution with the behavior of the stock market. Lepori (2009) states that air pollution affects negatively the stock yield. On the other hand, Guo and Zhang (2016) concluded that Shangai's stock market presented a better behavior, lower turnover and volatility when the air quality conditions are clearly good. In fact, Wu et al. (2018) studied the effect of air pollution on the stock market investor sentiment and states that there is a strong relation between them: when the air pollution is more severe and perfectly felt by the investor, the willingness to invest is clearly smaller due to a more pessimist state of spirit related to the possible earnings to be achieved. Moreover, the same author refers that there is a different investor behavior when the Air Quality Index (AQI) is below or above 300. Effectively, there is no negative reaction of the investors in the stock market until AQI reaches the index 300, value from which a pessimist feeling begins to invade the investors and the stock market yield starts dropping. This effect is not recent, but the negative influence of the air quality perceived in the stock market fluctuations have increased drastically since 2014. Furthermore, this effect is even more evident and difficult to reverse when we are talking about China, where the investors are living mainly in largely industrialized cities with huge population densities, having many more pollution sources and being difficult to drastically reduce or eliminate them in the short term. By this way, economic growth can be seriously affected given this pessimism wave. Thus, the effect of severe pollution should never be neglected regarding a development point of view, which can also affect the quality of life of the populations.

3.3.2 *Effects of Water Pollution*

Water covers roughly 70% of the Earth's surface, being essential for humanity, both in domestic and industrial terms however, perhaps due to this immense volume, discarding something to the water has become a somewhat common procedure. This way of thinking was deeply explored in the first half of the twentieth century, during the strong increase of the industrial business activities. Until pollution assumed alarming proportions, rivers, lakes and oceans were often used to discard difficult to manage chemicals without elevated costs or proper technology. An example of this practice can be seen in Fig. 3.6. Industries using chemicals in an intensive way, such as steelmakers, the pulp and paper industry, the textile industry, traditional metal coatings, among others, usually chose to install their plants close to rivers in order to collect fresh water and discard of contaminated water full of chemicals and other harmful products. Indeed, under a judicious analysis, a huge number of well-known products can be found in the rivers, lakes and oceans, since sunscreen to pesticides, oil or birth control pills. Thus, groundwater can be compared to a cocktail of chemicals. Moreover, the industrial world and domestic life are not the only ones contributing to water pollution: farms also usually use waterways to discard of animal excrements, nutrient-rich fertilizers and domestic sewage. Additionally, pesticides applied to soils are subsequently washed away by rainwater, penetrating deeper into

Fig. 3.6 Industrial wastewater discharge. *Credits* pingphuket/Shutterstock.com

the ground. Later, when wells or bore holes are dug for water sources, pollution levels should be checked to determine the degree of contamination. Furthermore, acid rain downfall induced by air pollution is also conducted to waterways, contributing to the abovementioned cocktail. Another effect also reported is water oxygen depletion. In fact, aerobic and anaerobic micro-organisms are present in the water, if too much biodegradable matter is disposed of, micro-organism growth is encouraged, requiring a higher water oxygen consumption, simultaneously promoting the death of aerobic micro-organisms. Thus, the concentration of anaerobic micro-organisms rises, producing harmful toxins, such as ammonia and sulfides. Unfortunately, oil spills also cause serious problems as these events usually happen in the ocean, even though the effects stay confined to a given area. However, sea life is usually truly affected as well as birds in the area, commonly affecting their ability to fly.

Unfortunately, wastewater remains an unsolved problem, despite many efforts made essentially by developed countries. In an effort to improve the situation these countries fund programs with the objective of helping the industry adopt advanced technologies to treat wastewater before discarding it into rivers, lakes or even in the ocean. It is estimated that about 2 million ton of wastewater is discharged to the groundwater every day (Earnhart 2013) but the contribution of the industry to this value has been reduced over time. In the U.S.A., the industry was responsible for around 14% of the wastewater generated in the country in 1998, value which has certainly been improved in the last two decades. The costs are elevated, but they must be integrated in the product value chain and paid by the customer for a very important reason: our future. However, it is crucial that the top management of companies internalize this mindset. Otherwise, health problems will still remain, threatening our wellbeing. Conversely, developing countries are increasing their production instigated by developed countries, due to modest labor costs and absence of severe restrictions regarding environmental preservation, constituting now a serious concern. In addition, these countries are still trying to solve basic problems such as the adequate treatment of domestic sewage. Regarding this issue and the reality in East Africa, domestic sewage is contributing with 50% of all biological oxygen demand discharges, while industrial wastewater just represents 21% of the total value (Earnhart 2013).

Wastewater usually has health harming chemicals and other products. In an article published in 2013, Earnhart (2013) refers that in the year 2000, the consumption of unsafe water led to about 4 billion events of diarrhea globally. On the other hand, the numbers referred by Wang and Yang (2016) are a little different, considering the existence of 2.3 million of effected people all over the world, of which 2.2 in developing countries. Moreover, these authors have carried out a large study concerning the consequences of water pollution in China, concluding that this type of pollution has different physical and mental effects on people. It is estimated that about 60,000 people die yearly caused by injuries and other diseases directly related with water pollution and around 190,000 contract sicknesses (Tao and Xin 2014). Researchers have investigated and proven a direct relation between water pollution and severe water-borne diseases, such as cholera, dysentery, hepatitis, giardiasis, diarrhea, typhoid and cryptosporidiosis (Jalan and Ravallion 2003; Cutler and Miller 2005; WB-SCEA 2006; Roushdy et al. 2012). Furthermore, water pollution has been also connected to the development of various cancer diseases, as referred by Moralez-Suarez-Varela et al. (1995), Lin et al. (2000), Ebenstein (2012) and Lu et al. (2015). Additionally, it is estimated that 24% of mammals, around 12% of birds and about one third of amphibians worldwide are also affected due to the consumption of wastewater, being at risk of extinction.

The amount and type of wastewater discharges impact the environment differently. Discharges can be considered continuous or accidental based on the amount of pollutants discharged and preceding events. The first case assumes that companies do not have treatment facilities or are only able to partially treat pollutants. This situation is usually inspected by authorities and, depending on the legislation, companies may be forced to review their entire production process, including water-treatment stations. Sometimes, when pollutants can be treated in the same manner, companies are allowed to discharge wastewater into the municipal sewer systems to be treated along with domestic sewage. In developed countries, this situation is usually under the control of specific authorities and there are heavy fines to pay when the legislation is not enforced. In developing countries however, the enforcement of these laws is dependent on many factors, from political aspects to the economic impact on a company and consequent negative effects regarding employment. On the other hand, accidental wastewater releases are difficult to control. There are different techniques to avoid these situations, but accidents usually occur when regulations and precautions are not followed. There are companies which store their wastewater until they are able to treat it or convey it to other treating facilities. These storage, transfer and transport procedures are prone for mishaps and accidents, leading to potential leakages and contamination. In these cases, prevention is impossible and authorities are usually called after a disaster has happened only to register and provide guidance on how to remedy the situation. The constituents of wastewater are also very important as the pollutants usually found in larger amounts in waterways, lakes and oceans include organic pollutants (e.g. pharmaceutical, petroleum hydrocarbons, pesticides, solvents, domestic sewage), nutrients (phosphorus and nitrogen), and some heavy metals (e.g. mercury, lead, and cadmium). This situation will be developed later in an appropriate chapter however, a generic view needs to be done here. Nimkar (2018) refers in his work that the textile industry alone works with more than 8000 different chemicals, which can be discharged in the wastewater of these companies. However, water pollution does not consist solemnly of chemical products. In fact, wastewater can be divided into two main groups: fund pollutants and stock pollutants (Earnhart 2013), having a significant difference between them. Fund pollutants are, for example, degradable organic pollutants, for which waterways or the ocean have the ability to process, integrate and regenerate by itself. On the other hand, stock pollutants are those for which the waterways or oceans don't have the ability to embrace, as for example some heavy metals. In addition, these pollutants have a clear tendency to integrate the food chain.

Regarding the textile industry once again, Nimkar (2018) developed a reasoning concerning the expected pollution to be created by this industry. The world population is growing fast, from 1 billion in 1800 to an expected number of 11 billion people at the end of this century. As stated by the same author, the average consumption per capita of textiles and apparel yearly is about 7 kg, which, considering the current world populations (7 billion people) represents a total consumption of 49 billion kg of textile products per year. This consumption may be correlated with the production increase of chemicals, from 1 million ton in 1930 to 400 million ton in 2000. The concerns associated with this huge increase should strengthen the education about sustainable consumption and related principles. Additionally, the pollution empowered by the textile industry does not stay confined to the country of production as some chemical products remain saturated in the textiles after production. These textiles are then exported, with the harmful products only being removed after the first washing cycle, polluting the water used in this process and wide spreading it in the domestic sewage system (Greenpeace 2011).

The consumption of fossil-fuel based products has dramatically increased in the last decades, increasing further the need for oil extraction. This kind of activity also interferes drastically with surrounding water quality. A recent study carried out by Yusta-Garcia et al. (2017) regarding the Corrientes River (Peru) and taking into account a meta-analysis of chemical data provided by governmental institutions and oil company reports, has evidenced that the water found in the same formations as oil and gas, increased lead concentrations of heavy metals in the region. This fact is directly related with oil extraction. Moreover, this activity concentrated in the Corrientes and Tigre rivers, has also been connected to the intensification of dissolved sodium (Na) and chloride (Cl) flows in the Amazon River (Moquet et al. 2014). However, there are no reports referring oil releases on to the rivers of that area

(Peruvian Amazon), with exception of the study published by Reátegui-Zirena et al. (2014), as stated by Rosell-Melé et al. (2018). In this example, water and soil has become contaminated, affecting hunting and fishing subsistence activities usually carried out by indigenous populations living in the vicinity of the oil extraction facilities.

Preceding investigations have disclosed that products directly related with agricultural activities, such as fertilizers, fungicides, farm manures, agricultural run-off, as well as other wastewater from municipal and industrial sources like effluents from domestic and industrial sewage, waste incineration, urban and storm water run-off, refineries, land fill and petroleum industries, iron and steelmakers industries, pulp and paper industries, direct discharge of acid mine effluents on to waterways, lakes and ocean are the major causes of metal pollution in the aquatic environment (Marcotullio 2007; Kibria et al. 2010).

In fact, the marine and coastal areas of Bangladesh usually present poor water quality due to pollution providing from domestic, industrial, agrochemicals and oil and grease refineries/industries. Thus, these pollution sources, being the primary ones, constitute severe concerns regarding the registered water pollution levels in that country (Hossain and Islam 2006). The heavy metal contamination of waterways in Bangladesh was carefully studied by Kibria et al. (2016) through artificial mussels technology, finding a series of pollutant elements, such as Cadmium, Copper, Iron, Manganese, Nickel, Lead, Uranium and Zink in the country's main water courses, but with more incidence in the Buriganga River located in the in the capital, Dhaka, where higher concentrations of Iron, Copper Manganese, Lead and Nickel were identified. These large concentrations of heavy metals have as main origin untreated effluents discharges providing from textile industries and tanneries strategically located close to the river. Moreover, some authors state that the color of the Buriganga River is black (Saifullah et al. 2012), consequence of the level of pollution directly discharged to it. Also, the level of dissolved oxygen is extremely poor, usually in the range of 0.722–1.204 mL/L (Rahman and Al Bakri 2010). In other areas of Bangladesh, farming activities are mainly responsible for water pollution, due to the direct discharge of effluents providing from cattle breeding farms, agricultural and aquaculture activities, giving rise to the detection of some elements such as Copper, Zink and Uranium in the water. This kind of pollution may affect the biodiversity of ecosystems, abolishing the most sensitive species as well as causing reproduction problems and higher rates of illnesses in other species (Wu et al. 2007; Kibria et al. 2012). In fact, some authors are somewhat pessimistic regarding the quality of the water and environment in general regarding Bangladesh, because many important activities for the national economy and population's way of living are impacted by the quality of water, such as aquaculture, fishing, seafood nurseries, due to the migration of some species caused by the elevated levels of pollution detected in its waters (Kibria et al. 2016). Thus, researchers are claiming for data registers and appropriated legislation in order to diminish the level of pollution in the water, namely the one caused by heavy metals (Kibria et al. 2016). As severe as in Bangladesh, the water pollution in China is also a huge concern. Jian (2012) states in his work that 40% of Chinese rivers are extremely polluted. The same author refers yet that 20%

of the Chinese rivers were so polluted with toxic products that it is dangerous to even touch the water. Moreover, regarding information providing from the Ministry of Environmental Protection of China, more than 1700 occurrences of water pollution are reported yearly in China (Xie 2011), most of them caused by improper and non-authorized direct industrial discharges into rivers and lakes. These facts constitute permanent stress on populations, with more than 70% of the people feeling embarrassed with this situation. The law adopted in China in April of 2014 regarding Environmental Protection was understood as a first step in the right direction, reverting the concerning previously trodden path. Though, in addition, it is necessary to create the adequate conditions to oblige offenders to abide by the law (Yang 2014). Thus, offices and inspectors need to be deployed to the ground in order to enforce respect for the law and report incidences. Regarding this aspect, Dong et al. (2011) understood that environmental complaints from local people were heard more intensively and induced the allocation of inspections resources in the affected area, reinforcing the surveillance. Thus, a new recommendation came from that work, to encourage environmental education, leading to the involvement of more people and a higher surveillance. Miao et al. (2015) also supports that the endowment of people with the necessary knowledge and culture concerning the environmental harm caused by industrial and agricultural activities will allow for a better monitoring of injurious situations, helping to act more rapidly and avoid prolonged damage to the environment. The discharge of wastewater presupposes the existence of water courses, lakes or oceans nearby, while the concentration of pollutants results from the amount of municipal/industrial wastewater discharged and volume of freshwater available for the discharges. In some areas, the flow of municipal and industrial effluents is not proportional to the available freshwater, exacerbating the situation. This case is reported in the work of Hu and Cheng (2013), based on a prior work developed by Cheng et al. (2009) regarding the Chinese spatial distribution of people and industrial activities, relatively to the volume of freshwater available. Indeed, around one third of the population and industrial activity in China is mainly located in the Northern area, usually called North China Plain. However, this area only has about 8% of the Chinese water resources. On the other hand, the Southwest of China contributes just 1% to national GDP but this region is endowed of about one fifth of the total Chinese water resources. Moreover, China is a good example on how the difficulties in supplying water with the necessary quality are substantially increasing. The drastic rise in population in the last decades with corresponding discharges, the effluents originating from agricultural farms which have intensified the use of agrochemicals, as well as a dramatic escalation of industrial activities without adequate implementation of environmental regulations, allied to a reduction of available fresh water due to climacteric reasons, has undermined water quality. Indeed, in the last three decades, it is estimated that around 150–200 million people have migrated from villages to cities, leaving rural activities to adhere to the industrialization (Cheng and Hu 2010). This high concentration of people in cities have led to a dramatic rise of freshwater consumption and wastewater discharges, having the last ones increased at an annual rate of 6%. Thus, any gains achieved by the treatment of industrial wastewater has been offset by the increase of municipal wastewater generation, decreasing the overall quality of freshwater. Effectively, around the main cities, the capacity of wastewater treatment has not been improved proportionally to the exponential population growth (Hu and Cheng 2013). Moreover, due to the abundant existence of lakes and rivers in southern China, industrial and municipal untreated wastewater is thrown into the river, polluting and impoverishing water quality. Likewise, rural activities in China are also increasing their pollution contribution as the use of fertilizers rose drastically over the period of 1978 and 2004, reaching 2004 roughly at a rate of 420 kg/ha/year (World Bank 2006). The use of pesticides also rose in the same period, worsening further the situation. Due to new eating habits, rising of livestock and poultry production was also reported, with respective consequences in terms of dejects discharge. This case is also more severe in China as it has received much less attention than when compared to the pollution generated through industrial activities. Moreover, agricultural activities are less prone to be regulated and controlled than industrial ones. However, the pollution providing from agriculture cannot be neglected as it represents a Chemical Oxygen Demand (COD) of 11.86 million ton and 47.7% of the total wastewater produced in China, following the data available in 2011 (MEP 2012). Regarding the release of ammonium, the total discharges produced by agricultural activities ascends to 0.83 million ton (MEP 2012), 31.8% of the total released in China. The work of Hu and Cheng (2013) describes in detail the main sources of water pollution in China, which was largely induced by the dramatic industrial growth, being the small and family run companies the largest contributors to this situation, due to the use of old processes and equipment. However, they are not alone in this contribution and some sectors present particular responsibilities. The paper industry is the major contributor to water pollution in China, being responsible for 18% (in volume) of generated wastewater, 23% of the COD and one of the main ammonium contributors (NH_4^+ -N). The industry of raw chemical materials and related products is the leader in terms of ammonium releases (35.3%), being also producer of hydrocarbons, COD and wastewater. On the other hand, the core source of heavy metals is provided by the non-ferrous metal processing industry (27.5%), along with the leather and fur industry (19.4%), the metalworking industry (17.7%) and the non-ferrous metal ore mining industry (14.0%). Also, the leading source of oil hydrocarbons is coal mining and washing industry (17.1%) and the ferrous metal processing and manufacturing industry (16.1%). The summary of these numbers can be seen in Fig. 3.7. It is believable that wastewater discharges have been kept constant in the last decade and, due to a greater attention to the environment, effluents such as COD or ammonium have been fading progressively throughout the last decade.

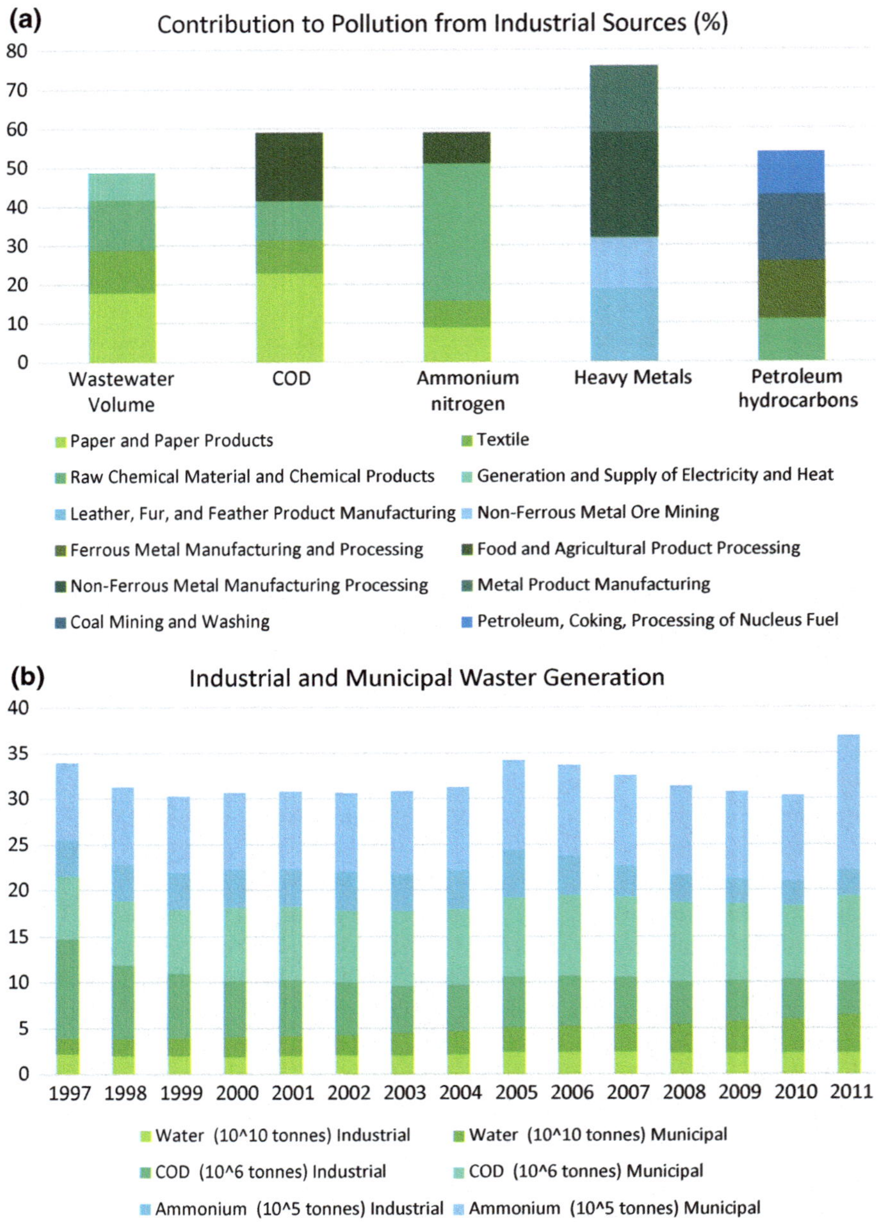

Fig. 3.7 Data referring to pollutants discharges providing from industrial and municipal sources in China: **a** main industrial sectors contributing to wastewater generation and pollutant releases in 2011 and **b** discharges of industrial and municipal wastewater, as well as COD and Ammonium pollutants over the period of 2001–2011. In each bar, the dark color corresponds to industrial source and the light color to municipal one. Adapted from Hu and Cheng (2013)

References

W.A. Alarcon, *Elevated Blood Lead Levels Among Employed Adults—United States, 1994–2013. Centers for Disease Control and Prevention* (2016), https://www.cdc.gov/mmwr/volumes/63/wr/mm6355a5.htm. Retrieved on 27 Jan 2018

N.E. Alexis, C. Carlsten, Interplay of air pollution and asthma immunopathogenesis: a focused review of diesel exhaust and ozone. Int. Immunopharmacol. **23**, 347–355 (2014)

A. Analitis, K. Katsouyanni, K. Dimakopoulou, E. Samoli, A.K. Nikoloulopoulos, Y. Petasakis, G. Touloumi, J. Schwartz, H.R. Anderson, K. Cambra, Short term effects of ambient particles on cardiovascular and respiratory mortality. Epidemiology **17**, 230–233 (2006)

D.W. Anderson, K. Pothakos, J.S. Schneider, Sex and rearing condition modify the effects of perinatal lead exposure on learning and memory. Neurotoxicology **33**, 985–995 (2012)

S.C. Anenberg, L.W. Horowitz, D.Q. Tong, J.J. West, An estimate of the global burden of anthropogenic ozone and fine particulate matter on premature human mortality using atmospheric modeling. Environ. Health Perspect. **118**, 1189–1195 (2010)

AQICN.ORG, *World Health Organization: 2014 Air Pollution Ranking* (2015), http://aqicn.org/faq/2015-05-16/world-health-organization-2014-air-pollution-ranking/. Retrieved on 30 Dec 2017

K.E. Barrett, S.M. Barman, S. Boitano, H. Brooks, *Ganong's Review of Medical Physiology*, 25 edn. (McGraw-Hill Education, 2015). ISBN 978-0071825108

C. Beaucham, E. Page, W.A. Alarcon, G.M. Calvert, M. Methner, T.M. Schoonover, Indoor firing ranges and elevated blood lead levels-United States, 2002–2013. Morb. Mortal Wkly Rep (MMWR) **63**, 347–351 (2014)

D. Bellinger, A. Leviton, J. Sloman, Antecedents and correlates of improved cognitive performance in children exposed in utero to low levels of lead. Environ. Health Perspect. **89**, 5–11 (1990)

M.L. Block, A. Elder, R.L. Auten, S.D. Bilbo, H. Chen, J.C. Chen et al., The outdoor air pollution and brain health workshop. Neurotoxicology **33**, 972–984 (2012)

E. Boldo, S. Medina, A. Tertre, F. Hurley, H.-G. Mücke, F. Ballester, I. Aguilera, Apheis: health impact assessment of long-term exposure to PM2.5 in 23 European cities. Eur. J. Epidemiol. **21**, 449–458 (2006)

J. Burney, V. Ramanathan, Recent climate and air pollution impacts on Indian agriculture. Proc. Natl. Acad. Sci. **111**(46), 16319–16324 (2014)

Centers for Disease Control and Prevention, *Lead* (2017), https://www.cdc.gov/nceh/lead/. Retrieved on 27 Jan 2018

Centers for Disease Control and Prevention, Take-home lead exposure among children with relatives employed at a battery recycling facility—Puerto Rico, 2011. Morb Mortal Wkly Rep (MMWR) **61**(47), 967–970 (2012)

W.L. Chameides, H. Yu, S.C. Liu, M. Bergin, X. Zhou, L. Mearns et al., Case study of the effects of atmospheric aerosols and regional haze on agriculture: an opportunity to enhance crop yields in China through emission controls? Proc. Natl. Acad. Sci. **96**, 13626–13633 (1999)

H. Cheng, Y. Hu, Planning for sustainability in China's urban development: status and challenges for Dongtaneco-city project. J. Environ. Monit. **12**, 119–126 (2010)

H. Cheng, Y. Hu, J. Zhao, Meeting China's water shortage crisis: current practices and challenges. Environ. Sci. Technol. **43**, 240–244 (2009)

R. Chowdhury, S. Ebelt Sarnat, L. Darrow, W. McClellan, K. Steenland, Mortality among participants in a lead surveillance program. Environ. Res. **132**(Suppl. C), 100–104 (2014)

A. Ciucci, I. D'Elia, F. Wagner, R. Sander, L. Ciancarella, G. Zanini, W. Schöpp, Cost-effective reductions of PM2.5 concentrations and exposure in Italy. Atmos. Environ. **140**, 84–93 (2016)

K. Crane, Z. Mao, *Costs of Selected Policies to Address Air Pollution in China* (RAND Corporation, Santa Monica, California, U.S.A., 2015), https://www.rand.org/content/dam/rand/pubs/research_reports/RR800/RR861/RAND_RR861.pdf. Retrieved on 29 Jan 2018

D.M. Cutler, G. Miller, The role of public health improvements in health advances: the twentieth-century United States. Demography **42**, 1–22 (2005)

S. Deguen, D. Zmirou-Navier, Social inequalities resulting from health risks related to ambient air quality—a European review. Eur. J. Pub. Health **20**(1), 27–35 (2010)

K.N. Dietrich, K.M. Krafft, R.I. Bornschein, P.B. Hammond, G. Berger, P.A. Succop, M. Bier, Low-level fetal lead exposure effect on neurobehavioral development in early infancy. Pediatrics **80**, 721–730 (1987)

Y.L. Dong, M. Ishikawa, X.B. Liu, S. Hamori, The determinants of citizen complaints on environmental pollution: an empirical study from China. J. Cleaner Prod. **19**(12), 1306–1314 (2011)

C.G. Douglas, J.S. Haldane, J.B.S. Haldane, The laws of combination of haemoglobin with carbon monoxide and oxygen. J. Physiol. **44**, 275–304 (1912)

S.D. Dubowsky, H. Suh, J. Schwartz, B.A. Coull, D.R. Gold, Diabetes, obesity, and hypertension may enhance associations between air pollution and markers of systemic inflammation. Environ. Health Perspect. **114**(7), 992–998 (2006)

D. Earnhart, Water pollution from industrial sources, encyclopedia of energy, in *Natural Resource and Environmental Economics*, vol. 3, ed. by J. Shogren (Elsevier's Science and Technology, Oxford, UK, 2013), pp. 114–120. ISBN 978-0-08-096452-2

A. Ebenstein, The consequences of industrialization: evidence from water pollution and digestive cancers in China. Rev. Econ. Stat. **94**, 186–201 (2012)

EPA, *Review of National Ambient Air Quality Standards for Ozone Final Staff Paper, Human Exposure and Risk Assessments and Environmental Report* (2007), https://www3.epa.gov/ttn/naaqs/standards/ozone/data/2007_01_finalsp_factsheet.pdf. Retrieved on 24 Jan 2018

EPA, *US EPA. Integrated Science Assessment for Ozone and Related Photochemical Oxidants (Final)* (National Center for Environmental Assessment (NCEA), 2013), pp. 1251

EPA, *Criteria Air Pollutants* (2018a), https://www.epa.gov/criteria-air-pollutants. Retrieved on 4th Jan 2018

EPA, *Basic Information About Ozone* (2018b), https://www.epa.gov/ozone-pollution/basic-information-about-ozone#what where how. Retrieved on 23rd Jan 2018

A.A. Ettinger, H. Lamadrid-Figueroa, M.M. Tellez-Rojo, A. Mercado-Garcia, K.E. Peterson, J. Schwartz et al., Effect of calcium supplementation on blood lead levels in pregnancy: a randomized placebo-controlled trial. Environ. Health Perspect. **117**(1), 26–31 (2009)

N. Fann, A.D. Lamson, S.C. Anenberg, K. Wesson, D. Risley, B.J. Hubbell, Estimating the national public health burden associated with exposure to ambient PM2.5 and ozone. Risk Anal. **32**, 81–95 (2012)

E. Fattore, V. Paiano, A. Borgini, A. Tittarelli, M. Bertoldi, P. Crosignani, R. Fanelli, Human health risk in relation to air quality in two municipalities in an industrialized area of Northern Italy. Environ. Res. **111**, 1321–1327 (2011)

T. Fawcett, R. Moon, P. Fracica et al., Warehouse workers' headache: Carbon monoxide poisoning from propane fueled forklifts. J. Occup. Med. Toxicol. **34**, 12–15 (1992)

Z.W. Feng, M.H. Jin, F.Z. Zhang, Y.Z. Huang, Effects of ground-level ozone (O_3) pollution on the yields of rice and winter wheat in the Yangtze River Delta. J. Environ. Sci. **15**, 360–362 (2003)

K. Feng, S.J. Davis, L. Sun, X. Li, D. Guan, W. Liu, Z. Liu, K. Hubacek, Outsourcing CO_2 within China. Proc. Natl. Acad. Sci. U.S.A. **110**, 11654–11659 (2013)

Z. Feng, E. Hu, X. Wang, L. Jiang, X. Liu, Ground-level O_3 pollution and its impacts on food crops in China: a review. Environ. Pollut. **199**, 42–48 (2015)

F. Forastiere, M. Stafoggia, C. Tasco, S. Picciotto, N. Agabiti, G. Cesaroni et al., Socio-economic status, particulate air pollution, and daily mortality: differential exposure or differential susceptibility. Am. J. Ind. Med. **50**(3), 208–216 (2007)

M.W. Frampton, J. Bausch, D. Chalupa, P.K. Hopke, E.L. Little, D. Oakes, J.C. Stewart, M.J. Utell, Effects of outdoor air pollutants on platelet activation in people with type 2 diabetes. Inhalation Toxicol. **24**, 831–838 (2012)

J.J. Gershanik, G.G. Brooks, J.A. Little, Blood lead values in pregnant women and their offspring. Obstet. Gynecol. **119**, 508–511 (1974)

GISS.NASA.GOV, *NASA Research Finds 2010 Tied for Warmest Year on Record (2011)* (2017), https://www.giss.nasa.gov/research/news/20110112/. Retrieved on 28 Dec 2017

Government Publications Office (2018), https://www.gpo.gov/fdsys/pkg/CFR-2011-title29-vol6/pdf/CFR-2011-title29-vol6-sec1910-1025.pdf. Retrieved on 28 Jan 2018

Z.J. Graff, M. Neidell, The impact of pollution on worker productivity. Am. Econ. Rev. **102**(7), 3652–3673 (2012)

P. Grandjean, P.J. Landrigan, Developmental neurotoxicity of industrial chemicals. Lancet **368**(9553), 2167–2178 (2006)

Greenpeace, *Dirty Laundry: How Big Brands are Making Consumers Unwitting Accomplices in the Toxic Water Cycle* (2011), https://www.greenpeace.org/archive-international/Global/international/publications/toxics/Water%202012/DirtyLaundryReloaded-ExecSum.pdf. Retrieved on 9th Feb 2018

C. Guerriero, L. Chatzidiakou, J. Cairns, D. Mumovic, The economic benefits of reducing the levels of nitrogen dioxide (NO2) near primary schools: the case of London. J. Environ. Manage. **181**, 615–622 (2016)

Y.J. Guo, Y.H. Zhang, Will air quality affect the stock market? J. Financ. Res. **2**, 71–85 (2016)

S. Guo, M. Hu, M.L. Zamora, J.F. Peng, D.J. Shang, J. Zheng, Z.F. Du, Z. Wu, M. Shao, L.M. Zeng, M.J. Molina, R.Y. Zhang, Elucidating severe urban haze formation in China. Proc. Natl. Acad. Sci. U.S.A. **111**(49), 17373–17378 (2014)

S. Gwini, E. MacFarlane, A. Del Monaco, D. McLean, D. Pisaniello, G.P. Benke et al., Cancer incidence, mortality, and blood lead levels among workers exposed to inorganic lead. Ann. Epidemiol. **22**(4), 270–276 (2012)

Y. Hao, L. Balluz, H. Strosnider, X.J. Wen, C. Li, J.R. Qualters, Ozone, fine particulate matter, and chronic lower respiratory disease mortality in the United States. Am. J. Respir. Crit. Care Med. **192**, 337–341 (2015)

S. Havard, S. Deguen, D. Zmirou-Navier, C. Schillinger, D. Bard, Traffic-related air pollution and socio-economic status: a spatial autocorrelation study to assess environmental equity on a small-area scale. Epidemiology **20**(2), 223–230 (2009)

X. He, S. Pang, J. Ma, Y. Zhang, Influence of relative humidity on heterogeneous reactions of O3 and O3/SO2 with soot particles: potential for environmental and health effects. Atmos. Environ. **165**, 198–206 (2017)

HealthyPeople.com (2018), https://www.healthypeople.gov/2020/tools-resources/evidence-based-resource/low-level-lead-exposure-harms-childrena-renewed-call. Retrieved on 28 Jan 2018

E.G. Hertwich, G.P. Peters, Carbon footprint of nations: a global trade-linked analysis. Environ. Sci. Technol. **43**, 6414–6420 (2009)

G. Hoek, R.M. Krishnan, R. Beelen, A. Peters, B. Ostro, B. Brunekreef, J.D. Kaufman, Long-term air pollution exposure and cardio-respiratory mortality: a review. Environ. Health **12**, 43–57 (2013)

D. Hoffmann, I. Hoffmann, K. El-Bayoumy, The less harmful cigarette: a controversial issue. Chem. Res. Toxicol. **14**, 767–790 (2001)

S. Holgate, J. Grigg, R. Agius, J.R. Ashton, P. Cullinan, K. Exley, D. Fishwick, G. Fuller, N. Gokani, C. Griffiths, Every breath we take: the lifelong impact of air pollution, report of a working party. J. R. Coll Phys. Edinb. 1–94 (2016)

Y.C. Hong, J.T. Lee, H. Kim, E.H. Ha, J. Schwartz, D.C. Christiani, Effects of air pollutants on acute stroke mortality. Environ. Health Perspect. **110**, 187–191 (2002)

M.M. Hossain, M.M. Islam, *Ship Breaking Activities and its Impact on the Coastal Zone of Chittagong, Bangladesh: towards Sustainable Management* (Young Power in Social Action (YPSA), Chittagong, Bangladesh, 2006), p. 54, http://www.shipbreakingbd.info/report/Ship%20Breaking%20Activities%20and%20its%20Impact%20on%20the.pdf. Retrieved on 9th Feb 2018

Y. Hu, H. Cheng, Water pollution during China's industrial transition. Environ. Dev **8**, 57–73 (2013)

International Business Times (2015), http://www.ibtimes.com/china-air-pollution-smog-blankets-beijing-shanghai-day-after-christmas-2240081, 26 Dec 2015. Retrieved on 16 Jan 2018

S. Iqbal, J.H. Clower, S.A. Hernandez, S.A. Damon, F.Y. Yip, A review of disaster-related carbon monoxide poisoning: surveillance, epidemiology, and opportunities for prevention. Am. J. Public Health **102**, 1957–1963 (2012)

J. Jalan, M. Ravallion, Does piped water reduce diarrhea for children in rural India? J. Econ. **112**, 153–173 (2003)

Jancovici, *How do Greenhouse Gas Emissions Presently Evolve?* (2017), https://jancovici.com/en/climate-change/ghgs-and-us/how-do-greenhouse-gas-emissions-presently-evolve/. Retrieved on 28 Dec 2017

M. Janík, M. Ublov, S. Kucerova, P. Hejna, Carbon monoxide-related fatalities: a 60-year single institution experience. J. Forensic Leg. Med. **48**, 23–29 (2017)

M. Jerrett, R.T. Burnett, C.A. Pope, K. Ito, G. Thurston, D. Krewski, Y. Shi, E. Calle, M. Thun, Long-term ozone exposure and mortality. New Engl. J. Med. **360**, 1085–1095 (2009)

Y. Jian, *China's River Pollution 'A Threat to People's Lives'. People's Dly* (2012), http://english.peopledaily.com.cn/90882/7732438.html. Retrieved on 10 Feb 2018

H. Joo, J.H. Choi, E. Burm, H. Park, Y.-C. Hong, Y. Kim, E.-H. Ha, Y. Kim, B.-N. Kim, M. Ha, Gender difference in the effects of lead exposure at different time windows on neurobehavioral development in 5-year-old children. Sci. Total Environ. **615**, 1086–1092 (2018)

K. Kanemoto, D. Moran, M. Lenzen, A. Geschke, International trade undermines national emission reduction targets: new evidence from air pollution. Glob. Environ. Change **24**, 52–59 (2014)

J. Kasten-Jolly, D.A. Lawrence, Sex-specific effects of developmental lead exposure on the immune-neuroendocrine network. Toxicol. Appl. Pharmacol. **334**, 142–157 (2017)

J. Kasten-Jolly, N. Pabello, V.J. Bolivar, D.A. Lawrence, Developmental lead effects on behavior and brain gene expression in male and female BALB/cAnNTac mice. Neurotoxicology **33**, 1005–1020 (2012)

K. Katsouyanni, G. Touloumi, C. Spix, J. Schwartz, F. Balducci, S. Medina, G. Rossi, B. Wojtyniak, J. Sunyer, L. Bacharova, Short term effects of ambient sulphur dioxide and particulate matter on mortality in 12 European cities: results from time series data from the APHEA project. The BMJ **314**(7095), 1658–1663 (1997)

Y.O. Khaniabadi, R. Polosa, R.Z. Chuturkova, S.M. Daryanoosh, G. Goudarzi, A. Borgini, A. Tittarelli, H. Basiri, H. Armin, H. Nourmoradi, A.A. Babaei, P. Naserian, Air pollution health impact assessment on total, cardiovascular, and respiratory mortality in Khorramabad, Iran (The AirQ Approach). Process Saf. Environ. Prot. (2018). http://dx.doi.org/10.1016/j.psep.2017.07.018

G. Kibria, A.K.Y. Haroon, D. Nugegoda, G. Rose, *Climate Change and Chemicals: Environmental and Biological Aspects* (2010), p. 460, http://dx.doi.org/10.13140/2.1.4384.0963. ISBN 9789-38-0235-301. https://www.researchgate.net/publication/261216635_Climate_Change_and_Chemicals_Environmental_and_Biological_Aspects. Retrieved on 9th Feb 2018

G. Kibria, T.C. Lau, R.S.S. Wu, Innovative 'Artificial Mussels' technology for assessing spatial and temporal distribution of metals in Goulburn-Murray catchments waterways, Victoria, Australia: effects of climate variability (dry vs. wet years). Environ. Int. **50C**, 38–46 (2012)

G. Kibria, M.M. Hossain, D. Mallick, T.C. Lau, R. Wu, Monitoring of metal pollution in waterways across Bangladesh and ecological and public health implications of pollution. Chemosphere **165**, 1–9 (2016)

K.L. King, S. Johnson, I. Kheirbek, J.W.T. Lu, T. Matte, Differences in magnitude and spatial distribution of urban forest pollution deposition rates, air pollution emissions, and ambient neighborhood air quality in New York City. Landscape and Urban Planning **128**, 14–22 (2016)

M.A. Kioumourtzoglou, J.D. Schwartz, M.G. Weisskopf, S.J. Melly, Y. Wang, F. Dominici, A. Zanobetti, Long-term PM2.5 exposure and neurological hospital admissions in the Northeastern United States. Environ. Health Perspect. **124**(1), 23–29 (2016)

M.J. Kosnett, R.P. Wedeen, S.J. Rothenberg, K.L. Hipkins, B.L. Materna, B.S. Schwartz et al., Recommendations for medical management of adult lead exposure. Environ. Health Perspect. **115**(3), 463–471 (2007)

H. Kruize, P.P.J. Driessen, P. Glasbergen, K.N.D. van Egmond, Environmental equity and the role of public policy: experiences in the Rijnmond region. Environ. Manage. **40**(4), 578–595 (2007)

N. Künzli, M. Jerrett, R. Garcia-Esteban, X. Basagaña, B. Beckermann, F. Gilliland, Ambient air pollution and the progression of atherosclerosis in adults. PLoS ONE **5**(2), e9096 (2010)

B.P. Lanphear, R. Hornung, J. Khoury et al., Low-level environmental lead exposure and children's intellectual function: an international pooled analysis. Environ. Health Perspect. **113**(7), 894–899 (2005)

J. Lelieveld, C. Barlas, D. Giannadaki, A. Pozzer, Model calculated global, regional and megacity premature mortality due to air pollution. Atmos. Chem. Phys. **13**, 7023–7037 (2013)

J. Lepeule, F. Laden, D. Dockery, J. Schwartz, Chronic exposure to fine particles and mortality: an extended follow-up of the Harvard Six Cities study from 1974 to 2009. Environ. Health Perspect. **120**(7), 965–970 (2012)

G.M. Lepori, Environmental stressors, mood, and trading decisions: evidence form ambient air pollution. Working papers series. http://ssrn.com/421abstract=1284549. Retrieved on 18 Jan 2018

R. Levin, The attributable annual health costs of U.S. occupational lead poisoning. Int. J. Occup. Environ. Health **22**(2), 107–120 (2016)

S. Li, G. Williams, Y. Guo, Health benefits from improved outdoor air quality and intervention in China. Environ. Pollut. **214**, 17–25 (2016)

T.I. Lidsky, J.S. Schneider, Lead neurotoxicity in children: basic mechanisms and clinical correlates. Brain **126**(1), 5–19 (2003)

N.F. Lin, J. Tang, H.S.M. Ismael, Study on environmental etiology of high incidence areas of liver cancer in China. World J. Gastroenterol. **6**, 572–576 (2000)

M. Lippmann, Toxicological and epidemiological studies of cardiovascular effects of ambient air fine particulate matter (PM2.5) and its chemical components: coherence and public health implications. Crit. Rev. Toxicol. **44**, 299–347 (2014)

C. Liu, Q. Ma, Y. Liu, J. Ma, H. He, Synergistic reaction between SO_2 and NO_2 on mineral oxides: a potential formation pathway of sulfate aerosol. Phys. Chem. Chem. Phys. **14**(5), 1668–1676 (2012)

H. Liu, S. Liu, B. Xue, Z. Lv, Z. Meng, X. Yang, T. Xue, Q. Yu, K. He, Ground-level ozone pollution and its health impacts in China. Atmos. Environ. (2017, in press), https://doi.org/10.1016/j.atmosenv.2017.11.014

Y.L. Lu, S. Song, R.S. Wang, Z.Y. Liu, J. Meng, A.J. Sweetman, A. Jenkins, R.C. Ferrier, H. Li, W. Luo, T.Y. Wang, Impacts of soil and water pollution on food safety and health risks in China. Environ. Int. **77**, 5–15 (2015)

T.F. Mar, G.A. Norris, J.Q. Koening, T.V. Larson, Associations between air pollution and mortality in Phoenix, 1995–1997. Environ. Health Perspect. **108**, 347–353 (2000)

P.J. Marcotullio, Urban water-related environmental transitions in Southeast Asia. Sustain. Sci. **2**, 27–54 (2007)

K. Matus, K.-M. Nam, N.E. Selin, L.N. Lamsal, J.M. Reilly, S. Paltsev, Health damages from air pollution in China. Glob. Environ. Change **22**, 55–66 (2012)

Ministry of Environmental Protection, *State of Environmental Report* (in Chinese) (2012), http://www.mep.gov.cn/gkml/hbb/qt/201306/W020130604382172090857.pdf. Retrieved on 16 Jan 2018

X. Miao, Y. Tang, C.W.Y. Wong, H. Zang, The latent causal chain of industrial water pollution in China. Environ. Pollut. **196**, 473–477 (2015)

J.-S. Moquet, L. Maurice, A. Crave, J. Viers, N. Arevalo, C. Lagane, W. Lavado-casimiro, J. Guyot, Cl and Na fluxes in an Andean Foreland basin of the Peruvian Amazon: an anthropogenic impact. Aquat. Geochem. **20**, 613–637 (2014)

X. Morelli, C. Rieux, J. Cyrys, B. Forsberg, R. Slama, Air pollution, health and social deprivation: a fine-scale risk assessment. Environ. Res. **147**, 59–70 (2016)

M.M. Morales-Suarez-Varela, A. Llopis-Gonzalez, M.L. Tejerizo-Perez, Impact of nitrates in drinking water on cancer mortality in Valencia, Spain. Eur. J. Epidemiol. **11**, 15–21 (1995)

A. Namdeo, C. Stringer, Investigating the relationship between air pollution, health and social deprivation in Leeds, UK. Environ. Int. **34**(5), 585–591 (2008)
National Aeronautics and Space Administration (2018), https://www.nasa.gov/topics/earth/features/health-sapping.html. Retrieved on 17 Jan 2018
National Toxicology Program (NTP), *Health Effects of Low-Level Lead. NTP Monograph* (2012), https://ntp.niehs.nih.gov/ntp/ohat/lead/final/monographhealtheffectslowlevellead_newissn_508.pdf. Retrieved on 27 Jan 2018
M. Neuberger, A. Rutkowski, H. Friza, M. Haider, Grripe, Luftverunrei nigungund Mortalitat in Wien (in German). Forum Stadtehygiene **38**, 7–11 (1987)
U. Nimkar, Sustainable chemistry: a solution to the textile industry in a developing world. Curr. Opin Green Sustain. Chem. **9**, 13–17 (2018)
H. Orru, M. Maasikmets, T. Lai et al., Health impacts of particulate matter in five major Estonian towns: main sources of exposure and local differences. Air Qual. Atmos. Health **4**, 247–258 (2011)
Occupational Safety and Health Administration, *Commonly Used Statistics* (2017). https://www.osha.gov/oshstats/commonstats.html. Retrieved on 28 Jan 2018
B. Ostro, R. Broadwin, S. Green et al., Fine particulate air pollution and mortality in nine California counties: results from CALFINE. Environ. Health Perspect. **114**, 29–33 (2006)
C.M. Padilla, W. Kihal-Talantikite, V.M. Vieira, P. Rossello, G.L. Nir, D. Zmirou-Navier et al., Air quality and social deprivation in four French metropolitan areas—a localized spatio-temporal environmental inequality analysis. Environ. Res. **134**, 315–324 (2014)
L. Perez, C. Declercq, C. Iniguez et al., Chronic burden of near-roadway traffic pollution in 10 European cities (APHEKOM network). ERJ Express **42**, 594–660 (2013)
A. Peters, Ambient particulate matter and the risk for cardiovascular disease. Prog. Cardiovasc. Dis. **53**(5), 327–333 (2011)
C.A. Pope, D.W. Dockery, Health effects of fine particulate air pollution: lines that connect. J. Air Waste Manag. Assoc. **56**(6), 709–742 (2006)
C.A. Pope, R.T. Burnett, M.J. Thun, Lung cancer, cardiopulmonary mortality, and long-term exposure to fine particulate air pollution. J. Am. Med. Assoc. **287**, 1132–1141 (2002)
C.A. Pope, R.T. Burnett, M.C. Turner, A. Cohen, D. Krewski, M. Jerrett, S.M. Gapstur, M.J. Thun, Lung cancer and cardiovascular disease mortality associated with ambient air pollution and cigarette smoke: shape of the exposure-response relationships. Environ. Health Perspect. **119**, 1616–1621 (2011)
L.D. Prockop, R.I. Chichkova, Carbon monoxide intoxication: an updated review. J. Neurol. Sci. **262**, 122–130 (2007)
X. Pu, T.J. Wang, X. Huang, D. Melas, P. Zanis, D.K. Papanastasiou, A. Poupkou, Enhanced surface ozone during the heat wave of 2013 in Yangtze River Delta region, China. Sci. Total Environ. **603**, 807–816 (2017)
O. Raaschou-Nielsen, Z.J. Andersen, R. Beelen, E. Samoli, M. Stafoggia, G. Weinmayr et al., Air pollution and lung cancer incidence in 17 European cohorts: prospective analyses from the European Study of Cohorts for Air Pollution Effects (ESCAPE). Lancet Oncol. **14**(9), 813–822 (2013)
M.B. Rabinowitz, J.D. Wang, W.T. Soong, Dentine lead and child intelligence in Taiwan. Arch. Environ. Health: Int. J. **46**, 351–360 (1991)
A. Rahman, D. Al Bakri, A study on selected water quality parameters along the River Buriganga, Bangladesh. Iran. J. Energy Environ. **1**, 81–92 (2010)
J.A. Raub, Health effects of exposure to ambient carbon monoxide. Chemosphere Glob. Change Sci. **1**, 331–351 (1999)
E.G. Reátegui-Zirena, P.M. Stewart, A. Whatley, F. Chu-Koo, V.E. Sotero-Solis, C. Merino-Zegarra, E. Vela-Paima, Polycyclic aromatic hydrocarbon concentrations, mutagenicity, and Microtox® acute toxicity testing of Peruvian crude oil and oil contaminated water and sediment. Environ. Monit. Assess. **186**, 2171–2184 (2014)

R.A. Rohde, R.A. Muller, Air pollution in China: mapping of concentrations and sources. PLoS One e0135749 (2015)

A. Rosell-Melé, N. Moraleda-Cibrián, M. Cartró-Sabaté, F. Colomer-Ventura, P. Mayor, M. Orta-Martínez, Oil pollution in soils and sediments from the Northern Peruvian Amazon. Sci. Total Environ. **610–611**, 1010–1019 (2018)

F.J.W. Roughton, R.C. Darling, The effect of carbon monoxide on the oxyhemoglobin dissociation curve. Am. J. Physiol. **141**, 17–31 (1944)

R. Roushdy, M. Sieverding, H. Radwan, *The Impact of Water Supply and Sanitation on Child Health: Evidence from Egypt* (New York Population Council, New York, 2012)

S.W. Ryter, L.E. Otterbein, Carbon monoxide in biology and medicine. BioEssays **26**, 270–280 (2004)

A.S.M. Saifullah, M.H. Kabir, A. Khatun, S. Roy, M.S. Sheikh, Investigation of some water quality parameters of the Buriganga River. J. Environ. Sci. Nat. Res. **5**, 47–52 (2012)

J.A. Salomon, T. Vos, D.R. Hogan, M. Gagnon, M. Naghavi, A. Mokdad, N. Begum, R. Shah, M. Karyana, S. Kosen, Common values in assessing health outcomes from disease and injury: disability weights measurement study for the Global Burden of Disease Study 2010. Lancet **380**, 2129–2143 (2013)

E. Samoli, A. Analitis, G. Touloumi et al., Estimating the exposure-response relationships between particulate matter and mortality within the APHEA multicity project. Environ. Health Perspect. **113**, 88–95 (2005)

L. Schnaas, S.J. Rothenberg, M.F. Flores, S. Martinez, C. Hernandez, E. Osorio et al., Reduced intellectual development in children with prenatal lead exposure. Environ. Health Perspect. **114**(5), 791–797 (2006)

G.W. Schweimer, M. Levin, *Life Cycle Inventory of the Golf A4* (Volkswagen, Wolfsburg, Germany, 2000), http://www.wz.uw.edu.pl/pracownicyFiles/id10927-volkswagen-life-cycle-inventory.pdf. Retrieved on 21st Jan 2018

A.L. Sexton, Health and environmental implications of Americans? Time use responses to external stimuli essays on air quality alerts and daylight savings time. Int. J. Work Organ Emot. **1**(1), 87–102 (2012)

R.M. Shaffer, S.G. Gilbert, Reducing occupational lead exposures: Strengthened standards for a healthy workforce. Neurotoxicology (2017). https://doi.org/10.1016/j.neuro.2017.10.009

P. Sicard, C. Talbot, O. Lesne, A. Mangin, N. Alexandre, R. Collomp, The aggregate risk index: an intuitive tool providing the health risks of air pollution to health care community and public. Atmos. Environ. **46**, 11–16 (2012)

P. Sicard, A. Anav, A. De Marco, E. Paoletti, Projected global tropospheric ozone impacts on vegetation under different emission and climate scenarios. Atmos. Chem. Phys. **17**, 12177–12196 (2017)

E.K. Silbergeld, Lead in bone: implications for toxicology during pregnancy and lactation. Environ. Health Perspect. **91**, 63–70 (1991)

H. Simon, A. Reff, B. Wells, J. Xing, N. Frank, Ozone trends across the United States over a period of decreasing NOx and VOC emissions. Environ. Sci. Technol. **49**, 186–195 (2015)

S.I.V. Sousa, M.C.M. Alvim-Ferraz, F.G. Martins, Health effects of ozone focusing on childhood asthma: what is now known—a review from an epidemiological point of view. Chemosphere **90**, 2051–2058 (2013)

K. Steenland, S. Selevan, P. Landrigan, The mortality of lead smelter workers: an update. Am. Pub. Health Assoc. **82**(12), 1641–1644 (1992)

K. Steenland, V. Barry, A. Anttila et al., A cohort mortality study of lead exposed workers in the USA, Finland and the UK. Occup. Environ. Med. **74**, 785–791 (2017)

Swaniti Initiative, *Air Pollution in India Impact, Policy Analysis and Remedial Measures by Governments* (2015), http://www.swaniti.com/wp-content/uploads/2015/04/Air-Pollution-in-India.pdf. Retrieved on 14 Dec 2017

T. Tao, K.L. Xin, Public health: a sustainable plan for China's drinking water. Nature **511**, 527–528 (2014)

Y. Tao, W. Huang, X. Huang, L. Zhong, S.-E. Lu, Y. Li et al., Estimated acute effects of ambient ozone and nitrogen dioxide on mortality in the Pearl River Delta of Southern China. Environ. Health Perspect. **120**, 393–398 (2012)

The Telegraph, *Mapped*: *The world's Most Polluted Countries* (2017), http://www.telegraph.co.uk/travel/maps-and-graphics/most-polluted-countries/. Retrieved on 30 Dec 2017

S. Tong, A.J. McMichael, P.A. Baghurst, Interactions between environmental lead exposure and sociodemographic factors on cognitive development. Arch. Environ. Health: Int. J. **55**, 330–335 (2000)

G. Touloumi, Short term effect of ambient oxidant exposure on mortality: a combined analysis within the APHEA project. Am. J. Epidemiol. **146**, 177–185 (1997)

TreeHugger, *Driving Sustainability Mainstream* (2017), https://www.treehugger.com/fossil-fuels/world-energy-use-over-last-200-years-graphs.html. Retrieved on 28 Dec 2017

M.C. Turner, D. Krewski, C.A. Pope et al., Long-term ambient fine particulate matter air pollution and lung cancer in a large cohort of never-smokers. Am. J. Respir. Crit. Care Med. **184**, 1374–1381 (2011)

UNIDO, Industrial Development Report, *The Role of Technology and Innovation in Inclusive and Sustainable Industrial Development* (2016)

United States Department of Labor, *Occupational Safety and Health Administration* (2018a). https://www.osha.gov/pls/oshaweb/owasrch.search_form?p_doc_type=INTERPRETATIONS&p_toc_level=3&p_keyvalue=1910.1025&p_status=CURRENT. Retrieved on 28 Jan 2018

United States Department of Labor, *Occupational Safety and Health Administration* (2018b). https://www.osha.gov/pls/oshaweb/owadisp.show_document?p_table=STANDARDS&p_id=10641. Retrieved 28 Jan 2018

L.S. Van Winkle, K. Bein, D. Anderson et al., Biological dose response to PM2.5: effect of particle extraction method on platelet and lung responses. Toxicol. Sci. **143**, 349–359 (2015)

D.R. Varma, S. Mulay, S. Chemtob, Carbon monoxide: from public health risk to painless killer, in *Handbook of Toxicology of Chemical Warfare Agents*, ed. by R.C. Gupta (Elsevier, 2015), pp. 267–286. ISBN 978-0128001592

M. Vrijheid, D. Martinez, I. Aguilera, M. Bustamante, F. Ballester, M. Estarlich et al., Indoor air pollution from gas cooking and infant neurodevelopment. Epidemiology **23**, 23–32 (2012)

Q. Wang, Z. Yang, Industrial water pollution, water environment treatment, and health risks in China. Environ. Pollut. **218**, 358–365 (2016)

X. Wang, Q. Zhang, F. Zheng, Q. Zheng, F. Yao, Z. Chen, Z., et al., Effects of elevated O_3 concentration on winter wheat and rice yields in the Yangtze River Delta, China. Environ. Pollut. **171**, 118–125 (2012)

J. Wang, S. Wang, A.S. Voorhees, et al., Assessment of short-term PM2.5-related mortality due to different emission sources in the Yangtze River Delta. China. Atmos. Environ. **123**(Part B), 440–448 (2015)

G. Wang, S. Gu, J. Chen, et al., Assessment of health and economic effects by PM2. 5 pollution in Beijing: a combined exposure response and computable general equilibrium analysis. Environ. Technol. 1–8 (2016a)

G. Wang, R. Zhang, M. E. Gomez, L. Yang, Z. M. Levy, M. Hu, Y. Lin, J. Peng, S. Guo, J. Meng, Persistent sulfate formation from London Fog to Chinese haze. Proc. Nat. Acad. Sci. USA **113**(48), 13630–13635 (2016b)

T. Wang, L. Xue, P. Brimblecombe, Y.F. Lam, L. Li, L. Zhang, Ozone pollution in China: a review of concentrations, meteorological influences, chemical precursors, and effects. Sci. Total Environ. **575**, 1582–1596 (2017)

WB-SCEA, Pakistan Strategic Country Environmental Assessment. Main Report, pp. 1–66. Report no. 36946-PK World Bank (2006)

L.K. Weaver, Carbon monoxide poisoning. New Engl. J. Med. **360**, 1217–1225 (2009)

C.L. Weber, G.P. Peters, D. Guan, K. Hubacek, The contribution of Chinese exports to climate change. Energy Policy **36**, 3572–3577 (2008)

WHO, *Health Impact Assessment of Air Pollution in the WHO European region* (World Health Organization, 2001), http://www.euro.who.int/__data/assets/pdf_file/0013/91111/E75492.pdf. Retrieved on 24 Jan 2018

WHO, *Meta-Analysis of Time-Series Studies and Panel Studies of Particulate Matter (PM) and Ozone (O3): Report of a WHO Task Group* (World Health Organization, 2004), http://www.euro.who.int/__data/assets/pdf_file/0004/74731/e82792.pdf. Retrieved on 24 Jan 2018

WHO, Air quality guidelines for particulate matter, ozone, nitrogen dioxide and sulfur dioxide, in *Global update 2005*, ed. by W.H. Organization (World Health Organization, 2005)

World Health Organization, *Climate Change and Health* (2017), http://www.who.int/mediacentre/factsheets/fs266/en/. Retrieved on 21st Jan 2018

World Health Organization (2018), http://www.who.int/airpollution/en/. Retrieved on 16 Jan 2018

T. Wiedmann, R. Wood, M. Lenzen, J. Minx, D. Guan, J. Barrett, *Development of an Embedded Carbon Emissions Indicator–Producing a Time Series of Input–Output Tables and Embedded Carbon Dioxide Emissions for the UK by Using a MRIO Data Optimisation System (EV02033)* (Defra, London, 2008), https://www.sei.org/publications/development-embedded-carbon-emissions-indicator-producing-time-series-input-output-tables-embedded-carbon-dioxide-emissions-uk-using-mrio-data-optimisation-system/. Retrieved on 16 Jan 2018

T. Wiedmann, R. Wood, M. Lenzen, J. Minx, D. Guan, J. Barrett, The carbon footprint of the UK—results from a multi-region input–output model. Econ. Syst. Res. **22**, 19–42 (2010)

Williams-Steiger Act, *Occupational safety and health act of 1970*. Public Law, 596 (1970)

World Bank, *Water Quality Management—Policy and Institutional Considerations* (Environment and Social Development Department, Washington, D.C., U.S.A., 2006), http://documents.worldbank.org/curated/pt/644721468025218243/pdf/377520CHA01Wat1management001PUBLIC1.pdf. Retrieved on 1st Feb 2018

R.S.S. Wu, T.C. Lau, W.K.M. Fung, P.H. Koa, K.M.Y. Leung, An artificial mussel for monitoring heavy metals in a marine environment. Environ. Pollut. **145**, 104–110 (2007)

X. Wu, S. Chen, J. Guo, G. Gao, Effect of air pollution on the stock yield of heavy pollution enterprises in China's key control cities. J. Clean. Prod. **170**, 399–406 (2018)

Y. Xiao, J. Murray, M. Lenzen, International trade linked with disease burden from airborne particulate pollution. Resour. Conserv. Recycl. **129**, 1–11 (2018)

Y. Xie, H. Dai, H. Dong, T. Hanaoka, T. Masui, Economic impacts from PM2.5 pollution-related health effects in China: a provincial-level analysis. Environ. Sci. Technol. **50**, 4836–4843 (2016)

Q. Xie, Frequent water pollution incidents exposure loopholes in supervision and punishment. Nanfang Dly (2001). Available from: http://epaper.nfdaily.cn/html/2011-08/19/content_7000739.htm. Retrieved on 10 Feb 2018

Y.-F. Xing, Y.-H. Xu, M.-H. Shi, Y.-X. Lian, The impact of PM2.5 on the human respiratory system. J. Thorac. Dis. **8**(1), E69–E74 (2016)

W.W. Yang, H. He, Q.X. Ma, J.Z. Ma, Y.C. Liu, P.F. Liu, Y.J. Mu, Synergistic formation of sulfate and ammonium resulting from reaction between SO_2 and NH_3 on typical mineral dust. Phys. Chem. Chem. Phys. **18**, 956–964 (2016)

H. Yang, China must continue the momentum of green law. Nature **509**(7502), 535 (2014)

H. Yin, M. Pizzol, L. Xu, External costs of PM2.5 pollution in Beijing, China: uncertainty analysis of multiple health impacts and costs. Environ. Pollut. **226**, 356–369 (2017)

R. Yusta-Garcia, M. Orta-Martínez, P. Mayor, C. Gonzalez-Crespo, A. Rosell-Mele, Water contamination from oil extraction activities in Northern Peruvian Amazonian rivers. Environ. Pollut. **225**, 370–380 (2017)

Q. Zhang, J. Tian, Y. Bai, Z. Yang, H. Zhang, Z. Meng, Effects of sulfur dioxide and its derivatives on the functions of rat hearts and their mechanisms. Proc. Environ. Sci. **18**, 43–50 (2013)

S.Q. Zheng, X.N. Zhang, Y. Sun, The mechanism of the effect of air pollution on outdoor activities of urban residents. J. Tsinghua Univ. **56**(1), 89–96 (2016)

Chapter 4
Sustainable Consumption

Abstract The large technological evolution witnessed during the twenty-first century has created an alarming new problem: unsustainable consumption. This has generated an unbalanced cycle which, if unchanged, will inevitably lead to the depletion of essential resources. As this problem gains momentum, several organizations and committees have begun to discuss and take action to tip the scale in favor of the environment. Improvements can be had for both personal and industrial consumption/production patterns, needing just an attitude shift to start thinking of ourselves as planetary citizens. The environmental impact of consumption is addressed, discussing topics such as the ecological footprint and how each human being needs approximately 18,000 m^2 of land to sustain current consumption patterns. However, new trends and lines of thought are emerging that are based on sustainable consumption, transforming niche markets into mainstream businesses due to the growing demand of eco- and organic products as well as the creation of networks which promote reusing, upcycling, and borrowing as a way of helping products retain value and further extend their lifecycles. Also, researchers are investigating the influence that personal, intellectual, emotional, political, religious, moral, age range, and social traits have on sustainable behavior, going even as far as looking into the practice of mindfulness and meditation. Environmentally responsible citizens choose to avoid impulse buying and consume only the strictly necessary to live comfortably, while ethical companies are praised for their positive commitment to environmental preservation. Then again, when it comes to purchasing eco-friendly goods, the lack of information makes it harder for consumers to choose wisely. The introduction and standardization of eco-labels should be pushed by governmental entities and committees in order to force producers to account for the environmental impact of their products, from the amount of resources utilized during production and transportation (i.e., electrical energy, water, etc.), to the emissions that their product creates throughout its entire life cycle, allowing consumers to make informed purchase decisions on the fly (analogous to nutritional facts labels). These topics are discussed thoroughly throughout this chapter. Although there is still much to be done to overcome current challenges and promote a larger and widespread sustainable behavior among modern society, the key starting points are within everyone's reach.

F. J. Gomes da Silva and R. M. Gouveia, *Cleaner Production*,
https://doi.org/10.1007/978-3-030-23165-1_4

Keywords Sustainable consumption · Consumer needs · Consumer behavior · Consumption patterns · Consumption policies · Citizenship · Moral values · Consumption drivers · Sustainable consumption strategies · Mindfulness · Eco-labeling · Cross-generational consumption · Challenges in sustainable consumption

The National Research Council defines sustainability in the book Sustainability and the US Environmental Protection Agency (US EPA) as follows: "Sustainability is based on a simple and long-recognized factual premise: Everything that humans require for their survival and well-being depends, directly or indirectly, on the natural environment (Marsh 1864). The environment provides the air we breathe, the water we drink, and the food we eat. It defines in fundamental ways the communities in which we live and is the source for renewable and nonrenewable resources on which civilization depends. Our health and wellbeing, our economy, and our security all require a high-quality environment." (National Research Council 2011). On the other hand, Zukin and Maguire (2004) define consumption as "a social, cultural and economic process of choosing goods," which empowers citizens to shape and illustrate their personality.

The Commission on Sustainable Consumption was created by Mansfield College, Oxford, in April 1999. The main goal behind the creation of this committee was to promote the debate and encourage the necessary actions to achieve the desired sustainable consumption aims. The Commission has carried out extensive investigation, discussion, and analysis, supported by staff from the Oxford Center for Environment, Ethics, and Society (OCEES). The idea behind the creation of the Oxford Commission on Sustainable Consumption (OCSC) arose from developments previously carried out by OCEES on sustainable consumption and lifestyles. It became apparent that, despite a series of international workshops, conferences, and publications, efforts to develop the concepts, goals, and strategies that might lead us toward sustainable consumption have not gone far. In addition, it became clear that despite many conferences, meetings, publications, and other attempts in international terms, the desired results on the evolution and implementation of sustainable consumption concepts were not being achieved. The difficulty in advancing certain concepts related to sustainable consumption had as its main origins the ethics linked to consumption, as well as the power of the companies that drive consumption (Michaelis 2003).

The twenty-first century brought a new important topic for research programs and public concerns: Sustainable Consumption and Production (SCP). In fact, the initial ideas about this issue began to be discussed in the mid-1990s through reports launched by the United Nations, OECD, and the World Business Council for Sustainable Development (WBCSD), pointing out the need for minimizing the consumption of natural resources, doing more with less. This process was sped up by the 2002 World Summit on Sustainable Development which took place in Johannesburg, where delegates from United Nations Environment Programme and the United Nations Department for Economic and Social Affairs were encouraged to develop a 10-year framework program on Sustainable Consumption and Production. The process was afterward driven by the multi-stakeholders Marrakesh process, launched in 2003, which brought new insights for the development of the 10-year program abovementioned, which was finally formally established by the United Nations Conference on Sustainable Development—also known as Rio + 20—which took place in Rio de Janeiro, Brazil on 20–22 June 2012 (Brauch et al. 2016). Some nationally developed programs in Finland, UK, and Germany also contributed to the adoption of a set of Sustainable Development Goals (SDGs), which built upon the Millennium Development Goals (MDGs), and came together with the post-2015 development agenda (Geels et al. 2015). This program engages state members to promote national programs on Sustainable Development based on the principles agreed in the Conference (UN 2018). The scientific community also has largely promoted deep discussions about this issue and the policies around it, giving rise to some special issues in widespread prestigious journals such as Journal of Industrial Ecology in the years of 2005 and 2010, Journal of Cleaner Production in 2008, and Natural Resources Forum in 2010 (Geels et al. 2015).

Sustainable consumption can be defined as the one that allows an adequate balance between the expenses incurred and time, leading to the satisfaction of the basic needs of the human being and providing the appropriate well-being, also taking into account the welfare of future generations (Cornwell and Drennan 2004).

4.1 How Consumption Interferes with the Environment

The beginning of the new millennium conveyed the urgency of reversing the global warming process. Will it be possible to get a global consensus, leading the world to a solution for the greenhouse gas emissions problem? Are we completely aware of our dependence on fossil fuels? Unrelated to issues such these, we continue to ignore the increase in population and concentration in major cities. Moreover, people are instigated to consume goods by several means, such as marketing actions, bank credit offers, sales, and so on, trying to increase profits. Behind that, there are companies feeding the necessary products, eager to increase their turnover and profits. Furthermore, stock markets need positive news about the economic activity, which depends on the global consumption. In addition, governments are also eager to announce GDP improvements every year, generating more jobs and charging more taxes (Watkins et al. 2016). But, how can we create the necessary balance to create jobs for people, live a comfortable life, and not contribute to a continuous degradation of the environment? Is this not a good time to reassess the consumerist lifestyles? It is for these and other reasons that we need to answer a very important question: How can we reverse all the harmful actions that human activity is constantly inflicting to the environment? The global environmental crisis results from unsustainable patterns of production and consumption and requires a wide awareness mainly regarding the growing and uncontrolled waste generation, from household waste to the dangerous industrial waste. New patterns of production and consumption, along with the reduction of its impact on the environment are goals that, day by day, require changes in people's habits and attitudes, as well as lifestyles. It is impossible to reach these goals without the development of an education focused on the construction of a planetary citizenship able to promote quality of life, social justice (fair income distribution), and economic development respecting the environment. The current lifestyle requires constant consumption of natural resources, both renewables and nonrenewables, which then gives way to trash. Thus, it is necessary to mind about how much land is required for each world inhabitant to be provided with basic resources, such as food, energy, transport, and clothing. Thinking about that, Rees (1992) and Wackernagel and Rees (1996) have defined the ecological footprint as the impact that the consumption of each world inhabitant prints on Nature, providing thus a reflection on the waste production and pollution, as well as the capacity of absorption by the environment of that waste and pollution. Regarding that balance, it was concluded that each human need 18,000 m^2 (1.8 ha), the equivalent area of 2.5 football fields. If the current global consumption is already 25% above the bearable level of the planet, how we will be in 2050, when production of consumer goods will double? The ability of the Earth to renew its own resources and absorb wastes is already quite compromised by this pattern of life that, in addition to being unsustainable, it is unfair as, less than 20% of the population holds 80% of the world's natural resources, relegating the vast majority to an undeniable situation of poverty. Indeed, Michaelis (2003) points out three main factors which make it possible to clearly see that consumption trends are not oriented in the right direction: (a) It is often referred that increasing material consumption in developed countries has not contributed positively to an increase in the quality of life of citizens and to their happiness, and that the cycle "to make money—to spend money" only causes stress and unhappiness; (b) current patterns of consumption translate into social injustice, corresponding to very high wages for a few and a very low wages for the majority, which translates into very different purchasing power between different social classes, accentuating inequalities; (c) the increasing intensification in material consumption is leading to serious environmental problems, such as global warming, pollution from increased energy consumption, enlarged consumption of drinking water, and increased generation of waste and effluents, as well as the generation of harmful substances in certain production processes. Thus, the current consumption behavior trend is unfair and will lead us to serious problem in the future. Therefore, it is urgent to reverse this trend by using all possible efforts, knowing that the selfish commercial strategy of companies would hardly contribute to this battle.

The relationship between consumption and well-being can be realized through a study of human needs, which can explain many of the social and behavioral consumption aspects, as well as the theoretical meaning of our economic system (Jackson et al. 2004). However, the needs approach should take into account that resources are finite and to some extent universal (Jackson and Michaelis 2003). This theoretical framework defines the basis of many of the consumption problems, as well as the inefficiencies of the current economic systems to satisfy the basic needs, also allowing social development.

There are some worrying indications that society still does not show a clear trend toward the adoption of sustainable consumption principles. The number of inhabitants per house is one of them. This has led to more intense demand for household appliances and furniture, showing a consumption pattern more focused on the individual than on the family (Røpke 1999). Another important indicator is the number of people who are using in a systematic way their own car in detriment of the use of public transports. This decreases the availability of a more efficient transport network, throwing consumers into a greater dependence of their own car, thus reversing the tendency to adopt and increase the use of public transportation, reducing by this way unsustainable consumption. The lack of social involvement and more active collective decision making misleads the decisions considered as socially more appropriate. The incentives in terms of adopting a more sustainable behavior for the individual are still very fragile in most of the countries (Vercalsteren and Geerken 2003). Thus, as some individuals continue to increase their consumption, all others face a higher benchmark, which continues to pressure the whole society to higher levels of consumption (Briceno and Stagl 2006).

The exercise of consumerism may be drilled in a compulsive manner and even become a disease. Despite this, our society considers this practice with great naturalness, forgetting with remarkable lightness of spirit how that practice can put at risk global sustainability. Thus, it is essential that every citizen knows how to perfectly distinguish the essential from the superfluous, avoiding impulse purchases, the substitution of devices and households items in perfect state of operation, among many other small actions which, repeated constantly, generates the creation of unnecessary waste, forcing the production of new products, which will generate the need of new natural resources and energy, leading to pollution and an unsustainable system. Indeed, an important question can be brought to the discussion: Are we consuming to live or are we living to consume? It is important that society does not opt for quantity in detriment of quality, and, nowadays, there are societies where the feelings of power and happiness seem to be strictly related to the excess of food, goods, information, and even cultural events, then to a more moderate lifestyle based on quality. Effectively, early education should instill children now with the values we want to see in the future. The education in the last decades was strongly influenced by instigation to compulsive consumption as the best way to make people feel the sensation of power. Marketing and media are the ideal means to create illusions in people's mind which, when already exhausted by work, see shopping actions as a sense of satisfaction and reward for the work done. However, by purchasing more, more work is needed to feed this cycle. By delivering a good economic and social education or teaching drastic events, such as insolvency and what led to such situation, a new awareness can be created and lead people to an increased responsibility regarding sustainable consumption. A responsible citizen chooses to consume only the necessary, avoiding impulse buying. A great option is to choose bulk materials using recyclable packages, instead of single products with complex packages leading to a difficult recycling process. In short: plan purchases, reduce waste, avoiding the disposable, and, whenever possible, seek to reuse. Above all, it is essential to value companies presenting environmental and ethical commitments with society.

Several investigations are pointing out that the excessive purchase of goods and services carries a heavy burden on society and the environment and threatens the economic comfort and development of society in general (Brown et al. 2005; Sheth et al. 2011; Alexander 2011; Mont et al. 2014; Lee and Ahn 2016; Hüttel et al. 2018). Despite the large research on this subject, a vital issue tends to remain to only a small number of studies: At the moment of purchase, how can consumers assess the consequences of their action for their own well-being and for the society in general? Furthermore, how do economic motivations act in the moment of purchase? Thus, it is necessary to understand how sustainable decision-making patterns are influenced by economic constraints. Sustainable and non-sustainable consumption have completely opposite effects: Sustainable consumption can be seen as one that is carried out in favor of social welfare and the environment, while non-sustainable consumption is essentially supported by the superfluous consumption of resources as a preferential lifestyle, with contempt for the waste of resources as something pertaining to the whole society. However, problems of overconsumption, in addition to bringing environmental problems, can expose or cause psychological complications and imply future financial difficulties (Achtziger et al. 2015). The economic factor has a strong influence on buying habits. According to several authors, middle-class consumers tend not to consume unnecessarily products because they fear waste, lean toward avoiding the burden of ownership, and are afraid of accumulating debts and falling into poverty (Soman 2001; Tatzel 2002; Todd and Lawson 2003; Canova et al. 2005; Moeller and Wittkowski 2010; Alexander and Ussher 2012; Piscicelli et al. 2015). Even with regard to the economic aspect, consumers normally use strategies to reconcile their lifestyle with their income allowance, but they also attach great importance on reaching short- and long-term economic goals in order to overcome unforeseen situations and accumulate wealth for their retirement or other purposes. Thus, it can be realized that citizens shrink from exaggerated consumption by economic security precautions, searching for higher autonomy. Some authors point out that a more moderate consumption is a consequence of a more mature attitude, which reflects the person's well-being or a spiritual attitude (Elgin and Mitchell 1977; Craig-Lees and Hill 2002; Todd and Lawson 2003; Devaney et al. 2007). On the other hand, it is also underlined by other authors that the relationship between citizens could be affected due to problems related to the practice of borrowing, which ends up conditioning people, eventually impelling them to purchase the goods in detriment of borrowing (Tinson and Nuttall 2007; Dezsö and Loewenstein 2012). Several authors have given special attention to the factors that may be the source of consumption considered economically unsustainable. Most of those motivations are centered on fun, aesthetic enhancement, personal well-being, quality of life, independence, freedom awareness, or need for a control sensation (Holbrook and Hirschman 1982; Livingstone and Lunt 1992; Richins and Rudmin 1994; Babin et al. 1994; Arnold and Reynolds 2003; Richins 2011).

Business marketing strategies should meet the specific needs of consumers concerned with the environment. Some authors refer that there is growing demand for sustainable products (Tódero et al. 2011; Souza 2012), and companies have been following the emergence of these consumers, putting special care in the resources used, avoiding waste generation as much as possible (throughout all stages of the product's life cycle) and the utilization of polluting products. These concerns are somewhat old, because in 1970, Kinnear and Taylor (1973) warned that personal consumption decisions can contribute to the preservation or deterioration of the environment, while Hartmann and Ibáñez (2006) explain that the sustainable purchasing behavior is vital for social transformation. Other authors are pointing out that there are consumers with clear conscience about the impact of the products on the environment, searching specifically for environmentally friendly products, using their lifestyle as a way of deliberately showing how they are against unruly habits of consumption (Micheletti et al. 2004; Newman and Bartels 2010; Watkins et al. 2016), seeking a new way of life. However, this "open-mind" state can be socially and politically interpreted as reactive (Stolle et al. 2005). Effectively, Crockett and Wallendorf (2004) advocate that consumers are increasingly using their individual consumption patterns as a way to express and demonstrate their ideology, despite the era of increasing political apathy that we are crossing. Thus, consumption patterns of certain groups of citizens require careful analysis, both from marketing and from political perspective. When the consumption pattern is strongly influenced by political motivations, it is important to correlate the ideology with the main concerns that it is inducing, trying also to make the parallel between moral principles and political ideology, as both converge to a better consumption pattern. By searching for the main motivations, even under different political interests, it is possible to adopt further techniques to motivate more people in the same way.

In the social and economic fields, companies have increased responsibilities, mainly in two perfectly distinct, but coincident, vectors: (a) Companies should generally profit, which should be used to promote new investments, as well as compensate investors, which reintroduces wealth into the economy, generating more wealth; (b) the generation of wealth by the companies contributes to the national GDP, engendering taxes that aim to improve social conditions of a part of the country's population.

Although numerous studies have already been carried out on how consumer behavior is affected by their cultural level, social class, ethical values, intellectual and psychological properties, self-confidence, among many other personal characteristics, there is still a gap between the attitude they intend to convey to the society and the behavior they currently take in consumption acts (Ajzen 2001; Loureiro and Lotade 2005; Grankvist et al. 2007; Perrini et al. 2010).

4.2 Needs and Consumer Behavior

The happiness of an individual, a family, or a community is not merely dictated by an eventual abundance of material possessions. Moreover, it is necessary for companies and citizens to realize that a responsible consumer standard is surely the basis of a better quality of life now and in the future, allowing the necessary socio-economic balance and the desirable preservation of the environment. Sustainable development is a serious challenge for humanity, and consumption patterns are at the heart of this task. If consumer habits do not find the path to sustainability, technological advancement, growth in average life expectancy, modern production systems, and social policies to protect the most disadvantaged will not be enough to ensure a sustainable development. Perhaps emotions are too important factors in the continued adoption of sustainable consumption patterns. However, there are many other factors which, probably in a less significant way, also contribute to the fact that sustainable consumption is not always followed with the desired and necessary determination. Indeed, some authors have concluded that emotions overlap with cognition in the development of continued practices of sustainable consumption (Carrus et al. 2008; Meneses 2010; Swim et al. 2011; Kanchanapibul et al. 2014). The relationship between patterns of consumption and emotions has been largely investigated after 2010, and a very strong relationship has been established between some of these emotions and patterns of sustainable consumption. In fact, it is widely known that anger is an emotion perfectly associated with the punishment of those who damage the environment, as well as their acts. This feeling makes it possible to perceive that the individual, which feels anger when witnessing an attack on the environment, will not want to do the same and therefore will have a strong tendency to have an ongoing habit of sustainable consumption. On the other hand, pride is a feeling that is at the basis of a behavior of sustainable consumption, because it clearly indicates that the individual wants to defend the environment as it would defend its own heritage (Harth et al. 2013; Peloza et al. 2013; Antonetti and Maklan 2014). Thus, the knowledge regarding the emotions that govern the mind of an individual will allow to predict what kind of behavior will tend to be assumed on his habits of consumption.

However, the list of emotions that are directly correlated with human behavior linked to consumption and environment protection is much more extensive. Indeed, emotions can be divided into two main groups: emotions linked to an individual's own consciousness and basic emotions. Emotions linked to an individual's consciousness usually include pride, embarrassment, regret, shame, shyness, guilt, among others. These emotions are closely related to four moral principles: pride and respect in a positive way and guilt and anger in a negative manner. Pride and guilt are usually strictly connected to self-behavior, whereas respect and anger are normally correlated to other behaviors. On the other hand, basic emotions are well-known: love, liking, fear, disliking, anger, joy, and sadness. The emotions related to the conscience of an individual are the ones that are at the base of the moral that conducts their actions, allowing to predict in a much more exact way their conduct. These emotions are also directly related to the norms by which the individuals are governed, having regard for ethical principles resulting from the moral principles. Wang and Wu (2016) have dissected largely the different emotions affecting consumer behavior, demonstrating that emotions such as pride and guilt influence the decision to switch from appliances that consume more energy for others with better energy saving characteristics. Also, emotions of respect and anger have shown a very positive effect on the decision to opt for more favorable appliances in terms of energy consumption. However, it has also been shown that a negative emotion can override all positive emotions regarding a more environmentally friendly choice, which shows how pernicious the mental process of formulating an intention to acquire or exchange a good is. Although this study leaves worrying indications on how emotions govern the human mind and decisions in terms of consumption, the results obtained also allow us to perceive that by stimulating emotions such as respect, pride, guilt, and anger and avoiding the establishment of negative emotions that decisively affect the formulation of an opinion during the act of exchange or acquisition of goods, one may have a higher degree of confidence that individuals will essentially make favorable decisions concerning environment, guiding their attitudes of consumption toward principles of sustainability.

Consumption comes from needs, which were divided into nine main classes by Max-Neef (1992), as follows: survival, security, relaxation, membership, friendliness, self-determination, sympathetic, conception, and uniqueness. Excluding survival needs, all others can occur together, in a balance game, where they can complement or replace each other, depending on how these classes are being met. The feeling of satisfaction will depend on the type of culture of the citizen and the socio-economic context in which it is inserted, and this satisfaction can be measured according to four different criteria: doing, having, being, and/or interacting (Doyal and Gough 1991; Max-Neef 1992). It may also be noted that different human needs can be divided into three main vectors: physiological, social, and/or psychosocial, which determines the different types of satisfaction (Jackson and Michaelis 2003). Nonetheless, considering the currently implemented economic system, particular emphasis has been given to material needs, placing the focus primarily on satisfaction rather than necessity itself (Witt 2001). The reliance on certain conveniences to satisfy some of the more complex needs has made satisfaction indicators more inefficient. Studies have shown that the lack of satisfaction of one or more needs generates reactions that manifest through emotions such as the need for demonstration of power, dominance, violence, and exacerbated consumption of goods and services (Kamenetzky 1992). It has been verified that the non-satisfaction of given material indicators has been systematically covered by the consumption of material of goods, easily turning into exacerbated and unsustainable consumerism (Briceno and Stagl 2006), becoming a coping mechanism which allows to overcome feelings of sadness generated due to deprivation of some sort. It has been found that, in general, material needs can be adequately attained by material goods, but if there are social or psychological needs, these will not be adequately compensated by material means. Furthermore, the higher the consumption of materials goods in an effort to meet these needs, the less they are effectively met (Jackson and Michaelis 2003). If social relations and well-being are improved in the community where citizens are involved, the main psychological needs are directly or indirectly fulfilled, improving the global well-being and reducing material needs (Putnam 2000). In developed countries, the rooted consumption patterns seem to want to override social and humanistic needs that common citizens also require. In these countries, the involvement of most citizens in the current economic system takes out the capacity for the creation and development of their own structures capable of meeting basic social needs such as creativity, autonomy, and involvement in community causes. If these needs are not adequately met, they can be replaced by others of more material nature, giving place again to consumerism. Moreover, it is recognized that in Western countries more self-centered educational systems have been developed, incentivizing individualism as a way to feel more autonomous, free, and with full sovereignty on their choices (Michaelis 2003).

Sustainable consumer behavior studies have assumed different facets regarding the main concern of each study. The first approach is usually performed by psychological models which intend to determine and describe consumer behavior and the main drivers capable of shifting their usual options, based on concepts and assumptions, as deeply referred in the work of Jackson (2005) and Darnton (2008). However, this problem can be found in literature under different names, such as political consumption as studied by Micheletti (2003) or Micheletti et al. (2004), ethical consumption as discussed by Devinney et al. (2010) or Newholm and Shaw (2007), green consumption as referred by Peattie (2010) or Roberts (1996), voluntary simplicity as mentioned by Iwata (2006), mindfulness consumption as denoted by Sheth et al. (2011) or Fischer et al. (2017), concerned consumers as called by Connolly and Shaw (2006), or citizen-consumer when studied by Webster and Lusch (2013).

Consumer behavior has been the subject of innumerable theories, some based on the more cognitive side of the individual, others, more recently, integrating moral values and factors able to influence the consumer's choice, including governmental incentives. Examples of these developments are the following theories: Ajzen and Fishbein's Theory of Reasoned Action, Ajzen's Theory of Planned Behavior, Schwartz's Norm Activation Theory, Stern's Value-Belief-Norm Theory, Ölander and Thøgersen's Motivation–Opportunity–Abilities Model, Cialdini's Focus Theory of Normative Conduct, Stern's Attitude-Behavior-Context Model, Triandis' Theory of Interpersonal Behavior, and Bagozzi's Comprehensive Model of Consumer Action (Jackson 2005; Phipps et al. 2013; Piscicelli et al. 2015). The Stern's Value-Belief-Norm Theory is one of the most known theories because it brings together the moral norm activation theory argued by Schwartz (1973), the theory of personal values and also the New Environmental Paradigm (NEP), establishing that social pro-environmental behavior, as well as individual moral values would be the main courses for sustainable consumption behavior.

This theory is based on the principle that if consumers present values that are highly unselfish and focused on a better future for the planet, then these consumers will be better prepared to accept the new environmental paradigms, respecting the balance between existing natural resources and consumption needs. Moreover, the Values-Beliefs-Norms theory establishes something that seems to be perfectly current, since it indicates that there are two completely antagonistic currents of behavior: (a) one in which citizens assume that pro-environmental actions concern them, and (b) those who believe that it is up to others to change their behavior in relation to how they contribute to solving environmental problems (Stern 2000; Phipps et al. 2013). However, this theory may be limited by constraints previously pointed out by the Motivation-Opportunities-Abilities theory, which identifies some limitations to actions that the individual may want to take in a pro-environmental sense, such as buying products that comply with sustainability rules or a wish to recycle products but not have the economic or logistical means to do so (Thøgerson 2005; Koos 2011). There may also be other restrictions impacting the sustainable behavior of individuals, such as a temporary or permanent lack (for distribution reasons) of sustainable products in the region where they live and/or carry out their purchases (Tanner and Kast 2003).

Bandura (2001) established that there are three factors that influence the behavior of an individual concerning the environment: personal, behavioral, and environmental. However, Bandura (2001) also noted that these factors cannot be considered in isolation, but together, representing a means to formulate a pro-environmental attitude. Still, they are not sovereign of themselves. In fact, they only represent the potential for environmentally friendly behavior but can also be affected by social or situational issues such as observation and following of community behaviors, following of third-party advice, presence of family or friends at the time of purchase, among many other factors. Thus, individuals are producers of behaviors, being also a component of the environment in which they live, which represents a consequence of the attitudes that have been taken by that same individual in the past. This mutual involvement allows us to perceive that our behavior as consumers affects the environment in which we live, and the environment in which we live affects our lives. This is defined as "reciprocal determinism" by Bandura (2001). Therefore, the cause-effect relationship is bi-univocal and cannot be disaggregated, for the reasons previously mentioned. The translation of this relationship can be represented by extremely simple cases, such as the bag used to carry groceries from the supermarket to home. If the consumer prefers to use a reusable bag, he is producing a sustainable behavior, which stems from the awareness that many non-reusable bags have been used in the past and that they are seriously damaging the environment. However, this pro-environmental attitude may be determined by factors linked to the consumers' sense of well-being by themselves, or just to the more materialistic way of thinking. In fact, consumers can feel satisfied with the use of a reusable bag as they feel good about themselves having contributed to a more sustainable consumption with their behavior, or just to analyze the more rational side of the option: A reusable bag will prevent them from needing more reusable bags in the near future (since the cost of the non-reusable bag has to be borne by the customer or the supplier, being inevitable). Bi-univocal relationships between suppliers and citizens regarding consumption behavior can assume different facets, some of which are quite curious, such as the following example: If a company causes a serious environmental accident and is found guilty by negligence, the consumer can take (a) the attitude of repudiating products of that company because it feels that it is favoring a company that does not take care of the environment that belongs to all of us or (b) feel that they can also have unruly behaviors toward the environment, as the company had. The exemplary punishment of the company, in the case above-referred, may serve as a stimulus for the consumer to choose the first route (a) and not the second (b).

The consumers' behavior has also been the subject of several studies to establish metrics that allow to evaluate their behavior in terms of economic, social, and environmental aspects (Roberts 1996; Lastovicka et al. 1999; Kaiser et al. 1999; Sen and Bhattacharya 2001; Huneke 2005; Gilg et al. 2005; Iwata 2006; de Pelsmacker and Janssens 2007; Pepper et al. 2009; Balderjahn et al. 2013;). Later, Balderjahn et al. (2013) carried out a study intending to establish a new measurement scale able to put together all effects previously referred giving more emphasis on the economic factor of consumer behavior. In addition, as people become socially and culturally more autonomous, they lose certain ties to politics, religion, and habits of the community in which they live, tending to adopt given consumption patterns as a way of creating a certain identity that they want to convey (Jackson 2002). When there is an individualization of the citizen, much of the security transmitted by the previously existing spirit of community is lost, and the pattern of consumption tends to replace the previously existing security (Putnam 2000). Consumption as a hunt for comfort and personal stimulation has replaced some of the voids created by increasingly individualized societies. The lack of social relations and coordinated actions has a strong potential to intensify the demand for goods and services, thus increasing materialism (Briceno and Stagl 2006). Moreover, and as a consequence, the family and the community have lost dedication on the part of individuals who have devoted more time to work and commercial activities. In this context, community activities have lost strength, such as educational projects and infrastructure improvements, and the sharing of assets has become more restricted to small village communities who insist on maintaining this old tradition. With this, the market has become trans-border, more complex, and abstract, making it more difficult to reflect on consumption processes (O'Hara and Stagl 2001). However, social problems begin to take place that start to demand changes in consumption systems, so that they also integrate social responsibilities. For example, the distribution of wealth and social justice are very problematic aspects of the current consumption, which compromise both the short-term and the long-term social welfare and citizen security (Borghesi and Vercelli 2003).

Consumer behavior is influenced by a complex set of dynamics which are intensely entrenched in, being also profoundly influenced by social, psychological, and cultural contexts. Besides these aspects and their influence on purposeful choice regarding consumption patterns, other important factors such as habit, price, quality, availability, convenience, lack of information, overall perception of products, disbelief of product claims, available time to research for the right choice, and so on are also hardheaded walls to pursue sustainable consumption principles. Furthermore, people tend to interact with each other in order to renew their sense of being and well-being, thereby giving way to the establishment of their own consumption needs (Jackson and Michaelis 2003). Moreover, all of the barriers previously referred contribute to a well-known gap between the real attitudes of the consumers and their sustainable consumption mindset (Moisander 2007; McDonagh and Prothero 2014; Watkins et al. 2016).

Moral values are usually intrinsically linked to sustainable consumption, and this relationship has been deeply studied by several authors (Schwartz 1992; Triandis 1993; Karp 1996; Vermeir and Verbeke 2006), being pointed out that personal characteristics such as responsibility, freedom, benevolence, self-direction, universalism, altruism, and collectivistic spirit are usually related with the sustainable consumption whereas tradition, ambition, security, power, and hedonism as values normally connected to a contrary attitude. Moreover, moral foundations have also been defined by other authors focusing on codes and rules based on the absence or minimization of the stress and harm of the society in general (Singer 1993; Gert 2005; Graham et al.

2009). However, this will be a permanently unsolved topic in terms of definition, which will contribute to the discussion and mind exercises of consumer behaviorists, sociologists, philosophers, and other society researchers, as it is directly related to human behavior and society nature. Nevertheless, this concept can be also directly connected to the worst attitudes of the human behavior, such as excessive or compulsive consumption in general, without any concern about the natural resources, waste and pollution generated, along with unequal distribution of wealth. Some theories about moral foundations are pointing out that behavior significantly differs mainly due to five psychological characteristics: harm/care, fairness/reciprocity, in-group/loyalty, authority/respect, and purity/sanctity (Watkins et al. 2016).

The substantial impact that sectors such as food, transportation, housing, or clothing represent in terms of global consumption has made sustainable consumption a "hot issue" in terms of investigation, attracting researchers for developing algorithms and models able to describe consumer behavior (Tukker et al. 2010; Kenis and Mathijs 2012, Ivanova et al. 2015). The recent advances achieved in the study of consumer behavior allow facing with optimism the future in this field; however, deeper investigations should be carried out in order to perceive all motivations behind consumption options (Kaufmann-Hayoz et al. 2012; Reisch and Thøgersen 2015). Moreover, it is still necessary to overcome the stigma of the broad group of consumers who think that environmentally friendly products are less efficient than traditional ones (Luchs et al. 2010).

Strategy interventions that tend to guide the consumer may also change the way consumers act. These policies are based on the psychological knowledge of how human beings react to different stimuli, which then try to generate ways of removing existing barriers in the consumer's mind, in an effort to alter their consumption habits. It is well-known that the announcement of tax incentives, awards, or other forms of consumer rewards can lead to the decision to be made on its own, without the need for a strong direct awareness campaign. Another way to interact with consumers in order to influence their decision is providing them with the correct product information. In this case, eco-labeling can provide consumers with the key information they need at the right time to make their decision based on the information received. However, this action will only succeed if the consumer is properly predisposed to consult the available information and to make their choice based on that information. Because these actions are not permanent and the consumer is not continually pressured, these initiatives tend to lose effectiveness over time. Furthermore, it is well-known that consumers are not permanently aligned with the sustainable consumption principles (Blake 1999; Morris et al. 2012).

There is a significant difference between the perceptions of sustainable consumption by Western countries vis-à-vis developing countries: People from developed countries adopt sustainable consumption patterns based on the benefits it brings to the environment in general, realizing immediately that it is equally important for future well-being. In developing countries, there are other types of perception: People adopt sustainable consumption habits because their surroundings clearly show that this behavior is absolutely necessary, or else they will only act in this way due to regulations, restrictions, or financial incentives, among others.

4.2.1 Water Consumption and Sustainability

Pollution has adversely affected the climate, with drastic changes in Nature occurring through the emergence of more frequent extreme events, as well as glacial melting and rising sea levels as a result of global warming. Certain areas of the planet have been further hampered by these phenomena, creating rush floods in one area and prolonged drought in others, severely affecting certain crops. Combining all of this with world population growth, the level of drinking water consumption has become a growing concern. Drought situations that have severely hit some parts of the planet have been the focus of several of the world's leading climate change organizations, such as the EPA, which has developed programs for the rational use of water resources. The OECD also published a report on the subject as early as 1998 (OECD 1998), resulting from a workshop held at that time, highlighting the importance of proper management of water resources, emphasizing their importance on the health of individuals, economic growth, and the quality of the environment. The European Union has also introduced a directive (2000/60/EC) to rationalize water use, which was in the meantime extinguished, since it had effects until 2015.

The way in which restrictions on water consumption have been imposed differs in form and effect. In fact, US restrictions are based on encouraging sustainable water consumption practices with a primary focus on residential consumption, forcing consumers to adopt a timetable for water consumption, car wash scheduling, and many other restrictions. In addition, an educational program has been established with a view to the careful management of water by citizens, as well as advisory services for the management of water consumption by sectors such as agriculture, industry, and services. On the other hand, the European Union preferred to use price as a main deterrent for water consumption, both for domestic consumption and for the business sector. In fact, the Directive 2000/60/EU had a much broader scope for improving water quality both in inland and coastal waters. However, bold goals were also set for water management, which included a much greater care with the consumption of drinking water. This has resulted in the need to create a deterrent effect on water consumption, which has undergone a gradual increase in its processing price, linking it with the "polluter pays" principle. This had direct effects on water consumption, not only by personal environmental concerns, but also due to the additional charge that water consumption brought to domestic and industrial entities. Figure 4.1 shows the effect of these policies, allowing to relate the impact of these measures in Europe and North America.

Nowadays, these policies have made water substantially more expensive in Europe than in North America, leading to a gradual but steady reduction of drinking water consumption. In the meantime, consumption in North America remains very high at a much lower price, as can be seen in Fig. 4.1a. Figure 4.1b allows to observe a great difference among consumption patterns throughout European countries, taking into account the use of water for manufacturing processes or household purposes.

Water management by corporate or common consumers is a constant concern. Indeed, when a company chooses to produce a particular product, it must take into

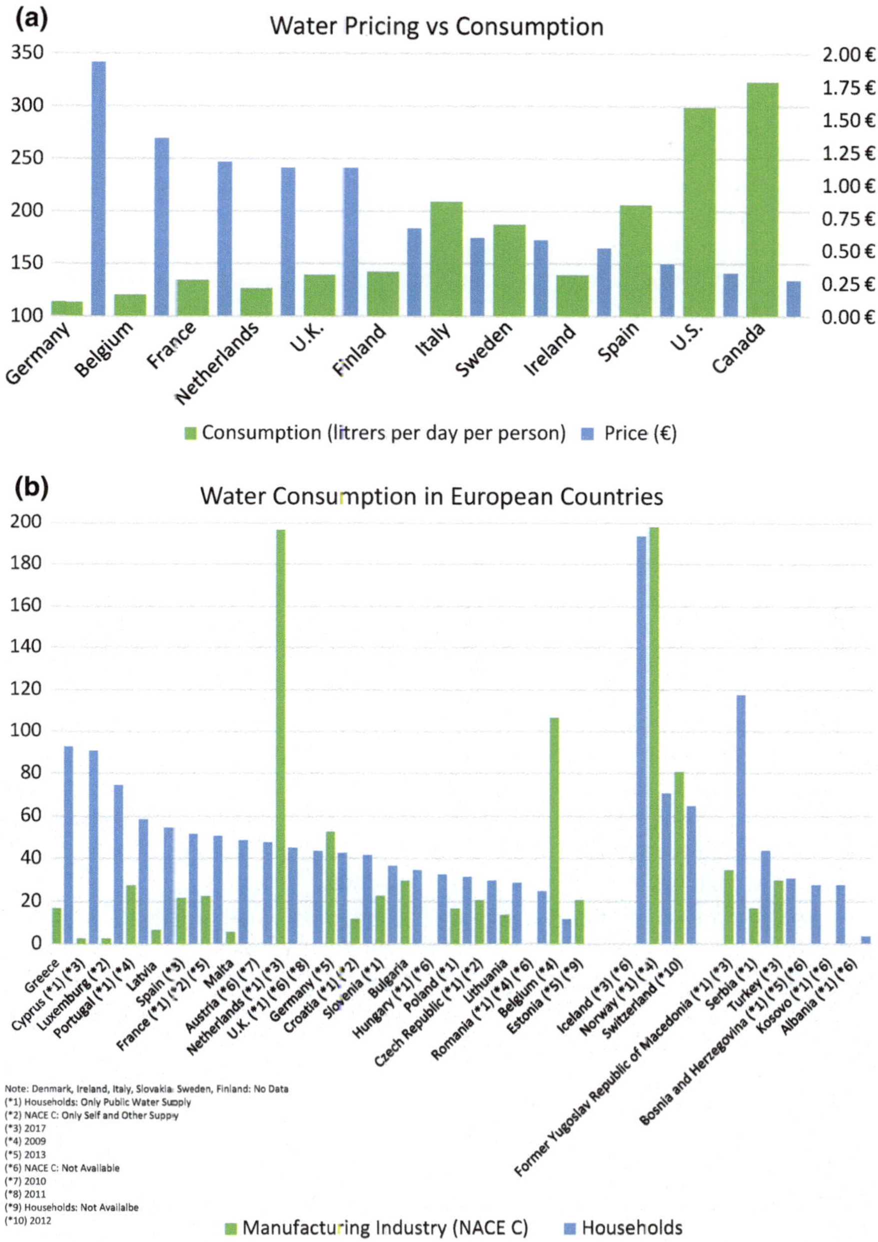

Fig. 4.1 **a** Water consumption versus water pricing in Europe, USA, and Canada; **b** water consumption in European countries regarding household and manufacturing consumption, taking into account the data available in 2015. Adapted from EC.EUROPA.EU (2015)

account the processes that will be integrated, as well as the water consumption which will entail. The domestic consumer should pay attention to the products he or she acquires and how they will consume more or less water. This concern should be taken into account before any decision is taken, as a poor decision will lead to a less responsible attitude toward the environment and, ultimately, a substantial increase of consumed water. However, these options, in addition to being predominantly related to the sustainable consumption of water, may also have a promotional character, calling for the purchase of products which have undergone a reformulation of their manufacturing processes, leading to significant water savings. This is the case of Levi's® (Levi's 2018a), which in 2011 launched jeans with a specific brand (Levi's water < less jeans), announcing a redesign of the product's finishing process, which allowed savings of around 170 million L of water annually during manufacturing. Furthermore, this company remains actively committed in improving the finishing processes in order to reach 80% of water savings by 2020, having 96% as their final target. Moreover, Levi's® is advising consumers to wash their products in a more moderate way, due to the type of finish utilized, which aims to achieve considerable water savings (Levi's 2018b).

Water saving has led to the launch of a number of other programs aiming at sensitizing producers and consumers to this need. WaterSense® is a voluntary partnership program sponsored by the US EPA, which gave rise to a seal, certifying products which comply with the directives issued by that Agency. The program aims to promote policies that ensure the future supply of freshwater in the USA, promoting the market entry of water-efficient products (EPA 2018). Products and services provided with WaterSense® label are certified when new models or products allow a saving of at least 20% of water, also reducing energy consumption and continuing to perform as well or better than previous versions. Essentially, this program aims to encourage innovation, leading producers, retailers, builders, and service providers to deliver more efficient and environmentally friendly products.

Sustainable water consumption should not only be seen as a state of mind that allows for proper care in water consumption in domestic activities, car washing, among other and should also encompass the concern of purchasing products that have consumed less water in their production, or which will save water during their use, essentially showing a proactive behavior for an effective reduction of water waste. This proactive attitude will incentivize companies to invest in rationalizing water consumption during manufacturing processes, or over the life-cycle of the products, to see their efforts distinguished by customers, forcing competitors to follow the same principles, thus helping to accelerate the process of reducing water consumption.

It is curious to note that, through a study carried out by Kang et al. (2017) on a large sample of Hispanic descent American citizens (825 people), it was possible to verify that the perception of extreme drought is perfectly ingrained in these citizens' minds but, when viewing their daily attitudes, this perception is not reflected, showing excessive water consumption in the most diverse daily tasks. Moreover, they do not feel any moral constraint with this relaxed attitude. If there are no such concerns to proactively generate water savings, it can also be expected that there will be no concerns toward the purchase of products that use less water throughout their manufacturing or life-cycle stages. This study corroborates the values shown in the graph of Fig. 4.1a where it is possible to observe that there is no concerted water saving policy in North America, leading to the excessive and unrestricted consumption of water by the vast majority of citizens and companies. Given the results of this study and knowing the differences in water supply policies between North America and Europe, it is clear that policies to discourage unnecessary water consumption in North America are lacking, a necessary need to avoid extreme situations in the future.

The rise in the price of water provided by local authorities is one way to contain water consumption. In Estonia, in the 1990s, subsidies for water supply were cut and prices updated, leading to a significant increase in the cost of water (EEA 2007). These measures promoted a 50% decrease of water consumption over a span of 15 years, which proves that, without regulatory measures that discourage unnecessary water consumption, the vast majority of consumers are not concerned regarding the savings of scarcening natural resources.

4.2.2 *Food Consumption and Sustainability*

Food is usually directly related to human health; thus, a sustainable food system can be defined as the one "that provides healthy food to meet current food needs while maintaining healthy ecosystems that can also provide food for generations to come with minimal negative impact to the environment" (American Public Health Association 2007).

Food consumption is a basic need, but it is also a phenomenon intrinsically linked to social, economic, and environmental factors. The sustainability of this type of consumption can be analyzed in two different ways: the type of consumption that is carried out, which is directly related to the quality of the products and their origin, and the quantity that is consumed, which is closely linked with the quantity of resources spent.

The type of consumption can also be divided into two main groups: conscious choice of products taking into account how they were produced (fair trade, organic products, etc.), or a choice based on diet, with a reduction in the quantity of some products that may have fewer appealing effects on health and the environment (lower consumption of meat, for example). Consumers seem to prefer to purchase meat products that comply with stricter sustainability principles than to radically eliminate meat from their diet (Vanhonacker et al. 2013), although the strategy used for each individual is closely linked to the way of thinking of each person and the social environment in which he/she is inserted (family, friends, and work group). The arrival of organic products has triggered a series of studies in this area, with an aim on perceiving the consumer receptivity this type of product, being the results a crucial factor for many companies to be able to make decisions on how to guide their strategies (Lockie et al. 2004; Bezençon and Blili 2011; Annunziata et al. 2011; Kareklas et al. 2014). In order to describe the selection of healthy and sustainable food by citizens in general, the Theory of Planned Behavior has been applied, which has shown high reliability in predicting behavior regarding the purchase of food products, incorporating the concepts of attitude, social norms, and perceived behavioral control (Dowd and Burke 2013; Verain et al. 2015). In relation to the acquisition of sustainable food products, there are reasons that positively influence this choice, namely an aim to be healthy, elegant, prefer natural products, or want to contribute to a more sustainable society, but there are also factors that negatively affect consumers choices regarding sustainable food products such as the unavailability of natural/organic products, their cost, or ambiguous information concerning their composition and production method (Mäkiniemi and Vainio 2014; Tobler et al. 2011). With regard to gender, carried out studies show that women have a greater affinity for organic food products than men (Aertsens et al. 2009; Hughner et al. 2007). Studies have also been carried out to establish correlations between socio-demographic characteristics and patterns of consumption of sustainable food products, but the results obtained did not allow to establish reliable relationships (Toma et al. 2011; Aertsens et al. 2009). If selection of organic food is an extremely important factor in making food sustainability a reality, it is also very important to pay attention to reducing the consumption of foodstuff categories which carry heavy burden for the environment (Garnett 2011; Foresight 2011). Indeed, it is recognized that meat-based menus entail far greater consumption of resources and levels of pollution than vegetable-based food (Westhoek et al. 2011; Garnett 2011; Tukker et al. 2011; Tilman and Clark 2014).

In Western societies, meat consumption is often understood as a sign of economic power and maleness. Hence, meat occupies a prominent place in the menus of numerous countries. In fact, according to FAO-OECD data, world meat consumption per capita was 34.2 kg between 1997 and 1999, rising to 44.3 kg between 2011 and 2013, and is expected to continue to grow to around 51.5 kg by 2022 (OECD-FAO 2013), but the greatest growth in the last two decades has been driven primarily by rising consumption in developing countries, not Western countries. Indeed, China, according to data from the US Department of Agriculture (USDA 2014), has been a major driver of meat consumption, having registered a consumption per capita of pork and poultry of 41.1 and 9.7 kg, respectively, values that can be compared with those registered in the USA also in 2013 of 27.2 and 43.0 kg also, respectively, for pork and poultry. Regarding recent data published by USDA (USDA 2017), the demand in East Asia remains high in terms of pork and poultry meats. Per capita consumption of beef in China has remained relatively low (4.4 kg) up until 2012, where after has registered a rapid growth, circa 50% from 2012 to 2018 (USDA 2017). Considering the values presented by Shimokawa (2015) in his study, taking into account other studies previously done by Li (2002) and Peters et al. (2007), it is easy to realize how the production and consumption of meat are unsustainable. The values can be seen in Table 4.1.

The above-mentioned values, together with the strong growth of meat consumption in China, have also forced the production of animal feed grain, devoting about a third of the Chinese grain production exclusively to animal feed for meat production. Moreover, China has become a grain importer as its internal production is not large enough to fill the demand in terms of direct population feed and animal feed.

Table 4.1 Feeding rate and land area needed for meat production regarding different types of animals

Type of meat	Ratio of kg of grains needed to achieve 1 kg of meat	Land area per 1000 kcal of meat production (m^2/1000 kcal)
Beef	5.0–7.0	7.7
Pork	2.8–4.5	10.6
Poultry (Chicken)	1.7–2.0	13.3

This increase in consumption of grains for livestock, particularly soybeans, implies the importation of these goods from other countries, such as the USA, Brazil, and Argentina, creating market price pressure due to high demand, which will contribute to a greater difficulty of accessing basic food necessities among the poorer members of the population (Shimokawa 2015).

It should also be noted that when countries do not have a maritime coast or preferential aquaculture areas, they have limited access to meat alternatives, such as fish. However, many Eastern countries have rooted cultures in which fish occupies a prominent place in their daily menus, namely island-based countries with large coastal perimeters and/or aquaculture tradition. Gradually, there has been a growing trend of citizens adhering to non-meat-containing diets, such as vegetarians or vegans. These citizens have been the subject of several studies, which have carefully analyzed the consumption of meat and its traditions, as well as the evolution in the trend toward a more sustainable diet (Ruby and Heine 2011; Hoek et al. 2011; Ruby 2012; Dagevos and Voordouw 2013; de Boer et al. 2014; Rothgerber 2014). Decreasing or even eliminating meat consumption may have religious, health, or even environmental motivations, depending on the education, culture, or even the knowledge path of individuals. The awareness that the consumption of animal meat has a greater burden on the environment will tend to weigh more and more on dietary decisions; however, it is crucial to provide adequate information and dietary education in the initial learning phases of children to deeply root these concepts and achieve true change of habits, with transition between generations. In recent years, the number of people concerned with animal welfare has also increased, avoiding anything that may arise and imply their sacrifice. This group includes vegans, which for the protection of animals, avoid the use of skins, as well as the consumption of meat, animal-derived products, among others. Not having a direct relationship with sustainable food consumption, this attitude contributes positively to a lower consumption of meat, which is advantageous for the environment (Fox and Ward 2008).

When sustainability is approached in terms of food, meat consumption usually takes a prominent place, as meat production within the food spectrum is considered to be one of the sectors that consumes more energy, generates greater negative environmental burdens, and raises greater concerns ethically. The graph presented in Fig. 4.2 depicts the meat consumption situation in several countries across the world, following the data provided by OECD regarding the year of 2016 (OECD 2018). In the last decade, meat consumption has been stagnant in developed countries, but has been consistently growing in developing countries, such as China and Brazil, with around a quarter of the world's meat production being consumed in China. Complete abstention of meat is usually related to health, moral principles, religious or social beliefs, concerns about the environment or animal welfare, and are usually options adopted by groups which follow the same line of thought, implying the use of very specific foods and the use of restaurants that are usually provided with these food options (Hoffman et al. 2013; Zur and Klöckner 2014; de Backer and Hudders 2015). On the other hand, the abundant and constant consumption of meat is usually associated with family or group habits, the need to exhibit above average status, lack of knowledge about the existence of meat substitutes and on how to cook them, and contempt for the environmental consequences of stockbreeding and slaughter (Vanhonacker and Verbeke 2009; Mäkiniemi and Vainio 2014). Some studies have made it possible to identify the characteristics of individuals or groups that voluntarily adhere to a reduction in meat consumption. Thus, it is found that people who are more advanced, more educated, better off, and provided with smaller houses are the ones who most easily adopt sustainable food principles (Schösler et al. 2012; de Boer et al. 2014; Hayley et al. 2015). Adoption of vegetarian or vegan diets is not the only method of reducing meat consumption. Reduction could be based on lowering the amount of meat associated with each meal, as well as a greater diversity of menus including meat substitutes, even if some continue to be of animal origin, such as cheese. However, the reduction in meat consumption can be achieved essentially at the expense of a larger number of meals made from fish or vegetables such as lentils, beans, chickpeas, coconuts, avocados, spinach, or nuts (Tobler et al. 2011; de Boer et al. 2014; Verain et al. 2015).

How certain groups of people face sustainability in food may differ significantly. Indeed, while some groups are essentially focused on the consumption of organic or "green" products, others focus primarily on reducing consumption of unsustainable products. Given this difference of approaches, it is also necessary that education, communication, and marketing strategies be properly segmented for different groups so that the message can be absorbed by all, in different ways, but ultimately aiming for the same goals (Abeliotis et al. 2010; de Jonge and van Trijp 2013).

Restaurants should also take a very active position in the defense of sustainability in food. Serving sizes should take into account the necessary food balance for a person of medium height, so that food is not wasted, or that the consumer is forced to ingest more than he should, according to the already established nutritionally standards. In addition, an appropriate mix between organic and other less sustainable products may deter consumers from consuming large portions of unsustainable products as long as this balance is properly studied and implemented. This strategy should also be applied gradually, so that consumer habits can be gradually changed without radicalism, which would lead the consumer to be able to take other less harmful options in environmental terms. The preparation of menus and options offered to consumers should be gradually approximated to more sustainable options, with a more attractive presentation and cost, sparking curiosity so that customers be willing to try new options without feeling that they had to change/condition their sense of pleasure and well-being, thus gradually leading consumers to more sustainable food

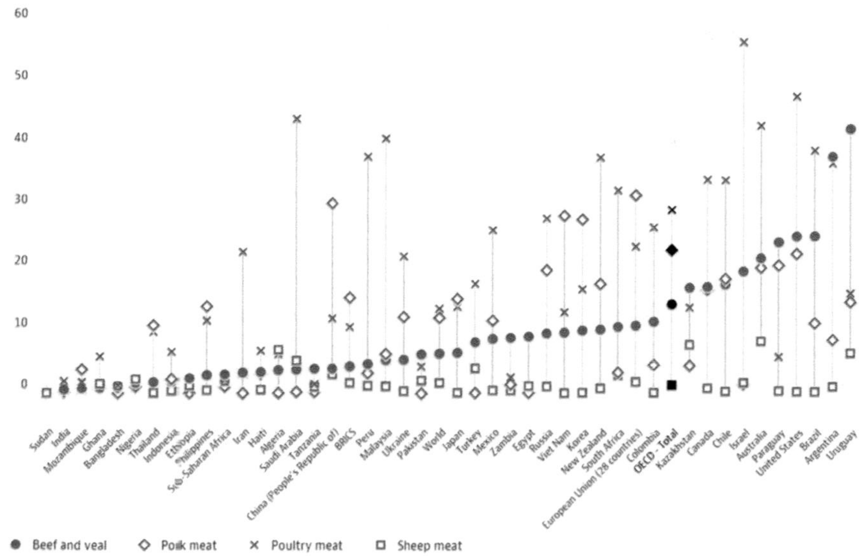

Fig. 4.2 World meat consumption following OECD regarding the year of 2016. Adapted from OECD (2018)

consumption habits. Due to the resistance of the majority of restaurants to adhere to this type of position, it would be very convenient to have regulations in this regard, i.e., through the obligation of presenting information regarding the environmental needs of the ingredients of a given dish, such as to act on product prices, in order to discourage consumption of unsustainable products.

In addition to restaurant menu policies and the predisposition of consumer consumption of meat, and viewing the recommendations of health-related organizations, the existing data is truly worrying. The World Cancer Research Fund or American Institute for Cancer Research recommended in 2007 a maximum intake of 0.3 kg of meat per week for each individual of medium height, but the per capita weekly intakes recorded in the USA and China are substantially higher: 1.44 and 1.03 kg, respectively (Shimokawa 2015). This abusive consumption of meat, in addition to all the environmental problems it entails, also has serious implications on consumer health, which will translate into diseases that are currently experiencing significant increases in patients, such as cancer, obesity, type 2 diabetes, hypertension, and cardiovascular problems. Death from some of these diseases is likely to reach alarming levels and is expected to account for more than 50% of deaths in China in 2025 (Shimokawa 2015).

4.2.3 Political Ideologies and Consumption Behavior

Given the different political options in existence since the end of the monarchy, consumption began to be seen as the product of individuals not committed with moral principles and as an ability to overcome everything (Ewen 1976; Lasch 1979). Often, moral foundations are interrelated with the political ideology of an individual. Political ideology is usually divided into two main groups: on one hand, left, liberals, or democrats, on the other hand right, conservatives, or republicans. Kidwell et al. (2013) and Graham et al. (2009) assume that individual behavior can be distinguished and directly correlated to one of these two groups, helping in predicting personal behavior. Effectively, political ideology can be assumed as "an interrelated set of moral and political attitudes that possess cognitive, affective, and motivational components. That is, ideology helps to explain why people do what they do; it organizes their values and beliefs and leads to political behavior" (Jost 2006), definition that is even extended by Kidwell et al. (2013), referring that political and moral positions describe how people should act in order to attain social justice and order.

According to the research performed by Watkins et al. (2016), the group of people usually connected to the "right" (conservative, republicans) are aligned with the five above-mentioned moral foundations because they characteristically struggle for high degrees of self-control and endorse a sturdy sense of responsibility, translated by respect for authority and selflessness for the rules imposed. Moreover, this group of people is usually linked to easy accepting of inequality in terms of economic power and quality of life, along with some difficulty to admit social changes (Carney et al. 2008). On the other hand, the main focus of the "left" (liberals, democrats) is promoting social justice, taking care, and preventing the damage of social equity (Graham et al. 2009) and are typically able to promote and accept the social changes but present thoughtful difficulties to accept social disparities (Carney et al. 2008). If these considerations are not directly related to sustainable consumption and moral foundations, they offer the possibility of understanding the bases on which the major part of purchasing decisions regarding goods or services are made. Furthermore, people connected to the "right" tend to assign more significance to some principles such as responsibility, traditionalism, and organization. On the other hand, the group linked to the "left" presents extra flexible behavior, added open mind, are more likely to accept different products, privilege the relationship with friends and acquaintances, and enjoy leisure as a complement to work.

In fact, many researches have been trying to correlate consumer behavior with political convictions linked to "right" or "left" groups, neglecting other factors that may be equally important in the moment of purchase or in the well-established mindset of the customer when a purchase is equated, namely moral foundations, individual ideology, impact of the group of people in which he/she is integrated and family influences. This relationship plays a crucial role in order to realize how the main motivations of consumers are influenced by their moral values as these are related to their political ideology. The consumption behavior induced by political convictions has also been profusely studied (Micheletti et al. 2004; Shah et al. 2007;

Schudson 2007; Glickman 2009; Nielson and Paxton 2010; Newman and Bartels 2010). Moreover, it was also stated by Crockett and Wallendorf (2004) that political motivations are capable of influencing not only the decision of what to buy, but also the place where the purchase will be made. Additionally, the group to the "left" are more likely to favor products that are truly innovative, allow new experiences and outcome-based consumption choices. Contrary, the "right" is more likely to prefer serviceable, conservative, and reasonable consumption options (Farmer 2014). Nevertheless, the study of Feinberg and Willer (2012) makes believe that outlining pro-environmental communication in an insistent way seems the most efficient strategy to dismantle some moral preconceptions or political ideologies associated with purchasing patterns which could avoid the adoption of a sustainable consumption way. Moreover, the change in consumption patterns can still take advantage of a high coherence between political ideology and the assertive communication, enhancing significantly the persuading efforts (Kidwell et al. 2013).

It is well-known that political ideologies usually exert significant influence on consumption patterns as a way to manifest their beliefs, and consumption choices can be used to transmit political messages to stakeholders in the universe of the political interests and market needs (Micheletti 2003; Stolle and Hooghe 2004; Willis and Schor 2012; Farmer 2014). It is also well-known that traditional political involvement is declining, as can be confirmed by the decreasing number of votes accounted in elections and lesser interest of people for political news. In contrast, the post-sharing on social networks, public petitions, and protests are growing, showing an increasing contribution to expressive actions (Gotlieb and Wells 2012). Thus, people are facing social concerns from an individual point of view, progressively abandoning the mass manifestations (Bajde 2007). Thus, the individual assumes preponderance in seeking for the solution of their problems, which can translate to a lack of hope in solving the problems together with the society that surrounds him and that shares the same concerns and desires. This shows that a gap is being engendered between governments and markets, political and commercial interests, and private and public sectors. Moreover, a concept of hybrid "consumer-citizen" emerges because the political orientation is often conducted taking into account market fluctuations and consideration (Newman and Bartels 2010). In any case, after empirical studies carried out in Europe, USA, and Canada by Willis and Schor (2012), it is clear that there is no relation between collective political accomplishment and the individualized political consumption behavior. However, when political coordination is focused on conscious consumption, better sustainable results can be achieved, because conscious consumption and political assignation tend to evolve hand in hand. However, political intervention in consumption is seen by the citizens as one of many other actions and strategies to increase sustainable consumption and reach a more balanced social justice. Individuals have a clear perception that their day-to-day consumption pattern is influencing the survival of companies, playing with their choice in order to lead to the kind of society of which they wish to be a part (Nielson and Paxton 2010).

In short, the people connected to "left" ideological principles are more prone to change their consumption patterns, adhering more easily to principles of sustainable choices regarding purchase options, as well as participating in actions regarding a higher engagement of the society with these principles. This is so as they are further predisposed to change and habitually feel some responsibility for their part in the common welfare of the society where they are inserted, believing that they can aid individually and collectively for a fairer social order, helping and protecting economically disadvantaged classes. Moreover, they are less attached to moral values, which give them an enlarged flexibility to change. On the other hand, people linked to the "right" present usually high moral values brought from the past, which they are not ready to contest, keeping them stuck to more retrograde consumption principles linked to an era where environmental problems were not a real concern. Furthermore, being usually an economically privileged class, it is common to see the use of consumption behavior as a manifestation of economic power, independently of the impact this attitude may have on the environment and society where they are inserted. Thus, due to their interest in national or international economic development and due to involvement in business, unregulated consumption is favorable for an improved money income, being an added reason to favor a more conservative behavior relatively to consumption pattern change. Moreover, this group is normally less prone to claim for political changes in terms of consumption patterns or be able to participate in programs leading to fairer resource distribution, preferring to preserve the already perfectly established models (Willis and Schor 2012). In the study carried out by Watkins et al. (2016), it is concluded that there is a straight connection between consumers' moral foundations, political alignment, and their individual and collective attentiveness to sustainable consumption as a factor for social transformation. As such, regarding further studies on how to act in order to modify the current consumption patterns, it is necessary to deeply understand how moral foundations and political orientations are affecting the capacity to change of individuals, along with the basis of the reasoning behind their consumption options. However, it is important to note that consumption is used for different purposes by customers, bringing aggregated different motivations. It is recognized that a good way to captivate the most skeptical for change in terms of sustainable consumption will be to educate the citizens in terms of energy savings and then create the necessary foundations for sustainable consumption to become rooted as a way of being.

4.2.4 Citizenship and Consumption Behavior

Concerns regarding sustainable consumption are somewhat already rooted in people's minds but are heavily dependent on the culture set up in each country and generally have a well-defined trend, which is formed by each individual through the reaction to the behavior of the society around him. Vassallo et al. (2016) refer that Italian people have food as their main sustainable consumption concern. On the other hand, Finish people are requiring tailored services in order to overcome their concerns about household sustainable consumption (Salo et al. 2016). However, the English people focus their concerns of sustainable consumption mainly on fashion products, based essentially on their moral values (Lundblad and Davis

2016). However, social influence does not affect all cultures in the same way. The Slovenian people are much more influenced by social pressures than the French people (Culiberg and Elgaaied-Gambier 2016). While in developed countries social pressure is one of the main factors influencing consumer behavior, in developing countries due to closer contact with environmental problems, this factor assertively determines consumption behaviors (Zhao et al. 2014). This has been proved through a study by Wang and Wu (2016), through the analysis of consumption patterns in the Chinese city of Hangzhou, where citizens' behavior in terms of consumption is strongly affected by values such as anger, guilt, respect, and pride, especially with regard to domestic consumption.

The approach carried out by Elkington (1997) emphasizes that sustainability is supported by three principal dimensions: environmental quality, social justice, and economic prosperity, which can be directly correlated to three important factors for the human being: planet, people, and profit. Moreover, Elkington (1997) also argues that the dimensions above-mentioned should be met, preferably, simultaneously. There is a growing recognition by business sectors that stakeholders are increasingly committed to environmental causes and sustainable consumption. Furthermore, sustainability principles are becoming a prerequisite for companies to start operating and remain competitive in the long run (Lubin and Esty 2010; Murphy et al. 2013). Following this evolution in business terms, consumers are expected to achieve greater social maturity, shaping their consumption patterns to a more sustainable development (Devinney et al. 2006; Caruana and Crane 2008). Moreover, it is recognized that consumers can play a key role as agents of change in corporate behavior and environmental preservation if they show increasing willingness to adapt their lifestyle to more sustainable consumption patterns (Balderjahn et al. 2013). By realizing that consumer behavior has changed, companies will thrive for producing parts with sustainable ingredients, components, and production process in an effort to meet customer demand and satisfaction. This consumer behavior can also push political forces to set more stringent requirements so that environmental targets can be effectively achieved or even exceeded. Thus, consumers can only assume the behavior and role of a true citizen when they effectively assume responsibility with the sustainability goals established for the society they are a part of. This includes the necessary pressure over the political system and companies in order to reach increasing levels of global sustainability.

As referred by Ricci et al. (2016), "the consumer as citizen is someone who makes purchasing choices in respect of the sustainable development of the world community." This is the concept of the perfect citizen, who sees in citizenship a responsible act of respecting the environment that surrounds him and has a vision of the future regarding the evolution of the planet. As everybody knows, our society is a mix of responsible and non-responsible citizens, and the last ones sometimes have no scruples when it comes to establishing their personal and financial interests above all.

On the one hand, companies should be obliged to inform consumers about their products (performance and how they were obtained), as well as the company's conduct in terms of social responsibility. On the other hand, each consumer, as a citizen, needs to be cultured in developing “lifestyle choices and social involvement that contribute to greater transparency, better policies and increased advocacy” (CCN 2005). Furthermore, the Consumer Citizen Network (CCN) has been associated to great organizations such as UNESCO, UNEP, and Consumers International, in order to foster consumers’ education, trying to reach higher levels of responsibility and sustainability from citizens in terms of consumption. The guidelines established by the CCN attach greater importance to the duties of citizens, in detriment of their rights, rebalancing the interactions between consumer rights and duties and establishing new roles on their behavioral structures based on moral principles and socially responsible consumption (Schrader 2007). Nevertheless, Schrader (2007) also argues that consumers’ rights should be extended as a principle to spread also their duties, as these rights can be easily transformed in duties. Taking into consideration that consumption should not be focused just on the fulfillment of citizen needs but it should also consider the social environment and corresponding sustainability, an extension of the consumers’ rights is necessary, beginning by the fundamental right to be correctly informed. This right allows consumers’ to better analyze product production chains and effects, yielding a more supported selection of products. This case has been explored successfully in the field of food products, where the effects of nutrition information help the consumer make a better decision regarding his health and the environment. This information reaches both the consumer who presents ethical concerns and willing to make sustainable choices and all others who remain indifferent to the call for sustainable consumption, hoping that one day their decision could be influenced by this message. Another right that must be safeguarded is the availability and diversity of environmentally sustainable products at prices similar to products without environmental care, pressing the consumer to make decisions based on the quality and environmental component of the product, and not on its price or availability. This example has been given by the organic food market in the food sector, where producers are gradually lowering the prices of their products and improving their dissemination through markets (Carrigan 2001).

Trivial social practices, such as moving, cooking, or laundering, can be taken as samples of citizen attitudes regarding the environment, allowing the characterization of individuals. If these practices are well understood and crossed with other personal traits, it will be possible to formulate the main rules that are governing personal conduct regarding a particular option or practice. Some decades ago, the lack of authority is argued by some authors as one of the main factors affecting consumption choices (Baudrillard 1998; Bauman 2008), because citizens had no decisive orientations and felt isolated in society, legitimating their social purchase selection by themselves, i.e., citizens use consumption as a means of affirmation in the society in which they are inserted. Nowadays, the citizen designs his lifestyle as a function of financial income and desires, adopting consumption strategies on a random basis. Beck (1992) refers that the permanent change of citizens’ lifestyle is a manifestation of dissatisfaction with himself and with society, showing how he struggles to establish for himself an identity, becoming a different person from what he is.

Independently of the amount of information that a consumer has and the number of actions carried out by various companies and organizations in order to positively influence the consumer to make the best decisions regarding environmental sustainability, the behavior of human beings presents certain fragilities difficult to overcome. This may mean that, independent of the number of actions a company or institution promotes in order to positively influence the consumer, their final choice may be completely influenced by other factors linked to the vulnerability of instinctual reasons at the time of acquisition (Carrigan 2001).

Nowadays, consumers are more capable of understanding the behind the scenes of product manufacturing processes and if companies are providing the correct information regarding product contents and production process. Moreover, they are also more attentive to lacks of information provided by manufacturers concerning the production process, which can translate the differences between the real characteristics of a product and the consumer's emotionally observed features through his/her own understanding and personal skills (Marinelli 2010; Menghini et al. 2011).

However, each culture presents its own characteristics, values, and social traits, which result in a multiplicity of behaviors with regard to consumption (Sharma and Jha 2017).

4.2.5 Influence of Moral Values on Consumption Behavior

Taking into account the notion of "values," Schwartz (1994) has established a widely used definition, which is described as "the concepts or beliefs about desirable end states or behaviors that transcend specific situations, guide selection or evaluation of behavior and events and are ordered by relative importance" (Schwartz and Bilsky 1987; Schwartz 1994). However, as referred by Rokeach (1973), values can assume a huge diversity of mindsets depending on regional and cultural dissimilarities, beliefs, political alignment, social arrangements, societal group, instruction, profession, and gender, which was later corroborated by Xiao and Kim (2009), when they refer: "The variations in individuals' personal, societal, and cultural experiences generate value differences, as well as the stability of values and value systems." Moreover, Schwartz has been very creative in studying the relationship between values and human behavior, being in the genesis of the conception of the Schwartz Values Scale (SVS) (Schwartz 1994), where a wide-ranging scale with 56 values is suggested. Nevertheless, this scale has been censured due to not including a factor able to distinguish different cultural characteristics. Then, Schwartz developed a new model, dubbed as Portrait Values Questionnaire, entailing 40 declarations, which represent 10 motivational values (Schwartz et al. 2001). Nevertheless, this new model would again be questioned due to the methods used, since other authors would prove through confirmatory factor analysis that there was a strong correlation between some of the motivational factors pointed out in the model proposed by Schwartz (Perrinjaquet et al. 2007; Knoppen and Saris 2009). Taking into account the Indian culture, Sharma and Jha (2017) decided to extend the Portrait Values Questionnaire, including some new motivational values, such as self-enrichment, compassion, self-evolution, and uprightness, tagging this new approach as Holistic Values Scale. Moreover, these last authors have discovered a new set of values that have positive implications on the behavior of citizens regarding sustainable consumption, such as tradition, acceptance, compassion, and universalism.

Kala and Sharma (2010) have argued that the connection to certain religions or ancestral traditions influences positively the individual or even society to enhance the linkage between man and Nature, which is directly related to an increased responsibility with the environment and the social order.

Interestingly, some studies point out that consumer care related to environmentally sustainable products is not directly related to moral values or typical characteristics of certain socio-economic classes. The most sustainable patterns of consumption are directly related to the influence of local producers, where there is a greater knowledge of the sustainable practices of companies. However, they can be negatively affected by the lack of time available to ponder what really needs to be purchased. There is also a significant tendency for people to assume they have internalized principles of sustainable consumption; however, the truth is that their actions are not in line with their values (Tanner and Kast 2003; Hume 2010).

4.2.6 Borrowing and Sharing Practices or Collaborative Consumption as a Sustainable Behavior

Most of the world's population has the perfect notion that current consumption levels are unsustainable. Another method that can help reduce or eliminate unnecessary use of resources and waste involves avoiding the acquisition of new goods by reusing existing products. Collaborative consumption is a socio-economic model that aims at the sharing of certain durable goods, allowing to extend the useful life of these products, making them profitable, thus avoiding the need to purchase new products, and clearly contributing to the decrease of waste. This can be perfectly achieved through exchanges, gifts, rents, negotiations, and loans for underutilized or unwanted goods among citizens (Prothero et al. 2011; Botsman and Rogers 2011; Heinrichs 2013). Such practices may or may not involve the assignment of property and may have money or other goods as a counterpart. A share usually does not contemplate an ownership transfer, while a gift implies an exchange of ownership, which is also usually valid when there is a transaction of a used good involving money or goods.

The proliferation of new technologies and easy access to them by any citizen has led to a significant increase in the number of platforms that promote the sharing or exchange of goods and services, becoming a major movement, quickly growing into business opportunities, and creating a disruption in the traditional way of doing business (Gansky 2010; Owyang et al. 2014). Some examples of these platforms have become a part of our daily lives, such as eBay, Uber, Cabify, Airbnb, among many others. The emergence and growth of this type of business have only become possible because it effectively fills a gap in users' needs, who are looking for goods that are still capable of meeting their requirements and that are no longer useful to others, bringing together two important factors: low cost and resource saving. The scale factor corresponding to these actions worldwide has a very significant effect on overall consumption, which contributes greatly to the improvement of the environment (Piscicelli et al. 2015). More recently, other platforms usually called social networks, such as Facebook, Instagram, and Twitter, have also served to share, cede, or sell low-value products, no longer useful for someone, but needed by others, which has increased the sharing and trade of used goods, with clear benefits for saving resources. Good examples of this are the commercialization, exchange or transfer of low-value goods, like clothes, footwear, esthetic products, among others.

Ozanne and Ozanne (2011) describe a very good example on how the consumption of products that may have limited temporal interest can be improved in order to reduce their environmental impact. In New Zealand, toy libraries have been set up where, through the voluntary collaboration of parents, toys are deposited and made available to other families in an effort to meet the temporary needs of children throughout their different growth phases. The toys are held for a preestablished period of time, in exchange for a fee that allows the library to keep running and the commitment to collaborate in its maintenance. When birth becomes a reality, parents feel that they can rely on such structures without the need to spend large amounts of money on the purchase of toys that are only useful during a passing phase of their children's lives and which usually lose any usefulness and even turn into difficult to recycle waste. Thus, reusing prevents harmful consequences for the environment and facilitates parenting. These libraries are usually located in the center of communities in order to facilitate family access, who may also find in these places a center where they can socialize with other families, thus increasing the sharing of experiences. In addition to all the already described advantages, this model allows the creation of a mentality of sharing in children, as well as a more responsible attitude toward sustainable behavior and the environment that surrounds them. These toy libraries also have the advantage that through the contact between families and exposure to the overall operating principles, learning and conditioning are promoted through behavioral observation and introspection of the values that underlie the library's activity. The exchange of toys, in addition to all the environmental advantages, also allows children to have a greater contact with a larger number of toys, increasing the possibilities of learning and development of skills. On the other hand, toy exchange avoids painful toys shopping rides as well as reducing the child's contact with the consumer market, moderately educating their children in a less consumerist way. This case of toy libraries can also serve to show how personal, behavioral, and environmental components are closely correlated in the development of sustainable consumption principles.

Sustainable behavior on the consumer's part may be based on four socio-economic drivers: (a) motivational, such as values, attitudes, and behavioral norms, (b) individual, for instance, social class or intrinsic knowledge and skills, (c) contextual, for example, advertising, availability, and legislation, and (d) historical, such as behaviors that are identified as usual (Stern 2000; Thøgersen 2006, Roos and Hahn 2017).

However, the practice of borrowing, being extremely devoted to environmental causes and altruism, can also be connoted with neoliberal approaches and practices that foster the unconditional alienation of assets, not having as main focus environmental sustainability concerns, as described by some researchers in this field (Rifkin 2000; Martin 2016). In fact, there are some causes that may justify the lack of adherence to a continued borrowing practice such as the values followed, habitual practices, and the behavior that citizens intend to default as their way of being (Barnes and Mattsson 2016).

The business sector and local political power have not been indifferent to these factors, essentially encouraging the sharing of transport. As far as cars are concerned, there are already common road-sharing networks, thus avoiding traffic overload, reducing air pollution, sparing existing energy resources, and also serving the interests of the people involved. More locally, within some cities, political entities have created shared systems for the use of bicycles within localities, which also contribute to a decrease in traffic in urban areas, avoiding the pollution caused by thermal engines driven by petroleum products and serving the interests of the population in terms of short distance mobility.

Borrowing of non-perishable goods can be seen as a sustainable consumption practice, since it avoids consumption on its own and makes a more effective use of existing goods. Moreover, Jenkins et al. (2014) describe the borrowing practice as expression of empathy, faith in the other, and interconnection. Therefore, the practice of borrowing can be a social action. On the contrary, there are also citizens who avoid borrowing as a way to avoid social dependence and obligations toward third parties (Belk 2010; Jenkins et al. 2014).

While there are still value barriers that tend to hinder the full acceptance of collaborative consumption and its diffusion, new business models will need to be developed that can be conveniently disseminated and have a sound economic base (add value for all stakeholders), as there has also been a collapse of many businesses in this branch of activity. Even when they deserve a high degree of consumer compliance, additional barriers may arise in terms of economic sustainability, appropriate service diffusion, adequate distribution of goods and services, appealing user's interface, and legal framework.

It should be noted that collaborative consumption can only exist based on technological platforms, leading to reduce distances between the diversity of existing needs and the huge supply of goods available for sharing, exchange, or acquisition in the used condition.

4.2.7 Influence of Mindfulness Practices in Consumption Behavior

Mindfulness is an issue that has attracted huge attention from researchers, as evidenced by the growth from 80 to more than 1450 publications performed in 2004 and 2015, respectively. This contrasts with the growth in the number of publications pertaining to sustainable consumption, which only grew 5 times over the same period, showing that this latter theme is more matured, while mindfulness is still to be heavily explored. Recent researches have re-embodied the concept of mindfulness, which is characterized by a thoughtful and diligent attention to the current moment, which includes complete mental availability for new challenges, ensuring a strong commitment with causes related to the well-being of the society (Kabat-Zinn 2003; Grossman 2010). This mindfulness is based on a relaxed attitude of paying the maximum attention to each moment, without letting it become a routine, and not allowing emotions to invade the moment and jeopardize future experiences (Fischer et al. 2017). This state of mind can be connected to Buddhist traditional culture, which proposes a deep reflection on how to face others, through simple gestures such as delicacy, compassion, generosity, and equality (Hyland 2011; Grossman 2013; Grossman 2015). It is whispered that this state of mind can also be improved through appropriate training, which can develop a more intense non-reactivity to emotions. Moreover, its practice can improve immunity to advertising stimuli and other marketing techniques, freeing citizens from the pressure to which they are permanently subjected by sound and visual inducements, giving them greater inner peace, which will result in a lesser tendency toward unnecessary consumption. This state of mind will allow a more assertive and thoughtful response to consumption incitements, which is contrary to the usual tendency of immediate adherence to consumption stimuli (Chambers et al. 2009). Furthermore, co-lateral positive personal properties have been also reported, such as an increase in self-esteem, higher self-acceptance, improved compassion, and stronger empathy with people and causes, factors which also positively contribute to a more relaxed state of mind, allowing individuals to contemplate consumption effects and be more vulnerable to sustainable consumption messages and policies (Shapiro et al. 1998; Chambers et al. 2009; Chiesa and Serretti 2009; Birnie et al. 2010; Bolz and Singer 2013). In addition, it is also believed that a more prolonged and attentive reflection by part of the consumer-facing consumption stimuli may reduce the gap between the expected behavior of certain groups of consumers and their actual behavior when stimulated to purchase with the usual means and subject to daily stress (Chatzisarantis and Hagger 2007). Moreover, the study published by Rosenberg (2005) refers that mindfulness increases the attentiveness of the cognitive-behavioral processes underlying consumption that have become rather instinctive, allowing for a higher freedom of choice, which is corroborated by other authors (Ericson et al. 2014; Bahl et al. 2016), referring that mindfulness tends to remove the materialistic propensity and reduces consumerist impetus. As above-mentioned, Rosenberg also proposes that susceptibility to specific promoting techniques and inducements "can be reduced when people are more mindful" (Rosenberg 2005).

Thus, and as a summary of what was referred above, the study carried out by Fischer et al. (2017) presents four topics which were identified as mindfulness relevant results regarding prior investigations relating mindfulness and sustainable consumption, as follows:

- Disruption of routines: It is recognized that mindfulness practice helps to break with installed habits of consumption and allows breaking routines in terms of con-

sumption, already regarded as automatic. This practice will help to curb reckless acts of consumption, especially when it comes as a non-sustainable behavior.
- Congruence: The sensitivity to the lack of attention given to routine consumption tasks was associated with the gap found between consumption expectations and unsustainable behavior. Mindfulness entails increased attention to routine tasks, with a constant concern to maintain sustainable consumption behavior.
- Non-materialistic values and well-being: The practice of mindfulness helps to clarify life values, putting aside the more materialistic values and treasuring the person's role in the society. Mindfulness practice helps people overcome three usual human characteristics in Occidental modern societies: greed, delusion, and aversion. These undesirable features can be counteracted through the practice of mindfulness, giving rise to much more beneficial behaviors for the entire society, such as mind-openness, bigheartedness, gentleness, and psychological clearness. Thus, mindfulness has positive effect not only on individual behavior and features, but also toward the whole society, by a more conscientious global behavior, mainly due to a less self-centered way of being.
- Pro-social behavior: Self-reflection mindfulness practice has showed to be beneficial to increase personal kindness and social attention, resulting in a clear benefit for the entire society, both through the widespread of good values and general well-being, as well as in the reflection on consumer choices, including lower-carbon lifestyles.

From the above-mentioned, it is clear that the practice of mindfulness is an extremely useful tool both for individual consumer behavior and improvement of society welfare conditions. More studies should be carried out in order to determine how mindfulness can help influence the more self-centered personalities in adopting more sustainable behaviors, as well as on how to propagated to non-practitioners, enlarging its beneficial influence on human behavior.

4.3 Eco-labeling as a Way to Influence Consumers

The information provided by companies and required by more environmentally attentive consumers is assumed by researchers and stakeholders involved in business as a very important factor for the increment of sustainable consumption practices. The information normally contained in these labels is usually the amount of water utilized, energy consumed, or amount of wastewater generated during the manufacturing of a given product (Jackson and Snowdon 1999). One of the ways to provide this information would be by eco-label. Eco-labels are intended to be a link between the producer and the consumer, showing to the latter the concerns that former one has had and is assuming regarding the product, in order to comply with the established environmental sustainability criteria. Basically, these labels aim to reduce the gap between the knowledge of the product by the environmentally concerned customer and the way it was thought and manufactured by the provider (Heinzle and Wüstenhagen 2012; Delmas and Grant 2014).

It is also expected that eco-labels can transmit to the consumer whether products were devised to have a longer life span, thus reducing the impact on the environment as less resources are needed to replace them in the short term. This may also have direct implications on the sharing of acquired good because, with longer life spans and after the initial owner no longer has use for a product, it can still be assigned or sold to third parties, thereby indirectly reducing environmental impact. In fact, this attitude must also start with manufacturers, which should aim to design and manufacture products with longer life cycles. On the other hand, it is also expected that consumers will have a favorable attitude toward products with longer duration, making them their product of choice. However, not all products may be subject to this type of life extension due to technological developments. Considering, for example, the case of mobile phones, the continuous increase of functions allowed by them has led to a significant decrease in useful life, estimated in about 50% from 2006 to 2009, implying an unsustainable consumption of this type of product, which also has direct implications on the environment (Nazzal et al. 2013). Indeed, many companies producing consumer goods have a clear strategy for innovation, constantly introducing renewed products, with ever more appealing functions and designs, in order to create customer loyalty and encourage premature consumption of goods, instead of exploring new receptive markets and focusing research efforts on increasing product life span (Bayus et al. 2003). It is well known that the electronics industry often launches intermediate technology products with the well-established objective of subsequently launching another product with a greater degree of innovation in order to stimulate consumption, which only undermines environmental sustainability. This problem can only be solved through agreements between governments and companies where product lifecycles are preestablished and updates can be controlled over time in order to limit unnecessary consumption. Although the current situation stimulates the market for used products, a large waste of resources continues to exist, seriously affecting the environment in the end.

Eco-labels can also indicate certifications that the producer has achieved, such as ISO 14001, ISO 26000, EMAS, EU Eco-label, EU Energy Star, and World Fair Trade Organization, among others. Nevertheless, eco-labels have also received some criticism because they are excessively focused on environmental issues and do not allow screen time to other equally important aspects as factors related to economic and social sustainability (Horne 2009). As referred by Nikolaou and Kazantzidis (2016), there are two different kinds of eco-labels: The first ones are focused on actions carried out by the companies themselves, without any other type of independent regulation, which aims to inform the customer of the reasons why the product may be more expensive due to the environmental practices employed (van Amstel et al. 2008); the second ones are usually called social-labels or eco-labels and have superior regulation in order to ensure that the information displayed on the product corresponds to the truth and is not a false advertising intended to deceive the customer, hinting that a company follows sustainable environmental practices when it is not true (Nilsson et al. 2004). The ISO standards present a different classification of labels, considering three types: (a) labels certified and periodically audited by certifying bodies taking into account all aspects related to the product life cycle,

(b) corporate responsibility labels usually indicating the products' degree of sustainability or recyclability, and (c) labels assigned by independent bodies that attest to a characteristic or several ecological characteristics of the product, such as carbon footprint (Gallastegui 2002; Nikolaou and Tsalis 2018). Eco-labeling has already been implemented, but the standardization process of the information is not properly uniformized, allowing the consumer to have access to vital information regarding the product in a fast, assertive, and uniform way, thus allowing him to make his purchase decision based on the information present on that label. Furthermore, the focus of the information presented usually changes from company to company, situation that is not favorable for a fast comparison of product information regarding environmental concerns. In addition, companies need to respect and show different aspects on how they are complying with the requirements to which they are bound by the various existing certifications, namely their environmental management system, such as ISO 14001, EMAS, or others, as well as direct environmental aspects, such as carbon footprints. The creation of social- and eco-labels aimed to show how producers usually follow the principles of social responsibility and are managing their production and/or operations under the principals of standards and had as main driver standardization improvement of information among producers (Loureiro et al. 2002). However, confusion in the information presented may exist as similar descriptions may be given by different companies. One company may only include information that corresponds to the way it manages its operations, while another may include the way production is managed, including product design. In other words, a company may wish to emphasize that its manufacturing processes follow the established environmental management standards, ensuring that the procedures followed do not harm the environment or want to demonstrate to customers that their product is environmentally friendly, and that all precautions with their products have been attained in order not to harm the environment, from its design stage to its distribution, including all stages of manufacture, avoiding negative impacts on the environment in terms of energy, materials usage, effluents, and future impact on natures after its use.

In addition, labels can take on a local or international character, having advantages and disadvantages in each of the systems. The local label allows for greater conformity between what the customer expects in a particular cultural environment, and what the producer is available to provide, in order to satisfy the client's need for information. On the other hand, especially for products with global distribution, information can be more confusing for the customer who is accustomed to reading it in a certain manner and will have to adapt to a different configuration of quantity and quality of the information.

The standardization of this label is particularly important when companies act in a global way, having different eco-labels depending on the type of product or country where they are operating and can have different levels of environmental management. Moreover, companies can present more than one certification, needing to attend to the requirements of the different standards. Thus, the information required in eco-labels can assume a certain complexity, which is not favorable for the promotion of customer sustainable behavior and selection through the rapid observation. However, the negative face of those labels is underlined by Bratt et al. (2011), who also refers that the proliferation of this type of label could cause confusion in the consumer, canceling out the desired positive effect, which consisted of providing useful information to the user, in order to allow a smarter and more conscious purchase.

There are not clear evidences on the willingness of stakeholders to pay higher prices for products to present proof of conformity with practices of environmental protection and sustainable consumption (Laroche et al. 2001; Tanner and Kast 2003; Tsakiridou et al. 2008; Gerpott and Mahmudova 2010), mainly due to the immense noise in the information and credibility that ends up being supplied to the consumer (Horne 2009). An example of this is the broadening of the categories for energy efficiency labeling. The expansion of category A into A+, A++, and A+++ categories has led to confusion in consumer reading, drastically reducing the initially intended effect of rapidly and assertively transmitting the energy consumption of certain appliances and other appliances in order to promote consumer choice toward those with higher energy saving ratings (Heinzle and Wüstenhagen 2012).

It can be clearly assumed that consumers only consult these labels if they are perfectly convinced that they have the necessary degree of confidence. Otherwise, they do not even ponder consulting them (Thøgersen 2002). Some authors tend to attribute this mismatch to lack of consumer information or lack of label credibility (Vermeir and Verbeke 2008; Vlaeminck et al. 2014). However, if the consumer really wants to have a responsible attitude at the time of purchase, he will certainly take due care in terms of information and then make an environmentally conscious decision. Anyway, it would be desirable for businesses and citizens to find a better match between the information provided by companies and the one that can be effectively perceived by consumers, so that labeling has real added value to the environment, benefiting companies concerned with environmental issues and helping consumers make more environmentally sustainable consumption decisions.

The certification body also assumes a relevant importance in the way the consumer interprets the information contained in labels. International or public certification bodies have greater credibility for consumers (Horne 2009). The complexity and lack of clarity revealed by certification bodies have also led to the loss of consumer confidence in the information contained in eco-labels.

More recently, and regarding the confusion surrounding eco-labels, Nikolaou and Tsalis developed an index, consisting of three subindices, aiming at the overall sustainability assessment of products and companies, taking into account the following aspects: (a) the evaluation of product's or company's economic contribution to the different stakeholders, (b) the assessment of the social impact of products and companies in the community, and (c) the estimation of the evolution of a given product or company in terms of compliance with environmental rules (Nikolaou and Tsalis 2018). The weight assigned to each of the subindices is established through consultation with consumer representative bodies as well as experts on corporate sustainability. In order to clarify the final classification of products, they should be divided by classes and the average classification for each sector should be determined. If a product or company stands out, or in other words, if they are above average, the classification should translate this clearly, thus helping the consumer to identify if a product or company performs above average, or if, conversely, they perform below average in terms of environmental sustainability. This classification, together with others considered as potentially useful, will help the consumer make more conscious choices of products and companies that behave more appropriately with their environmental posture.

4.4 Influences on Consumption Behavior of Younger People

Young people represent the future of our society, both as consumers, as well as creators of new trends, technologies, working methodologies, and even drawing new ways of business (Smola and Sutton 2002; Heaney 2006; McCrindle 2018). A deeper understanding of their behavior may allow us to anticipate trends in terms of how consumption will be viewed by these new generations in the near future. Given the social changes and economic growth in recent decades, it is quite understandable that the new generation mentality is shaped differently. Moreover, these generations were confronted early on with climate change, environmental disasters, excessive production of debris and hazardous materials, among other environmental hostilities, which would certainly promote a different way of looking at the future, feeling a greater need to contribute so that the trend of environmental degradation is reversed. In this context, the young generation, also dubbed Generation Y, is considered a population with "green thinking," growing in an era concerned with global environmental disasters linked to ecological attacks on our planet, a situation that is also enhanced by a broader and more constant access to information. However, it is also this media format that intoxicates these same generations of advertising, trying to create the necessary involvement so that they are attracted to the purchase of certain goods, even if unnecessary. However, the care put into acquiring so-called sustainable goods has grown significantly in importance in the market, largely due to this new generations' attitude. This is clearly shown by the percentage of eco-friendly products bought by young people, which ascended from 31% in 2009 to 36% in 2012, representing an increase of 16%, taking into account a survey carried out with a sample population of 4000 adults in the USA. This study had as focus natural/organic personal care products, such as shampoos, face creams, moisturizers, renewal creams, and beauty products, which registered a demand of 39% by the young people in 2012, compared with the demand of 27% registered in 2009 (The Guardian 2018).

Despite the above-mentioned, some authors are sure to point out that the new generations have a greater appetite for easy wealth creation, are exuberant in consumption and fun, think fast and have an easy career progression, and are also targets of marketing campaigns (Csikszentmihalyi 2000; O'Shaugnessy and O'Shaugnessy 2002; Hoey 2008). Indeed, new generations have a great appetite for new technologies, are very effective in their use, and are exposed to capitalist systems and rules in most Western countries, and the rapid growth of emerging economies such as China, Korea, Mexico, and Brazil, among many others, and the breakdown of most of the trade barriers, allowing to choose goods with high quality and very good design, having considerably fewer restrictions on consumption than their previous generations. Studies carried out on Belgian youth have allowed to conclude that the consumption of dairy sustainable products is influenced by social issues but is also conditioned by the daily routine and consumer availability (Vermeir and Verbeke 2008). Other authors, through studies previously carried out, are much more optimistic concerning the approach of the new generations to environmental problems, having as central concern consumption patterns. These authors point out that these groups of young people are socially and culturally more aware, and they are also more sensitive to environmental issues. However, do not fail to have a strong spirit of leadership and initiative, also enjoying capturing the attention of others both inside and outside their social group (Sheahan 2005; Heaney 2006; Sullivan and Heitmeyer 2008). However, due to the abundance of goods and services offered, greater freedom of movement through much more frequent travel to other countries and cultures, and the absence of barriers to trade, new generations are considered to be much more eager for consumption, surpassing clearly all other generations.

Modern society has steadily and progressively changed its consumption patterns, largely due to the broken barriers formerly existing between the different countries. The acquisition through online platforms has influenced in a notorious way our society. It should also be noted that young people, due to a greater degree of training and dynamics of the employment market, have incomes that allow them to have more money for nonessential consumption. Because of this, new generations are consuming more, being more demanding, and pursuing more eccentric lifestyles than their progenitors (Abela 2006).

The findings achieved by Te'eni-Harari and Hornik (2010) allowed to realize that there are five main factors able to influence young people in shopping situations: age, individual's idiosyncratic acquaintance, parental guidance, peers impact, and product class. The study carried out by Ginsberg (2004) suggests that approximately one-third of the world's population are young consumers. This represents a non-negligible market.

It is these young people who are also the target of huge marketing campaigns as companies try to win over customers, promoting product ranges suitable for different ages and loyalty campaigns, in an effort to maintain customer brand loyalty. However, companies essentially aim for profit and, to do so, need to sell. However, this is a vicious cycle. Companies show more profits by the end of a year, setting thereafter more ambitious goals for the next year, and so on. To do so, campaigns are needed, which aim essentially to increase sales. However, sales represent consumption, and companies which focus only on their increasingly ambitious goals usually lose focus on other important matters regarding the end-of-life and recycling of products, generating useless waste due to exacerbated and non-conscious manufacturing and consumption patterns. Despite the fact that these newer generations normally have a higher educational level than their parents, the need for group affirmation and young age make this range of consumers more vulnerable to marketeers, exploring their vulnerabilities. In order to avoid this situation in a closed cycle, it is necessary for companies to be committed on creating customer engagement from a very early age, which is essentially about sustainable consumption (Daub and Ergenzinger 2005). This agreement will only be possible when there is an absolute awareness by all the parties involved of the need to act quickly in this field, educating consumers and providing them with the necessary tools, so root behaviors can be developed that truly lead to a more sustainable way of living.

Other authors also point out that there are factors such as gender, education, and income that turn out to be less important than lifestyle or social values in consumption patterns exhibited by young people (Wellner 2000; Schewe and Meredith 2004; Lake 2006). Demographics also have a strong influence on the behavior of young people, since they are in different places, have access to different job opportunities and career progression, have diverse incomes, can find more or less development barriers, may have different access to technology and media, may be subject to different social habits, and have different opportunities to expand their knowledge and manifest their own personality. All this is reversed in dissimilar needs and opportunities to fill those needs through different manners, which leads to different ways of behaving in society and facing consumption, as well as the relationship between that consumption and the environment (Heaney 2006).

Greater contact with the environmental problems usually experienced in developing countries, such as China, induces a different mentality in young people in terms of consumption. In fact, the harm to the environment caused by previous generations is very evident for newer generations, arousing different behaviors in regard to environmental care, passing compulsorily by clear changes in their behavior with regard to consumption (Gandhi and Kaushik 2016). Moreover, it has also been confirmed that emotional and social values usually positively influence the decision of Chinese consumers in the early stages of their adult lives, leading them to opt for environmentally friendly products (Awuni and Du 2016). Some authors reinforce the idea that education for environmental concerns is also fundamental in teenage years, in order to root the principles that may be the basis of more responsible behavior when they reach adulthood, as well as family behavior (Zhu et al. 2013; Ollerer 2015; Scott et al. 2015). This idea is also emphasized by Jung and Jin (2016), referring that education is the first aspect to positively boost sustainable consumption among the younger population.

4.5 Current Situations and Future Perspectives

In terms of consumption, the alarming situation can be expressed in a simple sentence: "In less than eight months in each year, Humanity has already spent all the resources that nature will be able to regenerate in that year" (Global Footprint Network 2018). In 2017, the Global Footprint Network celebrated as usual the day of the year in which the resources consumed are equivalent to those that Nature will be able to regenerate in that given year, having happened on August 2, 2017. From that date and until the end of the same year, the world has spent resources that Nature is not able to regenerate, thus showing how wrong and unsustainable our behavior usually is in terms of consumption. The use of renewable resources is still very scarce, and, for example, in energy terms, there isn't enough area on Earth capable of producing sufficient biofuels to replace fossil fuels and meet current energy demand. Moreover, the same area of land is also becoming increasingly scarce to meet food production needs necessary to supply a growing population, which by having an increasingly better social condition, is leading to an even higher rate of consumption. Some rarer minerals, which are used intensively in new technologies, may very soon have higher demand than current extraction rates, causing serious mismatches between the needs and the actual existence of these ores for extraction. Adding to this, there are social problems that unsustainable consumption creates, and that will be accentuated in the future. Indeed, a citizen of a rich country consumes about 10 times more than another of a developing country. Between continents, there are also notable asymmetries: a European consumes about three times more than an Asian citizen and four times more than an African citizen. In the future, this situation will tend to aggravate cleavages between countries with higher and lower incomes, which may lead to insurmountable social problems. This situation, together with political instability, has led to a flow of people from the Middle East and Africa to Europe since 2015, looking for more opportunities and higher purchasing power. This situation, if it continues to worsen, will cause more and more social tensions, which can generate conflicts of unimaginable proportions.

Much remains to be done to make sustainable consumption a reality. Companies have realized that, after different approaches to the market with so-called ecological campaigns, the market is not yet sufficiently matured and focused on this type of products, so business is not yet profitable. Thus, and because immediate profit is the livelihood of companies, a good part has given up efforts in this field, since traditional consumption requires less effort and allows a greater immediate income (Wirtenberg et al. 2007; Mader 2008). Indeed, the mass communication policies of a culture of compliance with the most demanding norms in terms of preservation of the environment still do not present the rate of return that many would expect, because it requires a strong educational effort, in the adaptation or conversion of values and ways of being, which takes much longer than expected to be achieved.

4.6 Key Challenges in Sustainable Consumption

A bit of hope still exists for consumer trends to be reversed and makes the world more sustainable. A few years ago, Michaelis (2003) pointed out some ways to keep hope and reverse exacerbated consumption:

- Although the culture of individualism remains in Western countries, there seems to be a strong pressure in order to restore community relations through new, more engaging initiatives that are easily understood by each citizen of these communities;
- There are new entities emerging on the market that promote greater involvement of all stakeholders, strong links to the environment and consumer ethics, developing fair trade platforms, encouraging organic production and consumption;

- Movements of people begin to exist that tends to flee from large metropolis, valuing and giving preference to a more modest lifestyle in the silence and calmness of suburban and rural areas;
- There is a strong contrast between global trade and local production of goods and services, through a renewed offer, more focused on the culture of each region. This movement has found an echo in political power, which now delegates important decisions to local power;
- There is a greater perception and sensitivity for the risk to the business world and to the political class, so that transparency for them and between them needs to be evident and capable of developing dialogue between all stakeholders, respecting the communities' principles.

It is universally known that there is an urgent need for a change in consumption habits. There seem to be signs that it is possible and appropriately supported by the community. However, this change cannot be made abruptly or under extremely restrictive legislation. Only the involvement of companies, politicians, and consumers in policies following a coherent and well-understood strategy by stakeholders can result in the necessary change of habits.

More recently, many researchers have discussed the future of research consumer habits, pointing out three main challenges, as described below:

- Knowledge and attitude—This challenge is related to how the knowledge, the readiness to analyze the problem, and the attitude of citizens are related to their consumption behaviors, as well as the lack of linearity between knowledge and action, availability or attitude and behavior, because prior studies have concluded that consumption behavior is also strongly based on personal or familiar routines, along with social routines assumed as conventional (Kollmuss and Agyeman 2002; Spaargaren 2003; Fischer and Hanley 2007; Schäfer et al. 2012). Thus, it is necessary to reduce these gaps through new models and look for new policies able to lead citizens to a deeper reflection regarding their routines and make their behavior closer with sustainable consumption practices;
- Multi-disciplinary problem—Effectively, one of the main challenges for the discussion in the future is the multi-disciplinary character of the consumption behavior and sustainability, which allow for inputs from the most diverse sectors and novelties brought by deeper investigations (Di Giulio et al. 2014; Lorek and Vergragt 2015). Thus, a reinforcement of the work in the boundary fields is needed, to make more compatible speech and terminology, breaking the barriers sometimes erected between the knowledge in different disciplines;
- Global analysis—The last challenge concerns a poor overall analysis of the most relevant policies that are being implemented, particularly with regard to their results. This reveals a real gap between the installed policies and their implementation (Tukker et al. 2006; Heiskanen et al. 2014). It is necessary to internalize and consolidate the achievements already made, promoting the creation of a multifaceted database of knowledge capable of encouraging a solid implementation of sustainable consumption principles.

The integration of the dispersed knowledge as well as the establishment of bridges between different scientific fields remains as a global challenge for the scientific community interested in demystifying the blockages to a quicker understanding of all the factors that are behind the consumers' decisions, as well as the best way to modify their attitude facing consumption, adopting a sustainable behavior.

References

A.V. Abela, Marketing and consumerism. Eur. J. Mark. **40**(1/2), 5–16 (2006)

K. Abeliotis, C. Koniari, E. Sardianou, The profile of the green consumer in Greece. Int. J. Consum. Stud. **34**, 153–160 (2010)

A. Achtziger, M. Hubert, P. Kenning, G. Raab, L. Reisch, Debt out of control: the links between self-control, compulsive buying, and real debts. J. Econ. Psychol. **49**, 141–149 (2015)

J. Aertsens, W. Verbeke, K. Mondelaers, G. van Huylenbroeck, Personal determinants of organic food consumption. A review. Br. Food J. **111**(10), 1140–1167 (2009)

I. Ajzen, Nature and operation of attitudes. Annu. Rev. Psychol. **52**, 27–58 (2001)

S. Alexander, Property beyond Growth. Toward a politics of voluntary simplicity. Ph.D. Dissertation, University of Melbourne, Faculty of Law, Melbourne, Australia, 2011

S. Alexander, S. Ussher, The voluntary simplicity movement: a multinational survey analysis in theoretical context. J. Consum. Cult. **12**(1), 66–86 (2012)

American Public Health Association, *Toward a Healthy Sustainable Food System* (2007), https://www.apha.org/policies-and-advocacy/public-health-policy-statements/policy-atabase/2014/07/29/12/34/toward-a-healthy-sustainable-food-system. Retrieved on 24 Mar 2018

A. Annunziata, S. Ianuario, P. Pascale, Consumers' attitudes toward labelling of ethical products. The case of organic and fair trade products. J. Food Prod. Mark. **17**(5), 518–535 (2011)

P. Antonetti, S. Maklan, Feelings that make a difference: how guilt and pride convince consumers of the effectiveness of sustainable consumption choices. J. Bus. Ethics **124**(1), 117–134 (2014)

M.J. Arnold, K.E. Reynolds, Hedonic shopping motivations. J. Retail. **79**(2), 77–95 (2003)

J.A. Awuni, J.G. Du, Sustainable consumption in Chinese cities: green purchasing intentions of young adults based on the theory of consumption values. Sustain. Dev. **24**, 124–135 (2016)

B.J. Babin, W.R. Darden, M. Griffin, Work and/or fun: measuring hedonic and utilitarian shopping value. J. Consum. Res. **20**(4), 644–656 (1994)

S. Bahl, G.R. Milne, S.M. Ross, D.G. Mick, S.A. Grier, S.K. Chugani, S. Chan, S.J. Gould, Y.-N. Cho, J.D. Dorsey, R.M. Schindler, M.R. Murdock, S.B. Mariani, Mindfulness. The transformative potential for consumer, societal, and environmental well-being. J. Public Policy Mark. **35**, 1–13 (2016)

D. Bajde, *Consuming a Political Figure: A Post-Structuralist Reading of Four* (2007), http://www.acrwebsite.org/volumes/eacr/vol8/eacr_vol8_125.pdf. Retrieved on 23 Feb 2018

I. Balderjahn, A. Buerke, M. Kirchgeorg, M. Peyer, B. Seegebarth, K.-P. Wiedmann, Consciousness for sustainable consumption: scale development and new insights in the economic dimension of consumers' sustainability. AMS Rev. **3**, 181–192 (2013)

A. Bandura, Social cognitive theory: an agenetic perspective. Annu. Rev. Psychol. **52**(1), 1–26 (2001)

S.J. Barnes, J. Mattsson, Understanding current and future issues in collaborative consumption: a four-stage Delphi study. Technol. Forecast. Soc. Chang. **104**, 200–211 (2016)

J. Baudrillard, *The Consumer Society* (Sage Publications, London, UK, 1998). ISBN 0-7619-5691-3

Z. Bauman, *Does Ethics Have a Chance in a World of Consumers?* (Cambridge, MA, Harvard University Press, 2008). ISBN 978-0674-03351-1

B.L. Bayus, G. Erickson, R. Jacobson, The financial rewards of new product introductions in the personal computer industry. Manage. Sci. **49**(2), 197–210 (2003)

U. Beck, *The Risk Society, Towards a New Modernity* (Sage Publications, London, 1992). ISBN 0-8039-8346-8
R.W. Belk, Sharing. J. Consum. Res. **36**(5), 715–734 (2010)
V. Bezençon, S. Blili, Segmenting the market through the determinants of involvement. The case of fair trade. Psychol. Mark. **28**(7), 682–708 (2011)
K. Birnie, M. Speca, L.E. Carlson, Exploring self-compassion and empathy in the context of mindfulness-based stress reduction (MBSR). Stress Health **26**, 359–371 (2010)
J. Blake, Overcoming the "value-action gap" in environmental policy: tensions between national policy and local experience. Local Environ. **4**, 257–278 (1999)
M. Bolz, T. Singer (eds.), *Compassion. Bridging Practice in Science* (Max Planck Society, Munich, 2013). ISBN 978-3-9815612-0-3, http://www.compassion-training.org/en/online/files/assets/basic-html/index.html#1. Retrieved on 25 Feb 2018
S. Borghesi, A. Vercelli, Sustainable globalization. Ecol. Econ. **44**, 77–89 (2003)
R. Botsman, R. Rogers, 2011. *What's Mine is Yours: How Collaborative Consumption is Changing the Way We Live* (Harper Collins Publishers, London, UK, 2011). ISBN 978-0007395910
C. Bratt, S. Hallstedt, K.-H. Robèrt, G. Broman, J. Oldmark, Assessment of eco-labelling criteria development from a strategic sustainability perspective. J. Clean. Prod. **19**, 1631–1638 (2011)
H.G. Brauch, U.O. Spring, J. Grin, J. Scheffran (eds.), *Handbook on Sustainable Transition and Sustainable Peace* (Springer International Publishing, Switzerland, 2016). ISBN 978-3-319-43884-9
T. Briceno, S. Stagl, The role of social processes for sustainable consumption. J. Clean. Prod. **14**, 1541–1551 (2006)
S. Brown, K. Taylor, S.W. Price, Debt and distress: evaluating the psychological cost of credit. J. Econ. Psychol. **26**(5), 642–663 (2005)
L. Canova, A.M.M. Rattazzi, P. Webley, The hierarchical structure of saving motives. J. Econ. Psychol. **26**(1), 21–34 (2005)
D.R. Carney, J.T. Jost, S.D. Gosling, J. Potter, The secret lives of liberals and conservatives: personality profiles, interaction styles, and the things they leave behind. Polit. Psychol. **29**(6), 807–840 (2008)
M. Carrigan, The myth of the ethical consumer—do ethics matter in purchase behaviour? J. Consum. Mark. **18**(7), 560–577 (2001)
G. Carrus, P. Passafaro, M. Bonnes, Emotions, habits and rational choices in ecological behaviours: the case of recycling and use of public transportation. J. Environ. Psychol. **28**(1), 51–62 (2008)
R. Caruana, A. Crane, Constructing consumer responsibility: exploring the role of corporate communications. Organ. Stud. **29**(12), 1495–1519 (2008)
The Consumer Citizenship Network (CCN), Consumer citizenship education—guidelines, in *Higher Education*, vol. 1, ed. by W. Thoresen (2005). http://tuningacademy.org/wp-content/uploads/2014/02/CCN-Learning-Teaching-Guidelines-vol-1final-21-09-05-_2_.pdf. Retrieved 26 Feb 2018
R. Chambers, E. Gullone, N.B. Allen, Mindful emotion regulation. An integrative review. Clin. Psychol. Rev. **29**, 560–572 (2009)
N.L.D. Chatzisarantis, M.S. Hagger, Mindfulness and the intention-behavior relationship within the theory of planned behavior. Pers. Soc. Psychol. Bull. **33**, 663–676 (2007)
A. Chiesa, A. Serretti, Mindfulness-based stress reduction for stress management in healthy people. A review and meta-analysis. J. Altern. Complement. Med. **15**, 593–600 (2009)
J. Connolly, D.S. Shaw, Identifying fair trade in consumption choice. J. Strateg. Mark. **14**(4), 353–368 (2006)
B. Cornwell, J. Drennan, Cross-cultural consumer consumption research: Dealing with issues emerging from globalization and fragmentation. J. Macromark. **24**(2), 108–121 (2004)
M. Craig-Lees, C. Hill, Understanding voluntary simplifiers. Psychol. Mark. **19**(2), 187–210 (2002)
D. Crockett, M. Wallendorf, The role of normative political ideology in consumer behavior. J. Consum. Res. **31**(3), 511–528 (2004)
M. Csikszentmihalyi, Reflections and reviews: the cost and benefits of consuming. J. Consum. Res. **27**(2), 267–272 (2000)

B. Culiberg, L. Elgaaied-Gambier, Going green to fit in—understanding the impact of social norms on pro-environmental behaviour, a cross-cultural approach. Int. J. Consum. Stud. **40**, 179–185 (2016)

H. Dagevos, J. Voordouw, Sustainability and meat consumption. Is reduction realistic? Sustain. Sci. Pract., Policy **9**(2), 60–69 (2013)

A. Darnton, *GSR Behaviour Change Knowledge Review. Reference Report: An Overview of Behaviour Change Models and their Uses* (Government Social Research, London, 2008), http://www.civilservice.gov.uk/wpcontent/uploads/2011/09/Behaviour_change_reference_report_tcm6-9697.pdf. Retrieved on 5 Mar 2018

C. Daub, R. Ergenzinger, Enabling sustainable management through a new multi-disciplinary concept of customer satisfaction. Eur. J. Mark. **39**(9/10), 998–1217 (2005)

C.J.S. de Backer, L. Hudders, Meat morals. Relationship between meat consumption consumer attitudes towards human and animal welfare and moral behavior. Meat Sci. **99**, 68–74 (2015)

J. de Boer, H. Schösler, H. Aiking, "Meatless days" or "less but better"? Exploring strategies to adapt Western meat consumption to health and sustainability challenges. Appetite **76**, 120–128 (2014)

J. de Jonge, H. van Trijp, Meeting heterogeneity in consumer demand for animal welfare. A reflection on existing knowledge and implications for the meat sector. J. Agric. Environ. Ethics **26**(3), 629–661 (2013)

P. De Pelsmacker, W. Janssens, A model for fair trade buying behaviour: the role of perceived quantity and quality of information and of product-specific attitudes. J. Bus. Ethics **75**(4), 361–380 (2007)

M.A. Delmas, L.E. Grant, Eco-labelling strategies and pricepremium: the wine industry puzzle. Bus. Soc. **53**(1), 6–44 (2014)

S.A. Devaney, S.T. Anong, S.E. Whirl, Household savings motives. J. Consum. Aff. **41**(1), 174–186 (2007)

T.M. Devinney, P. Auger, G. Eckhardt, T. Birtchnell, The other CSR: consumer social responsibility. Stanford Soc. Innov. Rev. **4**(3), 30–37 (2006)

L. Dezsö, G. Loewenstein, Lenders' blind trust and borrowers' blind spots: a descriptive investigation of personal loans. J. Econ. Psychol. **33**(5), 996–1011 (2012)

A. Di Giulio, D. Fischer, M. Schäfer, B. Blättel-Mink, Conceptualizing sustainable consumption: toward an integrative framework. Sustain. Sci. Pract. Policy **10**, 45–61 (2014)

T.M. Devinney, P. Auger, G.M. Eckhardt, *The Myth of the Ethical Consumer* (Cambridge University Press, Cambridge, UK, 2010). ISBN 978-0-521-76694-4

K. Dowd, K.J. Burke, The influence of ethical values and food choice motivations on intentions to purchase sustainably sourced foods. Appetite **69**, 137–144 (2013)

L. Doyal, I. Gough, *A Theory of Human Need* (The Guilford Press, NewYork, 1991). ISBN 978-0898624137

EC.EUROPA.EU (2015), http://ec.europa.eu/eurostat/statisticsexplained/images/9/98/Water_use_by_households_and_the_manufacturing_industry%2C_from_public_water_supply_and_self_and_other_supply%2C_2015_%28m%C2%B3_per_inhabitant%29_V3.png. Retrieved 18 Mar 2018

European Environment Agency, *Policies and Measures to Promote Sustainable Water Use* (Report, Copenhagen K, Denmark, 2007)

D. Elgin, A. Mitchell, Voluntary simplicity. Plan. Rev. **5**(6), 13–15 (1977)

J. Elkington, Cannibals with forks: the triple bottom line of 21st century business. Environ. Qual. Manage. **8**(1), 37–51 (1997)

Environmental Protection Agency, *The WaterSense Label* (2018), https://www.epa.gov/watersense/watersense-label. Retrieved 20 Mar 2018

T. Ericson, B.G. Kjønstad, A. Barstad, Mindfulness and sustainability. Ecol. Econ. **104**, 73–79 (2014)

S. Ewen, *Captains of Consciousness: Advertising and the roots of Consumers* (Mc Graw Hill, New York, USA, 1976). ISBN 978-0465021550

A. Farmer, *Political Ideology and Consumer Preferences*. Theses and dissertations-marketing and supply chain. Paper 2. http://uknowledge.uky.edu/marketing_etds/2. Retrieved 22 Feb 2018
M. Feinberg, R. Willer, The moral roots of environmental attitudes. Psychol. Sci. **24**, 56–62 (2012)
A. Fischer, N. Hanley, Analysing decision behaviour in stated preference surveys: a consumer psychological approach. Ecol. Econ. **61**, 303–314 (2007)
D. Fischer, L. Stanszus, S. Geiger, P. Grossman, U. Schrader, Mindfulness and sustainable consumption: a systematic literature review of research approaches and findings. J. Clean. Prod. **162**, 544–558 (2017)
Foresight, *The Future of Food and Farming. Challenges and Choices for Global Sustainability* (The Government Office for Science, London, 2011), http://www.eracaps.org/sites/default/files/content/foresight_report.pdf. Retrieved on 24 Mar 2018
N. Fox, K. Ward, Health, ethics and environment. A qualitative study of vegetarian motivations. Appetite **50**(2–3), 422–429 (2008)
I.G. Gallastegui, The use of eco-labels: a review of the literature. Environ. Policy Gov. **12**(6), 316–331 (2002)
M. Gandhi, N. Kaushik, Socially responsive consumption behaviour—an Indian perspective. Soc. Responsib. J. **12**, 85–102 (2016)
L. Gansky, *The Mesh: Why the Future of Business is Sharing* (Portfolio Penguin, New York, 2010). ISBN 978-1-101-46461-8, https://informationdj.files.wordpress.com/2012/01/future-of-business-is-lisa-gansky.pdf. Retrieved on 17 Mar 2018
T. Garnett, Where are the best opportunities for reducing greenhouse gas emissions in the food system (including the food chain)? Food Policy **36**(S1), S23–S32 (2011)
F.W. Geels, A. McMeekin, J. Mylan, D. Southerton, A critical appraisal of sustainable consumption and production research: The reformist, revolutionary and reconfiguration positions. Glob. Environ. Change **34**, 1–12 (2015)
T.J. Gerpott, I. Mahmudova, Determinants of price mark-up tolerance for green electricity e lessons for environmental marketing strategies from a study of residential electricity customers in Germany. Bus. Strategy Environ. **19**(5), 304–318 (2010)
B. Gert, Moral arrogance and moral theories. Philoso. Issues **15**(1), 368–385 (2005)
A. Gilg, S. Barr, N. Ford, Green consumption or sustainable lifestyles? Identifying the sustainable consumer. Futures **37**(6), 481–504 (2005)
B. Ginsberg, *Purchasing for Sustainability—Young Consumers in the Green Marketplace: Why Youth? Why You?* (2004). http://www.popsustainability.org/popzine/issue3/trends3.htm. Retrieved 11 Feb 2017
L. B. Glickman, *Buying Power: A History of Consumer Activism in America* (University of Chicago Press, Chicago, U.S.A., 2009). ISBN 978-022-6298-67-2
Global Footprint Network, *Earth Overshoot Day* (2018), https://www.footprintnetwork.org/our-work/earth-overshoot-day/. Retrieved on 25 Mar 2018
J. Graham, J. Haidt, B.A. Nosek, Liberals and conservatives rely on different sets of morals foundations. J. Pers. Soc. Psychol. **96**(5), 1029–1046 (2009)
G. Grankvist, H. Lakedal, M. Marmendal, Values and eco- and fair-trade labelled products. Br. Food J. **109**(2), 169–181 (2007)
P. Grossman, Mindfulness for psychologists: paying kind attention to the perceptible. Mindfulness **1**, 87–97 (2010)
P. Grossman, Kindness and compassion as integral to mindfulness, in *Compassion. Bridging Practice in Science*, ed. by M. Bolz, T. Singer (Max Planck Society, Munich, Germany, 2013), pp. 192–207. ISBN 978-3-9815612-0-3
P. Grossman, Mindfulness. Awareness informed by an embodied ethic. Mindfulness **6**, 17–22 (2015)
N.S. Harth, C.W. Leach, T. Kessler, Guilt, anger, and pride about in-group environmental behaviour: different emotions predict distinct intentions. J. Environ. Psychol. **34**, 18–26 (2013)
P. Hartmann, V.A. Ibáñez, Green value added. Mark. Intell. Plan. **24**(7), 673–680 (2006)
J. Heaney, Generations X and Y's internet banking usage in Australia. J. Finan. Serv. Mark. **11**(3), 196–210 (2006)

H. Heinrichs, Sharing economy: a potential new pathway to sustainability. GAIA Ecol. Perspect. Sci. Soc. **22**(4), 228–231 (2013)

S.L. Heinzle, R. Wüstenhagen, Dynamic adjustment of eco-labeling schemes and consumer choice-the revision of the EU energy label as a missed opportunity? Bus. Strategy Environ. **21**, 60–70 (2012)

E. Heiskanen, O. Mont, K. Power, A map is not a territory—making research more helpful for sustainable consumption policy. J. Consum. Policy **37**, 27–44 (2014)

A. Hayley, L. Zinkiewicz, K. Hardiman, Values, attitudes, and frequency of meat consumption. Predicting meat-reduced diet in Australia. Appetite **84**, 98–106 (2015)

A.C. Hoek, P.A. Luning, P. Weijzen, W. Engels, F.J. Kok, C. de Graaf, Replacement of meat by meat substitutes. A survey on person- and product related factors in consumer acceptance. Appetite **56**(3), 662–673 (2011)

T. Hoey, How to turn generation Y into generation $. Sydney Morning Herald (2008) 15 Mar, http://www.smh.com.au/articles/2008/03/15/1205472147009.html. Retrieved 14 Nov 2009

S.R. Hoffman, S.F. Stallings, R.C. Bessinger, G.T. Brooks, Differences between health and ethical vegetarians, Strength of conviction, nutrition knowledge, dietary restriction, and duration of adherence. Appetite **65**, 139–144 (2013)

M.B. Holbrook, E.C. Hirschman, The experiential aspects of consumption: Consumer fantasies, feelings and fun. J. Consum. Res. **9**, 132–140 (1982)

R.E. Horne, Limits to labels: the role of eco-labels in the assessment of product sustainability and routes to sustainable consumption. Int. J. Consum. Stud. **33**, 175–182 (2009)

R.S. Hughner, P. McDonagh, A. Prothero, C.J. Shultz, J. Stanton, Who are organic food consumers? A compilation and review of why people purchase organic food. J. Consum. Behav. **6**(2–3), 94–110 (2007)

M. Hume, Compassion without action: Examining the young consumers consumption and attitude to sustainable consumption. J. World Bus. **45**, 385–394 (2010)

M.E. Huneke, The face of the un-consumer: an empirical examination of the practice of voluntary simplicity in the United States. Psychol. Mark. **22**(7), 527–550 (2005)

A. Hüttel, F. Ziesemer, M. Peyer, I. Balderjahn, To purchase or not? Why consumers make economically (non-) sustainable consumption choices. J. Clean. Prod. **174**, 827–836 (2018)

T. Hyland, *Mindfulness and Learning. Celebrating the Affective Dimension of Education* (Springer, Dordrecht, New York, 2011). ISBN 978-94-007-1910-1

D. Ivanova, K. Stadler, K. Steen-Olsen, R. Wood, G. Vita, A. Tukker, E.G. Hertwich, Environmental impact assessment of household consumption. J. Ind. Ecol. **20**, 526–536 (2015)

O. Iwata, An evaluation of consumerism and lifestyle as correlates of a voluntary simplicity lifestyle. Soc. Behav. Pers. **34**(5), 557–568 (2006)

T. Jackson, L. Michaelis, *Policies for Sustainable Consumption—A Report to the Sustainable Development Commission*, http://citeseerx.ist.psu.edu/viewdoc/download?10.1.1.322.6476&rep=rep1&type=pdf. Retrieved 3 Mar 2018

T. Jackson, W. Jager, S. Stagl, Beyond insatiability—needs theory, consumption and sustainability, in *The Ecological Economics of Consumption, Current Issues in Ecological Economics Series*, ed. by L. Reisch, I. Røpke (Edward Elgar Publishing, Cheltenham, UK, 2004), pp. 79–107. ISBN 978 1 84376 512 7

P. Jackson, K. Snowdon, Ecolabels: boon for the environment or confusion for the consumer? Eng. Sci. Educ. J. **8**(1), 20–22 (1999)

T. Jackson, Evolutionary psychology in ecological economics: consilience, consumption and contentment. Ecol. Econ. **41**, 289–303 (2002)

T. Jackson, *Motivating Sustainable Consumption: A Review of Evidence on Consumer Behaviour and Behavioural Change*. Report to the 'Sustainable Development Research Network', London. Available at: http://sustainablelifestyles.ac.uk/sites/default/files/motivating_sc_final.pdf. Retrieved 5 Mar 2018

R. Jenkins, M. Molesworth, R. Scullion, The messy social lives of objects: inter-personal borrowing and the ambiguity of possession and ownership. J. Consum. Behav. **13**(2), 131–139 (2014)

J.T. Jost, The end of the end of ideology. Am. Psychol. **61**(7), 651–670 (2006)
S.J. Jung, B. Jin, From quantity to quality: understanding slow fashion consumers for sustainability and consumer education. Int. J. Consum. Stud. **40**, 410–421 (2016)
J. Kabat-Zinn, Mindfulness-based interventions in context: past, present, future. Clin. Psychol. Sci. Pract. **10**, 144–156 (2003)
F.G. Kaiser, S. Wölfing, U. Fuhrer, Environmental attitude and ecological behaviour. J. Environ. Psychol. **19**(1), 1–19 (1999)
M. Kala, A. Sharma, Traditional Indian beliefs: a key toward sustainable living. Environmentalist **30**, 85–89 (2010)
M. Kamenetzky, The economics of the satisfaction of needs, in *Real Life Economics—Understanding Wealth Creation*, ed. by P. Ekins, M. Max-Neef (Routledge, London, New York, 1992), pp. 181–193. ISBN 978-0415079778
M. Kanchanapibul, E. Lacka, X. Wang, H.K. Chan, An empirical investigation of green purchase behaviour among the young generation. J. Clean. Prod. **66**, 528–536 (2014)
J. Kang, K. Grable, G. Hustvedt, M. Ahn, Sustainable water consumption: the perspective of Hispanic consumers. J. Environ. Psychol. **50**, 94–103 (2017)
I. Kareklas, J. Carlson, D.D. Muehling, 'I eat organic for my benefit and yours' Egoistic and altruistic considerations for purchasing organic food and their implications for advertising strategists. J. Advertising **43**(1), 18–32 (2014)
D.G. Karp, Values and their effects on pro-environmental behavior. Environ. Behav. **28**(1), 111–133 (1996)
R. Kaufmann-Hayoz, B. Brohmann, R. Defila, A. Di Giulio, E. Dunkelberg, L. Erdmann, D. A. Fuchs, S. Gölz, A. Homburg, E. Matthies, M. Nachreiner, K. Tews, J. Weiß, Societal steering of consumption towards sustainability, in *The Nature of Sustainable Consumption and How to Achieve it. Results from the Focal Topic "From Knowledge to Action—New Paths towards Sustainable Consumption* ed. by R. Defila, A. Di Giulio, R. Kaufmann-Hayoz (Oekom, München, 2012), pp. 113–142. ISBN 978-3865813022
A. Kenis, E. Mathijs, Beyond individual behaviour change: the role of power, knowledge and strategy in tackling climate change. Environ. Educ. Res. **18**, 45–65 (2012)
B. Kidwell, A. Farmer, D.M. Hardesty, Getting liberals and conservatives to go green: political ideology and congruent appeals. J. Consum. Res. **40**, 350–367 (2013)
T.C. Kinnear, J.R. Taylor, The effect of ecological concern on brand perceptions. J. Mark. Res. **10**(2), 191–197 (1973)
D. Knoppen, W. Saris, Do we have to combine values in the Schwartz' human values scale? A comment on the Davidov studies. Surv. Res. Methods **3**(2), 91–103 (2009)
A. Kollmuss, J. Agyeman, Mind the Gap: why do people act environmentally and what are the barriers to pro-environmental behavior? Environ. Educ. Res. **8**, 239–260 (2002)
S. Koos, Varieties of environmental labelling, market structures and sustainable consumption across Europe: a comparative analysis of organizational and market supply determinants of environmental labelled goods. J. Consum. Policy **34**(1), 127–151 (2011)
L. Lake, *The Power of Generation Marketing* (2006), http://marketing.about.com/od/demographics/a/generationmktg.htm. Retrieved 11 Mar 2018
M. Laroche, J. Bergeron, G. Barbaro-Foleo, Targeting consumers who are willing to pay more for environmentally friendly products. J. Consum. Mark. **18**(6), 503–520 (2001)
C. Lasch, *The culture of Narcissism* (Norton and Company, London, UK, 1979). ISBN 978-0-393-35617-5
J.L. Lastovicka, L.A. Bettencourt, R.S. Hughner, R.J. Kuntze, Lifestyle of the tight and frugal: Theory and measurement. J. Consum. Res. **26**(1), 85–98 (1999)
M.S.W. Lee, C.S.Y. Ahn, Anti-consumption, materialism, and consumer wellbeing. J. Consum. Aff. **50**(1), 18–47 (2016)
Levi's (2018a), http://www.levi.com/PT/en_GB/features/sustainability. Retrieved 17 Mar 2018
Levi's (2018b), http://store.levi.com/waterless/index.html. Retrieved on 17 Mar 2018
J. Li, *Introduction to animal science* (China Agricultural Press, Beijing, China, 2002). (in Chinese)

S.M. Livingstone, P.K. Lunt, Predicting personal debt and debt repayment: psychological, social and economic determinants. J. Econ. Psychol. **13**(1), 111–134 (1992)

S. Lockie, K. Lyons, G. Lawrence, J. Grice, Choosing organics. A path analysis of factors underlying the selection of organic food among Australian consumers. Appetite **43**(2), 135–146 (2004)

S. Lorek, P.J. Vergragt, Sustainable consumption as a systemic challenge: inter- and transdisciplinary research and research questions, in *Handbook of Research on Sustainable Consumption*, ed. by L.A. Reisch, J. Thøgersen (Edward Elgar Publishing, Cheltenham, UK, 2015), pp. 19–32. ISBN 978-1-78347-126-3

M.L. Loureiro, J. Lotade, Do fair trade and eco-labels in coffee wake up the consumer conscience? Ecol. Econ. **53**, 129–138 (2005)

M.L. Loureiro, J.J. Mccluskey, R.C. Mittelhammer, Will consumer pay a premium for eco-labeled apples? J. Consum. Aff. **36**(2), 203–219 (2002)

D.A. Lubin, D.C. Esty, The sustainability imperative—lessons for leaders from previous game-changing megatrends. Harvard Bus. Rev. **88**, 42–50 (2010)

M.G. Luchs, R.W. Naylor, J.R. Irwin, R. Raghunathan, The sustainability liability: potential negative effects of ethicality on product preference. J. Mark. **74**(5), 18–31 (2010)

L. Lundblad, I.A. Davies, The values and motivations behind sustainable fashion consumption. J. Consum. Behav. **15**, 149–162 (2016)

M.R. Gotlieb, C. Wells, From concerned shopper to dutiful citizen: implications of individual and collective orientations toward political consumerism. Ann. Am. Acad. Polit. Soc. Sci. **644**, 207–219 (2012)

B. Mader, *Is going green too expensive for most people?* (2008), http://www.contractormagcom/columns/editorial/going_green_expensive. Retrieved 11 Mar 2018

J.P. Mäkiniemi, A. Vainio, Barriers to climate-friendly food choices among young adults in Finland. Appetite **74**, 12–19 (2014)

N. Marinelli, Asimmetrie informative e sicurezza alimentare nei diritti del consumatore e nella competitività dei sistemi produttivi (in Italian). J. Agron. **4**, 13–21 (2010)

G.P. Marsh, Man and nature; or physical geography, as modified by human action. The New York Times. 25 July (1864). ISSN 0362-4331. Retrieved on 20 Nov 2017

C. Martin, The sharing economy: A pathway to sustainability or a nightmarish form of neoliberal capitalism? Ecol. Econ. **121**, 149–159 (2016)

M. Max-Neef, Development and human needs, in *Real Life Economics—Understanding Wealth Creation*, ed. by P. Ekins, M. Max-Neef (Routledge, New York, 1992), pp. 197–213. ISBN 978-0415079778

McCrindle Research, *Socially Conscious or Selfish and Materialistic?* (2018), http://www.mccrindle.com.au. Retrieved 10 Mar 2018

P. McDonagh, A. Prothero, Sustainability marketing research: past, present and future. J. Mark. Manage. **30**(11–12), 1186–1219 (2014)

G.D. Meneses, Refuting fear in heuristics and in recycling promotion. J. Bus. Res. **63**(2), 104–110 (2010)

S. Menghini, N. Marinelli, S. Fabbrizzi, *Il consumo responsabile di alcol dei giovani in Toscana: analisi delle caratteristiche del consumatore e studio sulla fattibilità di una strategia di marketing sociale (In Italian)*, in Alcol e Giovani. Disagio sociale, salute e competitività (Franco Angeli, Milano, 2011), pp. 105–172. ISBN 978-8856-84610-2

L. Michaelis, The oxford commission on sustainable consumption. J. Clean. Prod. **11**, 931–933 (2003)

M. Micheletti, A. Follesdal, D. Stolle, *Politics, Products, and Markets. Exploring Political Consumerism Past and Present* (Routledge, Taylor and Francis Group, New York, 2004). ISBN 978-1-4128-0552-0

M. Micheletti, *Political Virtue and Shopping. Individuals, Consumerism, and Collective Action* (Palgrave Macmillan, New York, U.S.A., 2003). ISBN 978-1-4039-7376-4

S. Moeller, K. Wittkowski, The burdens of ownership: reasons for preferring renting. Managing Serv. Qual. **20**(2), 176–191 (2010)

J. Moisander, Motivational complexity of Green consumerism. Int. J. Consum. Stud. **31**(4), 404–409 (2007)

O. Mont, A. Neuvonen, S. Lähteenoja, Sustainable lifestyles 2050: stakeholder visions, emerging practices and future research. J. Clean. Prod. **63**, 24–32 (2014)

J. Morris, M. Marzano, N. Dandy, L. O'Brien, *Forestry, Sustainable Behaviours and Behaviour Change: Setting the Scene* (The Research Agency of the Forestry Commission, 2012), https://www.forestry.gov.uk/pdf/behaviour_review_summary.pdf/$FILE/behaviour_review_summary.pdf. Retrieved 5 Mar 2018

P.E. Murphy, M. Öberseder, G.R. Laczniak, Corporate societal responsibility in marketing: normatively broadening the concept. AMS Rev. **3**(2), 86–102 (2013)

National Research Council, *Sustainability and the U.S. EPA* (The National Academies Press, Washington D.C., U.S.A., 2011). ISBN 978-0-309-21252-6

D. Nazzal, O. Batarseh, J. Patzner, D.R. Martin, Product servicing for lifespan extension and sustainable consumption: an optimization approach. Int. J. Prod. Econ. **142**, 105–114 (2013)

T. Newholm, D. Shaw, Studying the ethical consumer: a review of research. J. Consum. Behav. **6**(5), 253–270 (2007)

B.J. Newman, B.L. Bartels, Politics at the checkout line: explaining political consumerism in the United States. Polit. Res. Q. **64**(4), 803–817 (2010)

L.A. Nielson, P. Paxton, Social capital and political consumerism: a multilevel analysis. Soc. Probl. **57**(1), 5–24 (2010)

I.E. Nikolaou, L. Kazantzidis, A sustainable consumption index/label to reduce information asymmetry among consumers and producers. Sustain. Consumption Prod. **6**, 51–61 (2016)

I.E. Nikolaou, T. Tsalis, A framework to evaluate eco- and social-labels for designing a sustainability consumption label to measure strong sustainability impact of firms/products. J. Clean. Prod. **162**, 105–113 (2018)

H. Nilsson, B. Tuncer, Å. Thidell, The use of ecolabeling like initiatives on food products to promote quality assurance—is there enough credibility? J. Clean. Prod. **12**(5), 517–526 (2004)

S. O'Hara, S. Stagl, Global food markets and their alternatives: a socioecological economic perspective. Popul. Environ. **22**(6), 533–554 (2001)

J. O'Shaugnessy, N.J. O'Shaugnessy, Marketing, the consumer society and hedonism. Eur. J. Mark. **36**(5/6), 524–547 (2002)

OECD, *Water Consumption and Sustainable Water Resources Management*, Report (OECD Publishing, Paris, France, 1998). ISBN 978-9264160828

OECD (Organization for Economic Cooperation and Development), *Meat Consumption* (2018), https://data.oecd.org/agroutput/meat-consumption.htm. Retrieved on 25 Mar 2018

OECD-FAO (Organization for Economic Cooperation and Development-Food and Agriculture Organization), *Agricultural Outlook 2013* (2013), https://www.oecd.org/tad/agricultural-policies/OECD-FAO_Outlook_2013-2022.pdf. Retrieved 25 Mar 2018

K. Ollerer, Environmental education—the bumpy road from childhood foraging to literacy and active responsibility. J. Integr. Environ. Sci. **12**, 205–216 (2015)

J. Owyang, A. Samuel, A. Grenville, *Sharing is the New Buying. How to Win in the Collaborative Economy* (2014), http://www.visioncritical.com/sites/default/files/pdf/sharing-new-buying-collaborative-economy-report.pdf. Retrieved 3 Nov 2017

L.K. Ozanne, J.L. Ozanne, A child's right to plan: the social construction of civic virtues in toy libraries. J. Public Policy Mark. **30**(2), 264–278 (2011)

K. Peattie, Green consumption: behavior and norms. Annu. Rev. Environ. Resour. **35**, 195–228 (2010)

J. Peloza, K. White, J.Z. Shang, Good and guilt-free: the role of self-accountability in influencing preferences for products with ethical attributes. J. Mark. **77**(1), 104–119 (2013)

M. Pepper, T. Jackson, D. Uzzell, An examination of the values that motivate socially conscious and frugal consumer behaviours. Int. J. Consum. Stud. **33**(2), 125–136 (2009)

F. Perrini, S. Castaldo, N. Misani, A. Tencati, The impact of corporate social responsibility associations on trust in organic products marketed by mainstream retailers: a study of Italian consumers. Bus. Strategy Environ. **19**, 512–526 (2010)

A. Perrinjaquet, O. Furrer, J.C. Usunier, G. Cestre, P. Valette-Florence, A test of the quasi-circumflex structure of human values. J. Res. Pers. **41**(4), 820–840 (2007)

C.J. Peters, J.L. Wilkins, G.W. Fick, Testing a complete diet model for estimating the land resource requirements of food consumption and agricultural carrying capacity-The New York State example. Renew. Agric. Food Syst. **22**, 145–153 (2007)

M. Phipps, L.K. Ozanne, M.G. Luchs, S. Subrahmanyan, S. Kapitan, J.R. Catlin, R. Gau, R.W. Naylor, R.L. Rose, B. Simpson, T. Weaver, Understanding the inherent complexity of sustainable consumption: a social cognitive framework. J. Bus. Res. **66**, 1227–1234 (2013)

L. Piscicelli, T. Cooper, T. Fisher, The role of values in collaborative consumption: insights from a product-service system for lending and borrowing in the UK. J. Clean. Prod. **97**, 21–29 (2015)

A. Prothero, S. Dobscha, J. Freund, W.E. Kilbourne, M.G. Luchs, L.K. Ozanne, J. Thøgersen, Sustainable consumption: Opportunities for consumer research and public policy. J. Public Policy Mark. **30**(1), 31–38 (2011)

R. Putnam, *Bowling Alone: The Collapse and Revival of American Community* (Simons and Schuster Paperbacks, New York, U.S.A., 2000). ISBN 978-0-7432-0304-3

W.E. Rees, Ecological Footprints and appropriated carrying capacity: what urban economics leaves out. Environ. Urbanization **4**(2), 121–130 (1992)

L.A. Reisch, J. Thøgersen (eds.), *Handbook of Research on Sustainable Consumption* (Edward Elgar Publishing, Cheltenham, UK, 2015). ISBN 978-1786439277

C. Ricci, N. Marinelli, L. Puliti, The consumer as citizen: the role of ethics for a sustainable consumption. Agric. Agric. Sci. Proc. **8**, 395–401 (2016)

M.L. Richins, F.W. Rudmin, Materialism and economic psychology. J. Econ. Psychol. **15**(2), 217–231 (1994)

M.L. Richins, Materialism, transformation expectations, and spending: implications for credit use. J. Public Policy Mark. **30**(2), 141–156 (2011)

J. Rifkin, *The Age of Access: The New Culture of Hypercapitalism Where all of Life is a Paid-for Experience* (New York, U.S.A, Tarcher-Perigee/Putnam, 2000). ISBN 978-1585420827

J.A. Roberts, Green consumers in the 1990s: profile and implications for advertising. J. Bus. Res. **36**(3), 217–231 (1996)

M.J. Rokeach, *The Nature of Human Values* (Free Press, New York, U.S.A, 1973). ISBN 978-0029267509

D. Roos, R. Hahn, Does shared consumption affect consumers' values, attitudes, and norms? A panel study. J. Bus. Res. **77**, 113–123 (2017)

I. Røpke, The dynamics of willingness to consume. Ecol. Econ. **28**(3), 399–420 (1999)

E. L. Rosenberg, Mindfulness and consumerism, in *Psychology and Consumer Culture. The Struggle for a Good Life in a Materialistic World*, ed. by T. Kasser, A.D. Kanner (American Psychological Association, Washington DC, U.S.A., 2005), pp. 107–125. ISBN 978-1-59147-046-5

H. Rothgerber, A comparison of attitudes toward meat and animals among strict and semi-vegetarians. Appetite **72**, 98–105 (2014)

M.B. Ruby, S.J. Heine, Meat, morals, and masculinity. Appetite **56**(2), 447–450 (2011)

M.B. Ruby, Vegetarianism: A blossoming field of study. Appetite **58**(1), 141–150 (2012)

M. Salo, A. Nissinen, R. Lilja, E. Olkanen, M. O'Neill, M. Uotinen, Tailored advice and services to enhance sustainable household consumption in Finland. J. Clean. Prod. **121**, 200–207 (2016)

M. Schäfer, M. Jaeger-Erben, S. Bamberg, Life events as windows of opportunity for changing towards sustainable consumption patterns? J. Consum. Policy **35**, 65–84 (2012)

C.D. Schewe, G. Meredith, Segmenting global markets by generational cohorts: Determining motivations by age. J. Consum. Behav. **4**(1), 51–63 (2004)

H. Schösler, J. de Boer, J.J. Boersema, Can we cut out the meat of the dish? Constructing consumer-oriented pathways towards meat substitution. Appetite **58**, 39–47 (2012)

U. Schrader, The moral responsibility of consumers as citizens. Int. J. Innov. Sustain. Dev. **2**(1), 79–96 (2007)

M. Schudson, Citizens, consumers, and the good society. Ann. Am. Acad. Polit. Soc. Sci. **611**(1), 236–249 (2007)

S.H. Schwartz, W. Bilsky, Toward a psychological structure of human values. J. Pers. Soc. Psychol. **53**, 550–562 (1987)

S.H. Schwartz, G. Melech, A. Lehmann, S. Burgess, M. Harris, Extending the cross-cultural validity of the theory of basic human values with a different method of measurement. J. Cross Cult. Psychol. **32**, 519–542 (2001)

S. Schwartz, Normative explanations of helping behavior: A critique, proposal, and empirical test. J. Environ. Psychol. **9**(4), 349–364 (1973)

S.H. Schwartz, Universals in the content and structure of values: theoretical advances and empirical tests in 20 countries. Adv. Exp. Soc. Psychol. **25**(1), 1–65 (1992)

S.H. Schwartz, Are there universal aspects in the content and structure of values? J. Soc. Issues **50**, 19–45 (1994)

A. Scott, C. Oates, W. Young, A conceptual framework of the adoption and practice of environmental actions in households. Sustainability **7**, 5793–5818 (2015)

S. Sen, C. Bhattacharya, Does doing good always lead to doing better? Consumer reactions to corporate social responsibility. J. Mark. Res. **38**(2), 225–243 (2001)

D.V. Shah, D.M. McLeod, E. Kim, S.Y. Lee, M.R. Gotlieb, S.S. Ho, H. Breivik, Political consumerism: how communication and consumption orientations drive "Lifestyle politics". Ann. Am. Acad. Polit. Soc. Sci. **611**(1), 217–235 (2007)

S.L. Shapiro, G.E. Schwartz, G. Bonner, Effects of mindfulness-based stress reduction on medical and premedical students. J. Behav. Med. **21**, 581–599 (1998)

R. Sharma, M. Jha, Values influencing sustainable consumption behaviour: exploring the contextual relationship. J. Bus. Res. **76**, 77–88 (2017)

P. Sheahan, *Generation Y: Thriving and Surviving with Generation Y at Work* (Hardie Grant Books, Australia, 2005). ISBN 978-1740663175

J.N. Sheth, N.K. Sethia, S. Srinivas, Mindful consumption: a customer-centric approach to sustainability. J. Acad. Mark. Sci. **39**(1), 21–39 (2011)

S. Shimokawa, Sustainable meat consumption in China. J. Integr. Agric. **14**(6), 1023–1032 (2015)

P. Singer, *Practical Ethics* (Cambridge University Press, 1993). ISBN 0-521-43363-0

K.W. Smola, C.D. Sutton, Generational differences: revisiting generational work values for the new millennium. J. Organ. Behav. **23**, 363–382 (2002)

D. Soman, The mental accounting of sunk time costs: why time is not like money. J. Behav. Decis. Making **14**(3), 169–185 (2001)

J.V.R. Souza, *Verificação da relação entre os valores pessoais e a predisposição ao consumo sustentável (in Portuguese),* M.Sc. dissertation in Business Administration, Universidade Federal do Rio Grande do Sul—UFRGS. Porto Alegre

G. Spaargaren, Sustainable consumption: a theoretical and environmental policy perspective. Soc. Nat. Resour. **16**, 687–701 (2003)

P.C. Stern, Toward a coherent theory of environmentally significant behavior. J. Soc. Issues **56**(3), 407–424 (2000)

D. Stolle, M. Hooghe, The roots of social capital: attitudinal and network mechanisms in the relation between youth and adult indicators of social capital. Acta Polit. **39**(4), 422–441 (2004)

D. Stolle, M. Hooghe, M. Micheletti, Politics in the supermarket: political consumerism as a form of political participation. Int. Polit. Sci. Rev. **26**(3), 245–269 (2005)

P. Sullivan, J. Heitmeyer, Looking at Gen Y shopping preferences and intentions: exploring the role of experience and apparel involvement. Int. J. Consum. Stud. **32**, 285–295 (2008)

J.K. Swim, P.C. Stern, T.J. Doherty, S. Clayton, J.P. Reser, E.U. Weber, R. Gifford, G.S. Howard, Psychology's contributions to understanding and addressing global climate change. Am. Psychol. **66**(4), 241–250 (2011)

C. Tanner, S.W. Kast, Promoting sustainable consumption: determinants of green purchases by Swiss consumers. Psychol. Mark. **20**(10), 883–902 (2003)

M. Tatzel, "Money worlds" and well-being: an integration of money dispositions, materialism and price-related behavior. J. Econ. Psychol. **23**(1), 103–126 (2002)

T. Te'eni-Harari, J. Hornik, Factors influencing product involvement among young consumers. J. Consum. Mark. **27**(6), 499–506 (2010)

The Guardian, *The Guardian International Edition* (2018), https://www.theguardian.com/sustainable-business/blog/rise-generation-y-sustainable-marketplace. Retrieved 10 Mar 2018

J. Thøgersen, Promoting green consumer behavior with eco-labels, in *New Tools for Environmental Protection: Education, Information, and Voluntary Measures*, ed. by T. Dietz, P. Stern (National Academy Press, Washington, DC, 2002), pp. 83–104. ISBN 978-0309084222

J. Thøgersen, Understanding repetitive travel mode choices in a stable context: a panel study approach. Transp. Res. Part A **40**(8), 621–638 (2006)

J. Thøgersen, How may consumer policy empower consumers for sustainable lifestyles? J. Consum. Policy **28**(2), 143–178 (2005)

D. Tilman, M. Clark, Global diets link environmental sustainability and human health. Nature **515**(7528), 518–522 (2014)

J. Tinson, P. Nuttall, Insider trading? Exploring familial intra-generational borrowing and sharing. Mark. Rev. **7**(2), 185–200 (2007)

C. Tobler, V.H.M. Visschers, M. Siegrist, *Eating green.* Consumers' willingness to adopt ecological food consumption behaviors. Appetite **57**(3), 674–682 (2011)

S. Todd, R. Lawson, Towards an understanding of frugal consumers. Australas. Mark. J. **11**(3), 8–18 (2003)

M. Tódero, J. Macke, T.S. Biasuz, O consumo consciente e sua relação com as acções de responsabilidade social (in Portuguese). Revista de Gestão Social e Ambiental **5**(1), 158–175 (2011)

L. Toma, A. McVittie, C. Hubbard, A.W. Stotta, A structural equation model of the factors influencing British consumers' behavior toward animal welfare. J. Food Prod. Mark. **17**(2–3), 261–278 (2011)

H.C. Triandis, Collectivism and individualism as cultural syndromes. Cross-Cult. Res. **27**(3–4), 155–180 (1993)

E. Tsakiridou, C. Boutsouki, Y. Zotos, K. Mattas, Attitudes and behavior towards organic products: an exploratory study. Int. J. Retail Distrib. Manage. **36**(2), 158–175 (2008)

A. Tukker, M.J. Cohen, U. de Zoysa, E. Hertwich, P. Hofstetter, A. Inaba, S. Lorek, E. Sto, The Oslo declaration on sustainable consumption. J. Ind. Ecol. **10**, 9–14 (2006)

A. Tukker, M.J. Cohen, K. Hubacek, O. Mont, The impacts of household consumption and options for change. J. Ind. Ecol. **14**, 13–30 (2010)

A. Tukker, R.A. Goldbohm, A. De Koning, M. Verheijden, R. Kleijn, O. Wolf, I. Perez-Dominguez, J.M. Rueda-Cantuche, Environmental impacts of changes to healthier diets in Europe. Ecol. Econ. **70**(10), 1776–1788 (2011)

United Nations, *Sustainable Development Knowledge Platform* (2018), https://sustainabledevelopment.un.org/rio20. Retrieved 17 Feb 2018

USDA (U.S. Department of Agriculture), *Livestock and Poultry: World Markets and Trade* (2014), http://usda.mannlib.cornell.edu/MannUsda/viewDocumentInfo.do?documentID=1488. Retrieved 25 Mar 2018

SDA (U.S. Department of Agriculture), *Livestock and Poultry: World Markets and Trade* (2017), https://apps.fas.usda.gov/psdonline/circulars/livestock_poultry.pdf. Retrieved 25 Mar 2018

M. van Amstel, P. Driessen, P. Glasbergen, Eco-labeling and information asymmetry: a comparison of five eco-labels in the Netherlands. J. Clean. Prod. **16**, 263–276 (2008)

F. Vanhonacker, E.J. van Loo, X. Gellynck, W. Verbeke, Flemish consumer attitudes towards more sustainable food choices. Appetite **62**, 7–16 (2013)

F. Vanhonacker, W. Verbeke, Buying higher welfare poultry products? Profiling Flemish consumers who do and do not. Poult. Sci. **88**(12), 2702–2711 (2009)

M. Vassallo, M.L. Scalvedi, A. Saba, Investigating psychosocial determinants in influencing sustainable food consumption in Italy. Int. J. Consum. Stud. **40**, 422–434 (2016)

M.C.D. Verain, H. Dagevos, G. Antonides, Flexitarianism. a range of sustainable food styles, in *Handbook of Research on Sustainable Consumption*, ed. by L.A. Reisch, J.B. Thøgersen (Cheltenham: Edward Elgar Publishing, 2015), pp. 209–223. ISBN 978-1-78347-126-3

N. Vercalsteren, T. Geerken, *Need area "households", SusProNet Report WP7* (Tokyo, Japan, 2003)

I. Vermeir, W. Verbeke, Sustainable food consumption: exploring the consumer "attitude—behavioral intention" gap. J. Agric. Environ. Ethics **19**(2), 169–194 (2006)

I. Vermeir, W. Verbeke, Sustainable food consumption among young adults in Belgium: theory of planned behaviour and the role of confidence and values. Ecol. Econ. **64**, 542–553 (2008)

P. Vlaeminck, T. Jiangb, L. Vranken, Food labeling and ecofriendly consumption: experimental evidence from a Belgian supermarket. Ecol. Econ. **108**, 180–190 (2014)

M. Wackernagel, W.E. Rees, *Our Ecological Footprint: Reducing Human Impact on the Earth* (New Society Publishers, Gabriola Island, British Columbia, Canada, 1996)

J.M. Wang, L.C. Wu, The impact of emotions on the intention of sustainable consumption choices: evidence from a big city in an emerging country. J. Clean. Prod. **126**, 325–336 (2016)

L. Watkins, R. Aitken, D. Mather, Conscientious consumers: a relationship between moral foundations, political orientation and sustainable consumption. J. Clean. Prod. **134**, 137–146 (2016)

F.E. Webster, R.F. Lusch, Elevationmarketing: marketing is dead! Long live marketing! J. Acad. Mark. Sci. **41**(4), 389–399 (2013)

A.S. Wellner, Generational divide. Am. Demographics **22**(10), 52–58 (2000)

H. Westhoek, T. Rood, M. Van den Berg, J. Janse, D. Nijdam, M. Reudink, E. Stehfest, The Protein Puzzle. The consumption and production of meat, dairy and fish in the European Union. Eur. J. Food Res. Rev. **1**(3), 123–144 (2011)

M.M. Willis, J.B. Schor, Does changing a lightbulb lead to changing the world? Political action and the conscious consumer. Ann. Am. Acad. Polit. Soc. Sci. **644**, 160–190 (2012)

J. Wirtenberg, D. Lipsky, L. Abrams, M. Conway, J. Slepian, Empowerment, waste and new consumption communities. Organ. Dev. J. **25**(2), 11–21 (2007)

U. Witt, Learning to consume—a theory of wants and the growth of demand. J. Evol. Econ. **11**, 23–26 (2001)

G. Xiao, J.O. Kim, The investigation of Chinese consumer values, consumption values, life satisfaction and consumption behaviors. Psychol. Mark. **26**(7), 610–624 (2009)

H.H. Zhao, Q. Gao, Y.P. Wu, Y. Wang, X.D. Zhu, What affects green consumer behavior in China? A case study from Qingdao. J. Cleaner Prod. **63**, 143–151 (2014)

Q. Zhu, Y. Li, Y. Geng, Y. Qi, Green food consumption intention, behaviors and influencing factors among Chinese consumers. Food Qual. Prefer. **28**, 279–286 (2013)

S. Zukin, J.S. Maguire, Consumers and consumption. Ann. Rev. Sociol. **30**, 173–197 (2004)

I. Zur, C.A. Klöckner, Individual motivations for limiting meat consumption. Br. Food J. **116**(4), 629–642 (2014)

Chapter 5
Regulations and Environmental Management Systems

Abstract Regulations and close management of systems are essential to ensure that the best practices concerning environmental protection are being followed. From private groups such as Greenpeace to governmental entities, performing close monitoring of industries, processes, and procedures capable of environmental harm is essential to promote and maintain environmental safety. However, the range of these regulations can vary from country to country as they are dependent on governmental enforcement, giving rise to the concept of environmental management systems (EMS), which, due to their efficacy, have already been incorporated into the ISO 14000 family of standards. As consumers become more and more demanding, the search for certification also rises in an effort to showcase a company's will to align with the best practices. Besides improving corporate image, companies which implement EMS and work toward certification also see an increase in profit, having some studies shown that certified companies show an economic performance above the market's average. The close collaboration with external certifying entities, when done properly, allows for an optimal company restructure (especially during growing/expansion phases), leading to better process control, cost reductions, and lower environmental impact. This chapter discusses the concept of EMS, its tools and how it is implemented, different standards and regulations currently in practice and entities that enforce them, details on certification processes and how these affect corporate business as well as the concept of environmental indicators.

Keywords Environmental management system · EMS · Regulations · Environmental certifications · ISO 14001 · Environmental audits · Environmental standards · Environmental indicators

The environment can be protected in a coercive way by applying laws that oblige individuals, companies, state agencies, and non-governmental associations to follow established principles. If this holds true for all concerned, companies deserve a prominent place in this group, as their disastrous actions usually take on much larger proportions, requiring greater care, bearing in mind also that there are usually strong economic interests associated. For companies, therefore, standards have been established that limit their polluting or harmful activity to the environment. Thus,

F. J. Gomes da Silva and R. M. Gouveia, *Cleaner Production*,
https://doi.org/10.1007/978-3-030-23165-1_5

when companies comply with those standards, it is automatically ensured that the environment is protected due to the minimum safety parameters established by the standard.

In this way, this chapter will deal with various legislation-related issues but, given its diversity from country to country and constant updating, it does not make sense to dissect this topic, otherwise it will be outdated in the very short term.

5.1 Regulations

Environmental protection has been largely leveraged by the action of numerous environmental organizations, representing the part of society that is actually concerned about how weather conditions, severe climate events and ocean water level rise due to ice melting have been rapidly changing and affecting our lives. Obviously, the political class also shares the concerns usually raised by citizens, but it is a fact that properly organized groups of citizens can cause the necessary political constraints and denounce certain situations that are less obvious but still harmful for the environment, forcing their discussion, creating the necessary political "discomfort" for this class to feel compelled to respond. These organizations normally start with a very clear objective of denouncing or preventing any harmful action on Nature, thereby creating the necessary unity and force for further action in the future. One of the most recognized independent international organizations known for the impact of its actions is Greenpeace, which began in 1971 with a very focused action on the denunciation and influence of public opinion to avoid underground nuclear testing on Amchitka Island, which served as the last refuge to 3000 sea otters in danger of extinction, as well as birds and animals considered rare species. The campaign did not produce the immediate results intended, i.e., the non-detonation of the atomic bomb but created a wave of public opinion protests that was enough for the USA to end all testing on the island, essentially making it into a dedicated site for the preservation of rare birds. Greenpeace's media actions have made it a global organization, currently with more than 2.8 million members worldwide, being distributed throughout 41 countries with its headquarters in Amsterdam, the Netherlands. This organization, as well as other similar ones, tries to maintain their independence from governments and companies, subsidizing their actions via individuals and foundation donations, having as main motto the exposure of harmful activities to the environment, enacted by companies or political actions that fail to comply with established standards.

The creation of Greenpeace coincided with a time when the world began to pay more attention to environmental issues: the early 1970s. In 1972, the United Nations Environmental Program (UNEP) appeared, giving rise to the emergence of several other agencies and organizations which took on responsibilities in the area of environmental preservation. These organizations began serving as a regulating element of the, sometimes promiscuous, relations between governments and companies, while also attempting to independently establish the rules to be followed in environmental terms, through proper monitoring of situations and analysis of the evolution of environmental degradation. Nevertheless, although not having the power to legislate, these organizations play a decisive role in public opinion, in environmental preservation education and in the creation and definition of goals to be fulfilled, pointing to solutions so that these goals can be truly accomplished. It is then up to the political power to establish the necessary international agreements and impose rules, so that these same goals can be effectively achieved.

Several international organizations have been set up with the goal of creating the necessary conditions for solving the environmental problems that all of us face, including the following: European Environment Agency (EEA), US Environmental Protection Agency (US EPA), Earth System Governance Project (ESGP), Global Green Growth Institute (GGGI), Intergovernmental Panel on Climate Change (IPCC), International Union for Conservation of Nature (IUCN), Partnerships in Environmental Management for the Seas of East Asia (PEMSEA), and United Nations Environment Programme (UNEP), among many others. These organizations have worked in a multifaceted way, helping governments to establish the necessary legislation to comply with certain rules established worldwide and particularly in each country, working with consumers and companies to promote the information needed to understand everything that is involved, as well as promote educational actions enabling citizens and businesses to internalize basic principles, moving closer to the desired standards.

One of the agencies that has been working more consistently in the last decades in the interface between legislation and environmental education is EPA, promoting environment awareness sessions focused on specific sectors and issues, disclosure of the general and specific legislation for each sector, as well as divulging success cases concerning companies which are complying with the most updated requirements in environmental terms, as an incentive for others to follow the same steps. Also, EPA economically supports some initiatives in an effort to promote, streamline, and implement certain technological reforms which, while absolutely necessary, cannot always be fully supported by non-compliant enterprises. A dedicated "Laws and Regulations" folder can be found on the Agency's website, which provides the information necessary for companies to be fully aware of the rules they need to comply with. Given the technical character that some laws and regulations need to assume, regulatory power has been conferred to this agency by the American Congress as the agency is technically qualified to convert international agreements into federal regulations so that they can be correctly implemented. Following this step, the agency is still able to proceed to the NAICS (North American Industry Classification System), bringing together in each sector the legislation to be complied with, data on compliance with legislation by sector and cases of success. Thus, compliance with environmental rules is much easier for companies, which can also count on the technical support of the agency for this purpose.

The EEA website is organized differently but has a wide range of information and legislation on the different aspects of the environment, especially in the industry section tab. Being a union of different countries, this agency has cross-border information for the whole European Union, but it does not fail to contemplate the specific information corresponding to each country. It is also provided with a separator with the indicators and data corresponding to all the countries that integrate the European Union, as well as specific information of each country. The range of publications and reports is very broad, allowing the citizens of each country to be properly informed about their duties, as well as about the profile each country presents in meeting the targets set for each environmental parameter. As a positive aspect, the organization and the amount of information available should be highlighted. It should also be noted that the agency also includes countries which are not part of the European Community and which are cooperating countries. These countries, which are normally close to others in the European Union, or even candidates to join the European Union in the future, are already part of this agency and have already developed or are developing legislation and programs to bring them closer to the latest environmental preservation models. As a less positive aspect, it should be noted that there is no strong publicity about the actions developed by the agency, nor is there an adequate dissemination of successful cases of companies and/or countries that could stimulate others through their example. The agency therefore appears to play a much less intervening role in legislative policy and in the education of citizens, businesses and society at large, but rather analyzes, collects, and disseminates data on the environment in the European Union, leaving the legislative part to particular committees attached to the European Commission which, together with politicians, generate and manage legislation. In fact, environmental initiatives are essentially under the responsibility of the European Commission, which, through the Environmental Action Programs (the 7th program aims to achieve the desired results by the year 2020), promotes legislative, control policies and the necessary training/action so that the final objectives can be achieved as initially stipulated.

As previously stated, legislation on the preservation of the environment differs greatly from country to country and is usually more demanding in Northern European countries and has gradually spread to the rest of the European Union. However, this legislation, in compliance with global standards established by the European Union within its territory, or by the Federal Government in the United States of America, is very political in nature, as there is a need to balance income policies with environmental preservation policies. Thus, countries which are highly industrialized and with higher economic levels can establish more restrictive norms without causing serious problems relative to industrial or economic competitiveness. On the other hand, economically more fragile countries with weaknesses in attracting industry to their territory will tend to prefer more lenient legislation.

In global terms, this is even more noticeable since in recent decades there has been a migration of more polluting industries from developed countries, such as the USA or European countries, to countries in Asia and Africa where environmental legislation was less restricting, translating the willingness of politicians to sacrifice in the short term or to mortgage the health of their fellow citizens in the long term, in exchange for immediate economic benefits such as job creation and tax generation. The effectiveness of this type of policy is doubtful but has been followed by countless countries that are eager to develop, accepting to undermine the environment in exchange for faster economic development. However, if in political terms the idea could reap some benefits, the environment is highly impaired. In many cases, these situations would be easily avoidable if, for instances, a material change had been implemented, a product concept change had been done or a technological implementation had been coupled to production system, drastically reducing or even eliminating sources of pollution, allowing these countries to adopt the same technology and principles used in developed countries, where legislation does not allow the environment to be severely sacrificed.

Legislation in particular will not be addressed here, given its extremely volatile nature, as well as a huge dispersion in the degree of demand required by the different countries in the various regions of the world.

5.2 Environmental Managing Systems

Nowadays, for an increasing number of customers, consumers and investors, environmental responsibility for products and services is a basic market requirement. Organizations are expected to comply with environmental standards and demonstrate their commitment to the environment in all day-to-day actions. But these requirements can also be an opportunity for organizations to eliminate waste, risks, and unnecessary costs, while enhancing their values in protecting the environment (Nishitani et al. 2012; Bogue 2014).

Indeed, although environmental damage is caused by us all, the impact caused by companies is usually much more significant and dangerous, as it involves the transformation of materials that are proven to be detrimental to the environment and human health, representing continuous and comprehensive processes which lead to much more significant effects on the environment. In addition, the role of business can be much broader, covering a number of environmental protection stages, ranging from the product design phase to the environmental burden it brings throughout its life cycle, obviously also deeming the manufacturing and distribution phase of the product. More recently, the concept of environmental and social responsibility has been introduced, which has led many companies to redefine their mission and values. In fact, a company does not need to lose its focus on generating value in order to take a more responsible look at the environment in which it is inserted. On the contrary, the company should take this opportunity to reanalyze the development of its products, production, and distribution methods, attempting to eliminate waste, reducing risks, and avoiding waste generation as much as possible.

Thus, companies started to view compliance with environmental laws not as an obligation, but as a business opportunity, as when compliant with these directives, they show a more responsible behavior, which is welcomed by consumers who are more concerned with environmental preservation issues. In order to overcome this issue, the addition of labels, such as eco-labels, has been implemented, displaying the concerns taken during product design, creation, and distribution. However, a label reveals the environmental care taken for a particular product but may not translate the company's overall commitment to the preservation of the environment. To this end, the environmental management standards were created, which, when duly fulfilled, give rise to a compliance certification, effectively translating the company's global concern with the environment. The certification of environmental management imposes the fulfillment of an ever more extensive set of requirements, which have undergone changes in time, due to the extension of the vision created around the aspects of environmental preservation that, as previously mentioned, began by restricting the component of pollution generation in the process, to now include the entire product life cycle. This created the concept of EMS, which became governed in a very particular way by the ISO 14000 family of standards.

5.2.1 Environmental Managing Systems Concept

EMS is a flexible tool that establishes a set of procedures which define an organization's environmental policy and have arisen as a result of what has been called "corporate environmentalism" by Banerjee (2002), i.e., the recognition of environmental concerns by an organization, and the corresponding decision-making regarding the processes and procedures that it incorporates.

As stated by Testa et al. (2014), "An environmental management system (EMS) is a worldwide tool potentially applicable by any kind of organization in order to improve the management of their environmental aspects and reach a continuous improvement of environmental performance." The success achieved by any of these standards shows the high ease of application and the flexibility to adapt to any type of organization, regardless of its type of business, size or deployment.

An EMS is defined by EPA as "a set of processes and practices that enable an organization to reduce its environmental impacts and increase its operating efficiency" (EPA 2018a). The EMS models are based on a process of continuous improvement, which uses Shewhart's cycle designated by the Plan–do–Check–act (PDCA). The EMS should allow the company or organization to:

- Set an appropriate environmental policy for its case;
- Identify the environmental aspects arising from its activities, products, or services in order to determine the significant environmental impacts;
- Identify relevant legal and regulatory requirements;
- Identify priorities and set appropriate environmental goals and milestones;
- Establish a structure and program to implement and achieve the desired goals and milestones;
- Simplify the planning, control, monitoring, corrective actions, and audit tasks and review activities. For example, while ensuring that the policy is being fulfilled, also ensure that the environmental management system remains adequate;
- Be able to adapt to changes over time.

The adoption and implementation of an EMS by a given organization presupposes a significant increase in the quality of procedures related to environmental management, implying a much more accurate recording of all information, more assertive and conscious use of the same information and knowledge on the matter, the need for the implementation of authorizations, well-defined maintenance and operation plans and adequate training of all employees (de Oliveira et al. 2010; Franchetti 2011).

An EMS takes the form of a set of procedures that will allow a company to be aware of legislation issued by respective governments, avoiding penalties while helping maintain a positive image with consumers. However, the concept also encompasses situations as simple as energy saving, which is environmentally friendly and allows companies or organizations to save resources in the short and long term, thereby increasing their economic sustainability. The adoption of an EMS is also an opportunity to improve the reliability of processes, improving overall performance while increasing value generation and creating opportunities in more demanding markets (e.g., markets orientated toward the preservation of the environment) (Hunt and Johnson 1995; Fresner 1998; Wätzold et al. 2001; Hilson and Nayee 2002; Jasch 2003; Ghisellini and Thurston 2005; Thoumy and Vachon 2012). On the other hand, the change in organizational culture and values should allow for greater employee motivation. The adoption of an EMS is only possible if there is a strong motivation from the top management for this purpose, which will have to persuade and motivate all other employees to maintain focus and meet the expected results.

The implementation of EMS through the EMAS and ISO 14000 standards demonstrates the commitment and relevance that certain global structures are giving to environmental protection. Europe, through the European Commission, has strongly encouraged the adoption of this type of certification, giving certain benefits to companies adopting this type of certification, such as tax benefits, less supervision (as the adopted procedures ensure compliance) as well as the extension of the environmental permit (Wätzold et al. 2001).

5.2.2 *Environmental Managing Systems Implementation Steps*

In order to assist companies and organizations in the implementation of EMS, the EPA has defined a sequence of steps to be followed, which are described below (EPA 2018b):

a. *Outline the organization's main goals for EMS*: First of all, the organization must define the reasons why it wants to adopt an EMS. These objectives could only be achieved by complying with all existing environmental legislation, creating a more favorable image for consumers, wanting to expand their products or services to more demanding audiences, among others. The objectives should be clearly described, taking into account the initial situation, and these should often be revisited as progress is made in the process.
b. *Ensure that Top Management is committed to the project*: Getting the commitment of top management is one of the key pillars for successful implementation of an EMS. To this end, a SWOT analysis of the current situation should be made, highlighting the strengths and weaknesses of the organization in the current

context. Then, it is necessary to draw the attention of top management to the economic, environmental and image benefits that the implementation of an EMS can bring to the organization. The remaining hierarchy of the organization should know and realize that top management is committed to the implementation process.

c. *Selection of a project manager*: depending on the size of the company or organization, there may be one or more options for choosing who will lead the project. There are some characteristics that this project manager must have: leadership ability, power of persuasion, strong knowledge about the organization, and its constituents and preferably should have experience working with ISO 9000 and ISO 14000 standards. The project manager should also have strategic characteristics and deep thinking about how to delineate and implement systems. He/She will need to have unconditional support of the top management, time and availability enough to proceed with the structuring and implementation of the system, involving the remaining employees, and implicitly providing support for the project that has to be developed and executed.
d. *Selection of a team for implementation*: In order for the project to achieve its desired success, it is necessary to involve the all department managers of the organization, as they have a deeper knowledge of the utilized processes, which will be an asset in identifying and evaluating the depth of the existing problems, as well as enumerating opportunities for improvement and validating possible solutions. Whenever the size of the organization justifies it and, if possible, external entities with strong relations with the organization should be involved, helping to identify internal opportunities for improvement, as well as with external business partners. At an early stage, frequent meetings will be needed to get the project underway with the necessary direction and focus. If the team has multidisciplinary credentials, the results will be surely better, and implementation will be less likely to be unsuccessful, creating assertive and reliable procedures. This will help make the team feel more motivated and committed to the project.
e. *Promote the project launch meeting*: After the team is formed, it is time to hold an initial project launch meeting, which should clearly define the goals to be achieved and the expected benefits to the organization, from the outset defining the initial stages of the project. The different tasks should also be distributed among team members and their responsibilities. The presence of top management at this meeting will help the whole team realize the importance of the project and the commitment of top management to the implementation of the system, motivating the team. This is usually also a good opportunity to standardize knowledge on environmental management within the team, giving them some basic training for the early stages of implementation. The remaining collaborators should be aware of this initial meeting, in order to understand and collaborate and help make feasible the implementation of the system.
f. *Comparative analysis with standardized EMS*: The organization shall take as a comparison the self-assessment procedures provided by ISO 14001 to analyze the conformity of its organizational structure, suitability of its processes and procedures, environmental and social impacts, established policies for the preservation

of the environment and programs in this area. This preliminary analysis will be extremely useful in understanding exactly what needs to be done and what needs to be changed.

g. *Budgeting and timetable*: Based on the results of the preliminary assessment, the draft action, as well as the budget for implementation, should be prepared. The project must describe in detail all the phases by which it is constituted, as well as the actions to be taken, in which they are constituted and their respective drivers. The milestones should be carefully distributed over time, leaving some time lag. The project must be structured in such a way that some initial successes can be easily obtained, which can galvanize the team into the remaining missing work.
h. *Top Management commitment and approval*: After the plan is outlined, it must be submitted to top management for review and approval, as well as making available the necessary budget. This action will involve top management, strengthening the implementation of the plan. Whenever there is a possibility of external financing for the implementation of the project, it should be considered and requested.
i. *Involvement of employees*: Success and timely achievement of established goals will largely depend on employee involvement. Knowing the processes, environmental impacts, risks, safety problems and how to overcome them, the cooperation of employees is a key part of the process, as they can collaborate with new ideas, validate new procedures and improvement of overall effectiveness, due to the in-depth knowledge of the processes they hold.
j. *Monitoring and information on the process evolution*: the project should be built in order to contain monitoring processes, to check the progress made. Once each milestone is achieved, the results should be communicated to top management as well as to the entire team and collaborators so that they can feel a greater involvement with the process and an extra motivation to continue with the necessary changes.

All the previous steps just correspond to the PLAN step of the PDCA cycle. After that, it is time to pass to the action (DO step). The major actions are described as follows.

a. *Identification of legal requirements*: At the beginning of the process, it is necessary to survey all the legislation related to the products and processes that the organization incorporates so that all are contemplated and fulfilled through the EMS in implementation.
b. *Identification of sources of environmental concern*: All interactions with the environment must be identified and selected to be truly important. Then, one must look for the source of the problem, in order to be able to eradicate the problem at source, through an adequate management of the procedures and processes.
c. *Getting opinion from the stakeholders*: Stakeholders' opinion is very important in the EMS certification process, as it will empower third parties to emit an opinion on the organization's environmental performance, i.e., how those who relate to the organization, through business relations or the simple neighborhood, feel happy or affected by the performance of the company. Given that an assessment

has already been made of the legal and environmental aspects of the organization, there is nothing to fear, and the dialog will certainly be very constructive in terms of the present situation and future prospects.

d. *Establishing the environmental policy of the organization*: Based on the three previous steps, the organization should now be able to establish an environmental management policy, taking into account all steps previously taken. Thus, the organization must also create the appropriate mechanisms to monitor its own performance in environmental terms, based on the factors that are most important to its stakeholders.
e. *Responsibility distribution*: Once the organization's environmental policy is fully established and properly described through corresponding procedures, it is time to start allocating responsibilities to people, beginning with the top management. It will be necessary to establish who has management responsibilities, who leads operations and who belongs to the team responsible for the environment. After assigning roles and responsibilities, it is time to engage people, getting their contribution to setting goals and milestones.
f. *Establishing goals and milestones*: The objectives and goals to be achieved should be consistent with legislation, with the opinion of stakeholders and with any other analysis relating to the environmental performance of the company. Regarding the objectives set by the organization, adequate resources should be allocated to this objective, taking into account the profile of the allocated human resources and their motivation to achieve those goals. It will also be important to define which departments and sectors of the organization will have direct or indirect responsibility in pursuing these goals.
g. *Identifying the control parameters and monitoring needs*: As the various components of the environmental management process are implemented, it is now necessary to establish how and which parameters should be monitored. Through experience gained in identifying needs and establishing procedures, it will now be easier to understand what to control and when to do so. The proper establishment of milestones will also imply that ways of measuring performance are planned and whether these partial goals have been successfully achieved. Controlling and monitoring are crucial systems to evaluate the progress of the planned actions. It may also become necessary to create new control systems during the progress of the action plan, depending on the complexity and need for monitoring understood as necessary for the success of the project.
h. *Corrective actions, activity log and control documents*: The implementation reports generated by the EMS begin to accumulate and it becomes necessary to find an effective way to compile the gathered information. It is still natural that the program suffers setbacks or disruptions, which creates the need to establish contingency plans and corrective actions. Basically, it is about gathering all the information and making the system move forward, solving any problems that may arise.
i. *Monitoring procedures*: Taking into account the control parameters previously defined and the monitoring needs, it is now necessary to define the procedures that allow the operatives to obtain the data necessary for such monitoring. The

interviewing of such operators can provide the best insights on how to operationalize data collection for monitoring.

j. *Monitoring responsibilities*: Although the procedures for collecting information are already defined, it is very important to define who has responsibilities in their analysis on a managerial level, so that any deviation can be immediately identified and corrective actions taken.
k. *Structuring and inducing employee awareness*: For the quality management system to succeed, worker awareness is a crucial factor. Training should start with the environmental responsibilities of society, as well as with each one of us, evolving into the enumeration of the advantages of integrating an EMS, how an EMS works and what requirements it is subject to. Awareness should cover the entire workforce, but it is the responsibility of process managers to determine whether it is a priority for all, including temporary workers.
l. *Updating procedures*: The initially established procedures may no longer correspond to the evolving progression of the system. Therefore, it is necessary to review procedures so that they comply with the current requirements of the system, with the evolution of data generation and processing and with the greater maturity of the organization and its employees toward the EMS. At this stage, it is already possible to establish procedures for possible unforeseen situations, audits, detailed reports for top management, etc., which may be more demanding due to the greater training and increased predisposition of employees toward the system.
m. *EMS manual elaboration*: After all the preparation work previously developed, an EMS manual can be easily established, which will bring together the already reviewed procedures, describing the documentation that will result from the system, the different levels of responsibility and the employees assigned to them, as well as other elements important to this manual. The connections between the different elements of the system and the documentation must also be clearly defined. However, the manual should be as simple as possible, facilitating its analysis and making it useful to establish the interrelationship between the system's different interlocutors.
n. *System enhancement*: Once all procedures have been implemented, all documentation has been generated and the environmental management system has been put into action, it is now necessary to refine employee training so that they integrate the entire system in the best way possible, being able to perceive and collaborate in the different fronts of the process, integrating operational teams or monitoring the operation of the system. In this phase, employees should be able to test the operation of the EMS, as well as test its robustness, contributing systematically to its improvement and adapting to the natural evolution of the process.

After completing the DO phase, it is now necessary to go through the system test phase, looking for any gaps.

This stage involves internal audits to verify the effectiveness of the process. To this end, internal auditors capable of developing this type of activity should have been previously trained. The audits should comprehensively search the information and see if it has properly followed procedures. Auditors should also check the procedures to see if they are appropriate to the reality and needs of the organization. The initial audits should have a predominantly pedagogical function and should go up in level, increasing the level of demand and sophistication in the search for nonconformities. The eventual identification of nonconformities should give rise to due records, and the corresponding list of actions to be developed in a corrective and preventive sense so that the same situation does not repeat itself.

After the verification phase, it is time to act. Based on the results of the audits carried out in the previous phase, as well as based on the information generated by the EMS itself, it will be necessary to identify situations that do not yet correspond to the performance that the organization aspires to reach. Based on this analysis, the revisions to be made to the initially drawn plans should be listed in order to improve the system in the desired direction. Such revisions should be appropriately allocated to those already in the system, while maintaining consistency of functions. After reviewing all the situations detected as being improvable by top management, it is time to revisit the Planning steps, thus initiating a new iteration in the PDCA cycle.

5.3 Environmental Certifications and Audits

5.3.1 ISO Environmental Standards

Usually, a standard is a rule that must be respected and that allows to adjust certain conducts or activities. However, there are a lot of definitions for standard, which relate the term to a quality pattern or a model to be used in a comparative measurement. Nevertheless, a curious definition is given by Brunsson and Jacobsson (2000) regarding a standard: "pieces of general advice offered to a large number of potential adopters".

Standards can be viewed in a variety of ways, depending on context, political setting or market circumstances (Hilson 2003; Rao and Holt 2005; Khan 2006; Hicks and Dietmar 2007; Baas 2007; Gianetti et al. 2008; Calia et al. 2009; Shi et al. 2010; Klemes et al. 2012; Silva et al. 2013; Scandelari and Cunha 2013). In most cases, the adoption of standards is a voluntary act that is at the discretion of each entity or organization. However, there are also a large number of situations in which entities or organizations are forced to adopt certain standards by market or law enforcement. A third way is also quite common, when a large number of similar organizations adopt a certain standard, their competitors need to adopt the same rules in order to remain competitive and submit similar arguments.

ISO has established a technical committee to develop international standards on a wide range of aspects related to environmental management. This technical committee, ISO/TC 207, aims to develop and update the ISO 14000 series of standards, which covers the following areas: Environmental Management Systems,

Environmental Audits, Environmental Performance Assessment, Ecological Labeling, Life Cycle Analysis, Environmental Aspects in Product Standards, as well as Terms and Definitions. In this context, six Sub-Committees have been created with the associated ISO 14000 standards, as follows:

- TC 207/SC 1—Environmental Management Systems (ISO 14001 and ISO 14004);
- TC 207/SC 2—Environmental Audits (ISO 19011);
- TC 207/SC 3—Ecological Label (ISO 14020, ISO 14021, ISO 14024 and ISO 14025);
- TC 207/SC 4—Environmental Performance Assessment (ISO 14031 and ISO 14032);
- TC 207/SC 5—Product Life Cycle (ISO 14040, ISO 14041, ISO 14042 and ISO 14043);
- TC 207/SC 6—Terms and Definitions (ISO 14050).

The ISO 14000 series is a set of standards directed to the environmental management of companies of any level, size, or region. These standards have the main objective of creating an environmental management system in a company, reducing possible environmental damages. ISO 14000 standards are focused on curtailing environmental injuries. Its main goal is not to make a company in a fundamentalist way into an "ecological organization" but rather create a philosophy of continuous improvement in terms of environmental respect, through its EMS, and comply with all environmental policies and regulations. The development of an EMS varies accordingly to the company or organization, so two companies certified under the ISO 14000 standards may have differentiated performances, but both are qualified. The adoption of an EMS can be voluntary by a company/organization, or it may be due to market pressure or even public opinion, when there are facts that lead to the widespread concern of a large group of people affected by the practices of these companies or organizations. However, a company that has an ISO 14000 certificate gets significant advantages, either for the customer or for itself (Copeland and Le Clue 1999; Petroni 2001; Pun and Hui 2001; Raines 2002; Hilson 2003; Hillary 2004; Rao and Holt 2005; Khan 2006; Hicks and Dietmar 2007; Baas 2007; Sambasivan and Fei 2008; Gianetti et al. 2008; Fortunski 2008; Calia et al. 2009; Shi et al. 2010; Klemes et al. 2012; Silva et al. 2013; Hibadullah et al. 2013; Scandelari and Cunha 2013). Upon receiving an ISO 14001 certification, a company is logically associated with an international environmental management standard, which brings to the public a positive image of a clean and environmentally conscious company.

The reasons identified as most important for the endorsement of ISO 14001 as a means of certifying that companies are concerned with the environment and have their EMS performing perfectly, may be as follows (Whitelaw 2004; Noia and Nicoletti 2016):

- Gain or maintain market share by showing environmental concern, being confirmed by independent entities;
- Attract investments that are only possible due to the respect of values related to the environment;

- Reduce activity risk levels, lowering insurance premiums;
- Reduce the probability of environmental damage or breaching of laws/regulations and consequential financial finds;
- Reduce global costs, increasing competitiveness.

Nevertheless, many other motivations can be added so that companies choose to certify their EMS according to ISO 14001. Thus, companies also adhere to the ISO 14001 standard as they can attain financial incentives from governmental or public bodies, obtaining easier and better condition toward bank financing, easier access to permits, among others. Certification has been hugely successful and is particularly useful for organizations operating in environmentally sensitive fields, such as the pulp and paper industry, which have a significantly higher environmental footprint and whose repercussions in economic and image terms in the event of an environmental accident, are also much more severe (Hazudin et al. 2015). However, there is also the possibility that some companies may only consider certification as an attempt of creating a new brand image ("greenwashing"), without there being any real conviction that this process can contribute positively to the general well-being of humanity, and that this process, if well conducted, can bring significant efficiency gains to the organization, generating benefits that can perfectly cover the inherent processing costs. Other organizations that have narrower analytical capabilities understand that this type of certification only generates more records and circulation of information (often still on paper), without generating any economic benefit to the organization (Petroni 2001; Noia and Nicoletti 2016). If properly conducted, ISO 14001 certification can bring significant gains to companies, but to that end, it must be properly planned and executed. It is something that requires continuous action and, for its initial implementation, is a process that can take between 8 and 19 months to be successfully implemented (Babakri et al. 2004), requiring a significant investment in services of certifying entities. The documentation and training of collaborators is also necessary to better understand the main objectives of the initiative, to cooperate with it in the best way possible, and to allow the organization to function smoothly, complying with the laws and regulations of each country, enforcing data recording and continuous improvement of the EMS, with the objective of further improving results. Consequently, a rise of efficiency allows companies to save significant resources and a renew their image with customers, translating into competitive advantages over other players in the market. Obviously, these restrictions may be limiting for many small and medium enterprises (SMEs), being then unfeasible when weighing against the benefits that can be extracted from certification. However, if the system is implemented during the growth phase of a company, the organization will help manage this growth, allowing the company to focus on the essential while ensuring that growth is achieved in a sustainable manner.

However, companies need to meet certain requirements before they can proceed to the certification process. Thus, it is necessary for organizations to have: (a) the general requirements of an organization; (b) properly defined environmental policy; (c) Planning of improvement actions (EMS with objectives and goals duly traced);

(d) implementation and operation (as previously described for the EMS); (e) Ability to verify and promote the necessary corrective actions and possible review of EMS management.

5.3.2 EMAS and ISO 14001: Concurrence and Convergence

There are essentially two standards governing environmental management systems: the ISO 14001 standard, developed by a world-class private entity, the International Organization for Standardization, and another promoted by the European Union, known as the EMAS—Eco-Management and Audit Scheme, through Regulation EC 1221/2009. The first version of the EMAS standard was established in 1993, while the initial version of the ISO 14001 standard appeared only in 1996. ISO 14001 standard was developed having as base the BS 7750 standard issued by British Standard Institute in 1994, which laid down rules and procedures for the development, implementation and maintenance of EMS systems in order to ensure and demonstrate the conformity of its activity with the declaration of principles concerning its environmental policy, its objectives, and targets to be achieved. However, the launch of the ISO 14001: 2004 standard dictated the end of the BS 7750 standard, which became BS EN ISO 14001: 2004 (Campos et al. 2015).

Although they have a common focus, EMAS and ISO 14001 follow slightly different paths to achieving their intended purpose, i.e., establishing the principles of an EMS. Having these standards come out three years apart, at first, there was some rivalry between the adoption of one or the other standard (Morrow and Rondinelli 2002). This led to a European Union guideline which showed how to establish the link/equivalence between the two standards, with a view to facilitating the adoption of one or the other, fulfilling the requirements which the European Union demanded mainly for companies based on other continents that wanted to certify their companies located in Europe. However, in 2001 the European Commission officially recognized ISO 14001 as the standard for the regulation of EMS's, including in the revised version of EMAS (EMAS II) the entire text referring to ISO 14001 as an annex. In the revision of the EMAS I regulation for EMAS II, there was also an update regarding the scope of the standard, leaving the second version to be less focused on industrial aspects, moving into a more global scope, following what was already practiced with ISO 14001, making it applicable to any type of organization. The real convergence came with the second revision of the EMAS (EMAS III) standard in 2009, which now includes regulation for non-European countries. Despite these convergence efforts, there are fundamental differences that are difficult to blur, which can be summarized as follows (Daddi et al. 2010; Neugebauer 2012; Testa et al. 2014):

- **Nature of standards**: the ISO 14001 standard has been developed and managed by a private body, while the EMAS standard is managed and updated by the European Union in close cooperation with the competent national authorities, presenting by this way supra-organizational interests, while acknowledging the exemption that is absolutely necessary in these cases. However, the dynamics and the association of ISO with the standardization offices of each country have allowed the exemption of the evaluators to be made analogous, making the two standards comparable. Anyway, ISO is not supervised by any public body.

- **Area of application**: While the EMAS standard was only focused on the European Union until it was subject to the second revision (EMAS III), the ISO 14001 standard had since its first edition a universal character. This is reflected in the application of these standards, since the ISO standard is the only one to be applied in countries such as the USA or Japan and is also preferred by the overwhelming majority of multinational companies. The EMAS standard remains very regional in nature, usually linked to European companies, most of which are located in European space or, more rarely, in multinationals from Europe and that are located in countries that recognize this standard as equivalent to ISO 14001.
- **External communication**: EMAS imposes much stricter requirements in terms of external communication than ISO 14001, as well as continuous improvements in the environmental field, whereas ISO 14001 just imposes continuous improvement of the EMS. Organizations certified by the EMAS standard are obliged to update and make available annually a document called "Environmental Statement," which should reveal some indicators of perfectly defined performance, as well as the main environmental goals it proposes for the near future, in addition to other information considered relevant to the analysis of its EMS process. This data will have to be appraised by environmental entities duly accredited for this purpose, through audits. Thus, the EMAS scheme can also be considered as a tool that allows to make clear the organizations' environmental commitments to their stakeholders.
- **Scope extension**: The EMAS standard can be applied in practice not only within business organizations, but also in more generic applications, such as local authorities and groups of organizations, and an appeal is made in Article 37 for these groups to be encouraged on a regional basis.
- **Worldwide usage**: Taking into account the numbers published by the two certification bodies in 2016 and 2017, the use of the ISO 14001 standard is much more common, with around 346,147 certifications, 120,595 of which in Europe, while EMAS was only used as a reference 9140 times (EMAS 2017; ISO 2017). These numbers contrast with those known in 2008 and 2010 by the same institutions, which were 188,000 and 8000, respectively, showing that although both are rising, adherence to ISO 14001 continues to be stronger (Neugebauer 2012).

Since the end of the last century, a significant number of studies have focused on the motivations that may be at the basis of adopting certification as one of the ways to ensure the successful implementation and improvement of organizations' EMS, which include the increase financial gains, company image improvement, environmental performance, and operational efficiency enhancements (Klassen and McLaughlin 1996; Boudouropoulos and Arvanitoyannis 1999; Morrow and Rondinelli 2002; Melnyk et al. 2003; Bansal and Hunter 2003; Gonzalez-Benito and Gonzalez-Benito 2005; Chan and Wong 2006; Link and Naveh 2006; Boiral 2007; Gavronski et al. 2008; Lopez-Gamero et al. 2010; Campos 2012; Prajogo et al. 2012). The adoption of standards for EMS by organizations is based on internal and external factors. The main external factors could be the organizational state of the company and relationship with its stakeholders, the institutional environment in which it is inserted and the fact that if it is already certified by ISO 9001 or not, which gives companies a higher internal operation maturity, habituation to data recording and familiarity with the fulfillment of externally established requirements and audits. As internal factors, it is usually pointed out that certification can bring an increased market visibility, with the corresponding increase in market share, and the identification of problems and corresponding solutions, which can bring increased productivity gains and competitiveness. Above all, certification must be seen as a strategic decision to improve the effectiveness of the company in an increasingly competitive market. Nevertheless, it is very important to bear in mind that an EMS membership of an organization, by itself, does not ensure that the organization performs well in environmental terms. In order to ensure that this happens effectively, periodic monitoring must be carried out with the necessary depth, so that any gaps in the system can be detected and new targets can be established, in order to continue the process of continuous improvement established by the standards, through a PDCA methodology. To this end, companies should be periodically audited by accredited outside entities, ensuring exemption and quality of evaluation. It is at this point that certification assumes particular relevance by helping companies in this process and by externally ensuring that the company has its well-established environmental goals and is complying with them.

The number of studies on the perceived utility of the EMAS standard is much narrower, largely because the standard has been confined to a very limited geographical area by 2010, and because most organizations have preferred to use ISO 14001 standard (Hoffman et al. 2003; Iraldo et al. 2009). As in the case of studies concerning the perception of the usefulness of ISO 14001, the results did not follow a well-defined trend, having obtained very favorable results in some cases, and less positive in others (Wagner 2002; Daddi et al. 2011).

However, following the conclusions of Neugebauer (2012), it can be affirmed that companies are forced to be certified by the ISO 14001 standard mainly due to the pressure of their stakeholders, while the organizations that choose EMAS do it very strictly by their own conviction.

5.3.3 Perceptions Concerning Environmental Certification

The analysis of the certification processes has deserved countless studies in order to realize its true usefulness (Zenga et al. 2005; Newbold 2006; Gomez and Rodriguez 2011; Boiral and Henri 2012; Nishitani et al. 2012; Comoglio and Botta 2012; Zobel 2013). However, opinions have diverged significantly, and some of them do not support certification. Indeed, Boiral (2003) states that the intervention of certifying entities is merely symbolic, analyzing, and detecting possible gaps in the EMS organization but does not promote a better practice with respect to environmental issues.

On the other hand, Sandholtz (2012) states that organizations only seek certification to legitimize their policy and environmental management system. This conclusion is confirmed by other authors who, after having developed similar studies, verified that certification does not bring the desired added value (Welch et al. 2003; King et al. 2005; Christmann and Taylor 2006; Barla 2007; Boiral 2007; Boiral and Henri 2012). However, Castka and Balzarova (2018) found that there is also a proactive attitude by a significant number of companies that are not limited to a passive attitude toward the EMS certification process, seeking to take full advantage of this process through more precise enforcement of laws, but also aiming to reap benefits in terms of efficiency. This conclusion is corroborated by many other studies also developed in this area (Pun and Hui 2001; Melnyk et al. 2003; Potoski and Prakash 2005a; Goh-Eng et al. 2006; Russo 2009; Iwata et al. 2010; Gasbarro et al. 2012).

The environmental certification granted by an entity outside the organization usually has a very significant impact, since it independently certifies the commitment and practices of the company in this matter, since it is a voluntary submission to an evaluation carried out by third parties (Terlaak 2007; Brouwer and van Koppen 2008; Lozano 2012; Heras-Saizarbitoria and Boiral 2012; Castka and Balzarova 2018). Certifying entities are usually seen as a service provider and it is expected they have a high degree of professionalism, bringing real benefits to the organization, mainly through efficiency gains, but also through a renewed image that captures a greater number of customers and increases sales.

In a study developed by Heras-Saizarbitoria et al. (2011), on a sample of ISO 14001 certified companies, the authors found that these same companies have a commercial and economic performance above the market average in each sector. Moreover, Oliveira et al. (2016) consider that companies that have an ISO 14001 certified EMS have greater advantages in the implementation of CP programs and thus have a much more proactive position in the management of their business taking into account the principles of safeguarding the environment. In turn, Iraldo et al. (2009) detected a direct relationship between companies that had EMAS (Eco-Management and Audit Scheme) certification and increased consumer satisfaction. On the other hand, in a study carried out by Potoski and Prakash (2005b), these showed that some managers are typically skeptical about environmental certification by ISO 14001 as they do not believe that certification has a decisive impact on consumers, considering that the marketing effort carried out in parallel with the certification process is not enough to produce this effect. Indeed, certification can be advertised through marketing actions and labels placed on products however, consumers usually want more information. In fact, the environmentally conscious consumer wants to know how a company takes environmental issues into account, from product development to disposal, whether in landfill, incineration, reuse or recycling, and also to know if the organization has its structure set up for continuous improvement regarding environmental performance (Nicolau and Sellers 2002).

Certification by third parties may leave some doubts for the consumer and other regulatory authorities. In fact, if an external entity places many obstacles on the certification process, the company can dispense with the services of this certification entity and deliver its process to another entity that does not pose as many problems.

Therefore, it is necessary to standardize the performance of these certifying entities, through minimum levels of certification that are not ambiguous, allowing greater transparency in the whole process that, from the above, is permissive to external factors of pressure that may condition the results. In fact, especially in re-certification processes, there is supposed to be an effective improvement of the organization's behavior toward environmental issues. However, it is more difficult to accurately quantify the actions taken in the meantime, their effectiveness and efficiency, and whether it was possible to go much further or not (Lal 2004; Castka 2013; Castka and Balzarova 2018). In fact, Castka and Balzarova (2018) studied the relationship between organizations and certification bodies and found that contracted interventions and audits are the most common and most requested services by companies from certification bodies. During the certification process, the organization can assess how the certification body approaches the process in different strands, such as the quality of the audit report carried out by the certifying entity, as well as the ability of the certifying entity to adapt itself to align with the main expectations of the organization to be certified. For example, these expectations may be due to the coherence between reports from different auditors concerning different facilities of the same company working under the same type of procedures. Another way to evaluate certifying entities is to analyze the veracity of the statements produced in the reports. Indeed, if top management knows some of the company's operational flaws and the external audit report reveals that everything is fine, then surely the organization will lose confidence in the certifying entity.

5.3.4 Environmental Indicators

Environmental indicators are conceptual management tools that allow a more coherent assessment of the performance of organizations as individual entities or as a group. These tools enable academics and practitioners to assertively analyze available data and to make policy decisions on development to give to legislation and other types of regulations in order to achieve collectively defined performance levels. The indicators should be correctly defined by objectives, with a goal of facilitating further action. With regard to the environment, the more general indicator should reflect how sustainable the development is, being a global concern. Environmental performance indicators depend directly or indirectly on the quality of the environment, reflecting the environmental performance of companies in the vicinity (Tocchetto and Tocchetto 2004). These indicators allow, depending on the current situation and based on the historical records, to assess future trends and outline possible scenarios, which help to draw up strategies, outline policies and set short, medium- and long-term goals. In addition, they also make it possible to assess the effectiveness of the already implemented policies and compare their results with other organizations or groups of competing organizations.

In order to standardize the environmental performance indicators, ISO introduced in 1999 the ISO 14031—Environmental Management and Environmental Performance Evaluation Guidelines standard, which defines more than a hundred indicators that can be used for this purpose, establishing in detail the guidelines for the use of these indicators. These same indicators can be divided into two main groups, presenting very different scopes: (a) Environmental condition indicators, which are essentially intended to characterize the quality of the environment in a given location or region, usually starting from data automatically collected in strategically located stations and treated so that authorities can widespread information regarding the current environmental conditions, prompting interventive actions if necessary, and (b) environmental performance indicators, which are more oriented toward the evaluation of organizations. This type of indicators can still be divided into two subgroups: management indicators and operational indicators. Management indicators essentially aim to measure how management is contributing positively toward the performance of the organization as a whole, while the operational indicators are essential to assess how processes and operations are contributing to the environmental performance of the organization and how it is progressing.

Regarding the Environmental condition indicators, some of the world's most important entities, exclusively linked to the environment (e.g., EPA) or only as observers of the economic development of humanity (e.g., OECD), or others that study the development of the environment for scientific reasons or development of specific programs (e.g., NASA), conduct monitoring on a global or regional scale, providing data that becomes vital for the establishment of governmental and intergovernmental policies. In order to facilitate data comparison between regions or between perfectly established time periods, a standardization of the data is required, which can be perfectly ensured by the calculation of the indicators previously defined for this purpose.

References

L. Baas, To make zero emissions technologies and strategies become a reality, the lessons learned of cleaner production dissemination have to be known. J. Clean. Prod. **15**, 1205–1216 (2007)

K.A. Babakri, R.A. Bennett, S. Rao, W. Franchetti, Recycling performance of firms before and after adoption of the ISO 14001 standard. J. Clean. Prod. **12**, 633–637 (2004)

S.B. Banerjee, Corporate environmentalism: the construct and its measurement. J. Bus. Res. **55**, 177–191 (2002)

P. Bansal, T. Hunter, Strategic explanations for the early adoption of ISO 14001. J. Bus. Ethics **46**, 289–299 (2003)

P. Barla, ISO 14001 certification and environmental performance in Quebec's pulp and paper industry. Journal of Environmental Economics and Management **53**, 291–306 (2007)

R. Bogue, Sustainable manufacturing: a critical discipline for the 21st century. Assembly Autom. **34**(2), 117–122 (2014)

O. Boiral, J.F. Henri, Modelling the impact of ISO 14001 on environmental performance: a comparative approach. J. Environ. Manage. **99**, 84–97 (2012)

O. Boiral, ISO 9000: outside the iron cage. Organ. Sci. **14**(6), 720–737 (2003)

O. Boiral, Corporate greening through ISO 14001: a rational myth? Organ. Sci. **18**, 127–146 (2007)

I. Boudouropoulos, I. Arvanitoyannis, Current state and advances in the implementation of ISO 14000 by the food industries: comparison of ISO 14000 to ISO 9000 to other environmental programs. Trends Food Sci. Technol. **9**, 395–408 (1999)

M.A.C. Brouwer, C. van Koppen, The soul of the machine: continual improvement in ISO 14001. J. Clean. Prod. **16**(4), 450–457 (2008)

N. Brunsson, B. Jacobsson, The contemporary expansion of standardization, in *A World of Standards*, ed. by N. Brunsson, B. Jacobsson (Oxford University Press, Oxford, New York, U.S.A., 2000), pp. 1–19. ISBN 978-0-19-925695-2

R.C. Calia, F.M. Guerrini, M. de Castro, The impact of six sigma in the performance of a pollution prevention program. J. Clean. Prod. **17**, 1303–1310 (2009)

L.M.S. Campos, D.A.M. Heizen, M.A. Verdinelli, P.A.C. Miguel, Environmental performance indicators: a study on ISO 14001 certified companies. J. Clean. Prod. **99**, 286–296 (2015)

L.M.S. Campos, Environmental management systems (EMS) for small companies: a study in Southern Brazil. J. Clean. Prod. **32**, 141–147 (2012)

P. Castka, M.A. Balzarova, An exploration of interventions in ISO 9001 and ISO 14001 certification context—a multiple case study approach. J. Clean. Prod. **174**, 1642–1652 (2018)

P. Castka, Audit and Certification: what Do Users Expect? Joint Accreditation System of Australia and New Zealand (2013), https://www.bsigroup.com/LocalFiles/EN-AU/JAS%20ANZ%20Reports/ISO%2014001%20and%209001%20Report.pdf. Retrieved on 05 May 2018

E.S.W. Chan, S.C.K. Wong, Motivations for ISO 14001 in the hotel industry. Tour. Manag. **27**(3), 481–492 (2006)

P. Christmann, G. Taylor, Firm self-regulation through international certifiable standards: determinants of symbolic versus substantive implementation. J. Int. Bus. Stud. **37**, 863–878 (2006)

C. Comoglio, S. Botta, The use of indicators and the role of environmental management systems for environmental performances improvement: a survey on ISO 14001 certified companies in the automotive sector. J. Clean. Prod. **20**, 92–102 (2012)

A.M. Copeland, S. Le Clue, EMS as a mainstream business tool: Perspectives from Hong-Kong. Journal of Environmental Assessment Policy and Management **1**(2), 159–175 (1999)

T. Daddi, F. Testa, F. Iraldo, A cluster-based approach as an effective way to implement the ECAP (Environmental Compliance Action Program): evidence from some good practices. Local Environ. **15**, 73–82 (2010)

T. Daddi, M. Magistrelli, M. Frey, F. Iraldo, Do environmental management systems improve environmental performance? empirical evidence from Italian companies. Environ. Dev. Sustain. **13**(5), 845–862 (2011)

O.J. de Oliveira, J.R. Serra, M.H. Salgado, Does ISO 14001 work in Brazil? J. Clean. Prod. **18**, 1797–1806 (2010)

EMAS, Official Statistics of the European EMAS Helpdesk – Evolution of Organisations and Sites (Oct 2017). http://ec.europa.eu/environment/emas/emas_registrations/statistics_graphs_en.htm. Retrieved on 08 May 2018

Environmental Protection Agency, Environmental Management System (EMS) (2018). https://www.epa.gov/ems. Retrieved on 29 Mar 2018

Environmental Protection Agency, Guide to Developing an Environmental Management System – Plan (2018). https://www.epa.gov/ems/guide-developing-environmental-management-system-plan. Retrieved on 29 Mar 2018

B. Fortunski, Does the environmental management standard ISO 14001 stimulate sustainable development? Manag. Environ. Qual. Int. J. **19**(2), 204–2012 (2008)

M. Franchetti, ISO 14001 and solid waste generation rates in US manufacturing organizations: an analysis of relationship. J. Clean. Prod. **19**, 1104–1109 (2011)

J. Fresner, Cleaner production as means for effective environmental management. J. Clean. Prod. **7**(6), 171–179 (1998)

F. Gasbarro, F. Rizzi, M. Frey, The mutual influence of environmental management systems and the EU ETS: findings for the Italian pulp and paper industry. Eur. Manag. J. **31**, 16–26 (2012)

I. Gavronski, G. Ferrer, E.L. Paiva, ISO 14001 certification in Brazil: motivations and benefits. J. Clean. Prod. **16**, 87–94 (2008)

A. Ghisellini, D.L. Thurston, Decision traps in ISO 14001 implementation process: case study results from Illinois certified companies. J. Clean. Prod. **13**(8), 763–777 (2005)

B.F. Gianetti, S.H. Bonilla, I.R. Silva, C.M.V.B. Almeida, Cleaner production practices in a medium size gold-plated jewelry company in Brazil: when little changes make the difference. J. Clean. Prod. **16**, 1106–1117 (2008)

A. Goh-Eng, Z. Suhaiza, A.W. Nabsiah, A study on the impact of environmental management system certification towards firms' performance in Malaysia. Manag. Environ. Qual. Int. J. **17**, 73–93 (2006)

A. Gomez, M.A. Rodriguez, The effect of ISO 14001 certification on toxic emissions: an analysis of industrial facilities in the North of Spain. J. Clean. Prod. **19**, 1091–1095 (2011)

J. Gonzalez-Benito, O. Gonzalez-Benito, A study of the motivations for the environmental transformation of companies. Ind. Mark. Manag. **34**(5), 462–475 (2005)

S.F. Hazudin, S.A. Mohamad, I. Azer, R. Daud, H. Paino, ISO 14001 and Financial Performance: Is the Accreditation Financially Worth It for Malaysian Firms. Proc. Econ. Finan. **31**, 56–61 (2015)

I. Heras-Saizarbitoria, O. Boiral, ISO 9001 and ISO 14001: towards a research Agenda on management system standards. Int. J. Manag. Rev. **15**(1), 47–65 (2012)

K. Heras-Saizarbitoria, J.F.M. Molina-Azorin, G.P.M. Dicck, ISO 14001 certification and financial performance: selection effect versus treatment-effect. J. Clean. Prod. **19**, 1–12 (2011)

S.N. Hibadullah, N.F. Habidin, N.M. Fuzi, A.F.N.C. Desa, F.I.M. Zamri, ISO 14001 performance efforts and environmental performance in Malaysian automotive industry. J. Environ. Sci. Comput. Sci. Eng. Technol. **2**(1), 129–138 (2013)

C. Hicks, R. Dietmar, Improving cleaner production through the application of environmental management tools in China. J. Clean. Prod. **15**, 395–408 (2007)

R. Hillary, Environmental management systems and the smaller enterprise. J. Clean. Prod. **12**(6), 561–569 (2004)

G. Hilson, V. Nayee, Environmental management system implementation in the mining industry: a key to achieving cleaner production. Int. J. Miner. Process. **64**(1), 19–41 (2002)

G. Hilson, Defining "cleaner production" and "pollution prevention" in the mining context. Miner. Eng. **16**(4), 305–321 (2003)

E. Hoffman, K. Ankele, A. Ziegler, K. Rennings, J. Nill, The Influence of the EU environmental management and auditing scheme on environmental innovations and competitiveness in Germany: an analysis on the basis of case studies and a large-scale survey. ZEW Discussion Papers 03-14, ZEW - Zentrum für Europäische Wirtschaftsforschung/Center for European Economic Research (2003). https://ideas.repec.org/p/zbw/zewdip/959.html. Retrieved on 06 May 2018

D. Hunt, C. Johnson, *Environmental Management Systems: Principles and Practice* (McGraw-Hill, London, 1995). ISBN: 13:978-007 7079109

F. Iraldo, F. Testa, M. Frey, Is an environmental management system able to influence environmental and competitive performance? the case of the eco-management and audit scheme (EMAS) in the European Union. J. Clean. Prod. **17**, 1444–1452 (2009)

ISO, The ISO survey of management system standard certifications 2016, Geneva (2017). https://isotc.iso.org/livelink/livelink?func=ll&objId=18808772&objAction=browse&viewType=1. Retrieved on 08 May 2018

K. Iwata, T. Arimura, S. Hibiki, An empirical analysis of determinants of ISO 14001 adoption and its influence on toluene emission reduction. J. Contemporary Eur. Res. **62**, 16–38 (2010)

C. Jasch, The use of environmental management accounting (EMA) for identifying environmental costs. J. Clean. Prod. **11**(6), 667–676 (2003)

Z. Khan, Cleaner production: an economical option for ISO certification in developing countries. J. Clean. Prod. **16**(1), 22–27 (2006)

A.A. King, M.J. Lenox, A. Terlaak, The strategic use of decentralized institutions: exploring certification with the ISO 14001 management standard. Acad. Manag. J. **48**, 1091–1106 (2005)

R.D. Klassen, C.P. McLaughlin, The impact of environmental management on firm performance. Manag. Sci. **42**(8), 1199–1214 (1996)

J.J. Klemes, P.S. Varbanov, D. Huising, Recent cleaner production advances in process monitoring and optimization. J. Clean. Prod. **34**, 1–8 (2012)

H. Lal, Re-engineering the ISO 9001:2000 certification process. ISO Manag. Syst. **4**, 15–17 (2004)

S. Link, E. Naveh, Standardization and discretion: does the environmental standard ISO 14001 Lead to performance benefits? IEEE Trans. Eng. Manag. **53**(4), 508–519 (2006)

M.D. Lopez-Gamero, J.F. Molina-Azorin, E. Claver-Cortes, The potential of environmental regulation to change managerial perception, environmental management, competitiveness and financial performance. J. Clean. Prod. **18**, 963–974 (2010)

R. Lozano, Towards better embedding sustainability into companies' systems: an analysis of voluntary corporate initiatives. J. Clean. Prod. **25**, 14–26 (2012)

S.A. Melnyk, R.P. Sroufe, R. Calantone, Assessing the impact of environmental management systems on corporate and environmental performance. J. Oper. Manag. **21**, 329–351 (2003)

D. Morrow, D. Rondinelli, Adopting corporate environmental management systems: motivations and results of ISO 14001 and EMAS certification. Eur. Manag. J. **20**, 159–171 (2002)

F. Neugebauer, EMAS and ISO 14001 in the German industry e complements or substitutes? J. Clean. Prod. **37**, 249–256 (2012)

J. Newbold, Chile's environmental momentum: ISO 14001 and the large-scale mining industry e Case studies from the state and private sector. J. Clean. Prod. **14**, 248–261 (2006)

J.L. Nicolau, R. Sellers, The stock market's reaction to quality certification: empirical evidence from Spain. Eur. J. Oper. Res. **142**(3), 632–641 (2002)

K. Nishitani, S. Kaneko, H. Fujii, S. Komatsu, Are firms' voluntary environmental management activities beneficial for the environment and business? an empirical study focusing on Japanese manufacturing firms. J. Environ. Manag. **105**, 121–130 (2012)

E. Noia, G.M. Nicoletti, ISO 14001 certification: benefit, costs and expectations for organization. Studia Oeconomica Posnaniensia **10**(4), 94–109 (2016)

J.A. Oliveira, O.J. Oliveira, A.R. Ometto, A.S. Ferraudo, M.H. Salgado, Environmental management system ISO 14001 factors for promoting the adoption of cleaner production practices. J. Clean. Prod. **133**, 1384–1394 (2016)

A. Petroni, Developing a methodology for analysis of benefits and shortcomings of ISO 14001 registration: Lessons from experience of a large machinery manufacturer. J. Clean. Prod. **9**(4), 351–364 (2001)

M. Potoski, A. Prakash, Green clubs and voluntary governance: ISO 14001 and firms' regulatory compliance. Am. J. Polit. Sci. **49**(2), 235–248 (2005a)

M. Potoski, A. Prakash, Covenants with weak swords: ISO 14001 and facilities' environmental performance. J. Pol. Anal. Manag. **24**, 745–769 (2005b)

D. Prajogo, A.K.Y. Tang, K.H. Lai, Do firms get what they want from ISO 14001 adoption? an Australian perspective. J. Clean. Prod. **33**, 117–126 (2012)

K.F. Pun, I.K. Hui, An analytical hierarchy process assessment of the ISO 14001 environmental management system. Integr. Manuf. Syst. **12**, 333–345 (2001)

S.S. Raines, Implementing ISO 14001 – an international survey assessing the benefits of certification. Corporate Environ. Strategy **9**(4), 418–426 (2002)

P. Rao, D. Holt, Do green supply chains lead to competitiveness and economic performance? Int. J. Oper. Prod. Manag. **25**(9), 898–916 (2005)

M.V. Russo, Explaining the impact of ISO 14001 on emission performance: a dynamic capabilities perspective on process and learning. Bus. Strategy Environ. **18**, 307–319 (2009)

M. Sambasivan, N.Y. Fei, Evaluation of critical success factors of implementation of ISO 14001 using Analytic Hierarchy Process (AHP): a case study from Malaysia. J. Clean. Prod. **16**(13), 1424–1433 (2008)

K.W. Sandholtz, Making standards stick: a theory of coupled vs decoupled compliance. Organ. Stud. **33**(5–6), 655–679 (2012)

V.R.N. Scandelari, J.C. Cunha, *Ambidestralidade e desempenho socioambiental de empresas do setor eletroeletrônico (in Portuguese)*, RAE – Revista de Administração de Empresas **53**(2), 183–198 (2013)
H. Shi, M. Chertow, Y. Song, Developing country experience with eco-industrial parks: a case study of the Tianjin economic-technological development area in China. J. Clean. Prod. **18**(3), 191–199 (2010)
D.A.L. Silva, I. Delai, M.A.S. Castro, A.R. Ometto, Quality tools applied to cleaner production programs: a first approach toward a new methodology. J. Clean. Prod. **47**, 174–187 (2013)
A.K. Terlaak, Order without law? the role of certified management standards in shaping socially desired firm behaviors. Acad. Manag. Rev. **32**(2), 968–985 (2007)
F. Testa, C. Coetsiera, E. Carretiera, M. Ennahalib, B. Laborieb, P. Moulin, Recycling a slurry for reuse in chemical mechanical planarization of tungsten wafer: effect of chemical adjustments and comparison between static and dynamic experiments. Microelectron. Eng. **113**, 114–122 (2014)
M. Thoumy, S. Vachon, Environmental projects and financial performance: exploring the impact of project characteristics. Int. J. Prod. Econ. **140**(1), 28–34 (2012)
M.R.L. Tocchetto, A.L. Tocchetto, *Indicadores de desempenho ambiental para empresas com atividade galvânica (in Portuguese)*, in Congresso Acadêmico sobre o Meio Ambiente e Desenvolvimento do Rio de Janeiro, CADMA, Rio de Janeiro, Brazil, 9–10 Dec 2004
M. Wagner, The relationship between environmental and economic performance of firms and the influence of ISO 14001 and EMAS: an empirical analysis and implications for government policy. Paper presented at the 5th environmental management accounting network Europe (EMAN-Europe) annual conference, University of Gloucestershire, Cheltenham, UK, 2002, pp. 11–12
F. Wätzold, A. Bültmann, M. Eames, K. Lulofs, S. Schucht, EMAS and regulatory relief in Europe: lessons from national experience. Eur. Environ. **11**, 37–48 (2001)
E.W. Welch, S. Rana, Y. Mori, The promises and pitfalls of ISO 14001 for competitiveness and sustainability: a comparison of Japan and the United States. Greener Manag. Int. **44**, 59–73 (2003)
K. Whitelaw, *ISO 14001 Environmental Systems Handbook*, 2nd edn. (Elsevier Butterworth-Heinemann Ltd, Oxford, UK, 2004). ISBN 978-0-75-064843-1
S.X. Zenga, C.M. Tamb, V.W.Y. Tamb, Z.M. Dengb, Towards implementation of ISO 14001 environmental management systems in selected industries in China. J. Clean. Prod. **13**, 645–656 (2005)
T. Zobel, ISO 14001 certification in manufacturing firms: a tool for those in need or an indication of greenness? J. Clean. Prod. **43**, 37–44 (2013)

Chapter 6
Cleaner Production Tools and Environmental Management Practices

Abstract When Cleaner Production is sought after in the initial design phase of a product, all intrinsic and outcoming steps are optimized, allowing to create more environmentally friendly products, eco-designed products. Indeed, it is during the design phase that important decisions must be made, such as which processes and materials should be chosen, as these can affect the reusability/recyclability of a product. Recycling can be the revamping a product once its life cycle has ended, allowing it to perform, once again, similar functions as the ones it was created for. There are also other actions such as downcycling, upcycling, reusing, and so on, which have a similar goal as recycling but differ in the outcome. However, not all products are created equal or perform the same functions under the same conditions thus, their recyclability is affected. Tires, for instance, are a tough product to recycle due to materials and manufacturing processes employed. In the past, tires have been thrown in landfills in an "out of sight, out of mind" attitude; however, more recently a reverse phenomenon is happening where specific industries are reclaiming tires from landfills. This is because recycling of tires is gaining new outputs, allowing to incorporate them into other products or processes such as road asphalt, composite materials, or even energy production. Moreover, a process called pyrolysis shows great promise in this field as it enables the recovery of raw material and energy in an unprecedented way. Beyond tires, plastics and electronic products, which have become part of our day-to-day life, are also a large environmental concern due to the sheer volume that is produced yearly and lack of end-of-life solutions. So, when designing a product, and when possible, the designer/engineer should consider reusability option of the product once it completes its life cycle, with the goal of having the product perform exactly the same functions as before. This is seen in certain industries that collect water from rivers as part of their manufacturing process, which after its use, should strive to return it in the same or even better conditions as it was collected. Currently, the enforcement of a concept called environmental producer responsibility is helping achieve higher levels of recycling and reusing, enabling the commitment of industries to help minimize waste. In fact, following an eco-design and incorporating a Circular Economy strategy, industries can feed a part of their needs based on recycled materials and goods. This chapter details the concepts of recycling, recycling policies, drivers for recycling and examples; the concept of reuse and where it can be applied; environmental producer responsibility; and Circular Economy and eco-design.

F. J. Gomes da Silva and R. M. Gouveia, *Cleaner Production*,
https://doi.org/10.1007/978-3-030-23165-1_6

Keywords Cleaner Production tools · Environmental management practices · Recycling · Reuse · Reduce · 3R · Environmental producer responsibility · Circular Economy · Eco-design · Life cycle assessment · LCA · WEEE · Landfill

The methodology used to achieve Cleaner Production cannot be approached from the perspective of just limiting environmental damage during or after the manufacturing of products or provision of services. Environmental concerns should start as early as the product design stage and should extend all the way to the end-of-life phase of a product, encompassing all intermediate stages, including manufacturing, assembly, use, and even possible reuse and overwhelm of a product.

Figure 6.1 is a very simple diagram of the main input and output variables of a product manufacturing system. In this simple system, it is easily perceived that, when acting on input variables, the process and output variables will be conditioned, which may promote more or less favorable results in relation to the goals that are intended to be achieved. Therefore, a principle should be well established: If we start acting at the beginning of the system, taking care of situations such as excessive energy consumption, excess materials use, use of difficult-to-decompose materials, recycle or reuse, among many others, we can hardly come up with an optimal solution. Therefore, the first phase of the system, product design, is of particular importance in the product life cycle, being closely linked to how much a product can be environmentally friendly.

In Fig. 6.2, it is possible to view the existence of three main tools that can help to obtain Cleaner Production, which are the following: (a) source reduction, (b) recycling, and (c) product modification. The first case, source reduction, can be further subdivided into (a1) good housekeeping and (a2) process change, and this one can be achieved also following the different ways of: (a1) raw material change, (a1z) equipment modification, and (a1w) technology modification. On the other hand, recycling can be divided into (b1) on-site reuse or recovery and (b2) creation of useful by-products.

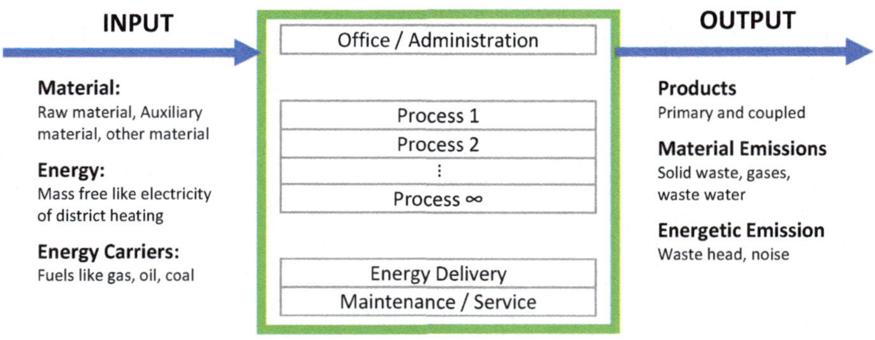

Fig. 6.1 Simplistic approach to a manufacturing process, its input, and output variables

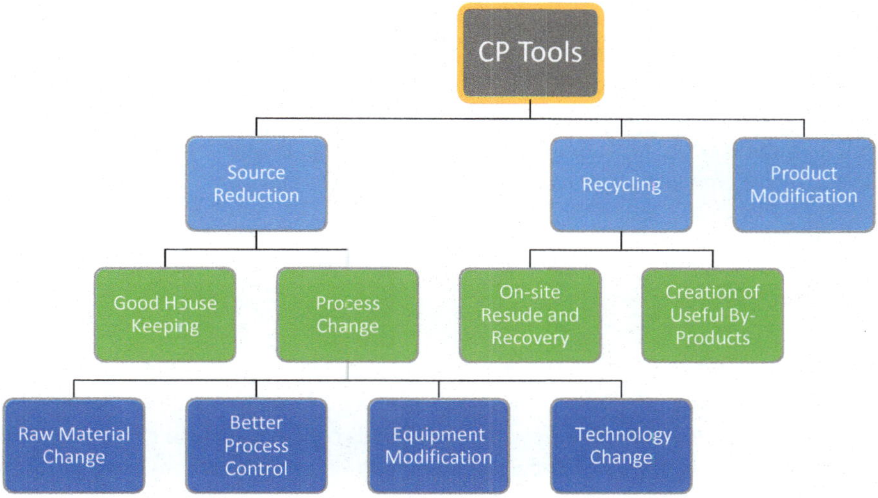

Fig. 6.2 Cleaner Production tools

Source reduction is essentially related to the means used to manufacture a product, not the product itself. Thus, it is essential that the productive means promote low resource waste, even when production systems are already optimized. However, poor equipment maintenance/management or the generation of waste during manufacturing will contribute negatively to the environment, a situation that should be avoided. On the other hand, if a process is not optimized, a change of processes should be considered in order to save resources and avoid the generation of effluents that are harmful to the environment. This situation often occurs in many types of industries where, after new technological developments occur, or only through the updating of equipment, it is possible to save energy, spend less water, reduce or eliminate the use of harmful chemicals for the environment, or radically change the approach to product creation, achieving a mix of all these situations. This is a situation that must be faced very assertively whenever there is marked technological progress in a given business sector, or when there are significant developments around the raw materials used and which directly affect the processes, or when a simple organization of productive processes allows efficiency gains that also translate into significant resource savings, changing only how the management of the production process is conducted. Better control of processes can also bring significant gains in terms of spent resources, or a significant reduction of waste released by a given process, acting only in the control. This may be a recurring situation in terms of equipment maintenance, but it may also be linked to the way equipment normally operates and can significantly increase its efficiency through more effective control.

Concerns about sustainability initially gave rise to policies designated as 3R, which consisted essentially of the following three components: reduce, reuse, and recycle. However, more recent studies have added another R, corresponding to the recover action. In the following chapters, emphasis will be placed on recycling, although the problem of reuse is also addressed, albeit with less emphasis. The recover action, due to its enormous diversity, will be treated in a slight way, with more emphasis on the reduce action, which is directly linked to product design, as well as some specific tools to act in this field, which in the view of the authors, can have a very significant importance in the decades to come.

6.1 Recycling

Recycling is also a complex subject as it involves very specific techniques for each type of product and material. Thus, this chapter will initially highlight the policies related to the incentive to recycling, then will focus on the attitude of citizens toward recycling, and only then will focus on some particular cases of recycling of more common products and materials.

6.1.1 Recycling Concept

Recycling essentially aims to recover materials that are contained in products which have reached their end-of-life and consequently are collected as waste in an organized or random manner (Worrell and Reuter 2014). Moreover, recycling is one of the central pillars of the so-called Circular Economy (Allwood 2014; Kirchherr et al. 2017; Grimaud et al. 2018). Products have a finite life cycle, which is defined as the time in which a product is able to accomplish the functions for which it was designed, without significant loss of performance. After the product's "end-of-life," there is a long way through the recycling chain, as shown in Fig. 6.3.

The recovery of materials can be classified in three different ways (Chini 2007; Calkins 2008): downcycling, recycling, and upcycling. In the case of downcycling, newer generation products are of inferior quality compared to those of previous generations, having a lower estimated life and so, less market attractiveness, negatively affecting resell prices of the reconverted material. The designation of recycling is mainly used when the new generation of material has the same or very similar charac-

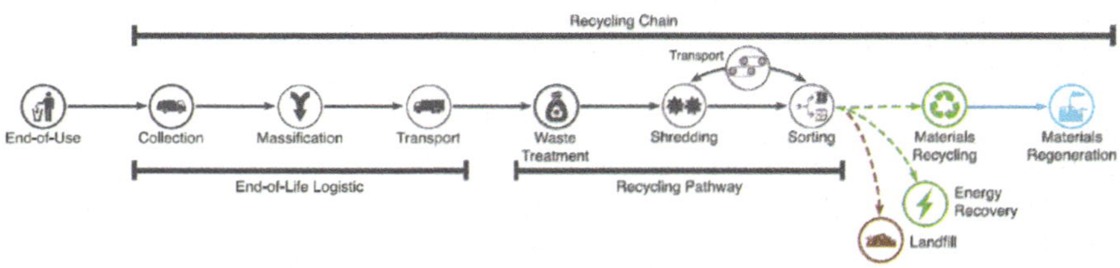

Fig. 6.3 Different steps of the recycling chain. Adapted from Grimaud et al. (2018)

teristics to those patented by the previous generation of the same material. Upcycling refers to situations in which the material, once reconverted, has better characteristics than in the previous generation, presents a longer life cycle, or has a higher market value. Nature is a good example of recycling, for instances, the sun contributes to the triggering of chemical reactions for the decomposition of organic materials into nutrients to be used for example as plant fertilizes. Another example of Nature's capabilities is physical transformations such as the cyclical reconversion of water. In this way, a considerable amount of waste can still be recovered, fueling the life cycle of other living creatures. Vefago and Avellaneda (2013) present in their work a very interesting classification on the different ways materials can be reused, dividing into four main categories:

- Recycled material: all materials which have undergone at least one chemical transformation or change in the physical state and still retain many of the initial properties shall be considered as recycled. It is not obligatory for the new function of the material to be the same as it was for the previous life cycle. This group includes metals, which, depending on the quality of the collected material, can be considered as almost applicable in all situations.
- Infracycled material: An infracycled material or element is one that has undergone at least a chemical transformation of its structure or has been altered in its physical state. Its initial properties are depreciated and do not have to perform the same functions performed in the previous life cycle. Polymers based on hydrocarbons are included in this category as they see initial properties degrade upon each recycling cycle.
- Reused: A material or element reused is one that does not suffer any chemical transformation or undergoes any change in its physical state. It should maintain its initial properties, although it does not necessarily have to perform the same functions it performed in its previous life cycle. Wood is often reused, going from larger to smaller shapes, but maintaining similar properties, although may perform completely different functions, e.g., wood that makes up pallets is often reused in other smaller applications.
- Infraused: An infraused material or element is one which neither undergoes any chemical transformation of its structure nor changes its physical state. However, their properties in the following life cycle can no longer reach the same level of the ones showed in the previous life cycle, so they will have to be used in different, less demanding applications. In this group are included several materials that lose characteristics and, for example, are incorporated into bituminous road mats, allowing even wear resistance and quite interesting water drainage functions.

Some biological/natural materials typically present a delayed infrautilization, such as wood containing resins, or ceramics, since after their first life cycle they are normally used in less important applications. However, there are other materials that mix natural and synthetic/technological components, called hybrids. An example of materials within this class is wood agglomerates, in which their natural base (wood) is bound by a resinous matrix (technological part), making its later treatment difficult. Plastics are also often infrautilized due to the loss of physical and mechanical properties at each recycling stage. Some materials, even natural ones, do not allow for easy reuse or recycling as they no longer can assume their initial state, leaving incorporation into aggregates as the only solution. Ceramic products are part of this group that, even though starting from natural resources, are subject to transformations that prevent them from returning to the initial state. On the other hand, there are other materials that allow a much wider choice of use, which may be carried out by reuse, recycling, or infrarecycling. Glass falls is in this category, and however, also presents some limitations. For example, the glass used in the optical application requires a high degree of quality and specific optical transmission properties, which can only be obtained with first-grade raw material. However, the glass used in food packaging and other less demanding applications can be from a recycled source.

As stated before, a given material that has reached the end of its first life cycle may be exploited in different ways, depending on its versatility of use. Figure 6.4 illustrates different ways to reuse wooden or metallic beams (Vefago and Avellaneda 2013), showing certain methods can be more profitable and environmentally friendly, while others may be more in line with what the local market is in need of. Usually, conversion costs, availability, and market demand determine the best route to follow.

Toward the end of a life cycle, the materials that make up a product can be reused, recycled, incinerated, or landfilled. Landfill disposal is the worst of all solutions as materials will still conserve their internal energy, not benefiting society in anyway, while simultaneously posing soil, water, and air contamination risks (Vefago and Avellaneda 2013). Moreover, landfills entail social costs, being initially estimated by Palmer et al. (1997) at 33 USD/ton, having then been updated to 5–9 USD/ton by Kinnaman (2006). Some materials, such as plastics, allow the partial release of energy through combustion, thereby allowing energy conversion. However, combustion may generate significant quantities of toxic solid wastes and gaseous pollutants. Thus, reuse and recycling routes emerge as the most environmentally advantageous. Reuse is the most environmentally favorable path because, after inputting a small amount of energy/work, it will be possible to readapt the product or material to new functions. This route is usually the one that generates less environmental impact. On the other hand, we also have recycling, which has been used for centuries mainly in the reconversion of metallic products (Fu et al. 2017). Recycling allows materials to be reclaimed through the intensive incorporation of energy and transformed into a perfectly usable raw material for new products (Björklund and Finnveden 2005). It should also be noted that the production of goods from end-of-life materials is a less intensive process in terms of emissions than the same process from virgin raw material (Acuff and Kaffine 2013). Moreover, the recycling process requires a considerable effort in the separation of products/families of materials, which is dependent on the citizens' goodwill, education and citizenship spirit, as well as on the transport and delivery logistics of products/materials in an organized way to centers where the recycling process can be carried out. The logistics necessary for this process is complex and generates in return other environmental burdens, namely transport pollution, energy consumption during the recycling process and waste and effluents also created during this process transformation. These factors need to be considered in terms of economic and environmental viability (Sultan et al. 2017).

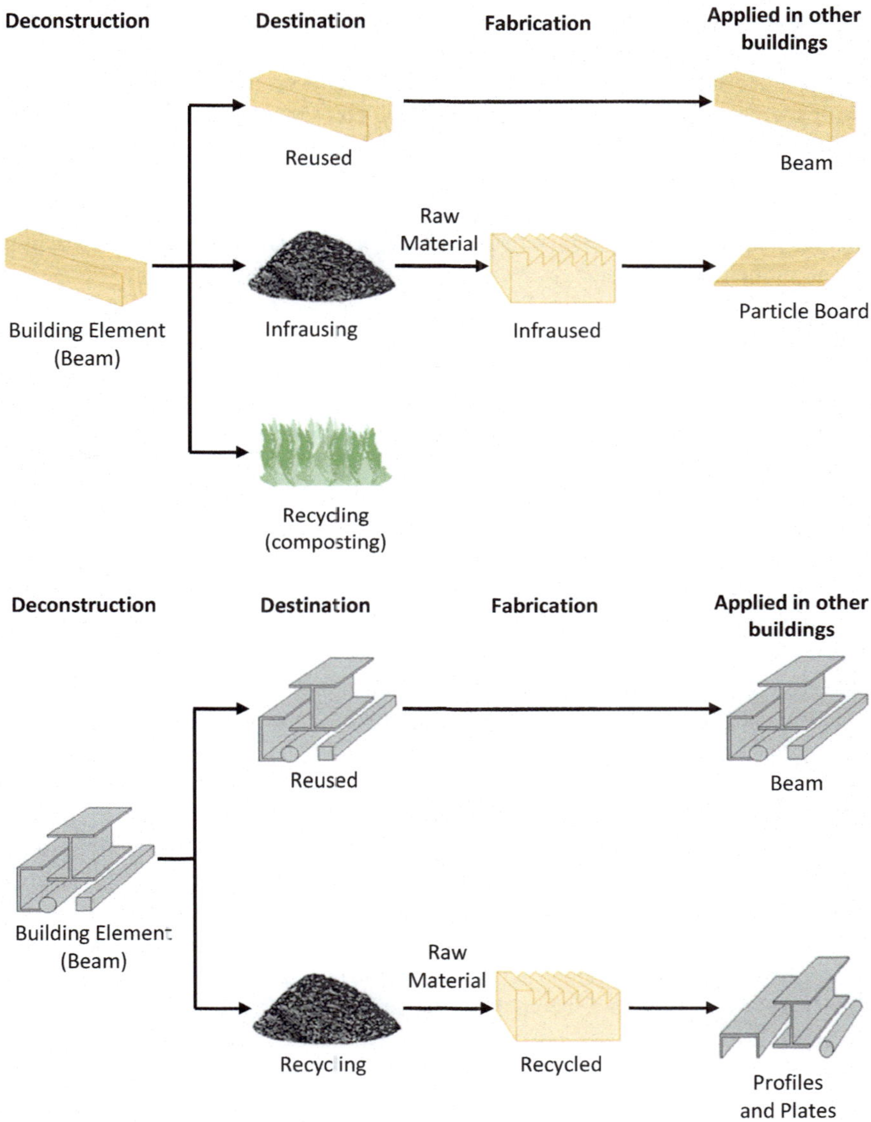

Fig. 6.4 Different ways of harnessing wooden or metallic beams. Adapted from Vefago and Avellaneda (2013)

Regarding metals, the characteristics of recycled material are the same as or very similar to those initially presented by the material in its new state, contrasting to other materials, that as discussed previously, may see their properties reduced (e.g., paper, glass, or thermoplastics). In Nature's biosphere, the same also happens. Materials considered biodegradable, gently spread into the biosphere without giving rise to toxic products. Braungart and McDonough (2008) designated these cycles as technical and biological, respectively. These authors created the "cradle-to-cradle" theory in which materials, when returned to their origins, gain a new life after reaching the end of their previous life cycle.

Furthermore, human activity is always associated with a burden on the environment, which may have a greater or lesser impact on the damage caused. Nature has the capacity of regenerating some of the negative impact caused, but that capacity has limits. The disorderly growth of the world's population, coupled with increasing consumption patterns and regional asymmetry of population concentration, has generated marked imbalances between Nature's regeneration capacity and the harm caused to it by human activity. The globalization of businesses has also contributed to the acceleration of energy consumption, raw material consumption, and consequent waste generation (Vefago and Avellaneda 2013).

The concept of recycling is to transform used objects into new products, enabling the preservation of the environment and favoring sustainable economic development. The incentive to recycling in Japan, as well as some appropriate legislation to this end, has had very significant effects in several areas. In fact, municipal waste deposited in landfills declined from 109 million tons in 2000 to only 51 million tons in 2009. During the same period, the recycled material gradually increased from 70 to 95 million tons, which proves the maturity of the population, as well as the effect of the legislation implemented in the meantime (Yamaguchi and Takeuchi 2016). Indeed, a citizen can produce usually between 1.8 and 5 kg of garbage daily. Adding to the entire world production, the numbers become frightening. As reported by Wan et al. (2015), the population in Hong Kong has grown about 36% in the last three decades. However, the amount of solid waste generated increased by 80%. The construction sector alone is responsible for the generation of 480 kg of garbage per person per year, with about 75% of this waste being deposited in landfills (Chandra 2005). Moreover, companies also contribute largely to the increase in these numbers when they produce goods which people need to consume. The degradation of Nature caused by litter has generated serious climate changes in the last years and can create serious problems for future generations. Taking into account population increase, today's world should be concerned with preserving what surrounds us: natural reserves, flora, and fauna. Nowadays, there are technologies for the treatment of waste and recycling of products, not only for the improvement of life quality as well as for the protection of Nature. Thus, it is necessary to be aware of the importance of recycling and waste avoidance, taking into account the practice of 3Rs:

- Reduce the amount of waste;
- Reuse materials;
- Recycle, separating what can gain a new life.

Using the resources that Nature offers us, we can have a cleaner and more sustainable environment, considering:

- For each ton of recycled paper, the slaughter of 20 trees is avoided.
- Plastic has an average decomposition time of 200–500 years.
- Aluminum takes 500 years to decompose.
- For each ton of recycled glass, about 1330 kg of sand is saved.
- Batteries and accumulators contain toxic materials which cause pollution in soil and groundwater.
- Organic waste is used in the production of organic fertilizers.

Separation is one of the major problems of recycling and can be approached from two different perspectives: the initial segregation, through separation from the operator, or the segregation that requires the disassembly and classification of each material. Complex products such as technological devices, often with a relatively short lifetime due to the permanent evolution of the features offered by manufacturers, make recycling much more difficult since they are made up of more than a dozen materials, from plastics to metallic materials (Hagelüken 2006; Sultan et al. 2017). The more varied materials a product contains, the more difficult its segregation becomes (Dahmus and Gutowski 2007). This is even more evident with the recycling of carbonated beverages cans, which are normally produced in a single material, making recycling much easier compared to the aforementioned technological products. If end-of-life products can be easily decomposed into different materials, making it easy to obtain new products with these materials, their market value will be much higher (Yam 2009; Despeisse et al. 2015). The added difficulty of segregating materials for recycling was mathematically modeled by Dahmus and Gutowski (2007), in order to parameterize the approach to the value of end-of-life products. However, the approach to the value of a particular product for recycling cannot be linearly established because there are different aspects that need to be considered, which can be listed in a simple way as follows (Morley and Eatherley 2008; Rybicka et al. 2016):

- Difficulty in the segregation process;
- Recycling technology cost and availability;
- Contamination state;
- Weight/volume ratio affecting the logistic process;
- Energy necessary for the reconversion of material into ready-to-market raw material;
- Short-term availability of pure material in nature;
- Current price of the pure raw material in the market;
- Current demand for this material in the market.

If some of the above items are easy to understand, others may be more complex. Indeed, certain products, because of the environment in which they were inserted, present increased recycling difficulties. Among many others are metals from nuclear power plants, the dismantling of nuclear submarines, or hospital products. In some of these cases, recycling can be a tremendous embarrassment for the later use of the material, which entails increased costs. Thus, the value of the material collected for recycling may or may not be economically feasible depending on the conjugation of these factors. However, in environmental terms, recycling has a very significant contribution to the saving of natural resources, and for any citizen or company, it is a duty to contribute with this tool to the preservation of resources.

However, when companies actually exploit recycling, they will need to be equipped with the necessary means to receive the returning products, as well as to provide all the reverse logistics of collecting the products to be recycled and delivery to their facilities. Reverse logistics, despite having costs, can be economically viable, if properly exploited (Silva et al. 2013; Giannetti et al. 2013). In fact, in the case of products whose end-of-life returns are systematic and continuous, transport can be monetized since there is cargo to transport in both directions, which could happen only in one direction if there were no return.

On the economic aspect, recycling provides more rational use of natural resources. On the social aspect, recycling provides a higher quality of life and preservation of Nature. However, as mentioned by Yamaguchi and Takeuchi (2016), recycling requires the incorporation of work and energy, and therefore, policies should be focused essentially on reducing the consumption of resources, avoiding all subsequent processes with a view to minimizing environmental damage. According to the same authors, it is verified that the instruments used to reduce waste on a source level are negatively affected by the policies established for recycling, as there is less market willingness to save resources on source level when there is the knowledge that material will still retain value through recycling.

A practical case of the application of recycling programs is presented in studies conducted by Andrews et al. (2013). Rush University Medical Center has expanded its recycling program constituting a team to implement sustainable practices. For this purpose, containers have been distributed in order to segregate materials according to the following categories: paper, plastics, aluminum, and glass. In 2010, this program avoided the deposit of 950,000 tons of trash in landfills. Moreover, the authors have concluded that by using different containers for different materials, the accuracy of the collecting process was increased. However, the addition of further trash containers should be done to avoid contaminations caused by mixed recycling. Regarding the situation in the USA in 2013, the authors also refer that a large investment in behavioral education is needed in order to improve citizen awareness for environmental sustainability practices.

6.1.2 Recycling Policies

Each country has its own culture, and it is the responsibility of the government of every given country to reflect the desires, interests, and values of its citizens in its policies (Ahlroth et al. 2011). A policy can be defined as a set of ideas that is transferred to legislation through its own institutions, that is, government or public bodies, which will promote a given direction of decisions and actions to be taken on a specific subject (Loomis and Helfand 2001). However, lawmakers are well aware that they should legislate to support the causes defended by most of their citizens, needing their support for the correct implementation of legislation (Rauwald and Moore 2002). The established policies can materialize through regulation, which can establish taxes for polluting activities, subsidies for Cleaner Production practices, incentives for the use of more environmentally friendly materials, mandatory use of recyclable materials in specific percentages, limitation of the use of certain polluting resources, among many other ways of guidance. In fact, these environmental targets are proposed by certain groups of countries, at regular interval meetings, where well-defined goals are set. Unfortunately, these goals have not always been successfully fulfilled (Murakami et al. 2015; Naess-Schmidt and Jensen 2015). Costa et al. (2010) and Sierzchula et al. (2012) point out that the policy of promoting tax reductions, or even their exemption regarding some specific environmental targets, has a significant impact on the behavior of populations and companies, inducing the latter to invest more in innovation. The number of studies dedicated to predicting the influence of a given variable on the outcome of a given policy is very limited, so it is not easy to predict accurately the expected results in setting a given policy (Field et al. 2012). However, policies always create barriers/requirements and condition business models, influencing strategies and having a direct influence on the labor market. These environmental constraints imposed by governmental policies and public bodies may cause market distortions of different natures: (a) Companies have different orientations and may be closer or further away from the imposed direction; (b) companies may be better or worse prepared to invest; (c) companies may be better or worse equipped with the necessary technological means to follow the new strategies, or, more importantly; (d) companies may or may not have the necessary human resources for the required changes imposed by new policies. Thus, there should be care by legislators to promote a smooth transition between different states of development, allowing the market to adapt accordingly. In fact, when more abrupt changes in policy are actually required, it may be necessary to inject public funds into the market in order to make the required policy change sustainable (da Cruz et al. 2012). Indeed, because of the pressure on governments to quickly do something for the environment, they have devoted very generous funds to support initiatives linked to renewable energy production, support for the transformation of processes into industrial enterprises, development of new and more environmentally friendly products, among many other initiatives. However, some disastrous public investments are also known, usually in conjunction with private initiatives, which lead to the loss of very large sums of money, part of which is the result of taxes paid by citizens (Olson 2014). Anyway, it is widely recognized that the most recent environmental policy guidelines have positively affected the way business is viewed, bringing broad social, economic and environmental performance benefits, as well as the need for a greater development of innovation related to technology (Mirata and Emtairah 2005; Popp 2006; Gibbs and Deutz 2007; de Abreu 2009; Costa et al. 2010; Testa et al. 2012; Geng et al. 2012; Yabar et al. 2013; Ribeiro and Kruglianskas 2013).

The study of the influence of environmental policies on improving recycling rates has been addressed by several authors, who essentially define two ways of action:

(a) implementation of new information strategies, which are essentially aimed at increasing citizen knowledge and education toward environmental issues and natural resources, in an effort to inform them about the main principles linked to the recycling process, what procedures to adopt and what are main benefits of recycling, inducing a greater willingness to participate. Or (b) implement strategies capable of acting in a structural way in the day-to-day context of citizens, influencing their behavior through the cost or benefit of certain actions related to the environment (Steg and Vlek 2009). Environmental policies may take different forms, depending on the objectives effectiveness to be achieved and on the time available for authorities to attain the required results. Thus, these policies may reach the citizen in one or more of the following ways (Elmore 1987; Wan and Shen 2013):

- Regulations: aiming essentially to enforce certain procedures and behaviors that the legislator intends to impose, implementing a set of rules that must be followed, and fines for non-compliance.
- Encouragements: aim to induce certain habits, behaviors, or procedures, in view of the purposes that the legislator intends to achieve. To that end, incentives can be created to promote more ambitious outcomes in a more assertive way.
- Direct investment: are aimed at securing the necessary financing for the pursuit of certain objectives, which could never be achieved in an accelerated manner due to economic constraints.

Recycling is a great business, which makes it possible to reduce costs for the environment, reuse materials by valuing waste while generating new industrial activities, promoting job creation. Recently, the concept of recycling has been incorporated as one of the possible phases of the Circular Economy, subject that will be treated more prominently in another subchapter.

Recycling, as well as other CP tools, has been the concern of many governments, which are setting priorities for their countries based on the behavior of their population, settled culture, consumption habits, greater or lesser ease of disposal, and reuse of certain materials, among many other factors. In fact, through several articles published, it is possible to perceive a great asymmetry between the policies adopted by each country, as is mentioned below. Tsai (2010) notes that Taiwan encourages the reuse of various materials, such as used plastic, textiles, cellulose, sugarcane bagasse, sawdust, and wood chips. In turn, Glew et al. (2012) stated that the UK has encouraged the recycling of springs. On the other hand, Malaysia supports the development of the chemical and biofuel sectors, with a view to minimizing environmental impacts (Ng et al. 2012). In the same sense, the Dutch Government promotes the improvement of environmental performance in the city of Rotterdam (Costa et al. 2013). In Canada, the government has invested primarily in training in business so that it has the necessary skills to behave more environmentally friendly (Granek 2011). In Germany, very specific objectives have been drawn which are linked to the percentage of recyclable material that end-of-life vehicles should possess, inducing a new wave of automobile design innovation (Blume and Walther 2013). It should be noted that around 8–9 million tons of waste are generated each year in the European Union, solely and exclusively due to end-of-life motor vehicles. Countries such as China, Portugal and the USA have encouraged the implementation of extended producer responsibility (EPR) policies to minimize the use of toxic or hazardous products in the development of new products and processes so that companies internalize the costs related to environmental preservation (Scheijgrond 2011; Wang and Chen 2013; Niza et al. 2014; Hickle 2014).

6.1.3 *Recycling Market Pressures and Company's Self-engagement*

Particularly in the industrial sector, companies are under increasing pressure regarding the need to recycle and contribute to a better environment. Market pressure is the starting point of these phenomena, as consumers are more attentive than ever to the environmental policies companies follow and employ (Beamon 1999; Walker and Preuss 2008; Walker et al. 2015). However, pressure is also applied by investors themselves, concerned with market trends and the positioning of their business going forward, as well as by their stakeholders, who also tend to pressure companies on improving their environmental performance (Trowbridge 2001; Walker et al. 2015; Guerrero et al. 2013). Furthermore, these investors are particularly concerned about the public image of a company since it directly interferes with its market share, influencing sales, and consequently profits. However, many other factors may put pressure on businesses for recycling and other environmental practices, such as market-driven pricing, increased demand for environmentally friendly products, increased citizen education for environmental preservation, government support for the development of sustainable products (eco-design), among many other factors (Galani et al. 2012; Fernández-Viñé et al. 2013).

On the other hand, there will have to be an effort by companies to motivate themselves for recycling, thus through an adequate relationship with their logistic chain, which plays an extremely important role in any recycling process (Saavedra et al. 2013). However, companies' efforts should be extended to the systematic search for new materials and processes through the systematic application of eco-design principles and life cycle assessment, through the creation of multi-disciplinary teams with a strong vocation for sustainability practices (Kiddee et al. 2013; Jabbour et al. 2013).

6.1.4 *Recycling in Practice: Some Examples*

Recycling should not be an objective, but an absolute necessity in the face of society's behavior toward consumption. The changes in consumption behavior in the last decades require that the scientific community and politicians find solutions to the recycling of certain products that were not consumed before, nor did they reach the end-of-life as quickly as some current technological products. Thus, there are sectors on which studies have been more focused, in order to respond to current consumption trends, which will trigger new demands on the treatment of end-of-life products in a few years' time. Electronic waste has gained some prominence in recent years, but there are other products that, although to a lesser extent, also deserve the attention of researchers.

6.1.4.1 E-Waste Recycling

The electronics industry is one of the most vigorous industries in the global market, interfering directly with our day-to-day. These technological products allow us to be networked, take advantage of our time and facilitate a great part of our daily tasks. It is an industry undoubtedly associated with high technology, which has changed our habits, way of managing time and even our social context (Li et al. 2015; Lu et al. 2015; Zeng and Li 2016). E-waste, also known as waste electrical and electronic equipment (WEEE), applies to all electrical and electronic products that reach their end-of-life and are disposed of in the trash for no longer being capable of carrying out desired functions. According to several authors, e-waste can be divided into several categories, which have known new groups over time, as follows (Gaidajis et al. 2010; Baldé et al. 2015): (a) televisions and monitors; (b) telecommunication equipment and small IT devices; (c) heavy equipment (washing machines, clothes dryers, among others); (d) heaters and refrigerators (refrigerators, freezers, air conditioners, among others); (e) small equipment (shavers, vacuum cleaners, toasters, among others); (f) lamps of all kinds; (g) medical apparatus; (h) toys; (i) sports equipment; (j) amusement equipment; and (k) automatic dispensers.

Within the aforementioned groups, mobile phones, and especially smartphones, deserve a prominent place. Indeed, the market share of these devices, which was unthinkable two decades ago, is reported as 88% in South Korea, 77% in Australia, 72% in the USA, 67% in Canada, and 58% in China, being that in Japan, the market share is only 39% (PEW 2015). China is the world's largest producer of such communication devices, accounting for 70.4% of the total volume of smartphones produced between January and November 2016, totaling 1.4 billion smartphones (Bai et al. 2018). However, mobile phone recycling globally accounts for only 10% of devices that have reached their end-of-life, so there is a huge scope here for something to be done in this domain (Welfens et al. 2016). In fact, the volume occupied by this type of devices is relatively small, so owners have a great tendency to preserve the device even though it is technologically outdated and has been already replaced by another device. Several studies have already been carried out to understand the attitude of consumers in this regard (Ongondo and Williams 2011; Li et al. 2012; Welfens et al. 2016; Yin et al. 2014), having found that the three main factors for this behavior are lack of reward for the delivery of the devices for recycling; awareness of how recycling works and recycling channels; and willingness to recycle.

Overall, e-waste reached more than 20 million tons in 2005 (Schwarzer et al. 2005), and it is supposed to have exceeded 40 million tons in 2014 (Baldé et al.

2015) and continues to increase at a rate of around 5–7% a year (He et al. 2006; Bigum et al. 2012), which is triple that of other types of waste (Rahmani et al. 2014). The case of China is particularly worrisome, as by 2020, e-waste is estimated to quadruple the values registered in 2007 (Yu et al. 2010). Furthermore, China is currently producing 9 million tons of e-waste per year (Zeng et al. 2016). It is also predicted that the waste from end-of-life mobile phones will increase sevenfold over the same period in China (Schluep et al. 2009). Additionally, predictions estimate that in 2020 China will surpass the USA in terms of e-waste generation (UNEP 2007). This sheer amount of e-waste is alarming and indicative of serious environmental problems to come if no efforts are done to minimize its impact (Chen et al. 2016; Hartard and Liebert 2015). Although the number of facilities devoted to recycling has grown significantly, the existing data points out to a strong inefficiency in the recycling process of this type of waste, given the values announced by registered operators: Only 13% of this waste is collected and recycled in centers proper for this purpose (Jiang et al. 2012). Given the danger presented by this type of waste and the value of certain materials incorporated in them, it is urgent that society cooperates to create a collection process for this waste. Also, the number of official recycling centers should be significantly expanded so that production of new products can be partially fed by the materials resulting from recycling, thereby reducing the demand for virgin raw materials and consequently generating energy and resource savings. The costs associated with the treatment of WEEE are strongly linked to the complete recycling cycle, which should include physical treatment through manual disassembly, mechanical treatment or heat treatment and the possible chemical recovery of some elements using hydrometallurgy and pyrometallurgy techniques (Le et al. 2014; Zeng et al. 2014, 2015a; Zeng and Li 2016).

E-waste includes toxic products and hazardous substances that, as of now, were never present in trash in such large quantities. These substances include, for example, polyvinyl chloride (PVC), or brominated flame retardants (BFRs) in the plastics class, or hexavalent chromium (Cr(VI)), lead or cadmium, in the group of metals (Xu et al. 2015; Kiddee et al. 2013; Patrick et al. 2012). Moreover, Namias (2013) reports that e-waste usually contains more than 60 different metal species. Along with all these toxic and dangerous substances are other materials that can be perfectly recycled, which are often called "mislaid resources," such as copper, aluminum, iron, silver, gold, and even platinum among other materials (Wibowo and Deng 2015; Cucchiella et al. 2015; Kumar et al. 2017). As pointed out by Cao et al. (2016) citing Schluep et al. (2009), one ton of scrap mobile phones contains, on average, about 30 kg of copper, about 3.5 kg of silver, and even about 340 g gold as well as about 140 g of palladium and traces of special metals such as antimony, indium and cobalt. The quantities of materials previously mentioned allow one to observe that it is much easier to extract precious metals from e-waste and its processed materials than to extract them from the mine and refine them by the usual processes (Hageluken and Meskers 2008). See, for example, the case of gold that in a ton of printed circuit boards has ten times more quantity of pure material than if extracted from a gold mine (Sun et al. 2011). The same is true for the copper, which the content printed on circuit boards is forty times higher than that in the copper mines (Qian 2010), given the same weight of raw material. Recycling of electronic waste can turn waste into wealth and minimize concerns about resource scarcity due to industrial development.

Based on the work already carried out by many researchers, Awasthi and Li (2017) summarized in a simple diagram how the recycling of WEEE can be conducted, regarding some of the most consumed devices that strongly contribute to the e-waste generated worldwide. Figure 6.5 shows this diagram.

The European Union (EU) and other countries have enthusiastically encouraged the EPR principle in the management of electronic waste (Gupt and Sahay 2015; Wang et al. 2014a). Generally, producers covered by the EPR pay the costs associated with recycling, which essentially depends on the economic value of the recycled products and the total cost of reprocessing. In order to encourage the EPR, the European Union has launched three main directives: WEEE (2012/19/EU) (EU 2018a), RoHS (2011/65/EU) (EU 2018b), and EU 2009/125/EU (EU 2018c). The first aims to consistently implement the EPR principle (Walther et al. 2010), which will be discussed later. The Directive on Electronic and Electrical Equipment (2012/19/EU) establishes two relatively recent targets in terms of recycling: The first one states that producers should be able to collect 45% of the average weight of EEE placed on the market by 2013, and the second one is even more demanding, putting the previous value in 65% by 2016. These rules are valid for any Member State of the Euro-

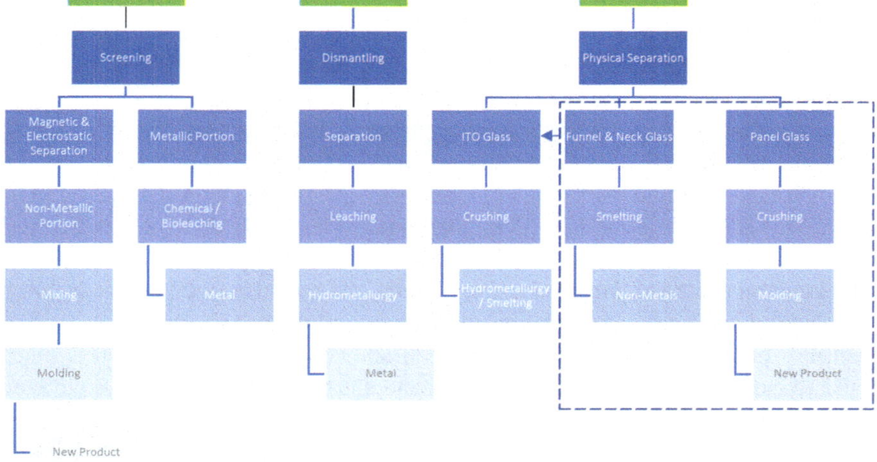

Fig. 6.5 Different ways to recycle WEEE. Adapted from Awasthi and Li (2017)

pean Union (Torretta et al. 2013). The Appliance Recycling Act was implemented in 2001 by Japan, stipulating that the manufacturer should take into consideration the recycling of products in the design and production of such products, adopting the regeneration of resources and the treatment of electronic waste to innocuous end products (Ogushi and Kandlikar 2007). Although there is no federal law in the USA, e-waste laws based on producer responsibility have been passed in 24 of its states since 2004 (Atasu et al. 2013).

6.1.4.2 Tires Recycling

Tires are composed of about 60–65 wt% rubber and 25–35 wt% carbon black, but still contain chemical elements such as accelerators and fillers, which are added during the manufacturing process. Tire rubber is comprised of a copolymer of long chain polymers, which includes elements such as isoprene, styrene, and butadiene, which are cross-linked by the addition of sulfur (vulcanization) (Antoniou et al. 2014). The elastomeric materials are present as C_xH_y, mixed with some fibers, and are considered thermoset polymers within the polymer family (Leung and Wang 1998). Moreover, tires consist of elastomeric materials, carbon black fillers, textile fibers, steel belt, and cord. The material used in tires commonly applied in motor vehicles is a blend of natural rubber, originating from the *Hevea* tree, with synthetic rubber derived from petroleum-based products such as butyl rubber and styrene-butadiene rubber (Shulman 2004; Martínez et al. 2013; Machin et al. 2017). Natural rubber is the main constituent of tires and given its unique elastic properties, allows a non-permanent deformation of its structure when subjected to external loads, presenting considerable mechanical resistance among the elastomer group of materials. Carbon black is an amorphous carbon species, with a very similar structure to graphite, which comes from the partial combustion of fossil hydrocarbons. The addition of carbon black is intended to increase the mechanical strength of the tire and give it greater resistance to abrasion. The remaining products that are added to this blend are essentially intended to make the rubber softer and workable for the vulcanizing process (Martínez et al. 2013). However, roughly speaking, it can be said that a tire is composed of about a hundred different products, including an organic sulfur compound as an accelerator, depending on the manufacturer, the function, and the target market of the tire.

Tires support much of the ground transportation activity, being an essential asset in the modern transport system created by society and indispensable for the movement of people and goods. In 2011, the tire production reported by the European Tyre and Rubber Manufacturers' Association (ETRMA 2011) in EU countries was about 1.5 billion units. Another source (Fredonia 2018), it is estimated that tire production is growing at about 4.1% per year in 2017–2019, reaching a consumption of about 3.0 billion units by 2019 in Europe. However, Asian countries and Oceania account for around 60% of world tire production. Data from 2010 indicates that China produced about 5.2 million tons of tires, while Japan produced about 1 million tons of tires in the same year. With economic growth in recent years, and with rising wages, economic activity is expected to generate more heavy-duty truck traffic and citizens will start to travel more, increasing the use of tires and requiring their replacement, which is in line with the forecasts described above. A recent study developed by Czajczyńska et al. (2017) based on a report launched by the European Tyre and Rubber Manufacturers Association reports that about 17 million tons of tires reach their end-of-life every year, of which about 3.6 million tons in Europe (European Tyre and Rubber Manufacturers Association 2016) and 4.0 million tons in the USA (Rubber Manufacturers Association 2016). Although the relationship is not direct, these numbers are directly linked to the growing number of vehicles circulating around the world. Considering just Europe, the number of cars in circulation in 2025 is expected to be 347 million, compared with 322 million in circulation in 2014, a growth of almost 8% in 11 years, showing that Europe is a relatively stable market. However, the situation is quite different in Eastern countries, where in 2025 the number of cars in circulation is expected to be 332 million in China and 69 million in India, which contrasts with the 107 million and 28 million in circulation in 2014 in the same countries, respectively, showing that growth here is very significant, almost tripling the number of vehicles in circulation in these countries (European Tyre and Rubber Manufacturers Association 2015). In addition to light-duty vehicles, the number of heavy-duty vehicles has also increased considerably, in response to the increasing need to move people and goods, as well as new construction and reconstruction of buildings. These tires have added durability characteristics, which makes them even less suitable for disposal without reconversion. It is estimated that approximately one billion of worn tires have been disposed of worldwide in 2015 (Freedonia 2014).

Due to the functions they perform on duty, tires have a number of properties that make them difficult to recycle because they are often highly resistant to abrasion, heat, conduction of electricity, humidity/water, and even bacteria or certain chemicals. This batch of properties is given to them through a set of chemical elements that are added to the rubber. In order to increase the mechanical strength of tires at high speeds, the rubber is further reinforced with metal wires and nonmetallic fibers, making tires a complex mix of materials. On the other hand, end-of-life tires, when stock-pilled for a long time, are a space-consuming waste, being as well a characteristic place for housing rodents and insects, and may be a source of problems for humans. If end-of-life tires are not conveniently disposed of, they can retain water, which is the ideal location for mosquito breeding and reproduction, promoting as well the spread of disease through these insects (Lin et al. 2008; Selbes et al. 2015; Derakhshan et al. 2017a). Due to the high flammability and high combustion power of tires, these can also constitute a danger in terms of initiation or propagation of difficult to extinguish fires (Martínez et al. 2013). Moreover, the combustion of tires releases intense fumes (dioxins, volatile organic compounds (VOCs), particulate matter, among others), oil, and other equally toxic effluents, which can contaminate the air, waterways, and soils if they are not properly controlled (Sciacca and Conti 2009; Sienkiewicz et al. 2012; Martínez et al. 2013; Derakhshan et al. 2017a, b; Dehghani et al. 2017). However, this thermal load contained in tires can still be very useful if it is properly utilized (Chyan et al. 2013; Antoniou et al. 2014; Naz et al. 2014), as referred ahead. End-of-life tires can also be recycled by incorporating them into the production process of other tires with less stringent requirements, in the form of road pavement mixtures, as building materials, motorway protections, artificial reefs for offshore fish farms, protective bumpers for docks and ships, and many other applications (Lin et al. 2008; Gupta et al. 2014; Derakhshan et al. 2017a). End-of-life tire management should be prioritized so that reuse must first be considered, then recycling, and finally reconversion into energy. These ways are accepted as the most environmentally and economically advantageous. Only as an extraordinary resource should landfilling be considered as an option, due to the slow degradation process, needing between 80 and 100 years. This is due to the fact that they are elastomers with a cross-linked structure, which hinders the separation of their chemical elements, making them unsuitable to suffer remelting processes (Martínez et al. 2013). In addition, landfill deposited tires may give rise to the accumulation of explosive gases, leading to dangerous situations. In this respect, a long and interesting path has been traversed in Europe due to the Waste Landfill Directive launched in 1999 (1999/31/EC) (EU 2018d) since in 2004 around 50% of end-of-life tires were deposited in landfills, a situation that has evolved to only 4% in 2010, showing a clear evolution in educational and environmental terms, demonstrating how science and environmental policies have achieved positive results in many fields (European Tyre and Rubber Manufacturers Association 2011). In the USA, the policy concerning end-of-life tires is not uniform across states, but a remarkable work has also been done in this area. In 1990, it was reported that there would be about one billion tires disposed in landfills; however, by 2015, 93% of these tires had already been removed from landfills and reused as a fuel complement or recycled (using as a blend of asphalt in pavements), depending on the emphasis given to each route by the respective state governments (California Environmental Protection Agency 2006; Rubber Manufacturers Association 2016).

End-of-life tires have known several applications, which essentially go through reconversion or incorporation into other products, as will be discussed later. When recycling is equated, one of the methods considered is the use of tires as fuel in compatible processes that are not affected by their combustion and do not impact the environment significantly. Thus, end-of-life tires are commonly used as a complement to the fuel used in cement kilns, as previously referred, or even as main fuel in bubbling fluidized bed reactors (BFBRs), used as combustors or gasifiers (Raman et al. 1981; Nakajime and Matsuyuki 1981; Giugliano et al. 1999; Scala et al. 2003; Mokrzycki and Uliasz-Bochenczyk 2003; Leung and Wang 2003; Pipilikaki et al. 2005). The use of end-of-life tires as complimentary fuel in thermal power plants reduces coal utilization and NOx emissions. The beginning of this route has been used since the early 1980s and, after validation by the competent environmental authorities, has been extended to other energy-intensive/energy-consuming types of industry, such as paper and pulp industry and thermal installations. Thus, the incineration of end-of-life tires is a possible solution to avoid its disposal in a landfill. Effectively, a life cycle assessment (LCA) was performed (Genan Business and Development 2009) regarding the use of waste tires in cement kilns, comparing the impact of this technique with other recycling streams. The LCA used as a unit the person equivalent (PE) to translate the impact caused by the treatment of a ton of tires in this way compared to the impact caused by a person during that same period of time. Through this study, it was found that recycling of used tires is advantageous in relation to incineration, saving around 0.07 to 0.31 PE, which makes it possible to realize that incineration, despite being an economically viable route, is not the best solution for the environment.

Other possible ways to deal with discarded tires is to turn rubber into high-value-added material (Song et al. 2018). The construction industry has also recycled the end-of-life tires as filler component of rubber blends used for sports flooring, roof insulation, children's playground, livestock mats, benches, vegetation protectors and windbreaks, garden hoses, and other domestic artifacts, in addition to the aforementioned mixture in bituminous asphalt coatings used on roads and motorways. As reported by Williams (2013), about 1.1 million tons of end-of-life tires are recycled to sports fields, playgrounds, roofing materials, and paving blocks, while 0.24 million tons are applied to embankments and foundations of motorways and railways. However, other methods of treating waste tires, such as gasification or pyrolysis, have been considered and developed in the last years. These processes have been exploited more intensively since roughly 2010, but given the EU's increasingly demanding environmental regulations, the energy crisis, the increasing cost of fuel, and an increasing number of investments in pyrolysis facilities, this appears to be the promising treatment of waste tires (Czajczyńska et al. 2017).

Pyrolysis, which is also commonly called thermal distillation or thermolysis, consists of a thermochemical treatment able to separate chemical bonds (Wampler 2006) using thermal decomposition under non-oxidative conditions (inert atmosphere or vacuum), acting on any gasification or combustion process as a first step, despite some dissonance with the term that gives rise to it as the Latin origin of the word refers to fire, which only exists in the presence of oxygen. Thus, the term thermolysis would be more appropriate than pyrolysis (Buekens 2006). Pyrolysis has also been referred to as reverse polymerization, thermal depolymerization, or polymer breaker. In terms of recycling, the great advantage that pyrolysis presents is the ability to manage difficult to treat waste relatively well, allowing the subsequent by-products to retain their initial properties and be considered for reuse (Martínez et al. 2013).

Most polymers are usually composed of carbon atoms, among others, which are combined through single or double bonds. The decomposition of rubber by pyrolysis has been the subject of research by several authors, who have studied and referred different mechanisms for their thermal decomposition. Basically, pyrolysis consists of heating the feedstock at temperatures above 400 °C in an inert, self-generated, deoxygenated atmosphere, causing volatilization and decomposition of the structural bonds of the feedstock. Under these conditions, the raw material undergoes several reactions, and several phenomena may occur such as dehydration, cracking, isomerization, dehydrogenation, aromatization, and condensation (Rezaiyan and Cheremisinoff 2005). However, other authors advocate that the best range of temperatures to optimize the char and gas production is between 500 and 600 °C (Williams et al. 1990; Laresgoiti et al. 2000). Pyrolysis gives rise to oil, which can be used directly as fuel or also utilized as a chemical raw material, as well as gases and char product, also allowing the recovery of the steel contained in the tires subjected to this decomposition process. The gases released by the pyrolysis process of end-of-life tires are usually H_2, C_1–C_4 hydrocarbons, CO_2, CO, and H_2S (Williams 2013). These gases have intrinsic energy enough to feed the heat necessary to feed the pyrolysis process in a continuous way. Otherwise, elemental analysis performed on oils providing from the pyrolysis process of end-of-life tire residues allowed to obtain a C/H ratio of 8/1 regarding their composition, indicating the presence of ring structures with molecular weight and heat energy, varying according to the parameters used in the process (between 27.8 and 44.8 MJ/kg) (Quek and Balasubramanian 2013; Antoniou et al. 2014). On the other hand, the char can be used as solid fuel, but also as carbon black filler in tire production or upgraded in order to give rise to activated carbon. Char is an activated carbon with an average adsorption capacity of between 400 and 700 m^2/g, which is seamlessly suited to the retention of liquid or gaseous phases of organic substances and is perfectly able to replace coal or biomass as a product ready to be marketed. The pyrolysis flowchart shown in Fig. 6.6 illustrates the expected products capable of being extracted from end-of-life tires, and how the process can be refed by self-produced energy.

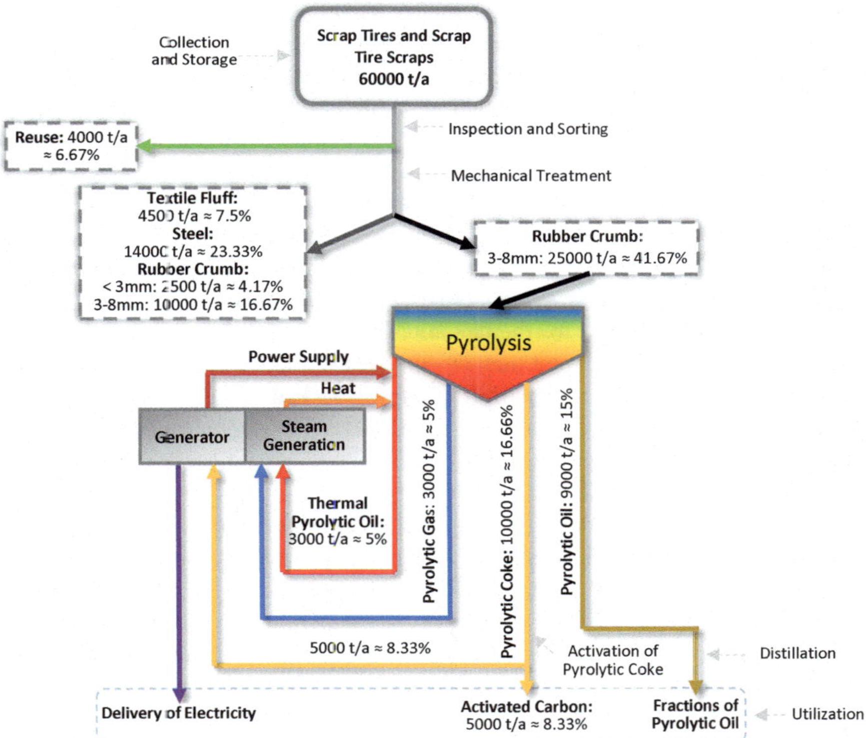

Fig. 6.6 Pyrolysis flowchart diagram

A practical case is described by Machin et al. (2017) regarding the information made available by the company Huayin Group (2018), with headquarters in China. This company produces industrial equipment for tire pyrolysis treatment, able to process from 3 to 10 tons per day of waste tires, using temperatures in the range of 400–450 °C, under oxygen scarce conditions, obtaining pyrolysis oil (45%), pyrolysis gas (10%), steel wire (15%), and black carbon (30%).

According to Scheirs (2006), pyrolysis of waste appears to be a very simple process, but the products to be treated, often being hydrocarbon-based products with a relatively low market value, have rather complex compositions. Therefore, in order for the process to be properly controlled and the resulting by-products to be those which are expected, it is necessary to have absolute control over the pyrolysis parameters, such as the type of atmosphere and the corresponding flow rate, the heat rate, the temperature at which the pyrolysis is carried out, the reaction time, among others. Chemical reactions regarding rubber decomposition have already been intensively studied by several authors, who have suggested different mechanisms for this decomposition. Thus, Dodds et al. (1983) found that the structure of rubbers is based essentially on double carbon bonds, which, when subjected to the thermal decomposition conditions established by the pyrolysis process, give rise to strongly reactive free radicals, which are considerably smaller than those presented by the molecule of rubber. On the other hand, Aguado et al. (2006) describe the thermal decomposition of rubber as a chain reaction process, in which there is a sequential transfer of hydrogen atoms, which is accompanied in parallel by a progressive breaking of the rubber's molecular chains. Based on this same principle, Groves et al. (1991) tried to study the phenomenon with particular detail, selecting for the study the pyrolysis of natural rubber at 500 °C through gas chromatography, suggesting as an explanation for the phenomenon, a monomer recombination, which will lead to the dimerization of rubber, through a process known as the Diels–Alter mechanism. However, the pyrolysis of rubber was also studied and explained by Mastral et al. (2000), which focused on another process for the decomposition of tire rubber waste. According to these authors, decomposition may be achieved through the depolymerization of polyisoprene and subsequent cyclization. However, Padkel et al. (2001) also suggested an alternative way for the process: Rubber polyisoprene can be thermally decomposed by a β-cleavage mechanism for intermediate radical isoprene, which in turn will be transformed into isoprene by depropagation, and then decomposed in a gas phase of dipentene, through dimerization. However, more recently published studies (Kwon and Castaldi 2009) indicate another type of decomposition process in the thermal degradation of rubber providing from end-of-life tires. This mechanism reveals that raising the temperature above a certain value leads to cleavage of the rubber monomer bonds, which is followed by hydrogenation and recombination, due to a gas phase reaction.

Based on the above-mentioned, it may be stated that the conditions under which end-of-life tire rubber pyrolysis is performed comprise the formation of intramolecular free radicals. Because it is difficult to accurately determine the result of each reaction, Li et al. (2004a) divided the rubber pyrolysis process into three different temperature levels, which may give rise to the formation of different reactions: pyrolysis reaction, which usually takes place between 250 and 520 °C, secondary reaction of the molecular chains volatiles, which occurs between 600 and 800 °C, and which has significant implications on the formation of benzenes, toluenes and xylenes, as well as the gasification reaction of pyrolytic carbon black with $CO_2/H_2O/O_2$ in the gases, which is carried out in the range of 750–1000 °C. It can also be seen that the influence of each of the temperature ranges previously considered is influenced by two main factors: temperature and reaction time, which has direct repercussions on the reaction rates, as can be determined by calculations, using the Arrhenius equation. These research examples show the large difficulty of reliably determining the outcome of a pyrolysis process, as it is not possible to describe the process by means of reaction equations. In this way, some more empirical approaches have been developed to determine the pyrolysis result, which use the principle of mass conservation to make the results more reliable and predictable (Buekens 2006). For this, end-of-life tire waste is considered to be converted into a solid carbon-based compound, accompanied by the formation of other volatile products. These solid compounds correspond to approximately 40% by weight of the organic components, while the volatile component corresponds to about 60% of the organic elements when also considering carbon black in this reaction. Equation 6.1 represents the above principle (Martínez et al. 2013). Thus, taking into account the enthalpy of the tire rubber reaction, it is possible to determine the energetic balance and calculate the volume of energy contained in the portion of volatile components.

Pyrolysis, when performed under the most appropriate conditions, promotes the elimination of problematic waste and also allows to obtain: (a) valuable chemicals, in a liquid state; (b) char of good quality (raw material for the production of activated carbon); (c) gaseous fuels, which provide sufficient energy to carry out the process and, as well as the production of electric energy. This raw material functions as an excellent fuel due to its high calorific value, comparable to other fuels such as crude oil or coal. The energy released by the burning of a medium-sized car tire is between 35 and 40 MJ/kg. Tires from motorcycles have lower available energy due to a lower content of volatiles and higher ash content, resulting in an average available energy rarely exceeding 30 MJ/kg (Islam et al. 2008).

Waste tire 100 wt% ↓	Solid fraction 40 wt% ↓	Volatile fraction 60 wt% ↓

$$C_{173}H_{165}O_3S_{1.5}N \rightarrow 1.1539C_{64.5}H_9S + 0.3453C_{285}H_{447}O_9N_3S \quad (6.1)$$

Elemental composition (daf) C = 88.20 wt% H = 69.7 wt% O = 2.21 wt% N = 0.59 wt% S = 2.02 wt%	Elemental composition (daf) C = 94.96 wt% H = 1.12 wt% S = 3.41 wt%	

The recycling of rubber from end-of-life tires has also known other developments than just pyrolysis. Indeed, the use of rubber waste from tires has also been added in the composition of epoxy-based composite materials. It is well known that the typical cross-linked structure of rubber as a result of the vulcanization process constitutes a problem because it restricts the movement of its molecular chains and limits the interaction forces between these chains and the typical structure of the matrices to which it can be added, forming composites. Thus, the mechanical properties of these composites can be compromised due to this lack of molecular interaction. In order to overcome this problem, it is desirable that the sulfur cross-linked molecular chains promoted by the vulcanization be broken, for which a devulcanization process is necessary (Naskar et al. 2001, 2002). Recently, rubber powder from end-of-life tires was used in the preparation of an epoxy-based composite by Yagneswaraan et al. (2008, 2013), in which the powder rubber was obtained by mechanochemical devulcanization. This epoxy–rubber waste composite showed good thermal stability and excellent dispersion and interaction with the epoxy matrix. Moreover, it was further detected that rubber waste has a catalytic effect when included in the composition of epoxy matrix composites, improving the curing reaction. However, previous studies performed by He et al. (1999) in which rubber tire used to obtain an acrylic matrix composite, allowed to realize that the fracture toughness of the composite was initially improved with the increasing addition of rubber waste, but from a certain content of this element, the fracture toughness began to decrease, indicating that there is an optimal value for the addition of rubber waste regarding fractural toughness. Studies were also developed to understand if rubber waste added to polyurethane-based matrices could bring benefits to the composite in terms of mechanical and thermal properties (Piszczyk et al. 2015). It was possible to verify through this study that the properties of the polyurethane foams are significantly altered by the introduction of rubber waste, improving compressive strength properties substantially, as well as the thermal stability, most probably due to a strong increase in the cross-linking density of the molecular chains. As can be understood from the aforementioned, devulcanization plays a preponderant role in the use of rubber waste from end-of-life tires. Hence, several studies have been carried out in the last two decades in order to find practical, economic, and effective solutions to achieve this desideratum. One of the routes studied was thermomechanical shearing devulcanization, which was explored by several authors, of which Sutanto et al. (2006), Zhang et al. (2010a, b), Kumar et al. (2002). This process imposes a drastic rise in temperature of the rubber, causing the structure to irreversibly degrade, and it has been used more and more frequently, using extruders to promote the above-mentioned effect (Tao et al. 2013). When a chemical component is added to this process, devulcanization can be said to have been obtained through a mechanochemical process, and various chemicals have been used to give rise to this type of devulcanization. However, this route is not the most appropriate since it uses and produces toxic products, which are harmful to the environment. Thus, an alternative was studied, based on the use of high-power ultrasonic electromagnetic radiation to promote the rubber devulcanization (Li et al. 2004b) since these waves, once they reach a certain level, become capable of breaking rapidly the reticulated three-dimensional rubber chains, while the process is also supported by appropriate pressure and temperature conditions. Extruders are again the most suitable equipment to carry out this process, producing the appropriate mechanical forces, which need to be complemented by the action of a powerful ultrasonic generator. Also, the use of microwaves has been studied as a devulcanization process for end-of-life tires, using their electromagnetic energy to promote the breakage of carbon–sulfur or sulfur–sulfur cross-linked chains, leaving intact carbon–carbon bonds. This electromagnetic energy is absorbed by the material's structure and transformed into heat, causing the temperature to rise rapidly. However, prolonged exposure to microwave radiation inevitably leads to the breakdown of some of the carbon–carbon bonds, which may affect the properties of the recycled material, consequently jeopardizing its reusability as add-in materials for composite matrixes for possible revulcanization processes. This is due to a lower-molecular-weight fraction presented by the devulcanized rubber (Karger-Kocsis et al. 2013; Garcia et al. 2015). However, this process has been used consistently and seems extremely promising since it presents some very attractive advantages such as (Adhikari et al. 2000): (a) the possibility of continuous recycling of rubber waste; (b) ease of adjustment of the parameters in function of the material to be treated; and (c) the possibility of treating large quantities of material in a relatively short period of time. This process also has the advantage of being environmentally friendly (Hong et al. 2015), as it does not need chemicals to improve its performance since the energy is accumulated by the material as a whole and the heat is developed locally, not by diffusion (Thostenson and Chou 1999). Moreover, Aoudia et al. (2017) recently tested a devulcanized rubber used in an epoxy matrix composite, comparing it to the same matrix filled with simple rubber residues, realizing that the flexural behavior of the devulcanized matrix epoxide composite showed better mechanical properties. These authors also verified that the amount of energy to be supplied to the process must be properly studied because excess energy deteriorates the properties of the devulcanized rubber.

More recently, another novel method has emerged capable of breaking the cross-linked molecular bonds created by sulfur in rubber. Although this type of rubber is not biodegradable under normal conditions, certain microorganisms have the ability to oxidize sulfur, thus breaking the rubber cross-linking and transforming sulfur into sulfate (Li et al. 2011; Yao et al. 2013), thus performing devulcanization. The powder generated by this devulcanization process can be added into the formulation of various composites without significant loss of properties (Presti 2013). However, this devulcanization process also entails some drawbacks, which can be divided into two main groups:

- The devulcanization process triggered by microorganisms is slow, which implies a prolonged immobilization of the tires, with the consequent problems of storage and waiting time.
- The microorganisms used in this process need to be neutralized at the end of the devulcanization process, which can only be accomplished through the addition of specific bacteria-killing products, a situation that can cause undesirable environmental impacts. In addition, a bacterial culture is difficult to control, which can lead to problems of another order.

Thus, this route of devulcanization has lost its application brilliance, being in disuse (Selbes et al. 2015).

On the other hand, biological technologies have also been tested and used successfully. Effectively, sequencing batch reactor (SBR) and fixed-bed sequence batch reactor (FBSBR) present an easy and flexible operation, using a compact structure, combining suspended sludge and biofilms, in an effective method for the treatment of effluents, due to its high efficiency in eliminating organic material from wastewaters, rapidly reducing biodegradable organic materials (Mahvi 2008; Maranon et al. 2008; Takdastan et al. 2009; Mahvi et al. 2009; Kulkarni 2013; Rodríguez et al. 2011; Santos and Boaventura 2015). The greater use of the FBSBR system is justified by the more stringent requirements for a higher quality of outgoing effluents after the end of the process, ensuring better preservation of water resources. Several studies have proved that FBSBR has very attractive properties, such as high biomass, high demand for oxygen in the chemical process (COD), robust acceptance loading, and no sludge bulking problem (Santos and Boaventura 2015). Furthermore, the FBSBR method is capable of taking full advantage of the sludge retention time in the biofilm, being also able to operate a suspended activated sludge system with a moderately tiny hydraulic retaining time, allowing as well the development of microorganisms possessing different sludge retaining times in a sole reactor. Taking into account the FBSBR process, it was found that it produces only about half of the sludge produced by the SBR process and that this sludge presents a much higher degree of stabilization, which potentiates its subsequent application as fertilizer. However, both bioreactors have shown excellent performance with the removal of organic substances, although the FBSBR bioreactor stands out positively, due to the higher capability of extracting biomass from the end-of-life waste tires (Derakhshan et al. 2017a).

6.1.4.3 Plastic Recycling

Following the data launched by the Plastics Europe Association (PlasticsEurope 2018), the world's global consumption of plastics reached 322 million tons, departing from 1.5 million tons in 1950, as shown in Fig. 6.7. The annual growth rate seen by the manufacturing industries of these types of products is higher than 3%, having been of 3.4% from 2014 to 2015. In the last decade, China has risen as the largest plastics producing country in the world, while all other countries or regions have lost a small part of their market share. Plastic consumption increase is due to the ease with which it replaces other materials, such as certain metal alloys, natural materials (wood), and even ceramics, as well as having the added characteristics of being easy to transform, with low densities, functional properties, and a low cost. However, this increase in consumption has triggered an increase in the volume of solid waste that needs to be landfilled or incinerated in conjunction with municipal waste, with truly harmful consequences for the environment (Patel et al. 2000; Subramanian 2000; Derraik 2002; Al-Salem et al. 2009; Hopewell et al. 2009).

As is well known, recycling of plastics is limited, which generates a huge amount of waste from non-recyclable or end-of-life plastic needing to be treated. When recy-

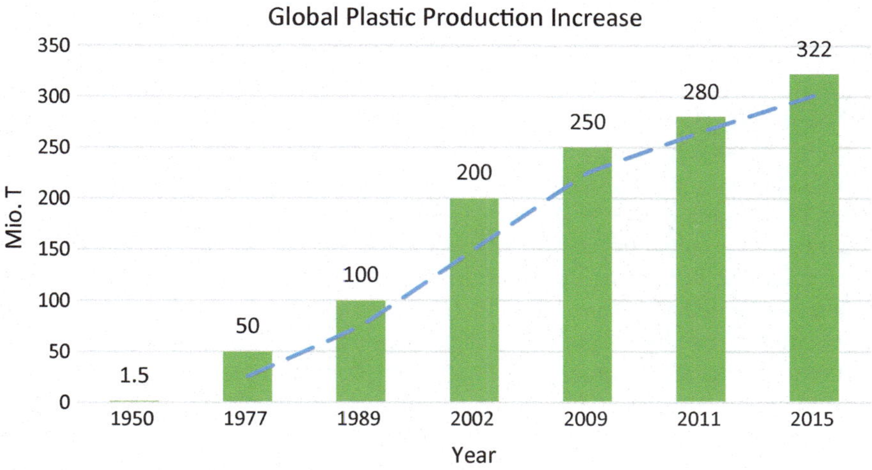

Fig. 6.7 Worldwide plastic production over the period 1950–2015. Data collected from PlasticsEurope (2018)

cling is not possible, plastic can be landfilled, incinerated, or decomposed through pyrolysis, as already explained with regard to the treatment of end-of-life tires, among other less utilized techniques. Both pyrolysis and incineration entail environmental problems, as pyrolysis is a highly energy-consuming process, although it also releases products with high-energy potential, while incineration requires the use of expensive, high-tech equipment to neutralize harmful products generated in the process, before releasing them in the neighboring environment. However, due to the gigantic amount of plastic that is not recycled, the disposal in the landfill of plastics takes up a lot of space, creating difficulties in their management. On the other hand, incineration entails severe environmental problems through the release of HCl gas, as well as chlorine-rich toxins, and ash containing cadmium and lead (Demirbas 2004; Achilias et al. 2007; Hopewell et al. 2009; Zhang et al. 2010c).

Basically, four different approaches can be considered regarding polymer recycling (García 2016):

- Main recycling process: remelting the polymer and remolding it;
- Mechanical process: polymer reprocessing, which requires pretreatment or decontamination, as well as repelletizing before reprocessing;
- Chemical or feedstock recycling: chemical breakdown of the molecular chains to lower-molecular-weight types, allowing for reuse. The process can reduce the chains to monomers (complete reversion) or oligomers (incomplete reversion).
- Energy recovery: incineration, allowing for energy recovering, taking advantage of the energy stored in the chemical bonds.

The benefits of polymer recycling are obvious: It is estimated that savings through primary recycling processes reach about 100 million BTUs of energy per ton of plastic, which is equivalent to the value of approximately 17 barrels of oil.

Thermoplastics allow recycling and can be remelted and molded again. However, thermosetting and elastomers cannot be reprocessed in the same way as the initial processing stage, as their reticulated microstructure does not allow for remelting and reworking. However, within the range of thermoplastics, there is an enormous diversity of polymers, with different compositions and structures, and also with different processing conditions, which implies increased care in the separation of the different types of plastics, with a view to its possible reutilization, thus avoiding the landfill. This problem occurs, for example, with polyethylene terephthalate (PET) when mixed with PVC, as the processing temperatures are incompatible, with no melting of the PET at the maximum PVC processing temperature. In addition, some degradation of the physical and mechanical properties as well as color of the plastics occurs during the second processing cycle, a situation which is even worse if there is a mixture of polymer types. This causes a noticeable price degradation of the recycled raw material, compared to virgin raw material.

One of the main concerns for the environment is that only a few polymers are recycled on a large scale. Polymers with the highest recycling rate are, in descending order of volume, high-density polyethylene (HDPE), PET, polypropylene (PP), and finally, polystyrene (PS). These polymers are essentially recycled using mechanical processes. Recycling implies that there is a maximum number of recycling cycles that can be imposed, as each cycle will promote a loss of ductility and molecular weight of the polymer. For example, ductility of PET falls from 310 to 218% only with the first cycle of recycling, and in the third cycle is just 2.9%, the reason why the polymer already has severe restrictions in terms of applicability. One of the main applications of PET is bottles (18% of the application worldwide, corresponding to about 900,000 tons). This polymer is recycled essentially for the production of fibers (50–77%), with only about 4–16% being reused for the same purpose—bottles (García 2016).

Thus, with respect to thermoplastics, the first difficulty in recycling is based on a correct separation of the different types of polymers making up a product, or in normally aggregated products. In fact, some of the principles that are already being used in construction, namely the design principles for deconstruction (DfD—Design for Deconstruction), should be widely applied in other areas of product manufacturing, so as to allow easier reuse or recycling of the materials initially employed, when the first phase of the product life cycle comes to an end. Over the last few decades, awareness of the recycling of plastics has increased significantly in the population, both in developed and developing countries, allowing to minimize environmental burden. In order to correctly separate the different types of plastics prior to recycling, different techniques have been developed, one of the most frequent being manual separation, which involves a large amount of labor and is a time-consuming task, providing poor conditions for workers involved (Zhao et al. 2018). Other techniques also used are triboelectrostatic separation and gravitational separation, but they have well-known limitations due to the similarity of properties among many of the thermoplastics available on the market for each of the most commonly used applications (Shent et al. 1999; Park et al. 2007; Gent et al. 2009; Malcolm Richard et al. 2011). On the other hand, the selective dissolution process involves the use of toxic organic solvents, which, besides presenting the typical drawbacks to the toxicity involved in the use of solvents, also presents a high price, which makes this process unattractive when compared to its competitors (Pappa et al. 2001; Miller-Chou and Koenig 2003). Thus, this method of polymer recycling has not seen significant developments in recent years, and it is expected that the problems posed are not easy to solve. However, the flotation separation method has presented good practical results, having proven to be a competitive and high-efficiency process (Jody et al. 2003; Wan et al. 2015). This separation method began in the mid-1970s by S. Izumi (Izumi and Tanaka 1975; Izumi and Saitoh 1978), and since then has been the subject of several developments, which have been based essentially on two very characteristic properties of polymers: low density and low surface energy (Wang et al. 2015). However, these properties are directly affected by other circumstances, inherent or not to the polymer itself, because different substances can be added to the polymers during its processing. Thus, chemical composition, plasticizers, degree of polymerization, crystallinity, and surface structure are factors which can directly affect the hydrophobicity of plastic surfaces, extremely important property regarding the flotation separation of plastics.

Although flotation separation is a very promising and competitive method, there are a number of factors that have to be studied in depth, leading for optimization of the process. In order for the polymer to be recycled, it first must be fragmented, which generates new surfaces on the obtained fragments. These fragments will then be mixed with original surface parts. Obviously, this is related to the cutting process, to which the molecular chains were subjected during fragmentation, causing the breakdown of macro-molecules. It also becomes obvious that the radicals formed in the sheared area will tend to oxidize. Therefore, there will be variations in the hydrophobicity of the polymer, presenting different oxidation degrees, being lower in the case of the new surfaces created by the fragmentation process. Other specific properties of polymers to be considered in this process are the free surface energy and the contact angle (Wang et al. 2015).

Most of the recycling processes begin with the reduction of size, from pieces or products, to waste of larger or smaller size, depending on the process of disaggregation and the time to which they are subject to this process. Then, the separation between the sinking and floating polymers is performed to separate the less dense plastics, such as polyolefins (PP and PE) and foam resins (PUR and expanded PS) from heavier plastics (ABS, PS, PET, PVC, and PC, etc.).

One of the key conditions for the flotation separation process to take place is related to the fact that its average specific gravity needs to be less than the density of the medium being used in the process. For this, it is still necessary to consider that bubbles are normally formed around plastic particles, so the specific average gravity has to consider the plastic bubble–particle aggregate. Following the theory formulated by Shen et al. (2001), the theoretical density of bubble–plastic aggregate is given by Eq. 6.2:

$$\rho_{\mathrm{a}} = \frac{\rho_{\mathrm{P}} * V_{\mathrm{P}}}{V_{\mathrm{P}} + nV_{\mathrm{B}}} \tag{6.2}$$

where ρ_a represents the density of bubble–particle aggregate; ρ_P and V_P characterize the density and volume of the plastic particles, respectively; V_B represents the average bubble volume; and n is the expected number of attached bubbles. Moreover, the shape of the particle impacts the specific surface area, which also influences the flotation ability of the particles/aggregates. Studies carried out by several authors (Marques and Tenório 2000; Shen et al. 2001) have demonstrated the validity of the previous expression, which, in practice, translates into a greater capacity of small particle flotation (greater specific surface area) with respect to larger ones. Particles of smaller volume and surface also do not need such an intense coverage of bubbles to float. Although some authors have concluded that plastic particles below 50 μm float in any condition, regardless of the type of plastic or flotation medium (Vogt 1981; Wang et al. 2015), states that the best range for the dimensions of plastic particles in view of this process should be between 2.00 and 6.00 mm, and this size is also influenced by the flotation device used in the process. It has also been found that hydrophobicity is affected by the surface roughness of the particles (Drelich et al. 1998; Oláh et al. 2005; Martin et al. 2011). In order to reduce the roughness and simultaneously to promote a cleaning of the particles, different chemical products were used, being verified that dilute alkaline solutions with very low alkaline contents promote a slight decrease in the surface roughness, while other alkaline treatments and ozonation, also initially used as cleaning products, promote an increase in surface roughness. Due to the typical hydrophobicity presented by a huge range of plastics, one of the main difficulties encountered in the plastic flotation process is to find efficient methods of selective wetting of plastics, which can be achieved by reducing the liquid–vapor surface tension (usually called gamma-type flotation), chemical conditioning, and treatment (Fraunholcz 2004; Alter 2005).

After a first rough division of the flotation process in three different groups (gamma flotations, physical conditioning and chemical conditioning) performed by Shent et al. (1999), it was verified that it was not sufficiently precise since there are chemical reactions that come from the physical condition. Thus, Wang et al. (2015) proposed another categorization for the flotation processes, which consist of four groups: gamma flotation, adsorption of reagents, surface modification, and physical regulation. These groups will be briefly described below:

- **Gamma flotation**: Regarding Fig. 6.8, it is possible to observe the balance between surface tensions. Zisman (1964) developed graphs that correlated the cos θ with the critical surface tension γ_c, the surface tensions of different liquids, and the experimental line extrapolated to cos $\theta = 1$. The abscissa value corresponding to the intersection is defined as the critical surface tension γ_c. If there is a sufficient difference between the critical surface tension of two hydrophobic materials, selective wetting may be achieved by reducing the surface tension $\gamma_{L/G}$ at the liquid/surface interface. For appropriate values of $\gamma_{L/G}$, the air bubble will adhere to particles with the lowest γ_c value, while the particles having the highest γ_c value will be sufficiently moistened to allow bubbles to be added thereto. This method is called gamma flotation, which is achieved by reducing the surface tension of the liquid to values comprised between the critical surface tension of the two

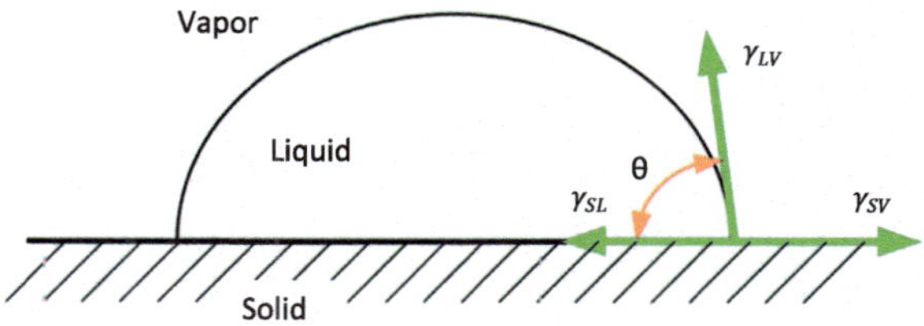

Fig. 6.8 Contact angle of a sessile drop (Alter 2005)

Table 6.1 Critical surface tension of different polymers

Polymers	γ_c (From)	γ_c (To)
PE	28	31
PS	33	36
PVC	39	
PET	40	43
PMMA	39	
PA6	42	42
PA11	33	42
PA6.6	42	46
PA10.10	32	32

Adapted from Wang et al. (2015)

plastics. The values computed for γ_c are shown in Table 6.1. Different liquids have been used in flotation to improve process efficiency. Shen et al. (2002a) studied the flotation behavior of seven different plastics using alkyl ethoxylated nonionic surfactant (15-S-7) as flotation mean and concluded that the surfactant effect is essentially felt in the reduction in surface tension. Regarding a problem previously referred, Buchan and Yarar (1995) carried out experiments in order to try to separate the PET of the PVC through the gamma flotation technique, concluding that PET becomes permeable to wetting after reducing the surface tension to appropriate values, while the PVC maintains its hydrophobicity unchanged under the same conditions. Gamma flotation also presents some drawbacks, namely the difficulty in selecting polymers with very close surface tension, especially if there are more than two types of plastics in the same selection process, and also the fact that some toxic substances used as flotation liquid are toxic and expensive, namely the aqueous solution of methanol, which makes the process uncompetitive both in financial and environment terms.

$$\gamma_{SV} - \gamma_{SL} = \gamma_{LC} \cos\theta \tag{6.3}$$

- **Adsorption of reagents**: The adsorption of hydrophilic reagents has been affirmed as an alternative method capable of selectively reducing hydrophobicity of plastics. Depressants tested for plastic flotation have been used to selectively wet plastics.

The usual composition of the wetting agents suitable for flotation of plastics contains molecular pieces capable of adsorbing on the plastic surface and molecular groups able to make plastic particles hydrophobic, such as lignosulfonates, tannic acid, or methylcellulose (Shibata et al. 1996; Shen et al. 2002b; Abbasi et al. 2010; Wang et al. 2014b). In addition to this method, many studies have been carried out in several directions, namely the study of the interaction mechanisms between the surface molecules of plastic fragments and wetting agent molecules (Stuckrad et al. 1997), a study of the absorption mechanisms of the wetting agent by the surface of the plastic fragments (Fraunholcz and Dalmijn 1997, 1998), an adsorption study of lignosulfonate by PET and PVC (Le Guern et al. 2000, 2001), studies of the influence of flotation agents on PS wettability (Basaróvá et al. 2005), investigation of the interaction between different phases involving hydrogen bonds, Van der Walls forces and Lewis acid–base interaction (Wang et al. 2011, Wang et al. 2013a, b), among others, which can be analyzed briefly in the work presented by Wang et al. (2015).

- **Flotation separation of plastics mixtures**: The flotation by dispersed air can be achieved by the introduction of a jet of air into the water, slightly decreasing air pressure until it reaches atmospheric pressure, forming a flotation column. However, flotation agents may also be used in order to more strongly wet one polymer over another or others, thereby promoting separation between them. A review presented by Wang et al. (2015) discloses an abundant list of reagents that may be used for this purpose, allowing an easier separation and recycling process. In this same work, the percentages of recovery efficiency of each of the polymers involved are indicated, and the recovery rates are quite high, which demonstrates the added value presented by this process in the separation of initially aggregated polymers. After successful tests on ABS and PS flaking through a tannic acid medium, as previously referred, it was found that the separation of these materials could also be carried out successfully by dispersed air flotation (Wang et al. 2012). Also, the complete split-up between PVC and PET was achieved by Saisinchai (2013) with PET being selectively moistening through a solution of calcium lignosulfonate after the optimization of the operational conditions. On the other hand, a process was developed that combines the sink–float methodology and the flotation technique in order to promote the separation of PE or PP mixed with PET from bottles through a separator drum, which, according to Dodbiba et al. (2002), resulted in a PET separation efficacy of about 99.995% based on a dodecylamine acetate or polyvinyl alcohol separating/wetting agent.
- **Surface modification**: The surface treatments of polymers in the separation process for subsequent polymer recycling assumes several aspects, and the first one to be carried out involves the removal of labels that are glued to the surface (a situation quite common in packaging), which can be done through an alkaline solution. A second stage is the surface treatment, which can be considered as a preparation of the polymers to facilitate the subsequent process of separation. Indeed, the surface modification mechanism relies essentially on a surface oxidation process, which results in the splicing of hydrophilic functionalities such as hydroxyl, carboxyl and carbonyl in the polymer chains located on the surface of the polymers. With this

treatment, it is intended to physically act on the surface of the polymers by modifying their wettability. This effect can be achieved through numerous approaches, such as plasma treatment, corona discharge and photografting (Fraunholcz 2004), wet oxidation (Wang et al. 2014b), flame treatment (Pascoe and O'Connell 2003a, b), and ozonation (Reddy et al. 2007, 2008, 2010; Okuda et al. 2007). Polymers have different sensitivities to different reagents, so it is necessary to know precisely what substances to use in the treatment, in order to act selectively on the polymer in which it is intended to act. For example, alkaline treatment may be used to promote selective PET wettability, while ammonia may be used to selectively act on PC (Wang et al. 2014b). Considering the developments carried out by Stückrad (1996), Pascoe and O'Connell (2003b), Okuda et al. (2007), and Reddy et al. (2007, 2008, 2010), the ozonation process showed better results than those revealed by the flame treatment process. A number of other surface treatment processes for polymers have been developed in order to modify their wettability and have been developed between 1988 and 1999 (Rånby 1992; Ozdemir et al. 1999; Allméar et al. 1988), flotation separation process, such as laser, electron beam, ultraviolet rays, corona discharge, which have proved to be efficient but have not seen significant developments in the last two decades.
- **Physical regulation**: Physical regulation appears as an alternative process to surface treatment, in which chemical reactions are no longer necessary, setting the process only in the physical properties of the materials involved. Much like surface treatment, it aims at creating the necessary conditions for subsequent separation by flotation. Several researchers have tested different alternatives for physical modification, which have included the addition of plasticizers to PVC to separate them more easily from PET (Guney et al. 2013), selective influence of boiling treatment on ABS or PS, or taking advantage of the difference in elasticity and mechanical resistance between the chlorinated plastics and other plastics, which gives rise to the selective twist formation for PVC and PVDC films because the chlorinated films retain bubbles more easily and in a more stable way (Reddy et al. 2010). Taking advantage that PA absorbs water more easily than other polymers, this fact can be used to easily separate it from other polymers since after being immersed in water for a few hours, and its buoyancy is affected compared to other plastics with which it is mixed (Fraunholcz 2004). When the density of the polymers to be separated is very similar and their separation by flotation is not feasible, the density of one of them may also be altered by the absorption of selected substances for this purpose. The density can be selectively altered through the absorption of a low-density solvent, thereby making PET–PVC separation viable, using after that flotation or float–sink methods (Grimm and Sehlmeyer 1986; Nugent 1991).

The diversity of methods used for the separation of polymers, as well as some incompatibilities between the results that have been obtained in laboratory works and what can be carried out on an industrial scale, clearly illustrates the difficulties in the effective separation of polymers that are normally collected together. Moreover, the same polymer may be provided with different additives, giving them different properties, also having a direct interference in the separation process. The flotation process has been the one that has deserved more attention by researchers, but one of the biggest difficulties has been to attach bubbles to the particles of the different polymers. Since bubble attachment to plastic particles is a sprightly process, it has become crucial to find an adequate physical parameter that can describe and characterize the floatability of particles. It should also be remembered that the dynamics of this process depends on factors such as the surface area of the polymer particles, hydrodynamic conditions, and size distribution of the bubbles generated. In spite of the numerous efforts of researchers to establish models capable of predicting the recoverability of a specified polymer in a given mixture, the values obtained so far have not been able to obtain sufficiently reliable results.

Another way that is particularly important in the recycling of plastics is the dissolution of polymers. Using different temperatures, the same solvent may be used to dissolve different non-separated polymers (Srivastava and Nauman 1993; Nauman 1993, 1994). This process consists of working a given mixture of polymers, usually from which the solvent will selectively act on one of the polymers, generating a solid phase which will contain polymers which are insoluble in the solvent at the initially used temperature and a solution phase. The solution phase will contain the polymer which has dissolved at the temperature which was originally used and is then drained by vaporization of the solvent, leaving behind pure polymer. The solvent is then sent back to the solid phase, where it is heated again, now at a higher temperature, to which another polymer is dissolved, again repeating the previous cycle. The temperature will have to be adjusted in each cycle so that each of the pure polymers can be dissolved (Narasimhan and Mallapragada 1998). In this process, it is found that the molecular weight of the polymer directly interferes with the dissolution rate because the higher the molecular weight, the lower the rate of dissolution achieved for the polymer. The solubility of the polymers is clearly conditioned by several marked physical and mechanical characteristics, such as glass transition temperature, wettability and surface tension, ultimate strength of the polymer, relation between thermal expansion and compressibility, or boiling point in the case of non-polar liquids (Miller-Chou and Koenig 2003). The process of dissolution of polymers in solvent entails two distinct stages: solvent diffusion and molecular chain deinterlacing. Several models have been proposed in order to clarify the physics behind the dissolution mechanisms of glassy polymers. Studies to model the dissolution of amorphous polymers can be classified as follows: (1) logical phenomena, (2) external mass transfer, (3) stress relief and molecular theories, (4) laws of scale and models of anomalous transport, and (5) molecular theories in continuous systems.

6.2 Reuse

Taking into account the existing literature, the term "reuse" does not have a consistent use. However, considering the recommendations issued by the European Commission on Waste Framework (Directive 2008/98/EC 2008), reuse refers to any operation by which products or components other than waste are used again for the same purpose for which they were designed. The repair, reconditioning, and remanufacturing of products are three activities that may lead to the reuse of a given good or material (Ijomah and Danis 2012; Bakker et al. 2014) and that can contribute significantly to avoid impacts, by deliberately generating waste when the product or material can still be reused (Geyer and Blass 2010).

In Fig. 6.9, two types of approach to the life cycle of products and materials can be seen, one more traditional (a) and one more advanced (b), both illustrating how reuse can be incorporated into the life cycle of products and materials.

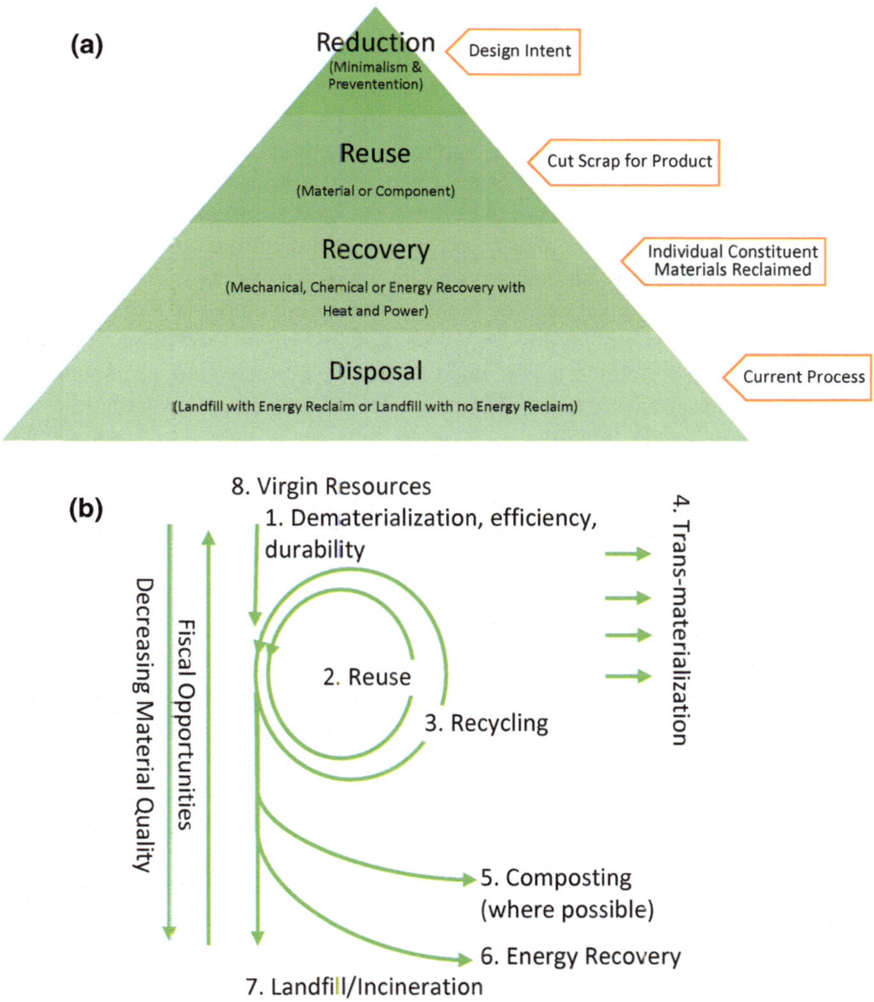

Fig. 6.9 Different approaches about the life cycle of the products: **a** traditional and **b** advanced. Adapted from Snudden et al. (2014)

The reuse of certain goods or equipment is a powerful tool that acts in a very positive way toward environmental preservation. Together with other tools, it can contribute in a very positive way to a drastic reduction in the consumer spirit deliberately taken on by some societies and help to preserve the natural resources with consequent benefits for the environment. Reuse is such a comprehensive topic that it can encompass items as simple as water, or as complex as industrial or electronic equipment. If well planned, this action may be one that spends the least amount of resources so that any good can regain a new life cycle, thus saving the environment. The number of studies on this subject is extensive, but not as much as the efforts devoted to recycling, for example, or more recently to the Circular Economy, although reuse fits perfectly in the spirit of Circular Economy, as can be seen later.

The reuse action can become effective in different ways, but always with the first aim of avoiding waste generation on the one hand, and avoiding the consumption of new resources by the other. Following are some practical examples of day-to-day actions where the will to reuse can be applied successfully: A torn piece of clothing can be sewn or made into another piece (pants can turn into shorts, for example); computers, printers, and monitors can be donated to social entities that will use them with people in need; pots and plastic bottles can be transformed into potted plants; single-sided paper sheets can be converted into sketch paper when using the blank side; old towels and blankets can be made into cleaning cloths or ballast for pets; coffee grounds can be a great ally for the healthy growth of your plants; the water used in children's swimming pools, when needing to be renovated, can be used for watering gardens: A broken piece of furniture does not have to go to waste (they can be repaired or donated); the water used for washing can be reused to wash the yard; the books used by children and young people in schools may be donated to others who use them in subsequent years; with creativity and packaging, sticks and plastic pots can be used to create several interesting toys; and resistant plastic bags can be used more than once.

It should be noted that the application of charges to the use of certain products, such as light plastic bags, acts as a disincentive to their use, but it is not enough by itself to promote a recycling policy. However, these rates can direct citizens' behavior toward reuse, which in itself is very positive for the environment. In some European countries, with an aim of increasing the reuse of the bags, taxes were implemented on the lighter plastic bags, trying to discourage their use to the detriment of stronger and more easily reusable plastic, paper, or fabric bags. This measure, when well understood by the population, prevents huge tons of plastic from being thrown into the environment/landfill, where they would take many years to decompose.

While recycling can be controlled by public authorities, making the process more demanding and dependent on the governmental funds or environmentally related public bodies, which can properly promote the process and encourage citizens to join such programs in a more or less routine and simple manner, the rooting of the principles of reuse implies other nuances. In fact, for reuse to be successful, it is vital that citizens have a much more developed environmental culture. This fact makes life harder for authorities as process efficiency is based on citizen behavior (Barr et al. 2001; Chen and Tung 2009).

Like recycling, the reuse of goods and materials implies a change in the habits of consumers, which should have certain previously nonexistent concerns (Henry et al. 2006). Nevertheless, as referred by some authors (Vermeir and Verbeke 2006; Mont and Plepys 2008), the implementation of these principles is much more difficult than those of recycling (which are already ingrained in the culture of some developed countries, such in Japan and many countries of the Northern Europe, among others), requiring different approaches, as well as a much more specific communication pathways (Ebreo and Vining 2001; Barr et al. 2001). The idea of reusing can be perfectly translated by a practical example: The water that we need to drink on a regular basis can be bought repeatedly using bottled water each time, or an own reusable bottle can be used. The change from the first to the last system will induce a significant saving in the generation of waste since countless small bottles will no longer be wasted or recycled because the water will be decanted day after day from a larger container to the bottle that can be carried with us and reused several times. This is just one example of the many situations where, with due care, we can make a daily contribution to reducing the burden on the environment by reusing certain products without any detriment to our well-being.

Next, some typical cases of the reuse of products or resources will be addressed, which, because they are of widespread use, will have a more detailed treatment.

6.2.1 Water

Water is probably the most essential asset for the survival of mankind, being responsible for much of the economic development of our society and in the way it is instituted. According to a Global Competitiveness Report of the World Economic Forum (2015), it is estimated that by the year 2030, there will be an effective water deficit in the order of 30% between the actual water needed and available freshwater for consumption. Based on this prediction, it is urgent to take measures to reduce consumption, but also to treat wastewater, generating a water management system that is appropriate for this emergent situation.

The shortage of drinking water has been widespread concern for several years and is considered as one of the biggest challenges for the next generations (UN-Water 2012). The substantial growth of the world population, coupled with the climatic changes that have occurred in the last decades, has accentuated the scarcity of water in more susceptible areas. In fact, the availability of water varies in space and time, but consumption continues to grow, driven by population growth and the development of the economic activity. In some regions, water demand clearly exceeds water availability. Molden (2007) estimated that 1.2 billion people live in physical watersheds situation, with another 1.6 billion living in areas of economic water scarcity, where affordable waterworks are not available. Moreover, the report issued by the UN in 2017 (UN 2017) refers that two-thirds of the world's population currently live in areas where water scarcity is really felt for at least one month a year, leading to realize that around 500 million people live in areas where water consumption surpasses the locally renewable water resources by a factor of two. The shortage of drinking water is also related to non-compliance with effluent treatment directives, polluting water that could be used in situations where its quality is absolutely necessary due to hygiene and public health issues. Fortunately, this situation tends to improve significantly in developed countries, and even in developing countries, so that the future can be looked at with much more optimism. Compliance with certain rules in this area is also imposed by environmental certifications, which are becoming more frequent and more valued by consumers, as mentioned above.

The water consumed in irrigation and food production represents a considerable part of freshwater consumption, and agriculture is responsible for the consumption of about 70% of the freshwater resources extracted from watercourses and river basins. In the case of countries strongly based on agriculture activities, this consumption represents more than 90% of the water withdrawn from freshwater courses and basins (FAO 2018).

The commitment of some countries to the mandatory consumption of biofuels also implies a greater consumption of water for the irrigation of the biological species that incorporate these fuels. Given that the EU's target is for about 10% of biofuels to be consumed by 2020 (European Commission 2009), this will bring an estimated 20% increase in water consumption. However, here one can consider that irrigation can be the final destination of much of the water previously consumed by the industry and later treated, that means, reused. Moreover, The International Energy Agency expects that within the next 20 years, the water needed to generate energy for population needs, including ground mobility, will have to increase by about 85%.

Reuse of wastewater in its current form is a concept that can be considered as innovative in global terms, although it has been practiced empirically for more than five millenniums in watercourses and river basins around the world. However, only in the twentieth century have studies and technology arisen which enable this process to be done in a perfectly controlled way, meeting current hygiene standards. This technology is based on the treatment of wastewater and is assumed as a practically unique way of satisfying the needs of populations where water scarcity is a constant.

Irrespective of other factors such as increasing water resources or treating contaminated water more effectively through the administration of nutrients in previously treated effluents, the possibility of reusing water has extremely positive benefits, which are the motivators for the implementation of water treatment programs and corresponding water reuse (EPA 2012). The benefits brought about by this reuse of water go through the improvement of agricultural production, treatment and distribution of drinking water and significant environmental benefits, such as a reduction in the amount of nutrients in the incoming water, since the latter, in the case of already treated wastewater, will bring a lower nutrient load (Fatta-Kassinos and Dionysiou 2016).

Unfortunately, water reuse has not yet assumed the desired proportions, and it is necessary to cope with the increasing demand for drinking water. However, there are studies that show that the panorama has changed quite favorably. The global volume for water reuse was expected at 33.7 GL/day in 2010, and it was predictable to reach

54.5 GL/day in 2015, with the highest growth in China, the USA, Middle East, North Africa, Western Europe, and South Asia (Chen et al. 2013).

Industry plays a leading role in water consumption, estimated to account for about 10% of global consumption in Asia, and about 57% in Europe, according to FAO's (2012) data. A large number of industries need large quantities of water to feed their industrial processes. Just as an example, the steel industry requires large amounts of water in the rolling process, among others. Water is so important for these types of industries that they are usually located along watercourses or in river basins, in order to facilitate the collection of water. If a few decades ago water was used and returned to the watercourse contaminated by the process, nowadays in developed countries, legislation obliges that the industries that consume water, return it to the same watercourse under similar conditions, allowing for its consumption regarding a variety of purposes, including household consumption. In this way, most industrially consumed water, depending on the process, can be returned to the place where it was collected under the same conditions, or sometimes in a better state than the one in which it was collected.

The water treatment process usually takes three phases: (1) removal of suspended materials that may affect the later stages of treatment; (2) primary sedimentation, which removes about 55% of the biodegradable solids; and (3) biological final treatment, capable of removing about 95% of biochemical oxygen demand (BOD), as well as heavy metals and organic compounds (Chon et al. 2012; Salgot et al. 2018). Usually, this is the last stage of conventional wastewater treatment. However, the different steps of the conventional treatment may not be enough to remove all the compounds present in the wastewater, and some microorganisms may survive this treatment (Al Aukidy et al. 2014). For water to be safely reused, a third treatment is necessary, which guarantees the complete removal of contaminants, such as microbial pathogens, particles or nutrients. If water reuse entails exceptional quality requirements, advanced treatment processes may be required (Malik 2014). Drewes et al. (2003) referred that essentially two different technologies exist to convert wastewater in drinking water: microfiltration and reverse osmosis. However, when using conventional techniques or even advanced technologies, there is still a risk that organic wastewater and/or pathogens may be found in treated water, which embodies a risk to public health. Moreover, Salgot et al. (2018), in a recent review, make a deep analysis of these and other techniques used to treat wastewater, converting it in reusable water for different purposes. Certain substances are often detected in sewage effluents, such as pharmaceuticals, which require special treatment, especially if water reuse includes domestic use (Jones et al. 2007). However, as stated by the National Research Council (2012), engineering has created technological systems capable to conveniently manage wastewater treatment processes, helping to mitigate all microorganisms and possible contaminants that wastewater could contain, in order to provide drinking water that meets the internationally established quality levels. Figure 6.10 attempts to describe the various possible flows between some forms of consumption and the respective reuse of water, verifying that, depending on the type of treatment applied, treated water can be destined to different types of consumption, a few more demanding than others, depending on the degree of purity ensured by each type of treatment.

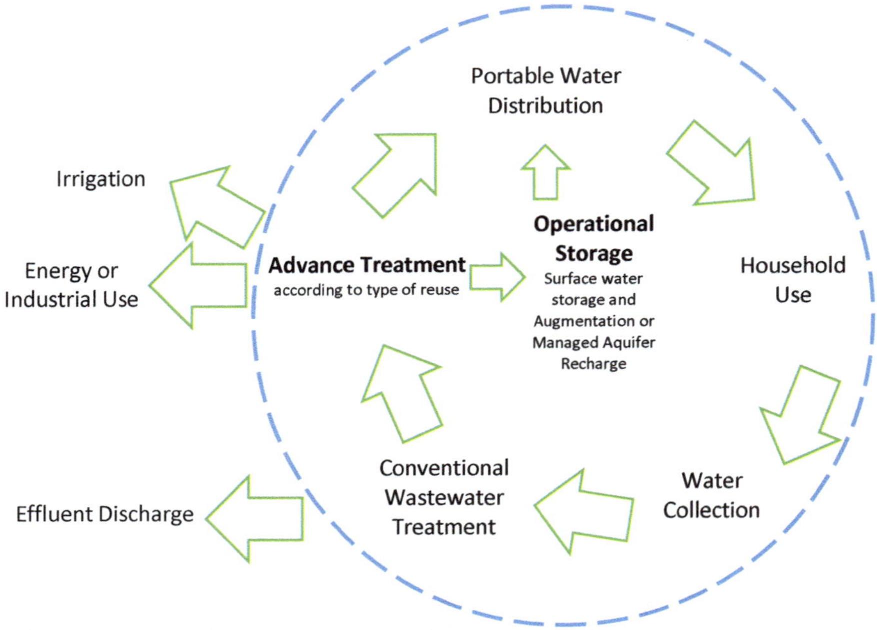

Fig. 6.10 Water reuse: possible direct and indirect flows. Adapted from Voulvoulis (2018)

In Southern European countries, such as Greece, Italy, Cyprus, or Spain (Bixio and Wintgens 2006), water reuse for irrigation has increased substantially over the past two decades and has become a common practice. In the case of islands and coastal zones, reuse of water allows to extend the life cycle of freshwater, improving the efficiency of its use and avoiding that it is prematurely sent to the sea. This reuse of water in agriculture has proved to be a policy of great utility for the environment. Taking an example from the island of Gran Canaria, according to a study carried out as early as 2007, about 20% of the water used in all sectors comes from treated wastewater, including irrigation of 5,000 hectares of tomatoes and 2,500 hectares of plantations of banana (MED-EUWI 2007). On the other hand, Cyprus set ambitious targets for 2014, in which it intended that 28% of the estimated consumption of water for irrigation would come from treated wastewater (European Environmental Agency 2012; WDD 2008). However, not all countries are conveniently exploiting water reuse. UK is an example of this, where the Environment Agency (2009) states that only about 0.16% of the 335,191,033 m^3 of treated water is actually used in irrigation of golf courses or for industrial purposes. In addition, water consumption that could potentially have as source treated water ascends to 22,000,000 m^3, of which 52% is collected in rivers and lakes, 11% is extracted from the underground water reservoirs, and 37% is from tidal water. In this last case, the water is essentially used for cooling systems.

6.2.2 *Automotive Parts*

The automotive industry represents one of the most important industrial sectors in the world, being a highly competitive and resource consuming industry, generating on the other hand a huge amount of trash which, if not taken care of properly, could pose serious problems for the environment. In this type of industry, recycling has been taken very seriously for a long time since the steel coming from the end-of-life structure of automobiles has been for many years one of the sources of raw material for steel mills. However, recycling represents a strong introduction of energy and work, to give a new life to a given material, when the structure of that material allows to be recycled. Certain end-of-life car parts assume an interesting commercial value, since they are in good condition, as manufacturers fail to produce certain parts after a few years afterward the end of the life cycle of each model, and the replacement of some parts in older vehicles starts to be compromised. Thus, a market of used parts is usually created independently of the automobile brands, feeding the needs of old car owners. Due to rapid technological development, certain car components tend to become obsolete during the life of a vehicle. In these cases, reuse becomes less appealing. However, because the age and different intensities of use, although still being in perfectly usable condition, these vehicles require periodic maintenance, making the used parts market appealing even if these parts seem obsolete by the current technology standards.

In order to analyze the capacity of a given country to recycle and reuse its resources, an indicator that assesses the use of natural resources (commonly referred to as "resource productivity") is defined as the ratio between the gross domestic product and the raw material used, including raw materials produced internally or imported and semi-finished and finished goods (Möller 2010).

As far as car production is concerned, China has become the world's largest producer of motor vehicles since 2012. Considering the effort China has been making to mitigate the environmental impacts that the high growth of its economic activity has provoked in its own country (National Development and Reform Commission of China 2010), the Chinese Government has recently devoted particular attention to the issues of remanufacturing and reuse of products and materials (Liu et al. 2017). The concepts of remanufacturing and reuse differ in their genesis, although they have common purposes. Remanufacturing can be defined as "a process of recapturing the value added to the material when a product was first manufactured" (Gray and Charter 2008). While remanufacturing entails taking a product already worn-out and recovering it, until it acquires functions similar to or better than those initially presented, reuse implies that the product or material that sees its first complete life cycle can be reused with functions very similar to the previous ones, with a minimal improvement intervention, or even without any improvement intervention. Some interesting studies have already been developed in order to analyze the energy gain and the savings of material resources obtained through the reuse of automobile engines (Liu et al. 2014) and truck injectors (Amaya et al. 2010). It has been possible to verify that, depending on the type of product, gains of between 20 and 80% can be obtained by remanufacturing or reusing certain components (Zhu et al. 2004). As is well known, the automotive industry is governed by very high quality and competitive standards, so remanufacturing or reuse policies will undoubtedly affect the key performance indicators usually used to assess the efficiency of this industry (Ng et al. 2015; Alves and de Medeiros 2015; Yang et al. 2015). A recent study conducted by Liu et al. (2018) has shown that remanufacturing and reuse is, in spite of everything, a viable alternative in the Chinese automotive industry, but for that, it is necessary that appropriate policies be implemented, such as legislation imposing such conduct, regulating the collection of components and technical barriers be broken, which avoid this practice in a current way. Moreover, the same authors draw some guidelines on how legislation should be amended, including:

- Evaluate the efficiency of the main industrial sectors in the country, analyzing performance indicators.
- Promote legislation and regulations to encourage the remanufacturing and direct reuse of products and materials.
- Promote the creation and organization of structures/sites for the collection and dismantling of end-of-life vehicles.
- Reinforce fiscal policy in order to encourage the development of remanufacturing and direct reuse companies.

Direct secondary reuse is something that has already been put into practice in Germany, notably by the Robert Bosch company since the mid-1960s, notably through components such as starters and generators (Robert Bosch GmbH 2011). However, given the quality standards that automotive companies need to exhibit due to the standardization and regulation they are subject to, reuse is confined to aftermarket only (McKenna et al. 2013). It should be noted that the aftermarket accounted for a turnover of EUR 20 billion only in Germany and only during 2008, so reuse could take advantage of this turnover to evolve very positively. Reuse at this level can also be strongly encouraged if, in political terms, specific legislation is created which obliges it to adopt this path as the solution to a better global environment. The EU has created specific legislation for some more critical sectors, such as the automotive industry (Commission Decision 2005/293/EC 2015) and electrical and electronic equipment, in order to achieve concrete results more quickly and efficiently. Regardless, other countries have created their own regulations, mainly dedicated to the reuse of products and materials, such as the German Recycling Oriented Product Development Directive (VDI 2243) or the UK Remanufacturing Standards (BSI 8887). These standards essentially aim to guide the thinking and management of companies, and also the market, in order to take care of the environment more assertively, as wells as through the saving of resources through reuse. If well applied, these standards could bring significant gains to companies in the automotive sector (Hatcher et al. 2011).

In the automotive industry, the use of fiber-reinforced plastics has grown very high in recent decades. However, the matrices of these reinforced plastics are essentially thermosets and are mixed with the fiber, which makes it unfeasible to recycle them. Due to this fact, its incorporation into cars began to be more difficult, in order to comply with the ratios of easy recyclable material that each type of car should incorporate demanded by legislation. This, however, runs counter to the aim of reducing vehicle weight with a goal of decreasing emissions, thus contributing to a better environment and complying with emission limits imposed in each region by internationally established agreements. In addition, electrically driven vehicles will need to incorporate relatively heavy batteries, which must be compensated by an equally rigid structure, but lighter. In addition, the autonomy of electric vehicles is strongly dependent on the weight they need to move, so composite materials with a polymeric matrix should continue to be one of the areas to be explored in the automotive industry. Given the difficulty of recycling these composites, reuse is the most viable path. However, for this it is necessary that the forms are compatible. Since the history of this type of components is usually known, it is easy to predict if they can be reused with the necessary safety (Snudden et al. 2014). Another factor that prevents easy reuse is the need to know the materials in question (matrix and reinforcement). Given that labeling is not feasible, this represents another handicap to the classification and reuse of the components in composite materials, making necessary the creation of accurate traceability. When reutilization cannot be integral, many composites can be crushed and reused in sandwich-like structures, or as low-cost raw materials for certain types of space filling.

According to statistics produced by the competent authorities or associations of the sector, in the USA 86% of the material used in end-of-life vehicles is recycled, reused or used for energy generation. Moreover, collecting these end-of-life vehicles, the Automotive Recyclers Association also helps the environment and the American economy, recovering about 30 million liters of gasoline/diesel, about 91 million liters of motor oil, around 30 million liters of engine coolant, about 19 million liters of windshield washer fluid, and 96% of all lead acid batteries (Automotive Recyclers Association 2018), all of which are harmful products for the environment and with commercial value. Regarding the European case, the number of end-of-life vehicles dropped from 9 million by 2009 to 6 million by 2015, being possible to observe that recycling and recovering rates are increasing steadily, in line with the EU Council expectations (EC-EUROPA-EU 2015). The values referring the reuse/recycling are within the targeted values, but reuse/recovering is still below the European political expectations.

6.2.3 *Waste from Electrical and Electronic Equipment (WEEE)*

The strong technological evolution that has been felt in the last decades has led many electric and electronic products to be discarded long before they are worn down by their intensive use. This situation has created enormous amounts of electronic and electronic waste (Yazici and Deveci 2015; Scruggs et al. 2016; Tansel 2017). The situation has been aggravated by global population growth and technological disruptions, such as the switch from cathode-ray tube to liquid crystal display or light-emitting diodes televisions, and from traditional keypad technology in mobile phones to touch-screen technology. Following the Global e-Waste Monitor 2017 report (The Global e-Waste Monitor 2017), a total WEEE (e-waste) volume of 44.7 million metric tons was recorded, expecting to reach 52.2 million metric tons in 2021, corresponding to 6.1 and 6.8 kg/inhabitant, respectively, as can be analyzed in the graph of Fig. 6.11. These values are different from those presented in the recycling section, trying to show in this way the evolution that has existed in the presented values and the diversity of sources that can be found with statistics on e-waste.

As the main producers of this type of waste are the countries where there is greater economic power, it is common to see the transfer of this waste from developed countries to developing ones. This garbage contains a bit of everything, from heavily polluting materials such as brominated and chlorinated flame retardants, as well as metals considered toxic such as lead, cadmium, chromium, and mercury, to precious metals such as gold, silver, and palladium (Awasthi et al. 2016). Much of this waste follows the traditional processes of treatment, which go through an initial phase of dismantling, followed by acid leaching and then open burning, which causes the generation and release of toxic fumes, as well as heavy metals. Some of these pollutants are released to the atmosphere and others to the soil, which may reach watercourses.

In the waste treatment hierarchy in general (EU 2008), and regarding the particular case of WEEE, reuse will always be the pathway that will lead to fewer environmental burdens, although it is not a final resolution of the problem, but rather a postponement, since issues related to the presence of harmful products to the environment, heavy metals and others, will remain. However, by prolonging the life of the equipment or its components, it is avoided that new equipment is manufactured in which other products harmful to the environment will be used (Ruediger et al. 2011).

Fig. 6.11 Global WEEE generated until 2016 and estimated from 2017 to 2021, as well as volume of WEEE generated per inhabitant in the same periods. Adapted from The Global e-Waste Monitor (2017)

As reported by Zlamparet et al. (2017), the best way to treat WEEE is to reuse it since the equipment is in a functional state. An alternative way, if the device is not fully functional, involves the individualization or extraction of the components that have the highest commercial value, with care due to the sensitivity of some components, and apply them in products of low cadence of manufacture or in the repair of damaged equipment (Kissling et al. 2012, 2013). When none of the above paths is valid, due to none of the conditions being met, the equipment should proceed to mechanical recycling which, according to a model developed by Zeng et al. (2015b) purposefully studied for mobile phones, could consist of two distinct phases: first manual dismantling followed by mechanical recycling. This approach proved to be quite valid when applied to computers and their peripheral devices, according to a study carried out by Oguchi et al. (2012).

Due to the very high technological evolution and characteristic specificity of some components, WEEE is not easy to reuse, so most of the studies carried out to date are essentially recycling (Parajuly and Wenzel 2017). The above reasons also mean that policies on what to do with the WEEE are not uniform. Taking China as an example, reuse is not considered in the existing regulations at the moment for the treatment of WEEE (Lu et al. 2018). However, in the EU there is a directive that has its own objectives for reuse, but integrated into the recycling regulations. Given the specificity of this sector again, it will be extremely important to analyze the evolution of reuse policies and draw the appropriate conclusions so as to try to standardize objectives and strategies to achieve them. Of the few existing studies on reuse of WEEE, which have essentially exploited the potential for reuse by product category (Bovea et al. 2016), it can be concluded essentially that it will be necessary to improve the management policy and implementation of the reuse of this type of waste (Hickey and Fitzpatrick 2016). Moreover, there are already some ways that define what reuse is, what care must be taken in the collection in order to ensure that the products properly retain their characteristics to be reused, among other important factors. The most relevant European standards in this field are EN 50625 (Collection, logistics and Treatment requirements for WEEE—Part 1: General treatment requirements), EN 50614 (Requirements for the preparation of waste electrical and electronic equipment), and EN 50574 (Collection, logistics and treatment requirements for end-of-life household appliances containing volatile fluorocarbons or volatile hydrocarbons). Specifically, in Germany, VDI 2343 establishes the main concepts related to the reuse of this type of equipment, but also describes recommendations for the preselection of which equipment can be included in the reusable group, separating them from those that cannot be reusable and will have to be recycled. Preparation for reuse should be seen as an indispensable step in this process (Seyring et al. 2015), and recycling and reuse should be definitively distinguished in the future through well-defined procedures, and established clear rates for each kind of waste treatment, definitively encouraging the route of reuse of WEEE as a priority.

6.3 Environmental Producer Responsibility

The EPR has essentially two distinct objectives: (1) improvement of product and system design and (2) improvement of the end-of-life treatment system, ensuring high levels of product collection, treatment, reuse, or recycling, while preserving the environment and society adequately (Manomaivibool 2009). However, companies have devoted much more attention to the second strand, i.e., how the product is treated when the end-of-life phase is reached, than to ensure during the design stage that the manufacturing could be environmentally friendly and that the product could certainly be reused or recycled in its end-of-life, without being a heavy burden for the environment. This may be due to the heavy fines which companies will have to bear if they do not guarantee a proper life end for the product which is not liable to cause harm to the environment.

The EPR should be considered by companies as a method of significantly improving their efficiency in the consumption of resources, being adopted strategically as a market stance. Based on this business policy, the environmental burden is reduced, as companies are given financial and physical responsibility over the products they produce, taking into account the product's life cycle and its end-of-life (Choi 2017). In fact, the EPR blames the producer for the protection of the environment in each step of the life cycle of the products that the company conceives and produces (Lifset et al. 2013). In essence, the product impact has to be internalized by companies in order to encourage producers to increase their care with the entire process of product design, production, and logistics, minimizing their environmental burden. Otherwise, they have to pay higher costs due to the damage caused by their products to the environment. The efforts made by each country to promote the EPR have not been uniform, essentially based on a voluntary basis, but whose regulations have evolved remarkably, with more than seventy regulations issued by the USA in the last two decades (Nash and Bosso 2013).

Environmental producer responsibility has been extensively studied and debated, involving different stakeholders, but concentrating its perspective essentially on one of the stakeholders: the producing companies or the market. In fact, there has been a line of thought that normally assigns higher costs to companies that take on a greater role in terms of environmental responsibility. This is an old way of thinking, which believes that this type of management can destroy wealth without any advantage (Palmer et al. 1995). However, this view is not consensual because there is another chain of thought that advocates that companies able to assume true environmental responsibility can take advantage of this, achieving greater market shares through a recognition by this market of their way of acting and, with that, larger volumes of business (Porter and Kramer 2011). In more recent work, Wang et al. (2016) argue that a more active position of companies in terms of sustainability with better economic performance is perfectly compatible, generating together greater wealth both for private companies and for the public sector. It is a consensus that stakeholders value the fact that the advances made in the environmental performance of companies are always closely intertwined with the financial performance component so that the balance between the costs generated by compliance with the environmental rules is duly compensated by the generation of a larger volume of revenues, better market recognition and, eventually, higher profit margins (Freeman 1984: Latan et al. 2018). Other authors argue that the environmental performance of companies is closely linked to economic performance since the company, when complying with the legislation, is not under the authority's focus, will not be subject to fines for non-compliance, will no longer require to consume money for its own defense in legal actions and can guarantee to reduce the number of complaints, boycotts, and customer returns, significantly improving its image in the market, gaining advantage (Cordeiro and Tewari 2015; Wang et al. 2016). However, as stated by Boesso and Kumar (2009), companies cannot satisfy all the interests of their stakeholders, due to the limitations of ever-existing resources, with managers having to prioritize and comply with the requirements that represent a greater risk potential for the company in terms of compliance with legal requirements and may be contained within the company's investment capacity (Dal Maso et al. 2018). Although the principles of this process are extremely clear, recent studies in the area of electrical and electronic material production and tires have shown, through examples, that some companies, although voluntarily adopting programs of this nature, did not fully understand the real objectives to be achieved with this type of accountability, that is, they have not been able to analyze the product since its design to its entire life cycle, in order to minimize the environmental impacts (Rodrigues et al. 2016; Richter and Koppejan 2016). If recycling entails high-energy costs and intensive manual labor, it is also true that there is a significant saving in the preparation of virgin material and that the resource environment is spared. This was also verified by Choi (2017) in a work carried out based on EPR applied to the manufacture of carpets in the USA.

As mentioned earlier, the extended producer responsibility aims to internalize the costs with the environmental impact on the part of the producer (Forslind 2009; OECD 2001). This was the first step toward putting the polluter pays principle into practice (Wang et al. 2017). Indeed, if a producer has several alternatives to design a product and opts for the one that brings the greatest burden on the environment, they should contribute to the process of recovering that material or to minimize the damage caused by that product. However, this new philosophy imposed on producers aims to use other available tools, such as eco-design, a tool that is one of the pillars of the Circular Economy. In fact, the EPR also aims to ensure that the environmental costs of a given product are not generating fees to cover its post-treatment, but rather encourage producers to implement CP principles from the product design so that they will have a much lower environmental impact at the end of their useful life. In part, this prevents governments from collecting fees that they will later have to spend on investments to minimize the environmental impacts of certain products. By trying to get producers to invest in materials that will be easier to treat at the end of the product life, using recycling, reusing or other processes, lowers the burden put on the Nature. Thus, the main concern of companies should be focused on how to project a product well and to think about which materials at the end of the product life will ensure better conditions to be reused or recycled using the lowest possible energy level and spending the least amount of other resources. Knowing and properly thinking about the materials to use early in the design phase of the product will greatly help companies avoid the need to pay extra fees for recycling organizations to do so.

However, EPR policies will need to be duly considered, as they should combine and establish (Hickle 2014):

- The responsibilities of each stakeholder, including governments, producers, distributors, and consumers;
- The financing mechanisms of programs, both by means of internalization of costs and through eco-taxes;
- Plans that help identify those companies that are capable of integrating programs that lead to the implementation of EPRs;
- Establishment of indicators to help quantify the evolution of plans and define the levels of recycling and reuse to be achieved.

As a country that has experienced a remarkable industrial increase in the last decades, China has been the target of many stakeholders in trying to understand how it is following the latest trends in environmental protection. Since the year 2000, the Chinese Government has published immense legislation for the protection of the environment, and in 2017, it has finally published regulations on EPR for companies operating in China, namely the Implementation of extended producer responsibility system (Peng et al. 2018). In other countries where the EPR is already properly implemented, manufacturers must take responsibility for the end-of-life of their products following different pathways. In the case of California, for example, producers need to pay a recycling fee to a government fund, which is responsible for the operationalization of this recycling program. In Japan, there is an obligation to recycle all collected WEEE, but the program is supported through fees paid by the last users of the product, system that is shown schematically in Fig. 6.12. In South Korea, each producer is responsible for collecting and processing a certain share of the e-waste that is produced in the country during a given year. In the European Union, the WEEE Directive (2002/96/EC), as well as the Swiss and Norwegian programs, attribute a very broad degree of responsibility to producers, although the transposition of these regulations for each of the 27 member states varies considerably (Sander et al. 2007; Huisman et al. 2007; Manomaivibool 2009). The evolution of EPR systems in some developed and developing countries can be seen in detail in Wang et al. (2017).

However, companies to successfully adhere to an EPR system need to fulfill some prerequisites, which have been enumerated by several authors who have studied the implementation of this type of system, namely

- Companies should have a thorough knowledge of the principles governing environmental protection (Zhao and Zhang 2008).
- The way companies manage their production systems must respond appropriately and timely to the environmental requirements that are being established (Peng et al. 2018).
- Companies should design products that meet environmental requirements from the design stage to the end-of-life stage (Peng et al. 2018).

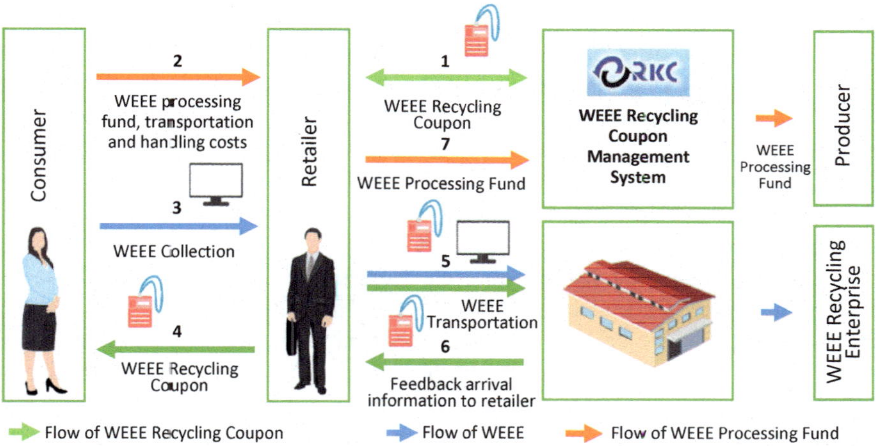

Fig. 6.12 Scheme on how the EPR system works in Japan regarding the WEEE. Adapted from Wang et al. (2017)

- The supply of environmentally friendly raw materials which are effectively recycled is also important components of the EPR system (Zhou et al. 2017). Moreover, materials with regenerative capacity technologically improve the product, reducing its cost, and adding economic benefits to the company (Atasu and Subramanian 2012).

Thus, rather than unconsciously adopting an EPR system, companies must first prepare themselves properly, both in terms of management and in terms of training their human resources so that adherence to one of these programs can be successfully implemented.

6.4 Circular Economy

Circular Economy can be defined as a new economic model operating in closed loops, catalyzed by innovation along the entire value chain, which is considered as an alternative solution to minimize material consumption and energy losses. Thus, it is intended that through design, materials can be reduced, reused, or recycled, keeping the value and usefulness of products, components, or materials, always at their highest level (EMAF 2015). The Circular Economy is based on a strategic concept aimed at the reduction, reuse, recovery, and recycling of resources in general, and in particular materials and energy, i.e., aiming at help to pursue the environmental sustainability goals (Homrich et al. 2018). In this philosophy, the end-of-life concept of a product faced as it usually was in the conventional economy and is replaced by new circular flows of reuse, remodeling, and renovation in an integrated process. This new philosophy, which integrates several previously developed tools, is seen as a key element to promote the decoupling between economic growth and the increase in the consumption of resources, a relation hitherto seen as impossible to break. It can be said that the concept of Circular Economy was inspired by the mechanisms of natural ecosystems, which manage the long-term resources in a continuous process of regeneration. The Circular Economy promotes a reorganized economic model, through the coordination of production and consumption systems in a closed loop. It is characterized by a dynamic process, which requires technical and economic compatibility, through the reconciliation of capacity and productive activities, but also requires a social and institutional framework since it must be based on a fairly well-defined base of social values. The Circular Economy goes beyond the scope and stern focus of the actions of waste management and recycling, intending to have a much broader field of action. In this way, and following what will happen with the EPR systems implementation, the action of the Circular Economy must be based on the reformulation of processes, rethinking of products and redesigning of business models, culminating in the optimization of resource utilization, making them circulate in a closed loop and in an efficient way. The concept of the Circular Economy also emphasizes the use of clean energy, namely solar, wind, biomass, and waste energy reuse as an energy source capable of feeding the entire value chain of the product (Rashid et al. 2013).

As discussed by Stahel (2016), the concepts of Circular Economics began to be successfully applied in Denmark in the 1990s on a controlled scale, starting with eco-industrial parks such as the Symbiosis Kalundborg. However, the principle was also successfully applied by large US manufacturing companies, namely through xerox, which began selling equipment modules as part of the assistance services, or by Caterpillar, which began to refurbish and sell used diesel engines.

The application of the concepts of Circular Economy should give rise to the development of economically viable and ecologically efficient new products and services, based on ideally everlasting cycles of upstream and downstream reconversion. It is expected that these actions materialize in minimizing the use of resources, trying to maximize reuse and increase the overall efficiency of companies (EC 2014). In Fig. 6.13, the way in which the different CP tools are articulated can be analyzed, forming a closed cycle that aims to achieve the aforementioned purposes.

Looking at Fig. 6.13, it may also be noted that the genesis of the concept is to avoid the consumption of raw materials through internal cycles, extending the useful life of the resources by using the 4Rs tool (reduce, reuse, recycle, recover), which should be articulated with other tools with the same purpose, namely eco-design, so that the design phase is also considered in the same perspective (Korhonen et al. 2018).

The concept of Circular Economy arises from the awareness that society has begun to have concern about the form and speed at which resources are being consumed, essentially taking into account two factors: the growth of the world population and greater availability of financial resources for an increasing number of people. Given that just in 2010 around 65 billion tons of raw materials needed to be introduced into the economic system, and that 82 billion tons of raw materials are expected to be needed by 2020, it is possible to realize that, at this rate, the world population

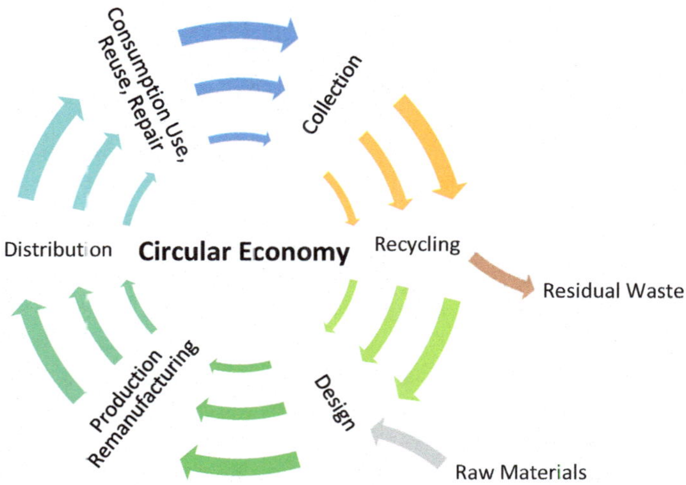

Fig. 6.13 How circular economy incorporates several Cleaner Production tools

may suffer from a shortage of certain types of raw materials in the medium term. In addition to the problems that global society may encounter in the medium term, the overconsumption of resources may also contribute to a greater susceptibility to markets, as the dependence on certain raw materials may subject companies or a particular country to speculation which may interfere decisively with the market and economy of a particular country or region.

The main benefits of the Circular Economy are:

- Promote eco-innovation.
- Create new business opportunities, as well as new products and services.
- Extend the life cycle of products and keep materials in the economic circuit for as long as possible.
- Design new business approaches.
- Reduce the dependence on fossil fuels.
- Decrease carbon emissions.
- Minimize waste generation as much as possible.
- Preserve Nature, respecting the natural periods of renewability.
- Contribute to the fight against climate change.

The concept of Circular Economy has been strongly encouraged by the European Union and followed by many other countries, such as Japan, Canada, China, UK, Finland, Sweden, the Netherlands, among others. According to reports from the European Commission, the shift from the concept of conventional to Circular Economy could generate gains in the order of 600 billion euros per year, taking into account only the EU industrial sector (EU 2014a, 2015). Another European Union study states that the implementation of the Circular Economy could reduce CO_2 emissions by 48% in Europe, which would generate an economic benefit estimated at 1.8 trillion Euros and the generation of around 2 million new jobs by 2030 (EU 2014b). A study by seven European countries concluded that implementation of the Circular Economy in these countries would help reduce greenhouse gas emissions by 70% and increase jobs by about 4% (Wijkman and Skanberg 2016). The studies, announcements, and appeals of the European Union and some other developed countries in order to stimulate the adoption of the Circular Economy have been noted (Lacy and Rutqvist 2016), always referring to very high economic and political—social benefits, but the transition to practice has not reflected this commitment to the cause (Ghisellini et al. 2016; Stahel 2016). However, China was the first country to turn into effective legislation the purpose of adopting the principles of Circular Economy through the Circular Economy Promotion Law of the People's Republic of China in 2009, having even anticipated it relatively to Europe. Thus, the principles of Circular Economy have been imposed in China through regulations (Geng et al. 2012; Geissdoerfer et al. 2017).

In fact, the concept of Circular Economy has been developed predominantly on a political and economic basis, but the ecological issue has been lacking since the facet of economic gains and the generation of other business models has superimposed on the main factor that should preside over the concept of Circular Economy: the preservation of natural resources and environment. In fact, the Circular Economy should be analyzed according to three different aspects: (a) maintenance or improvement of economic performance, (b) environmental sustainability, and (c) sustainable social model preservation (Korhonen et al. 2018).

Indeed, the Circular Economy may seem like an easy concept to absorb and put into practice, but there is a huge complexity behind it, since Nature is equally complex and the network created by man to meet his needs is also very intricate. The example pointed out by Korhonen (2001) is very illustrative of this complexity: When forest residues from tree cutting are used for the production of energy, we are avoiding the consumption of fossil fuels on the one hand, but at the same time we are removing from the forest parts of trees, barks, and others, all rich in nutrients, which will affect ecosystem balance, since they contribute positively to the balance of the forest and biodiversity. In addition, the recovery of these forest residues consumes economic and material resources, as it requires equipment, materials, and energy to do so. In addition, there is the release of by-products, such as lubricants and others. Therefore, each case needs to be properly studied in order to analyze the advantages and drawbacks, verifying if the balance is really positive for the environment and economically sustainable.

It is assured that the Circular Economy intends to conduct the flow of raw materials in a different way, allowing them to be reused as many times as possible, before being deposited in a landfill (if that is the case), as shown in Fig. 6.14, which is an evolution of Fig. 6.13. Given the multiplicity of raw materials and the different treatments that each need, and the way in which it is incorporated into the product, logistics in terms of raw materials can change profoundly, and the extraction or synthesis of raw materials may need to decrease significantly, while employment around collection, services for reuse, or recycling may suffer a significant increase. All of these changes require some time to adapt in order to keep society prepared for the jobs needed at each moment

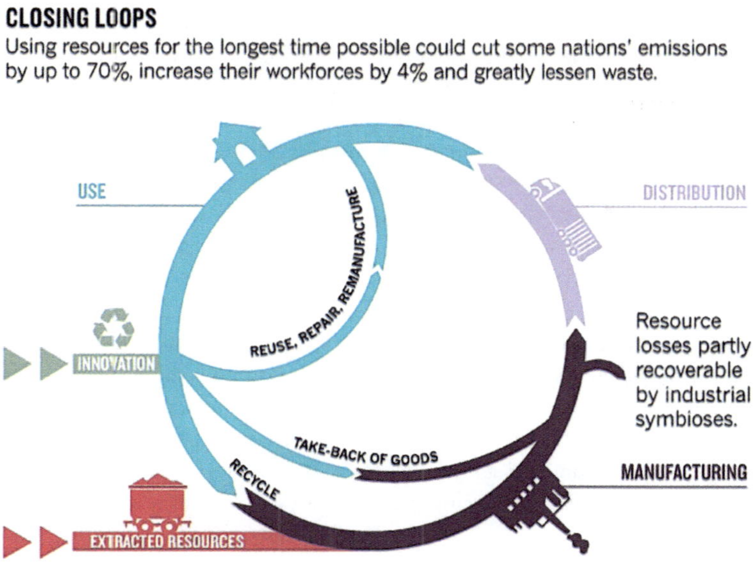

Fig. 6.14 Diagram about the proposed flows which are the core idea of the Circular Economy. Adapted from Stahel (2016)

since the change of functions requires adaptability and training by the employees. Moreover, technology will certainly evolve significantly around reuse, recovery, or recycling, forcing new facilities and work procedures to appear, which also requires time and preparation of society for this purpose. The logistics around these new materials flows will be different and will require facing business in a different way. Material collecting and segregation, as well as the recycling process, will be done in a different way from the usual supply of raw materials because the provenance will be much more diversified, the quantities may be significantly different, and the quality procedures will need to be more accurate, which will lead to some uncertainty in how all these flows will proceed until they can be optimized. The complexity of the situation may also increase if specialization is implemented, segmenting the flows, which will imply new algorithms for managing this type of business. Thus, it is logical to say that there is still an enormous degree of uncertainty on how the Circular Economy can function optimally in the future. Logistics also plays an extremely important role in these issues since the movement of raw materials and products needs to allocate human and material resources, in addition to consuming energy, which, at present are mostly from fossil origin. Thus, it will be necessary to carry out studies in order to optimize flows in this new reality. However, this seems to be a sure path to success in terms of reducing the consumption of virgin raw materials and a very assertive way of minimizing the effects of world population growth and consumerism on the environment.

In addition to all the positive effects that the Circular Economy may have, there are also some problems that need to be addressed. When a developed country decides to increase Nature conservation and eco-efficiency through legislation, the effects will be felt in the poorest or least well-prepared border countries, as migration from the most pollutants to these countries, due to the pressure exerted by fees and the obligation to comply with certain regulations (Mayer et al. 2005). If this migration makes economic sense, in social terms it has very harmful effects, since, despite apparently generating more employment, it also generates greater concerns in terms of population health, with future costs in terms of repairing these negative effects. Therefore, it can be seen that if the Circular Economy is implemented hastily and politically, the side effects can be very harmful in social terms. However, because the principle associated to Circular Economy is not merely the improvement of the environment, but also society harmonization, allowing the necessary stability for an improved lifestyle, for this to be a reality, the implementation will have to be carried out gradually, well explained and, essentially, well understood by society, without there being a usurpation of the principles of this philosophy by the large economic groups, disturbing the natural development of society in this process.

Despite all the advantages pointed out before, there are extremely skeptical researchers regarding the concept of Circular Economy (Skene 2017; Korhonen et al. 2018). Contrasting the sound numbers presented by the political class in relation to the gains obtained with the implementation of this philosophy, Korhonen et al. (2018) argue that:

- The disorderly development of countries where the population grows more strongly, or where the quality of life becomes better (which increases the consumption almost linearly) and the urbanization of more areas in these countries, will offset the gains expected to be achieved by developed countries through the application of Circular Economy. This argument, although valid, also shows that if nothing is done, the situation would become much worse, not making sense to think that way.
- When a country makes efforts to improve its environmental performance and moves more pollutant companies to poorer neighboring countries, it will contribute to a degradation of biodiversity in the poorest country to the detriment of the richest country, but the gains are smaller than the losses.
- It will always be difficult to make prevail a project based on Circular Economy principles with stakeholders if it does not have significant economic, social, and environmental benefits at the same time. These circumstances arise from the natural resistance that operators have in changing procedures and business models that, given the previously existing paradigm, worked perfectly.

In a recent study, Kirchherr et al. (2018) identified some non-technical barriers for the implementation of the Circular Economy, which are essentially linked to cultural factors. In a first approach, two non-expected barriers were recognized: (a) lack of consumer interest and awareness and (b) uncertain company culture. Based on this conclusion, it is obvious that a lot of fieldwork is lacking in terms of society's awareness regarding this issue, as well as research and development work concerning the study of new models around the one initially proposed is much less important. Other barriers that intersect with the previously focused ones were also identified as possible causes of the low application of the concepts of Circular Economy, namely low price of raw materials and high initial investment. Indeed, the low price of raw materials leads to disinterest in the use of end-of-life material, leading to a lack of interest and awareness of stakeholders, as previously noted. On the other hand, the need for a high initial investment leads to the great majority of companies preferring to operate in the Linear Economy system, rather than switching to the Circular Economy system.

As reuse is the most sustainable model of extending the life of a product within the tools developed so far, involving a new form of sharing that may not add value to the economy since it is not a business and can be done through nonprofit electronic platforms, creates a natural resistance in business players, who will try to find alternatives that can generate added value. This will be a not negligible resistance factor at all since current market paradigms show a strong marketing influence on consumption patterns.

Stahel (2016) argues that the future of Circular Economy must contain important achievements in two ways:

- Regarding the materials, it is needed to deepen the knowledge on how to separate/divide molecules of materials at the atomic level, so that there can be greater flexibility in their reuse, without the need to extract more natural resources. In this study, the automobile chassis is mentioned, for example, where more than a dozen different alloys of steel and aluminum are used, and it is necessary to identify them properly and reuse them in the simplest possible way.
- It is recognized that the main efforts of the Circular Economy are being carried out by large companies so far and only very weakly by SMEs. It is clear that SMEs do not have enough structure or know-how to act at certain levels, so it will be imperative that the development and knowledge developed in universities be transferred to these companies, substantially extending the application of the Circular Economy in the market. Hiring graduates with know-how on business management and technical aspects, SMEs can evolve in the right direction much more rapidly.
- Through the application of fees, it is incumbent upon governments to discourage companies from continuing to practice the traditional economic system, by directing them through legislation for Circular Economy practices. The value-added tax (VAT) should be charged only to activities involving resource extraction, construction, or manufacturing, exempting from fees the activities of reuse, repair, and reconversion. Activities that reduce the need to use fossil fuels should be compensated.
- It is incumbent on science to unlock some of the remaining limitations in reusing materials, creating new avenues, and simplifying processes to achieve this. Thus, there is the need to develop new technologies able to depolymerize, de-alloy, delaminate, devulcanize, decoat, and so on.

- Communication to society of these principles must be done in a clear and massive way, through the means that people usually use to inform themselves, and not only through scientific journals, which are consulted only by a small part of society. Communication has to be massive and overlap with established economic interests.

Given the balance that needs to be achieved in the near future, it will have to be understood how the gains obtained from saving materials and others through the Circular Economy can be channeled toward the development of sustainable consumption habits. As stated by Korhonen et al. (2018), one of the main challenges of the Circular Economy is to broach tools that can act in consumption so that the current standards can be significantly altered and society acquires patterns of sustainable consumption. Otherwise, the Circular Economy may become a simple technical tool, which will have a modest contribution to sustainability, considering that the economy as we know it today is based on unsustainable paradigms.

Although utopian, it will be necessary to change the paradigms of society in general, moving from a philosophy in which happiness is essentially based on the generation of wealth, which, in turn, allows the acquisition and ownership of material goods and services, to be based on a perspective of global social balance and with Nature, guaranteeing a better future for the next generations, without affecting their quality of life at present. However, many generations will be needed to put in practice this type of economy and society, and it will inevitably lead to a homogenization of human ambition, a situation that can hardly be achieved without another of the cataclysms that marked the history of the twentieth century, and which made society reorganize itself in an almost global term. This time, the situation may again be generated by Man, but in an indirect way, acting against Nature in such an aggressive way that it forces society to rethink its interaction with the planet.

6.5 Eco-design

Eco-design is the systematic integration of environmental considerations into the product design process, which should be understood as goods and services. The main goal of eco-design is to develop products that contribute to sustainability by reducing its environmental impact throughout the life cycle, along with requirements such as functionality, quality, safety, cost, ease of production, ergonomics, and aesthetics (ISO 2011). The definition of eco-design can be considered as the European equivalent of the Design for the Environment (DfE) used in the USA, which aims to reduce the environmental impact of products and services throughout their life cycle. According to the European Union's vision of eco-design (Charter and Tischner 2001), it is to improve the environmental performance of products throughout their life cycle (production, use, and end-of-life) by systematically integrating environmental aspects into an early design phase of a product. The concept of eco-design or eco-design reflects a worldwide trend that integrates the areas of architecture, engineering, and design with the aim of reducing the use of non-renewable resources and minimizing their environmental impact. Eco-design, in addition to its technological function of resource optimization, also includes a strong didactic component on consumer awareness regarding the negative impact that the productive process can have on the environment, providing alternatives on how to minimize these impacts, leading the consumer to choose products, systems, and services that are environmentally friendly. However, eco-design also induces recognition by society that we must recreate a closer proximity to Nature and draw the proper inferences from the processes of regeneration that Nature itself possesses, applying them in the best possible way to the material world created by man.

The relationship between design and ecology dates back to the 1970s and is commonly attributed to Victor Papanek, an Austrian emigrant in the USA, a designer and a professor of the profession. He was a visionary man, misunderstood in his day, who from the beginning defended the existence of an intrinsic relation between the design of products and the natural surroundings. His biggest contribution to the promotion of the concept of eco-design was the book "Design for the Real World," where he expresses his concern regarding the relationship between man and Nature as well as the role of design in the production of goods. There are, of course, other important names in the field, such as Buckminster Fuller, an American designer, architect, inventor, and author (Vieira et al. 2013).

It can be said that the most ambitious goal of the new trends in ecology, when applied to economy, is to make society think and act in a truly sustainable way in all its aspects, which implies a change in consumption habits, production concepts, and even postconsumer behavior—leading to the concept of sustainable design, i.e., the application of these principles to the conception of products. However, this change in behavior seems to be still quite utopian in relation to the standards that citizens and the economy adopt in their daily lives, and what is currently being discussed and tentatively applied as the concept of eco-design, an intermediate level toward sustainable design. Such radical change in the paradigm of society in which we live can only come from the creation of new lines of thought and innovation, being impossible to achieve them based on the current paradigms on which our social and economic life is based, and which has led us in a gradual but consistent way, to the environmental crisis and climate changes that we are currently feeling. About 80% of the environmental impact caused by a product is conditioned by the way it is designed, which includes the materials that it is built out of, as well as the manufacturing processes foreseen for its production (McAloone and Bey 2009). Thus, it is believed that by acting during the design phase, a product can be much more environmentally friendly throughout its life, as well as when it reaches its end of life cycle (InEDIC 2011).

All products have environmental impacts, which can occur at any stage in their life cycle: raw material extraction, manufacturing, distribution, use, and end-of-life. Lifecycle impacts can range from reduced to significant and from short-term to long run and can occur at local, regional, or global levels. Integrating environmental considerations from the beginning of the product development process is the most effective way to introduce changes that positively affect a given environmental profile at all lifecycle stages. Opting for eco-design means that environmental criteria will also have to be taken into account when it is necessary to make decisions throughout the creative process of a new product or reformulation of an existing one, with the ecological factor being an additional factor to all others that are usually taken into account (costs, quality, functionality, aesthetics, safety, and so on).

It is extremely important to have a complete overview of the entire product life cycle so as to clearly identify all the inputs and outputs of the process that have a direct influence on environmental degradation, and this analysis should not be confined only to the production process itself, but must consider all stages of the product life cycle from conception to the end-of-life of the product. The next step is to minimize the amount of required resources (materials, water, and energy), as well as minimize anything that can disrupt the environment at each stage of the product life cycle (emissions, effluent, waste, landfilling, etc.), i.e., finding the best balance possible in order to minimize the overall impact of the product throughout its entire life. This requires a great involvement of the top management of the companies that go down this route, being necessary that the entire structure of the company be focused on the same objectives.

According to the model proposed by Rozenfeld (Rozenfeld et al. 2006; Rozenfeld 2007; Amaral and Rozenfeld 2007), the management of eco-design practices in the product development process comprises the following stages: product, information design, conceptual design, detailed design, production preparation, product launch, product tracking, and monitoring, as well as other generic activities. The management of eco-design implementation processes in practice usually encompasses twelve well-defined and well-known areas:

- Eco-design drivers and identification of new technologies and process development;
- Technological strategy and behavior of technologies in environmental terms;
- Development of eco-design support and training processes;
- Awareness-raising, communication, and incentives to use eco-design principles;
- Strategies for the manufacturing, packaging, distribution, and end-of-life of products;
- Strategic management of eco-design implementation;
- Portfolio management, environmental performance assessment, and trends;
- Concept and product requirements, eco-design guidelines and trade-off management;
- Provision of services, legislation, and standards;
- Procedures related to the value chain and corresponding strategy;
- Management and integration of eco-design principles, benchmarking and selection of the most appropriate tools;
- Environmental viability, stage evaluation, and eco-design performance measurement.

Adoption of eco-design principles in practical terms is usually aggregated into three different levels of action: (a) strategies; (b) guidelines; and (c) design options. Indeed, the strategy used in the development of a new product could: minimize the use of raw materials, use 100% recyclable materials, lower the energy that the product will consume in service or during its useful life, increase the life span of a product, or determine ways of reusing the product, as well as definition of the target market and possible suppliers for raw materials. As regards for guidelines, they may include shape optimization, materials constraints to be used, and restrictions in terms of packaging. Finally, the design options will be able to correlate the materials to be used with the necessary manufacturing processes and the energy they will consume in order to minimize harmful effects on the environment. By implementing eco-design in the development of their products, companies can reap benefits in various sectors of their production process, such as:

- Raw materials: Significant savings can be achieved on necessary raw materials as products are designed to last longer and to be reused after the end of their first life cycle. Less need to extract materials from Nature leads to a decrease in impact on the environment.
- Energy: As mentioned above for raw materials, eco-design usually has a positive effect on the energy consumption necessary for each product. The simple fact that one can reuse certain equipment has the effect that all the energy that would be necessary for its manufacturing is no longer necessary. Moreover, to admit the possibility that a product can regain a new life cycle, through reuse, remanufacturing, or recycling strategies, instead of being deposited in a landfill, allows to recover the energy incorporated in it.
- Transport: Some eco-design strategies require special logistics in order to return the product after the end of the first life cycle. However, the reduction in the consumption of raw materials regarding new products conceived under the eco-design principles makes unnecessary many vital transport trips, such as international transport, thus compensating for the increase in logistics caused by the collection of discarded material.
- Waste: The implementation of eco-design concepts will allow a drastic reduction in waste generation, taking into account the strategies considered in product development. Generally, the strategy is to discard only what has no potential to be reused.

However, there are also well-identified barriers to the implementation of eco-design by companies, namely (InEDIC 2011; Brones and Carvalho 2015)

- Difficulties in understanding eco-design by the majority of producers, customers, and users of products;
- Poor perception of the environmental impact of products by producers and corresponding stakeholders;
- Belief that eco-design necessarily implies a high investment;
- Resistance to changing current production and consumption practices;
- Lack of training in environmental aspects and eco-design;
- Gaps in environmental studies;
- Belief that eco-design implies an increase in costs and human resources;
- Technical difficulties in adapting to new developments;
- Difficulties in creating and putting into operation interdisciplinary teams.

Although the concept of eco-design dates back to the 1970s, and its strong evolution has been in the 1990s, more than two decades later there is still some inertia in its implementation. Researchers and scholars have devoted huge efforts in terms of research to this subject and have consistently attributed the reasons for failure to two different strands: (a) the so-called hard eco-design, which is essentially related to the technical component on how to apply eco-design tools and (b) the so-called soft eco-design, which is almost strictly related to the management and motivation factors of human resources for the adoption of these practices (Boks 2006; Brones et al. 2017).

Since the 1990s, many researchers, whether from schools or companies, have developed integrated tools in the concept of eco-design in order to parameterize the different factors that must be taken into account when developing products. As a result, around two dozen tools have been developed, with evaluation systems that are not always coherent, allowing the project teams to have something more tangible to take into account for environmental aspects of product definition. Next, some of these tools are identified, segmented according to the type of analysis they allow (Byggeth and Hochschorner 2006; Schöggl et al. 2017):

- Qualitative analysis: ABC Analysis, Philips Fast Five Awareness, Dominance Matrix or Paired Comparison, eco-design checklist, E-concept spiderweb, Environmental Objectives Deployment (EOD), LiDS-Wheel and the Morphological Box, Strategy List, Ten Golden Rules, Volvo's White, Grey and Black Lists, The ten United Nations Global Compact (UNGC) principles, Method for Sustainable Product Development, and Checklist for Sustainable Product Development (CSPD);
- Semi-qualitative analysis: eco-design checklist, eco-design pilot, UNEP Design for Sustainability impact profile;
- Quantitative analysis: Environmentally Responsible Product Assessment Matrix (ERPA), Ford of Europe's Product Sustainability Index (PSI), Streamlined LCA framework for early decision making in vehicle development and Funktionskosten;
- Qualitative and quantitative: Materials, Energy, Chemicals, and Others (MECO), and MET-Matrix (Materials, Energy, and Toxic emissions).

Byggeth and Hochschorner (2006) concluded that, although these tools allow for a rapid and structured analysis of information, only nine allowed for adequate justifications for compromise situations, but these justifications were not consistent as necessary to select a given solution as the most feasible, taking into account the multiplicity of necessary factors during the development of a new product. The disappointment of these authors was essentially due to the fact that the assessment of the different factors related to the environment was not adequately reasoned, and therefore suggested that this assessment should take into account the full life cycle of the product and create a framework for sustainability so that in situations of conflict of interest between features, the option for certain factors, taking into account not only sustainability, but also aspects, such as cost, service, social impact, among others, would be clear. Spangenberg et al. (2010) state that the integration of the three dimensions mentioned above can be achieved through a further developed tool, Design for Sustainability (DFS), which can contribute to a more consistent transition to a society based on sustainable development. This tool does not only refer to the function of assisting in the discovery of the most sustainable solution for a certain product in development, but also to help designers and engineers to meet the real customer needs, always through iterations that promote growing sustainability of products (Schöggl et al. 2017). However, Davidson et al. (2007) point out an additional challenge to designers and engineers because they are not properly trained to effectively solve problems related to the environmental and social impact caused by a particular product. Thus, the concerns previously referred were taken up by Bovea and Pérez-Belis (2012), who extended the number of tools analyzed, and reperformed a critical analysis of the various tools available in this area. The idea was to use multi-criteria analysis to combine the technical requirements imposed on the product by the function it will perform, with the ecological aspects it must safeguard. The tools added by Bovea and Pérez-Belis (2012) were as follows: matrix element checklist for ERP, environmental product life cycle matrix (EPLC), streamlined life cycle assessment (SLCA), product investigation, learning and optimization tool (PILOT), environmental indicators, oil point method (OPM) and LCA, among others. Despite the virtues and limitations of all these tools, starting with the qualitative ones, then moving to the semi-quantitative ones and ending with the quantitative ones, LCA has been one of the tools that has collected more interest with researchers, and will be independently detailed. The importance of the LCA came to be mirrored in ISO 14040-44: 2006, which will also be mentioned later. Taking into account the different tools mentioned above, Bovea and Pérez-Belis (2012) drew a diagram in which they tried to establish the degree of difficulty and the time required to apply some of these tools in practice, versus the main objectives to be achieved with these tools, indicating also through different colors if the method is qualitative, semi-quantitative or quantitative, thus making a very useful summary for those interested to perceive very quickly and effectively which tool is the most suitable to each particular case. This diagram is shown in Fig. 6.15, and the integration of the different tools is shown in Table 6.2. There are many studies, in most cases theoretical, and others conducted in partnership with big companies, but not in a regular manner, where environmental considerations are effectively taken into account in the development phase of the product. However, unfortunately, these cases are still the exception, not the rule. The three main vectors that should be used in order to effectively use an eco-design tool are: integration of environmental aspects as a mandatory requirement in the design and development process of a product; adequate study of the life cycle of the product, taking into account how the product may affect the environment during the different life stages; and multi-criteria approach. However, although there are methodologies that integrate the criteria, their implementation will always be dependent on the way each company faces the process of developing its products.

Eco-design continues to be a tool that is not applied as vastly as it should (Rodrigues et al. 2017), and the negative aspects pointed out by several investigators who have carried out surveys to designers have been immense, trying to analyze the reason behind the resistance offered by designers to use eco-design tools in a more massive manner. In fact, one of the main negative causes mentioned is that the

Table 6.2 Tools for integrating environmental requirements into the product design process

	Methodology for environmental requirement evaluation	Product requirement (in addition to environmental)								
		Customer	Technical	Function	Cost	Cultural	Company	Health	Safety	Legal
Requirements matrix	Semi-qualitative matrix			✓	✓	✓				✓
DFE matrix	Semi-qualitative matrix							✓	✓	
House of ecology (HOE)	Environmental indicator	✓			✓					
Environmental-QFD (E-QFD)	Screening LCA	✓	✓							
Readiness assessment for implementing DFE strategies (RAILS)	MET-Matrix, screening LCA, LCA	✓					✓		✓	
Environmental objective deployment (EOD)	Design matrix, screening LCA, LCA	✓	✓							
Integrated approach to sustainable product development	LCA	✓	✓		✓					
Green-QFD (GQFD)	LCA	✓	✓		✓					
Life cycle house of quality (LC-HoQ)	LCA	✓	✓							✓
Quality function deployment for environment (QFDE)	Environmental indicators	✓	✓							
Eco-redesign methodology	LCA	✓	✓		✓					
Eco-value analysis (Eco-VA)	LCA	✓		✓	✓					

(continued)

Table 6.2 (continued)

	Methodology for environmental requirement evaluation	Product requirement (in addition to environmental)								
		Customer	Technical	Function	Cost	Cultural	Company	Health	Safety	Legal
Environmental impact and factor analysis (EIFA)	Environmental indicators			✓						
Environmental failure mode effects analysis (E-FMEA)	Environmental indicators			✓						
Eco-FMEA	Environmental Indicators/LCI	✓		✓					✓	
Environmental conscious QFD (ECQFD) and (LCA)	LCA	✓	✓		✓		✓			
Life cycle environmental cost analysis (LCECA)	LCA				✓					
Life cycle planning (LPC) TRIZ	LCA			✓	✓					
Early stage of environmental conscious design	KANO	✓			✓					
Environmental performance strategy map	Environmental indicators				X				X	

Adapted from Bovea and Pérez-Belis (2012)

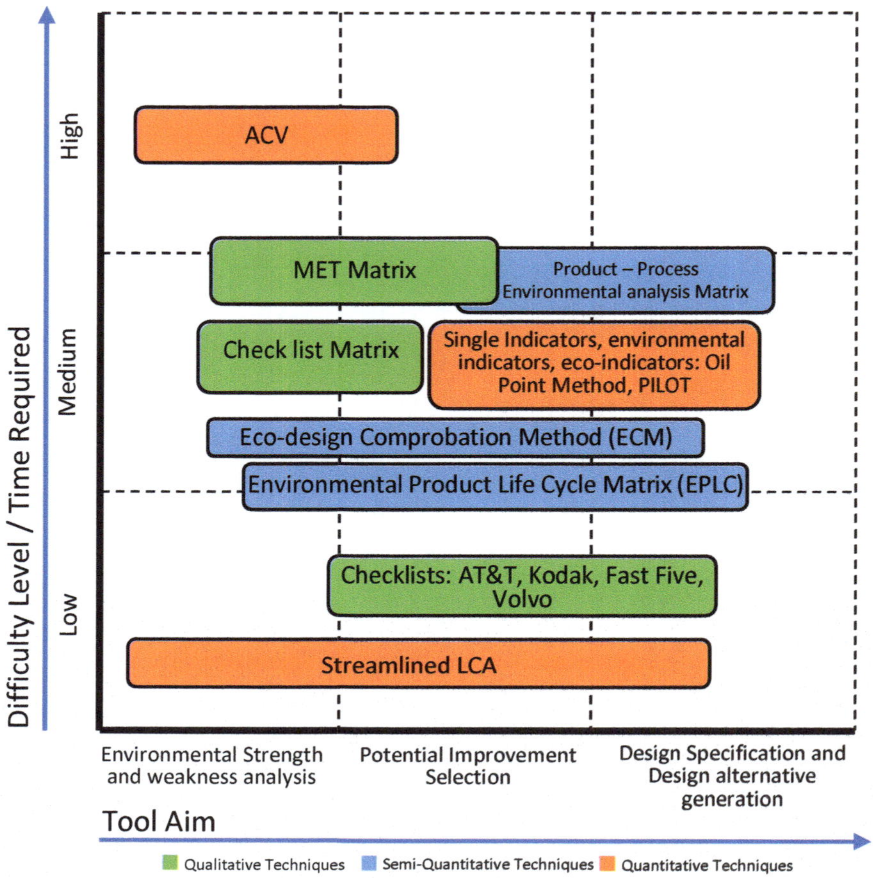

Fig. 6.15 Classification of eco-design tools according to their application and the way they consider environmental issues. Adapted from Bovea and Pérez-Belis (2012)

tools which have been developed are much more aimed at managers than designers (Walker 1998). Moreover, it has also been identified that most eco-design tools do not take into account the industrial designers' own culture and how they would like to approach eco-design (Sherwin 2000). According to a study carried out by Bundgaard et al. (2015), it can be seen that companies and industrial designers have put a very large focus on the issue of energy savings both in the manufacturing of products and in the consumption that the products will generate during their estimated useful life, leaving aside all other factors equally important when developing a new product or reformulating an existing product.

However, the mismatch of eco-design tools has also been well dissected and should have taken into account the five lines of thinking normally followed by industrial designers: content, style, time, service, and culture (Lofthouse 2006). In fact, studies show that most designers feel that eco-design is not a priority of the companies they work for, but just another aspect to be taken into account; however, the already existing tools, procedures, and checklists do not help them overcome practical issues they face in product development (Dewberry 1996). In addition, designers often complain that existing tools do not show them how to apply eco-design principles in a clear way. On the other hand, it is also pointed out by Tischner and Charter (2001) as a gap, that the existing knowledge regarding eco-design is very dispersed, which makes the search very time-consuming. Based on interviews with designers, it seems clear that they crave a much more assertive list of items to take into account in the development of new products, leaving perfectly clear the main issues to be safeguarded and what goals to achieve, avoiding wasted time with complex tools that provide results that are not always easy to decipher, given the initially defined project requirements (Lofthouse 2006). Still according to Lofthouse (2006), it is necessary to create a framework that is constituted by seven main aspects considered as essential for success in the proper use of the principles of eco-design: inspiration, information, education, guidance, visual, dynamic access and accessible, and non-scientific language. A correct combination of these factors can make eco-design tools easier for industrial designers to meet their own culture and achieve a much easier adherence to the principles it is intended to follow.

More recently, the eco-design maturity model (EcoM2) was developed, based on principles of product development management, built on a step-by-step process, through a systematic approach, which aims to assist integration of eco-design into product development processes (Pigosso et al. 2013). The model is based on the best eco-design practices, taking into account its characteristics and target (Pigosso et al. 2014), being divided into three components: eco-design management practices (EMP), operational eco-design practices and methods, and tools. Some researchers (Boks and Stevels 2007; McAloone 1998; Pigosso 2012; Pigosso et al. 2013) also studied the degree of maturity that companies can present with respect to the practice of implementing eco-design principles, as defined in the EcoM2 model, defining five levels that are briefly presented:

- The company does not have the conditions to apply a sustainable management practice in terms of eco-design.
- The company applies sporadically and, under unorganized manner eco-design principles, mainly with the purpose of overcoming an explicit concern or solving a specific problem.
- The company has already started a process of almost systematic application of eco-design practice in product development, having already defined the necessary documentation, internal organization, allocation of responsibilities and has already assigned resources for the continued practice of applying these principles.
- At this stage, the company has already defined specific performance indicators for monitoring the application of eco-design principles over time, with a perfect notion of how the organization is applying these principles.
- At this level, the company no longer intends only to monitor the application of the principles of eco-design and also intends that this routine be permanently

improved. Therefore, it allocates resources so that the process is progressing positively on a permanent basis.

In fact, the decisions made by designers and engineers in the design phase of the product will decisively impact the recovery capacity of the product after its end-of-life, the energy consumption during the manufacturing and distribution phase, the energy consumption due to its use during the life span of the product, in addition to other impacts that the same product can produce. In most cases, designers and engineers still do not have data that allow them to make sustainable decisions regarding the environmental and social impact of the product. One example is the fact that very high-performance composite materials are emerging on the market, which do not yet have a track record capable of providing the data needed for designers to make decisions (Schöggl et al. 2017). Thus, two different situations can occur: (a) The designer is inhibited from using new materials that he does not yet have enough data to support his decision and may not be moving in the right direction at the desired speed due to lack of initiative by the designer, or (b) the designer assumes responsibility for the application of new material, but there are risks of producing harmful effects on the environment and to society than with choosing other solutions equally available and with already proven data. The decision may not be consensual, nor is there a standard procedure for these cases, but the delay in acquiring relevant information concerning materials may affect decisions on a product that might be on the market for several years. Here, common sense will be vital, as well as the definition of the tests to be performed so that a final decision is not compromising.

From the above, it is clear that, despite the many tools developed and regulation established so far, the conditions are not yet created for eco-design to be applied technically, in a consistent and effective way. This opinion is expressed by several researchers who have recently studied this theme, noting that there is still much to be done in integrating concepts and in the appropriate structured study of the different alternatives during the design stage, based on consistent data on environmental performance of the different options (Ardente and Mathieux 2014; Dalhammar 2015; Bundgaard et al. 2015, 2017). It will certainly be necessary to legislate with mandatory character so that eco-design becomes a more present reality of products in our daily lives. Although there is a diversity of legislation in the making, especially in the EU, it is required that it be easily and legally enforceable and that appropriate preparation of stakeholders be addressed to ensure that the evolution of eco-design truly corresponds to environmental preservation.

However, there are EU regulations that set out very concrete values to be respected by manufacturers regarding the application of recoverable or recyclable materials. These values are shown in Table 6.3 and require that some of the most important industrial sectors, that is, of greater consumption by the world population and containing harmful products, follow the principles appropriate to be within the legally established values, regardless of the type of tools they wish to use. Thus, whether through eco-design tools or others, policies have converged to oblige manufacturers to obey standards that respect the environment, a situation that should continue

Table 6.3 EU rules for the recovery of some specific products

EU rules related to collection, recycling and landfilling				
Waste stream	Target years	Minimum recovery	Minimum recycling (%)	Collection rate
Packaging	2008	60%	55	
Cars	2015	95%	85	100%
Electronics	2006	75%	50	Min. 4 kg/inhabitant; 85% by 2020
Batteries	2011	50–75% (efficiency)		
	2012			25%
	2016			45%
Tires	2006	Zero tires going to landfill		
Biowaste diverted from landfills	2006	Reduction to 75% of the 1995 level		
	2009	Reduction to 50% of the 1995 level		
	2016	Reduction to 35% of the 1995 level		
New targets in waste framework directive	2015	Separate collection: at least paper, metal, plastic, and glass		
	2020	50% of municipal waste		
	2020	70% construction and demolition waste		

Adapted from Dalhammar (2015)

to be reviewed in the future, increasing the respect for the environment, making manufacturers and the population more aware about these concerns.

6.6 Life Cycle Assessment

One of the products that triggered the environmental impact assessment was, curiously, the beverage packaging. Given that this product can be manufactured in different materials and using different manufacturing processes, which will also imply different consumption of resources and generation of different quantities and types of waste at the end of their use. With different manufacturing approaches and environmental repercussions on their reuse or recycling, Coca-Cola was a pioneer in launching and funding a study on resource consumption and the effect on the environment of beverage containers (Jensen et al. 1997). This first initiative took place in 1969 and was followed by a similar one in European terms by Ian Boustead in 1972, in UK, which studied the energy consumed in the production of various beverage containers, which included different materials such as steel, aluminum, glass, and plastics. In 2018, a survey carried out by a specialist company concluded that there is a greater awareness about environmental issues in the packaging industry, but also revealed that around 55% of the more than 1000 respondents do not yet have reliable measurement and analysis mechanisms regarding the environmental impact of their products (Greenerpackage.com 2018).

At the UN Earth Summit in 1992, LCA methodologies were rediscovered and gained prominence in the view of stakeholders, being considered one of the most promising new tools for environmental management tasks. In 1993, interest in the LCA was essentially centered on the scientific community, but it was beginning to show signs of industrial adoption. Although there is a lot of research in this area today, there is still a lack of practical training, which prevents further progress on the ground. The lack of practical development of this methodology has also been associated with two main factors: exaggerated expectations and excessive advertising.

LCA is a complete approach, i.e., from the source of the raw material to the end of the life cycle of the product, which aims to evaluate the environmental performance of products, processes, industrial systems, and so on. This global analysis aims to analyze the cycle from the stage of extraction of raw material from Nature to create the product, until the return of the material to Nature, including stages such as material processing, production processes, logistics, useful life cycle, and maintenance, as shown in Fig. 6.16. The LCA evaluates all stages of a product's life under the assumption that they are interdependent, meaning that one operation may have implications on the next. The LCA allows the estimation of the accumulated environmental impacts resulting from all stages of the product's life cycle and, as a result, allows for the selection of the path or process that is environmentally preferable. In a short way, the LCA methodology aims to take into account the environmental, economic, and social impacts of a given product or range of products, throughout its entire life cycle.

The LCA can help decision makers select a product or process that induces the least impact on the environment. This information can be used in conjunction with cost and performance data to select a product or process. LCA data identifies the possibility of linking between different types of environmental impacts, such as by eliminating atmospheric emissions replacing them by wastewater effluent and/or a life cycle stage by another one, such as, for example, choosing between reusing a given material or the use of virgin raw material. If an LCA is not carried out, the transfer may not be recognized and included correctly in the analysis because it is out of the scope or typical focus of product selection processes. Due to the integrated view that the LCA allows, it is possible to perceive the harmful effects that the manufacture of a given product can present with respect to another, which is very important in its

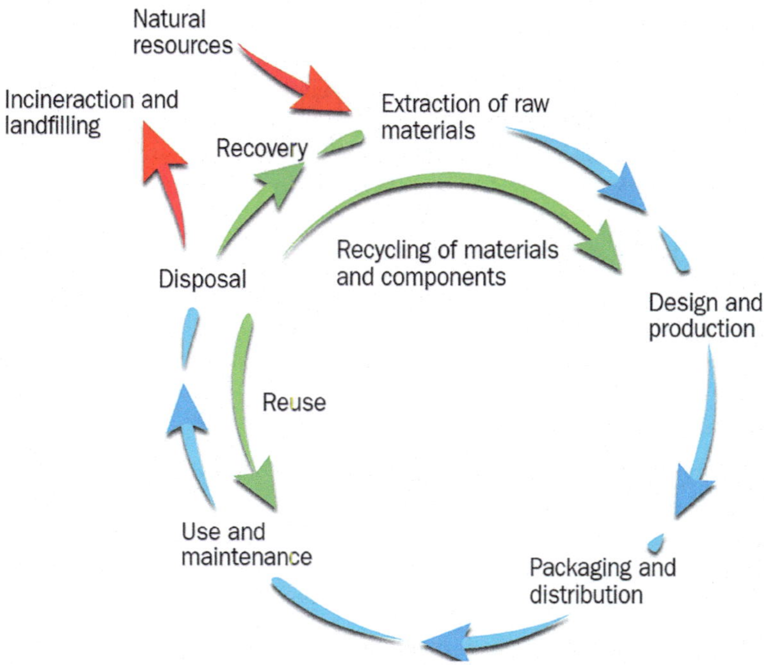

Fig. 6.16 Diagram of the materials and product flows in a "cradle-to-grave" system. Adapted from Nikkakyo.org (2018)

selection. This is particularly relevant if one wants to choose between two similar products, one of which produces a lot more solid waste, compared to another that produces less solid waste, but releases harmful chemicals into the atmosphere as well as effluents. Focusing only on solid waste would be a mistake as the second product, in overall terms, turns out to be more harmful to the environment. This overview is an added advantage of the use of the LCA tool. By undertaking an LCA, a work team can:

- Have an accurate, real-time notion of the environmental impacts caused by a given product, by systematic monitorization.
- Get the environmental trade-offs related to one or more specific products/processes in order to influence stakeholders as a strategic action.
- Quantify the release of harmful products into air, water, and land, taking into account each stage of the life cycle and/or process, in order to be able to act specifically on the product or process, minimizing adverse effects on the environment.
- Assist in the identification of significant changes in environmental impacts produced by the different phases of the life cycle of a product.
- Assessment regarding the effects on citizens' health and ecosystems of the consumption of virgin raw materials and the release of chemicals, harmful gases, and

effluents into the environment, which have a direct impact on local communities, the region, and the world.
- To be able to compare the health impacts on people and ecosystems of two or more competing products/processes, identifying the specific advantages or problems generated by each one, also identifying the phase of their useful life or process in that the situation is more serious, allowing a reflection on how to surpass the problem in a surgical way;
- Allow the identification of environmental impacts in one or more specific areas of interest.

Given that the methodology is primarily intended to assess the environmental impact of products or processes, there are different approaches that can considered, addressing for example the whole life cycle of a product or just a specific part. Thus, some partial approaches may be considered:

- Cradle-to-grave: from raw material extraction to product disposal, including manufacturing, transport and use;
- Cradle-to-gate: from raw material extraction to the gate of the factory;
- Gate-to-gate: analysis between two points of the time line, considering the life cycle of a given product.

The LCA methodology is commonly used to compare environmental impacts resulting from:

- Similar products;
- Different waste treatment processes;
- Different stages of manufacture of a given product;
- Different ways to use in the conception of a new product.

The methodology can be applied in a simplified manner considering four phases (Nikkakyo.org 2018):

- Consideration of input and output flows, taking into account important aspects of the system, as specified later. This phase usually includes the following tasks: (a) definition of product functions, as well as the reason and scope of the study; (b) understanding of who is interested in the study and how the results will be used; (c) creation of system boundaries, defining it in the ways already mentioned, that is, "cradle-to-cradle," "cradle-to-grave," or "gate-to-gate;" (d) identification of the required data; (e) recognition of study limitations; and (f) establishment of temporal and geographical references.
- Collection of data required for system analysis, with identification and quantification of predictable environmental impacts, associated with each of these flows. This data needs to be treated according to each group (inputs and outputs), categorizing them, so that the final analysis can be duly substantiated, and improvement solutions can be more easily studied.
- Environmental impact assessment: The data previously collected needs to be transformed into actual environmental impact data, through databases that correlate the quantity of each of the input and output variables of the system with an effective

impact on the environment, ecosystem, health of people, in the welfare of society, among others.
- Interpretation of the results obtained in relation to the impact assessment: The set of results and calculations previously obtained now needs to be effectively understood so that plans can be established for the implementation of improvements. To this end, the results should be verified by a panel of external evaluators with a view to results-free validation and collection of ideas on how to further reduce identified impacts.

After defining the life cycle or part of the life cycle to be studied, the boundaries of the considered system will have to be defined, so that the input and output variables of that system can be analyzed. These variables can take on extremely different contours since in a productive process it is necessary to analyze variables such as the raw material consumed, the energy required during the production, distribution and use of the product, the need for production facilities, as input variables, and it will be necessary to account for all output variables, such as waste materials, effluents, emissions, among others. However, if we consider agricultural tasks, the input variables will already be quite different, such as seeds, water, pesticides and fertilizers, among others, with the output variables being the residues that will enter watercourses (resultant of the chemicals used). All these factors have an impact on the environment and therefore need to be taken into account in the calculation of the LCA (Qi et al. 2018). In general terms, regarding the Environmental Protection Agency (EPA) report (EPA 2006) and considering the three components of the system, one can consider:

- Inputs: raw materials, energy;
- System within its boundaries: raw materials acquisition, manufacturing, use/reuse/maintenance, recycling/waste management;
- Outputs: atmospheric emissions, waterborne wastes, solid wastes, coproducts, and other releases.

In terms of standardization, the LCA is governed essentially by two ISO standards, each defining some of the main parameters of the methodology:

- ISO 14040: 2006—"Environmental management—Lifecycle assessment—Principles and framework;"
- ISO 14044: 2006—"Environmental management—Lifecycle assessment—Requirements and guidelines."

These standards have replaced other previously existing ones, such as ISO 14041:1999, ISO 14042:2000, and ISO 14043:2000, keeping the requirements and technical aspects unchanged, but improving the reading and understanding of them, with a view to making their use simpler (ISO 2018).

In order to have a comparison between different products in terms of environmental impact, there must be easily collectable, well-established metrics that use measurable and reliable parameters. However, the diversity of existing products, as well as the diversity on how to respect or attack the environment, can assume so many forms, that summarizing the metrics to a couple of factors has become an impossible task. Thus, there is a multiplicity of indexes that can be calculated and compared, which are described in some detail in the document issued by the EPA—Environmental Protection Agency, designated Life Cycle Assessment: Principles and Practice (EPA 2006), which was revised in August 12, 2015. Thus, based on that document, the International Council of Chemical Associations (ICCA 2018) revealed some indexes, which are following presented in alphabetical order, in order to only stimulate the curiosity of the most interested in this topic. Each index is calculated regarding a single functional unit of the product.

- Acidification potential (AP): Index intended to illustrate the relative effect of total emissions of acid gases (e.g., sulfur oxides (SO_x), nitrogen oxides (NO_x), hydrochloric acid (HCl), hydrofluoric acid (HF), and ammonia (NH_4)) on air quality throughout the life cycle of a product, including end-of-life waste management. Exposure to these emissions can acidify soils and watercourses and may cause corrosion on roofs and buildings facades.
- Conscious water footprint (CWF) and water emission footprint (WEF): Translates the total amount of water needed to produce a given product and must include the water needed to treat it at the end of its life cycle. This index can be subdivided into freshwater, brackish water or seawater, taking into account the type of source (river, lake, well, sea);
- Cumulative energy demand (CED): total energy consumed throughout the product life cycle, from the production stage to the waste management phase;
- Cumulative fossil energy consumption (CFED): Corresponds to a subset of the CED and relates to the total energy consumption of fossil fuels during the product life cycle, including the final disposal phase of the product. Fossil fuels include coal, crude oil and its by-products, and natural gas.
- Cumulative renewable energy demand (CRED): Integrating the CED as a subset indicates the portion of renewable (non-fossil) energy that is consumed by the product during its entire life cycle (including end-of-life waste management). Renewable energy includes hydro, wind, solar, photovoltaic, geothermal, and wave energy.
- Direct land use change (LUC): This is a very important indicator, especially in developing countries, which translates the rate at which lands are being converted from one kind of use into another one, or in other words, from an original state, which could be forest, pasture, degraded land, to another land utilization, which can be land devoted to agricultural production or forest products, with the aim of increasing the production of biofuel feedstock. This change produces significant changes in the quantity and type of emissions of GHG and carbon stocks in that land.
- Eco-toxicity and human toxicity assessment: Toxicity assessment indices, such as the USEtox™ model, allow the determination of specific chemical characterization factors that quantify the environmental fate of chemical emissions and their impact on human health and ecosystems, in terms of exposure to these chemicals, as well as their effect.

- Eutrophication potential (EP), also known as potential nutrition (NP): index that is intended to reflect the effect of excessive growth of algae caused by emissions of limiting nutrients (compounds containing nitrogen or phosphorus) directly or indirectly to surfaces or courses of water, such as lakes, rivers, estuaries, and so on, and to the soil during the life cycle of a product, including the management of its waste at the end of its useful life;
- Global warming potential (GWP), also known as carbon footprint: index that aims to convey the impact caused by manufacturing, use throughout its useful life and management of its waste after the end of its useful life, in climate change for a given period of time, usually taken as 100 years, in terms of total greenhouse gas emissions such as carbon dioxide (CO_2), methane (CH_4), and nitrous oxide (N_2O);
- Indirect land use change (ILUC): This is a significant indicator which is currently under development which represents the secondary land use change resulting from a primary land use change and promoting a commercial crop displacement, which will be consequently grown in a different place, leading to further land being altered from one state to another. This conversion will result in fluctuations in GHG emissions and carbon stocks on that land, which will produce different environmental impacts.
- Ozone depleting potential (ODP): Index which intends to transmit the effect caused by the total emissions of gases which are destroying the stratospheric ozone layer during the life cycle of a product, including waste management at the end of the product's useful life. Stratospheric ozone is a gas layer that naturally forms in the upper atmosphere and protects living cells from overexposure to ultraviolet (UV) solar radiation. Overexposure to UV radiation can cause numerous health and environmental problems, including skin cancer and crop yield impairment.
- Photochemical ozone creation potential (POCP): An index which intends to convey the relative effect of the total emissions of VOCs and nitrogen oxides throughout the life cycle of a product, including the management of its waste at the end of its useful life. Emissions of VOCs in the presence of nitrogen oxides and sunlight can induce chemical reactions that form ozone (O_3) on the Earth's surface and are commonly known as photochemical smog.

Some of these indexes and others can also be found in a comprehensive fashion in the work of Ingwersen et al. (2014). The application of LCA has been used successfully by some reputable companies, such as Procter and Gamble (Saouter and van Hoof 2002), which has begun studies at an early stage regarding the industry's interest in this tool and has maintained studies in this area. More recently, another company-based study has revealed how the company uses the LCA to study the environmental impact of its products, given that they are used by billions of people worldwide (Agarwal and Thiel 2015). Table 6.4 shows a perspective of the acceptance and depth of analysis carried out by 20 multinational companies regarding the application of the LCA methodology, showing the situation in 2010. Certainly, there have been developments, but there is surely still a lot to be done so that the application of the LCA becomes an indispensable routine in the development of new products and processes in companies.

Table 6.4 Engagement of 20 multinational companies regarding the implementation of LCA methodologies

	Energy Consumption	Climate Change	Acidification	Eutrophication	Material Depletion	Phochemical Ozone Formation	Ozone Depletion	Waste Problem	Eco-Toxicity	Human Toxicity	Water Reserve Impacts	Land Use	Biodiversity
BASF													
Bombardier Transportation													
Continental													
Daimler													
Electrolux													
GE													
GlaxoSmithKline													
Interface													
KONE													
Nestle Waters													
Procter & Gamble													
Siemens													
Unilever													
Vattenfall													
Vestas													
Xerox													

Adapted from Nygren and Antikainen (2010)

The evolution of the LCA has meanwhile developed other tools, such as life cycle costing (LCC), social life cycle assessment (S-LCA), Design for the Environment (DfE), and eco-labeling, which have been treated in an integrated way through another tool, called life cycle management (LCM). This tool is an integrated approach that seeks to quantify and minimize impacts on the environment, economy, and society of a product, process or service throughout its life cycle.

References

M. Abbasi, M.M. Salarirad, I. Ghasemi, Selective separation of PVC from PET/PVC mixture using floatation by tannic acid depressant. Iran. Polym. J. **19**, 483–489 (2010)

D.S. Achilias, C. Roupakias, P. Megalokonomos, A.A. Lappas, E.V. Antonakou, Chemical recycling of plastic wastes made from polyethylene (LDPE and HDPE) and polypropylene (PP). J. Hazard. Mater. **149**, 536–542 (2007)

K. Acuff, D.T. Kaffine, Greenhouse gas emissions, waste and recycling policy. J. Environ. Econ. Manag. **65**, 74–86 (2013)

B. Adhikari, D. De, S. Maiti, Reclamation and recycling of waste rubber. Prog. Polym. Sci. **25**, 909–948 (2000)

R. Agarwal, M. Thiel, P&G: Providing sustainable innovative products through LCA worldwide. South Asian J. Bus. Manag. Cases **2**(1), 85–96 (2015)

J. Aguado, D.P. Serrano, J.M. Escola, *Catalytic upgrading of plastic wastes*, in *Feedstock Recycling and Pyrolysis of Waste Plastics: Converting Waste Plastics into Diesel and Other Fuels*, ed. by J. Scheirs, W. Kaminsky (Wiley, New Jersey, 2006), pp. 73–110. ISBN 978-0-470-02152-1

S. Ahlroth, M. Nilsson, G. Finnveden, O. Hjelm, E. Hochschorner, Weighting and valuation in selected environmental systems analysis tools—suggestions for further developments. J. Clean. Prod. **19**(2–3), 145–156 (2011)

M. Al Aukidy, P. Verlicchi, N. Voulvoulis, A framework for the assessment of the environmental risk posed by pharmaceuticals originating from hospital effluents. Sci. Total Environ. **493**, 54–64 (2014)

K. Allméar, A. Hult, B. Rårnby, Surface modification of polymers. I. Vapour phase photografting with acrylic acid. J. Polym. Sci., Part A: Polym. Chem. **26**, 2099–2111 (1988)

J.M. Allwood, Squaring the circular economy: the role of recycling within a hierarchy of material management strategies, in *Handbook of Recycling, State-of-the-art for Practitioners, Analysts and Scientists* (Elsevier Inc., Amsterdam, 2014), pp. 445–477. ISBN 978-0123965066

S.M. Al-Salem, P. Lettieri, J. Baeyens, Recycling and recovery routes of plastic solid waste (PSW): A review. Waste Manag **29**, 2625–2643 (2009)

H. Alter, The recovery of plastics from waste with reference to froth flotation. Resour. Conserv. Recycl. **43**, 119–132 (2005)

J.L.S. Alves, D.D. de Medeiros, Eco-efficiency in micro-enterprises and small firms: a case study in the automotive services sector. J. Clean. Prod. **108**, 595–602 (2015)

D.C. Amaral, H. Rozenfeld, Integrating new product development process references with maturity and change management models, in *Proceeding of the 16th International Conference on Engineering Design, Paris, France, 28th–31st July 2007*. ISBN 1-904670-02-4 (Proceedings on CD)

J. Amaya, P. Zwolinski, D. Brissaud, Environmental benefits of parts remanufacturing: the truck injector case, in *17th CIRP International Conference on Life Cycle Engineering* (2010). https://hal.archives-ouvertes.fr/hal-00519559/document. Retrieved on 10th of Aug 2018

A. Andrews, M. Gregoire, H. Rasmussen, G. Witowich, Comparison of recycling outcomes in three types of recycling collection units. Waste Manag. **33**, 530–535 (2013)

N. Antoniou, G. Stavropoulos, A. Zabaniotou, Activation of end of life tyres pyrolytic char for enhancing viability of pyrolysis—Critical review, analysis and recommendations for a hybrid dual system. Renew. Sustain. Energy Rev. **39**, 1053–1073 (2014)

K. Aoudia, S. Azem, N.A. Hocine, M. Gratton, V. Pettarin, S. Seghar, Recycling of waste tire rubber: Microwave devulcanization and incorporation in a thermoset resin. Waste Manag. **60**, 471–481 (2017)

F. Ardente, F. Mathieux, Identification and assessment of product's measures to improve resource efficiency: the case-study of an energy using product. J. Clean. Prod. **83**, 126–141 (2014)

A. Atasu, O. Ozdemir, L.N.V. Wassenhove, Stakeholder perspectives one-waste take-back legislation. Prod. Oper. Manag. **22**, 382–396 (2013)

A. Atasu, R. Subramanian, Extended producer responsibility for e-waste: individual or collective producer responsibility? Prod. Oper. Manag. **21**(6), 1042–1059 (2012)

Automotive Recyclers Association, http://a-r-a.org/what-we-do/industry-statistics/ (2018). Retrieved on 10 Aug 2018

A.K. Awasthi, J. Li, An overview of the potential of eco-friendly hybrid strategy for metal recycling from WEEE. J. Clean. Prod. **126**, 228–239 (2017)

A.K. Awasthi, X. Zeng, J. Li, Environmental pollution of electronic waste recycling in India: A critical review. Environ. Pollut. **211**, 259–270 (2016)

H. Bai, J. Wang, A.Z. Zeng, Exploring Chinese consumers' attitude and behavior toward smartphone recycling. J. Clean. Prod. **188**, 227–236 (2018)

C. Bakker, F. Wang, J. Huisman, M.D. Hollander, Products that go round: exploring product life extension through design. J. Clean. Prod. **69**, 10–16 (2014)

C.P. Baldé, V. Forti, V. Gray, R. Kuehr, P. Stegmann, *The Global e-Waste Monitor* (2017), https://www.itu.int/en/ITU-D/Climate-Change/Documents/GEM%202017/Global-E-waste%20Monitor%202017%20.pdf. Retrieved on 11 Aug 2018

C.P. Baldé, R. Kuehr, K. Blumenthal, S.F. Gill, M. Kern, P. Micheli, E. Magpantay, J. Huisman, *E-waste Statistics: Guidelines on Classifications, Reporting and Indicators*. United Nations University, IAS—SCYCLE, Bonn, Germany (2015)

S. Barr, A.W. Gilg, N.J. Ford, Differences between household waste reduction, reuse and recycling behaviour: a study of reported behaviours, intentions and explanatory variables. Int. J. Environ. Waste Manage. **4**(2), 69–82 (2001)

P. Basaróvá, L. Bartovska, K. Korinek, D. Horn, The influence of flotation agent concentration on the wettability and flotability of polystyrene. J. Colloid Interface Sci. **286**, 333–338 (2005)

B.M. Beamon, Designing the green supply chain. Logist. Inf. Manag. **12**(4), 332–342 (1999)

M. Bigum, L. Brogaard, T.H. Christensen, Metal recovery from high-grade WEEE: a life cycle assessment. J. Hazard. Mater. **207–208**, 8–14 (2012)

D. Bixio, T. Wintgens, *Water reuse system management manual AQUAREC* (Official Publications of the European Communities, European Commission, Brussels, Belgium, 2006). ISBN 978-9279019340

A. Björklund, G. Finnveden, Recycling revisited—life cycle comparisons of global warming impact and total energy use of waste management strategies. Resour. Conserv. Recycl. **44**(4), 309–317 (2005)

T. Blume, M. Walther, The end-of-life vehicle ordinance in the German automotive industry e corporate sense making illustrated. J. Clean. Prod. **56**, 29–38 (2013)

G. Boesso, K. Kumar, Stakeholder prioritization and reporting: evidence from Italy and the US. Accounting Forum **33**, 162–175 (2009)

C. Boks, The soft side of eco-design. J. Clean. Prod. **14**, 1346–1356 (2006)

C. Boks, A. Stevels, Essential perspectives for design for environment. Experiences from the electronics industry. Int. J. Prod. Res. **45**, 4021–4039 (2007)

M.D. Bovea, V. Pérez-Belis, A taxonomy of ecodesign tools for integrating environmental requirements into the product design process. J. Clean. Prod. **20**, 61–71 (2012)

M.D. Bovea, V. Ibáñez-Forés, V. Pérez-Belis, P. Quemades-Beltrán, Potential reuse of small household waste electrical and electronic equipment: methodology and case study. Waste Manag. **53**, 204–217 (2016)

M. Braungart, W. McDonough, *Cradle to cradle: remaking the way we make things*, 2nd edn. (Vintage Books, Cambridge, UK, 2008). ISBN 13: 978-0865475878

F. Brones, M.M. Carvalho, From 50 to 1: integrating literature toward a systemic ecodesign model. J. Clean. Prod. **6**(9), 44–57 (2015)

F.A. Brones, M.M. Carvalho, E.S. Zancul, Reviews, action and learning on change management for ecodesign transition. J. Clean. Prod. **142**(1), 8–22 (2017)

R. Buchan, B. Yarar, Recovering plastics for recycling by mineral processing techniques. JOM J. Miner. Metals Mater. Soc. (TMS) **47**, 52–55 (1995)

A. Buekens, Introduction to feedstock recycling of plastics, in *Feedstock recycling and pyrolysis of waste plastics: Converting waste plastics into diesel and other fuels*, ed. by J. Scheirs, W. Kaminsky (Wiley, USA, 2006), pp. 3–41. ISBN 978-0470021521

A.M. Bundgaard, A. Remmen, K.O. Zacho, *Ecodesign Directive Version 2.0: from Energy Efficiency to Resource Efficiency*. Report. Danish Environmental Protection Agency, Miljostyrelsen (2015). ISBN 978-87-93283-56-5

A.M. Bundgaard, M.A. Mosgaard, A. Remmen, From energy efficiency towards resource efficiency within the Ecodesign Directive. J. Clean. Prod. **144**, 358–374 (2017)

S. Byggeth, E. Hochschorner, Handling trade-offs in Ecodesign tools for sustainable product development and procurement. J. Clean. Prod. **14**, 1420–1430 (2006)

California Environmental Protection Agency, *Technology Evaluation and Economic Analysis of Waste Tire Pyrolysis, Gasification, and Liquefaction, Contractor's Report to the Board* (2006). http://www.calrecycle.ca.gov/publications/Documents/Tires/62006004.pdf. Retrieved on 28 May 2018

M. Calkins, *Materials for Sustainable Sites: A Complete Guide to the Evaluation, Selection, and Use of Sustainable Construction Materials* (Wiley, Hoboken, 2008). ISBN 978-0-470-13455-9

J. Cao, B. Lu, Y. Chen, X. Zhang, G. Zhai, G. Zhou, B. Jiang, J.L. Schnoor, Extended producer responsibility system in China improves e-waste recycling: Government policies, enterprise, and public awareness. Renew. Sustain. Energy Rev. **62**, 882–894 (2016)

S. Chandra, Conference report. Cement Concr. Compos. **27**, 738–741 (2005)

M. Charter, U. Tischner, *Sustainable Solutions: Developing Products and Services for the Future* (Routledge, New York, 2001). ISBN 978-1351282468

M.-F. Chen, P.-J. Tung, The moderating effect of perceived lack of facilities on consumers' recycling intentions. Environ. Behavior **42**(6), 824–844 (2009)

Z. Chen, H.H. Ngo, W. Guo, A critical review on the end uses of recycled water. Crit. Rev. Environ. Sci. Technol. **43**(14), 1446–1516 (2013)

M. Chen, O.A. Ogunseitan, J. Wang, H. Chen, B. Wang, S. Chen, Evolution of electronic waste toxicity: trends in innovation and regulation. Environ. Int. **89–90**, 147–154 (2016)

A.R. Chini, *General issues of construction materials recycling in USA,* In *Sustainable Construction. Materials and Practice: Challenges for the New Millennium* (IOS Press, Amsterdam, 2007). ISBN 978-1-58603-785-7

T. Choi, Environmental impact of voluntary extended producer responsibility: The case of carpet recycling. Resour. Conserv. Recycl. **127**, 76–84 (2017)

H.-S. Chon, D.-G. Ohandja, N. Voulvoulis, Assessing the relative contribution of wastewater treatment plants to levels of metals in receiving waters for catchment management. Water Air Soil Pollut. **223**(7), 3987–4006 (2012)

J.-M. Chyan, D.-B. Senoro, C.-J. Lin, P.-J. Chen, I.-M. Chen, A novel biofilm carrier for pollutant removal in a constructed wetland based on waste rubber tire chips. Int. Biodeterior. Biodegrad. **85**, 638–645 (2013)

Commission Decision 2005/293/EC. https://eur-lex.europa.eu/legal-content/EN/TXT/PDF/?uri=CELEX:32005D0293&from=EN (2015). Retrieved on 11 Aug 2018

J.J. Cordeiro, M. Tewari, Firm characteristics, industry context, and investor reactions to environmental CSR: a stakeholder theory approach. J. Bus. Ethics **130**, 833–849 (2015)

I. Costa, G. Massard, A. Agarwal, Waste management policies for industrial symbiosis development: case studies in European countries. J. Clean. Prod. **18**(8), 815–822 (2010)

A. Costa Jr., K. Pasini, C. Andrade, Clean development mechanism in Brazil: an instrument for technology transfer and the promotion of cleaner technologies? J. Clean. Prod. **46**, 67–73 (2013)

F. Cucchiella, I. Adamo, S.C.L. Koh, P. Rosa, Recycling of WEEEs: an economic assessment of present and future e-waste streams. Renew. Sustain. Energy Rev. **51**, 263–272 (2015)

D. Czajczyńska, R. Krzyżyńska, H. Jouhara, N. Spencer, Use of pyrolytic gas from waste tire as a fuel: a review. Energy **134**, 1121–1131 (2017)

N.F. da Cruz, P. Simões, R.C. Marques, Economic cost recovery in the recycling of packaging waste: the case of Portugal. J. Clean. Prod. **37**, 8–18 (2012)

J. Dahmus, T. Gutowski, What gets recycled: an information theory based model for product recycling. Environ. Sci. Technol. **41**, 7543–7550 (2007)

L. Dal Maso, F. Mazzi, M. Soscia, S. Terzani, The moderating role of stakeholder management and societal characteristics in the relationship between corporate environmental and financial performance. J. Environ. Manage. **218**, 322–332 (2018)

C. Dalhammar, Industry attitudes towards ecodesign standards for improved resource efficiency. J. Clean. Prod. **123**, 155–166 (2015)

C.I. Davidson, H.S. Matthews, C.T. Hendrickson, M.W. Bridges, B.R. Allenby, J.C. Crittenden, Y. Chen, E. Williams, D.T. Allen, C.F. Murphy, S. Austin, Adding sustainability to the engineer's toolbox: A challenge for engineering educators. Environ. Sci. Technol. 4847–4850 (2007). https://pubs.acs.org/doi/pdf/10.1021/es072578f. Retrieved on 23 Aug 2018

M.C.S. de Abreu, *How to define an environmental policy to improve corporate sustainability in developing countries*. Bus. Strategy Environ. **18**(8), 542–556 (2009)

M. Dehghani, L. Keshtgar, M.R. Javaheri, Z. Derakhshan, G. Oliveri Conti, P. Zuccarello, M. Ferrante, The effects of air pollutants on the mortality rate of lung cancer and leukemia. Mol. Med. Rep. **15**(5), 3390–3397 (2017)

A. Demirbas, Pyrolysis of municipal plastic wastes for recovery of gasoline range hydrocarbons. J. Anal. Appl. Pyrol. **72**, 97–102 (2004)

Z. Derakhshan, M.T. Ghaneian, A.H. Mahvi, G.O. Conti, M. Faramarzian, M. Dehghani, M. Ferrante, A new recycling technique for the waste tires reuse. Environ. Res. **158**, 462–469 (2017a)

Z. Derakhshan, M.H. Ehrampoush, M. Faramarzian, M.T. Ghaneian, A.H. Mahvi, Waste tire chunks as a novel packing media in a fixed-bed sequence batch reactors: volumetric removal modeling. Desalin. Water Treat. **64**, 40–47 (2017b)

J.G. Derraik, The pollution of the marine environment by plastic debris: a review. Mar. Pollut. Bull. **44**, 842–852 (2002)

M. Despeisse, Y. Kishita, M. Nakano, M. Barwood, Towards a circular economy for end-of-life vehicles: a comparative study UK–Japan. Procedia CIRP **29**, 668–673 (2015)

E.L. Dewberry, *Eco-design—Present attitudes and future directions*. Doctoral Thesis (Open University, Milton Keynes, 1996)

Directive 2008/98/EC, 2008. http://ec.europa.eu/environment/waste/framework/. Retrieved on 08 Aug 2018

G. Dodbiba, N. Haruki, A. Shibayama, T. Miyazaki, T. Fujita, Combination of sink–float separation and flotation technique for purification of shredded PET-bottle from PE or PP flakes. Int. J. Miner. Process. **65**, 11–29 (2002)

J. Dodds, W.F. Domenico, D.R. Evans, L.W. Fish, P.L. Lassahn, W.J. Toth, *Scrap tyres: a resource and technology evaluation of tyre pyrolysis and other selected alternative technologies*. US Department of Energy Report EGG-2241 (1983)

J. Drelich, T. Payne, J.H. Kim, J.D. Miller, R. Kobler, S. Christiansen, Selective froth flotation of PVC from PVC/PET mixtures for the plastics recycling industry. Polym. Eng. Sci. **38**, 1378–1386 (1998)

J.E. Drewes, M. Reinhard, P. Fox, Comparing microfiltration-reverse osmosis and soil aquifer treatment for indirect potable reuse of water. Water Res. **37**(15), 3612–3621 (2003)

A. Ebreo, J. Vining, How similar are recycling and waste reduction? Future orientation and reasons for reducing waste as predictors of self-reported behavior. Environ. Behav. **33**(3), 424–448 (2001)

EC-EUROPA-EU, http://ec.europa.eu/eurostat/statistics-explained/index.php?titleFile:Recovery_and_recycling_rate_for_end-of-life_vehicles,_2015_(%25).png (2015). Retrieved on 11 Aug 2018

European Commission, *Communication from the Commission to the European Parliament, the Council, the European Economic and Social Committee and the Committee of the Regions* (2014). http://ec.europa.eu/environment/circular-economy/pdf/circular-economy-communication.pdf. Retrieved on 14 Aug 2018

R.F. Elmore, Instruments and strategy in public policy. Rev. Policy Res. **7**(1), 174–186 (1987)

EMAF, Allen MacArthur Foundation, *Growth Within: A Circular Economy Vision for a Competitive Europe*. Report (2015). https://www.ellenmacarthurfoundation.org/assets/downloads/publications/EllenMacArthurFoundation_Growth-Within_July15.pdf. Retrieved on 26 May 2018

Environment Agency, *Water for People and the Environment: Water Resources Strategy for England and Wales* (Environment Agency, Bristol, 2009). http://publications.environmentagencygov.uk/pdf/GEHO0309BPKXE-E.pdf. Retrieved on 09 Aug 2018

Environmental Protection Agency, *Life cycle Assessment: Principles and Practice*. EPA/600/R-06/060, May 2006 (U.S. Environmental Protection Agency, Cincinnati, Ohio, USA, 2006)

EPA, *Guidelines for Water Reuse* (2012). https://www3.epa.gov/region1/npdes/merrimackstation/pdfs/ar/AR-1530.pdf. Retrieved on 25 Aug 2018

EU, European Parliament, *Directive 2008/98/EC the European Parliament and of the Council of 19 November 2008 on Waste and Repealing Certain Directives* (2008)

European Union—European Commission, *Towards a circular economy: a zero waste programme for Europe*. Communication from the Commission to the European Parliament. The Council, the European Economic and Social Committee and the Committee of the Regions, Brussels, Belgium (2014a)

European Union—European Commission, *Study on Modelling of the Economic and Environmental Impacts of Raw Material Consumption* (2014b). http://ec.europa.eu/environment/enveco/resource_efficiency/pdf/RMC.pdf. Retrieved on 16 Aug 2018

European Union—European Commission, *Closing the loop—An EU Action Plan for the circular economy*. Communication from the Commission to the European Parliament. The Council, the European Economic and Social Committee and the Committee of the Regions (2015)

European Union, *Directive 2012/19/EU of the European Parliament and of the Council* (2018a). https://eur-lex.europa.eu/legal-content/EN/TXT/PDF/?uri=CELEX:32012L0019&from=en. Retrieved on 27 May 2018

European Union, *Directive 2011/65/EU of the European Parliament and of the Council* (2018b). https://eur-lex.europa.eu/legal-content/EN/TXT/PDF/?uri=CELEX:32011L0065&from=pt. Retrieved on 27 May 2018

European Union, *Directive 2009/125/EU of the European Parliament and of the Council*, (2018c). https://eur-lex.europa.eu/legal-content/EN/TXT/PDF/?uri=CELEX:32009L0125&from=en. Retrieved on 27 May 2018

European Union, *Directive 1999/31/EC of the European Parliament and of the Council* (2018d). https://eur-lex.europa.eu/legal-content/EN/TXT/PDF/?uri=CELEX:31999L0031&from=PT [ONLINE], retrieved on 28th of May 2018

European Comission, *Directive 2009/28/EC of the European Parlamient and of the Council on the promotion of the use of energy from renewable sources and amending and subsequently repealing Directives 2001/77/EC and 2003/30/EC* (2009)

European Environmental Agency, *Climate change, impacts and vulnerabiy in Europe—an indicator-based report*. EEA report no 12/2012. European Environmental Agency (2012)

European Tyre and Rubber Manufacturers Association, 2011, ETRMA, 2011. End of Life Tyres: *A Valuable Resource with Growing Potential*. 2011 Edition, European Tyre and Rubber Manufacturers Association, Brussels, Belgium

European Tyre and Rubber Manufacturers Association, *European tyre & rubber industry. Statistics*, 2015 Edition, Brussels, Belgium (2015).

European Tyre and Rubber Manufacturers Association, *End-of-life Tyre Report 2015* (Belgium, Brussels, 2015)

European Tyre and Rubber Manufacturers Association, *End of life tyres. End life tyres—a valuable resource with grow potential*, Brussels, Belgium (2011).

Food and Agriculture Organization of the United Nations (FAO), Clean Water and Sanitation. Progress on Level of Water Stress 2018. Global baseline for SDG indicator 6.4.2 (2018). http://www.unwater.org/app/uploads/2018/08/642-progress-on-level-of-water-stress-2018.pdf [ONLINE]. Retrieved on 15 Aug 2018

FAO, *Irrigation Water Requirement and Water Withdrawal by Country*. FAO AQUASTAT Reports (2012). http://www.fao.org/nr/water/aquastat/water_use_agr/IrrigationWaterUse.pdf. Retrieved on 08 Aug 2018

D. Fatta-Kassinos, D.D. Dionysiou, *Advanced treatment technologies for urban wastewater reuse*, in *The Handbook of Environmental Chemistry* (Springer International Publishing, Switzerland, 2016). ISBN 978-3-319-23885-2

M.B. Fernández-Viñé, T. Gómez-Navarro, S.F. Capuz-Rizo, Assessment of the public administration tools for the improvement of the eco-efficiency of small and medium sized enterprises. J. Clean. Prod. **47**, 265–273 (2013)

A. Field, J. Miles, Z. Field, *Discovering statistics using SPSS* (SAGE Publications, London, 2012). ISBN 978-1-4462-4917-8

K.H. Forslind, Does the financing of extended producer responsibility influence economic growth? J. Clean. Prod. **17**(2), 297–302 (2009)

N. Fraunholcz, W. Dalmijn, *Wetting mechanisms in the flotation of plastics*. In: Proceedings of the XX International Mineral Processing Congress. GMDB, Gesellschaft für Bergbau, Metallurgie, Rohstoff- und Umwelttechnik (1997). ISBN 978-3980486040

N. Fraunholcz, W. Dalmijn, Selective wetting of polymers by surfactant adsorption during froth flotation. J. Dispers. Sci. Technol. **19**, 859–873 (1998)

N. Fraunholcz, Separation of waste plastics by froth flotation—a review, part I. Miner. Eng. **17**, 261–268 (2004)

The Fredonia Group, *Demand Will Be Driven by Growth in the Number of Vehicles in Use in Developing Regions, Fueling Demand for Tires, and by Increases in Average Annual Vehicle Mileage Worldwide, Boosting Tire Replacement Rates, Report* (2018). https://www.freedoniagroup.com/industry-study/world-tires-3357.htm. Retrieved on 27 May 2018

The Freedonia Group, *World Tires, Industry Study 3105* (2014). https://www.freedoniagroup.com/brochure/31xx/3105smwe.pdf. Retrieved on 21 July 2018

R.E. Freeman, *Strategic management: a stakeholder approach* (Cambridge University Press, Cambridge, 1984). ISBN 978-0521151740

X. Fu, S.M. Ueland, E. Olivetti, Econometric modeling of recycled copper supply. Resour. Conserv. Recycl. **122**, 219–226 (2017)

G. Gaidajis, K. Angelakoglou, D. Aktsoglou, E-waste: environmental problems and current management. J. Eng. Sci. Technol. Rev. **3**(1), 193–199 (2010)

D. Galani, E. Gravas, A. Stavropoulos, Company characteristics and environmental policy. Bus. Strategy Environ. **21**(4), 236–247 (2012)

P.S. Garcia, F.D.B. de Sousa, J.A. de Lima, S.A. Cruz, C.H. Scuracchio, *Devulcanization of ground tire rubber: physical and chemical changes after different microwave exposure times*. eXPRESS Polymer Lett. **9**, 1015–1026 (2015)

J.M. García, Catalyst: design challenges for the future of plastics recycling. Chem **1**(6), 813–819 (2016)

M. Geissdoerfer, P. Savaget, N.M.P. Bocken, E.J. Hultink, The circular economy—a new sustainability paradigm? J. Clean. Prod. **143**, 757–768 (2017)

Genan Business and Development A/S, Comparative life cycle assessment of two options for waste tyre treatment: material recycling vs. co-incineration in cement kilns. *Executive summary* (2009). http://www.genan.eu/wp-content/uploads/2017/04/LCA_Material_recycling_vs._civil_engineering_applications.pdf. Retrieved on 21 July 2018

Y. Geng, J. Fu, J. Sarkis, B. Xue, Towards a national circular economy indicator system in China: an evaluation and critical analysis. J. Clean. Prod. **23**(1), 216–224 (2012)

M.R. Gent, M. Menendez, J. Toraño, I. Diego, Recycling of plastic waste by density separation: prospects for optimization. Waste Manage. Res. **27**, 175–187 (2009)

R. Geyer, V.D. Blass, The economics of cell phone reuse and recycling. Int. J. Adv. Manuf. Technol. **47**(5–8), 515–525 (2010)

P. Ghisellini, C. Cialani, S. Ulgiati, A review on circular economy: the expected transition to a balanced interplay of environmental and economic systems. J. Clean. Prod. **114**, 11–32 (2016)

B.F. Giannetti, S.H. Bonilla, C.M.V.B. Almeida, An energy-based evaluation of a reverse logistics network for steel recycling. J. Clean. Prod. **46**, 48–57 (2013)

D. Gibbs, P. Deutz, Reflections on implementing industrial ecology through eco-industrial park development. J. Clean. Prod. **15**(17), 1683–1695 (2007)

M. Giugliano, S. Cernuschi, U. Ghezzi, M. Grosso, Experimental evaluation of waste tires utilization in cement kilns. J. Air Waste Manag. **49**, 1405–1414 (1999)

D. Glew, L.C. Stringer, A.A. Acquaye, S. McQueen-Mason, How do end of life scenarios influence the environmental impact of product supply chains? Comparing biomaterial and petrochemical products. J. Clean. Prod. **29–30**, 122–131 (2012)

Global Competitiveness Report of the World Economic Forum, http://www3.weforum.org/docs/gcr/2015-2016/Global_Competitiveness_Report_2015-2016.pdf (2015). Retrieved on 08 Aug 2018

F. Granek, Business value of toxics reduction and pollution prevention planning. J. Clean. Prod. **19**(5), 559–560 (2011)

C. Gray, M. Charter, *Remanufacturing and Product Design: Designing for the 7th Generation* (The Centre for Sustainable Design, Farnham, 2008), http://www.cfsd.org.uk/Remanufacturing%20and%20Product%20Design.pdf. Retrieved on 11 Aug 2018

Greenerpackage.com, https://www.greenerpackage.com/metrics_standards_and_lca/new_packaging_study_reveals_complexities_sustainability (2018). Retrieved on 25 Aug 2018

G. Grimaud, N. Perry, B. Laratte, Development of an evaluation tool for engineering sustainable recycling pathways. Procedia CIRP **69**, 781–786 (2018)

M.J. Grimm, T.R. Sehlmeyer, *Method for the separation of a mixture of polyvinyl chloride and polyethylene terephthalate* (1986). US Patent 4617111

S.A. Groves, R.S. Lehrle, M. Blazsó, T. Székely, Natural rubber pyrolysis: study of temperature and thickness-dependence indicates dimer formation mechanism. J. Anal. Appl. Pyrol. **19**, 301–309 (1991)

L.A. Guerrero, G. Maas, W. Hogland, Solid waste management challenges for cities in developing countries. Waste Manag **33**(1), 220–232 (2013)

A. Guney, M.I. Poyraz, O. Kangal, F. Burat, Investigation of thermal treatment on selective separation of post consumer plastics prior to froth flotation. Waste Manag **33**, 1795–1799 (2013)

Y. Gupt, S. Sahay, Review of extended producer responsibility: a case study approach. Waste Manage. Res. **33**, 595–611 (2015)

V.K. Gupta, A. Nayak, S. Agarwal, M. Chaudhary, I. Tyagi, Removal of Ni (II) ions from water using scrap tire. J. Mol. Liq. **190**, 215–222 (2014)

C. Hagelüken, C. Meskers, Mining our computers—opportunities and challenges to recover scarce and valuable metals from end-of-life electronic devices, in *Electronics Goes Green*, Stuttgart (2008). https://www.researchgate.net/profile/Christian_Hagelueken/publication/270686794_Mining_our_computers_opportunities_and_challenges_to_recover_scarce_and_valuable_materials/links/550051390cf2d61f820d6870/Mining-our-computers-opportunities-and-challenges-to-recover-scarce-and-valuable-materials.pdf. Retrieved on 26 May 2018

C. Hagelüken, Improving metal returns and eco-efficiency in electronics recycling metals smelting and refining, in *Proceedings of the 2006 IEEE International Symposium on electronics and the environment* (IEEE, Scottsdale, AZ USA, 2006), pp. 218–223

S. Hartard, W. Liebert, *Competition and conflicts on resource use* (Springer International Publishing, Switzerland, 2015). ISBN 978-3-319-10953-4

G.D. Hatcher, W.L. Ijomah, J.F.C. Windmill, Design for remanufacture: a literature review and future research needs. J. Clean. Prod. **19**(17–18), 2004–2014 (2011)

J. He, D. Raghavan, D. Hoffman, D. Hunston, The influence of elastomer concentration on toughness in dispersions containing preformed acrylic elastomeric particles in an epoxy matrix. Polymer **40**, 1923–1933 (1999)

W. He, G. Li, X. Ma, H. Wang, J. Huang, M. Xu, C. Huang, WEEE recovery strategies and the WEEE treatment status in China. J. Hazard. Mater. **136**, 502–512 (2006)

R.K. Henry, Z. Yongsheng, D. Jun, Municipal solid waste management challenges in developing countries—Kenyan case study. Waste Manag. **26**(1), 92–100 (2006)

S. Hickey, C. Fitzpatrick, *Reuse Potential-Evaluation of Reuse Opportunities within WEEE Compliance Schemes* (2016). https://www.itu.int/en/ITU-T/climatechange/resources/Documents/Global-Portal/Step_GP_Reuse%20Potential_final.pdf. Retrieved on 12 Aug 2018

G.T. Hickle, Moving beyond the "patchwork:" a review of strategies to promote consistency for extended producer responsibility policy in the U.S. J. Clean. Prod. **64**, 266–276 (2014)

A.S. Homrich, G. Galvão, L.G. Abadia, M.M. Carvalho, The circular economy umbrella: trends and gaps on integrating pathways. J. Clean. Prod. **175**, 525–543 (2018)
Y.J. Hong, K.M. Jeong, P. Saha, J. Suh, J.K. Kim, Processing and characterization of microwave and ultrasonically treated waste-EPDM/LDPE polymer composites. Polym. Eng. Sci. **55**, 533–540 (2015)
J. Hopewell, R. Dvorak, E. Kosior, Plastics recycling: challenges and opportunities. Phil. Trans. R. Soc. B Biol. Sci. **364**, 2115–2126 (2009)
Huayin Group, http://www.huayinenergy.com/products/HY-6th_Waste_to_Oil_Pyrolysis_Machine/ (2018). Retrieved on 21 July 2018
J. Huisman, F. Magalini, K. Ruediger, M. Claudia, S. Ogilvie, J. Poll, C. Delgado, E. Artim, J. Szlezak, A. Stevels, *Review of Directive 2002/96 on Waste Electrical and Electronic Equipment (WEEE)*. DG ENV. Study Contract No. 07010401/2006/442493/ETU/G4 (2007)
International Council of Chemical Associations, *How to Know If and When It's Time to Commission a Life Cycle Assessment—An Executive Report* (2018). https://www.nikkakyo.org/sites/default/files/ICCA_LCA_Executive_Guide_en.pdf. Retrieved on 25 Aug 2018
W. Ijomah, M. Danis, *Refurbishment and reuse of WEEE*, in Waste Electrical and Electronic Equipment (WEEE) Handbook, ed. by V. Goodship, A. Stevels (Woodhead Publishing, Cambridge, 2012), p. 145. ISBN 978-0857096333
InEDIC, *Innovation and EcoDesign in Ceramic Industry (in Portuguese),* Manual de Ecodesign, LNEG/UCPS (2011). http://www.lneg.pt/download/12237/InEDIC_MANUAL_PT.pdf. Retrieved on 19 Aug 2018
W. Ingwersen, H. Cabezas, A.V. Weisbrod, T. Eason, B. Demeke, X. Ma, T. Hawkins, S.-J. Lee, J.C. Bare, M. Ceja, Integrated metrics for improving the life cycle approach to assessing product system sustainability. Sustainability **6**, 1386–1413 (2014)
M.R. Islam, H. Haniu, M.R.A. Beg, Liquid fuel sand chemicals from pyrolysis of motorcycle tire waste: product yields, compositions and related properties. Fuel **87**, 3112–3122 (2008)
ISO, *EN ISO 14006:2011—Environmental Management Systems—Guidelines for Incorporating Eco-design* (International Organization for Standardization, Geneva, 2011)
International Standard Organization, *ISO Standards for Life Cycle Assessment to Promote Sustainable Development* (2018). https://www.iso.org/news/2006/07/Ref1019.html. Retrieved on 25 Aug 2018
S. Izumi, K. Saitoh, 1978. *Method for separating mixed plastics*. US Patent 4119533
S. Izumi, H. Tanaka, *Flotation method of separation of mixture of plastics* (1975). US Patent 3926791
C.J.C. Jabbour, F.C.A. Santos, S.A. Fonseca, M.S. Nagano, Green teams: understanding their roles in the environmental management of companies located in Brazil. J. Clean. Prod. **46**, 58–66 (2013)
A.A. Jensen, L. Hoffman, B.T. Møller, A. Schmidt, *Life Cycle Assessment—A Guide to Approaches, Experiences and Information Sources*. Environmental Issues Series, 6 (European Environment Agency, 1997)
P. Jiang, M. Harney, Y. Song, B. Chen, Q. Chen, T. Chen, G. Lazarus, L.H. Dubois, M.B. Korzenski, Improving the end-of-life for electronic materials via sustainable recycling methods. Procedia Environ. Sci. **16**, 485–490 (2012)
B. Jody, J. Pomykala, E. Daniels, Cost effective recovery of thermoplastics from mixed scrap from mixed-scrap plastics. Mater. Technol. **18**(1), 18–24 (2003)
O.A. Jones, P.G. Green, N. Voulvoulis, Questioning the excessive use of advanced treatment to remove organic micropollutants from wastewater. Environ. Sci. Technol. **41**(14), 5085–5089 (2007)
J. Karger-Kocsis, L. Mészáros, T. Bárány, Ground tyre rubber (GTR) in thermoplastics, thermosets, and rubbers. J. Mat. Sci. **48**, 1–38 (2013)
P. Kiddee, R. Naidu, M.H. Wong, Electronic waste management approaches: an overview. Waste Manag. **33**(5), 1237–1250 (2013)
T.C. Kinnaman, Examining the justification for residential recycling. J. Econ. Perspect. **20**(4), 219–232 (2006)

J. Kirchherr, D. Reike, M. Hekkert, Conceptualizing the circular economy: An analysis of 114 definitions. Resour. Conserv. Recycl. **127**, 221–232 (2017)

J. Kirchherr, L. Piscicelli, R. Bour, E. Kostense-Smit, J. Muller, A. Huibrechtse-Truijens, M. Hekkert, Barriers to the circular economy: evidence from the European Union (EU). Ecol. Econ. **150**, 264–272 (2018)

R. Kissling, C. Fitzpatrick, H. Boeni, C. Luepschen, S. Andrew, J. Dickenson, Definition of generic re-use operating models for electrical and electronic equipment. Resour. Conserv. Recycl. **65**, 85–99 (2012)

R. Kissling, D. Coughlan, C. Fitzpatrick, H. Boeni, C. Luepschen, S. Andrew, J. Dickenson, Success factors and barriers in re-use of electrical and electronic equipment. Resour. Conserv. Recycl. **80**, 21–31 (2013)

J. Korhonen, A. Honkasalo, J. Seppälä, Circular economy: the concept and its limitations. Ecol. Econ. **143**, 37–46 (2018)

J. Korhonen, Four ecosystem principles for an industrial ecosystem. J. Clean. Prod. **9**(3), 253–259 (2001)

P. Kulkarni, Nitrophenol removal by simultaneous nitrification denitrification (SND) using *T. pantotropha* in sequencing batch reactors (SBR). Bioresources Technology **128**, 273–280 (2013)

C.R. Kumar, I. Fuhrmann, J. Karger-Kocsis, LDPE-based thermoplastic elastomers containing ground tire rubber with and without dynamic curing. Polym. Degrad. Stab. **76**, 137–144 (2002)

A. Kumar, M. Holuszko, D.C.R. Espinosa, E-waste: an overview on generation, collection, legislation and recycling practices. Resour. Conserv. Recycl. **122**, 32–42 (2017)

E. Kwon, M.J. Castaldi, Fundamental understanding of the thermal degradation mechanisms of waste tires and their air pollutant generation in a N_2 atmosphere. Environ. Sci. Technol. **43**, 5996–6002 (2009)

P. Lacy, J. Rutqvist, *Waste to Wealth: The Circular Economy Advantage* (Palgrave Macmillan, London, 2016). ISBN 978-1-137-53068-4

M.F. Laresgoiti, I. de Marco, A. Torres, B. Caballero, M.A. Cabrero, M.J. Chomón, Chromatographic analysis of the gases obtained in tyre pyrolysis. J. Anal. Appl. Pyrol. **55**, 43–54 (2000)

H. Latan, C.J.C. Jabbour, S.F. Wamba, M. Shahbaz, Effects of environmental strategy, environmental uncertainty and Top Management's commitment on corporate environmental performance: the role of environmental management accounting. J. Clean. Prod. **180**, 297–306 (2018)

H.-L. Le, E. Yamasue, H. Okumura, K. Ishihara, Improving sustainable recovery of metals from waste printed circuit boards by the primary copper smelter process. J. Mater. Cycles Waste Manage. **16**, 298–305 (2014)

C. Le Guern, P. Conil, R. Houot, Role of calcium ions in the mechanism of action of a lignosulphonate used to modify the wettability of plastics for their separation by flotation. Miner. Eng. **13**, 53–63 (2000)

C. Le Guern, P. Baillif, P. Conil, R. Houot, Adsorption of a lignosulphonate polymer onto PVC and PET surfaces: evaluation by XPS. J. Mater. Sci. **36**, 1547–1554 (2001)

D.Y.C. Leung, C.L. Wang, Kinetic study of scrap tyre pyrolysis and combustion. J. Anal. Appl. Pyrol. **45**, 153–169 (1998)

D.Y.C. Leung, C.L. Wang, Fluidized-bed gasification of waste tire powders. Fuel Process. Technol. **84**, 175–196 (2003)

S.-Q. Li, Q. Yao, Y. Chi, J.-H. Yan, K.-F. Cen, Pilot-scale pyrolysis of scrap tires in a continuous rotary kiln reactor. Ind. Eng. Chem. Res. **43**, 5133–5145 (2004a)

J. Li, M. Liang, S. Guo, Y. Lin, Studies on chain scission and extension of polyamide 6 melt in the presence of ultrasonic irradiation. Polym. Degrad. Stab. **86**, 323–329 (2004b)

Y. Li, S. Zhao, Y. Wang, Microbial desulfurization of ground tire rubber by *Thiobacillus ferrooxidans*. Polym. Degrad. Stab. **96**, 1662–1668 (2011)

B. Li, J. Yang, X. Song, B. Lu, Survey on disposal behavior and awareness of mobile phones in Chinese university students. Procedia Environ. Sci. **16**, 469–476 (2012)

L. Li, X. Zeng, A. Stevels, Ecodesign in consumer electronics: past, present and future. Crit. Rev. Environ. Sci. Technol. **45**, 840–860 (2015)

R. Lifset, A. Atasu, N. Tojo, Extended producer responsibility. J. Ind. Ecol. **17**(2), 162–166 (2013)
C. Lin, C.-L. Huang, C.-C. Shern, Recycling waste tire powder for the recovery of oil spills. Resour. Conserv. Recycl. **52**, 1162–1166 (2008)
Z. Liu, T. Li, Q. Jiang, H. Zhang, Life cycle assessment-based comparative evaluation of originally manufactured and remanufactured diesel engines. J. Ind. Ecol. **18**(4), 567–576 (2014)
W. Liu, J. Zhang, M. Jin, S. Liu, X. Chang, N. Xie, Y. Wang, Key indices of the remanufacturing industry in China using a combined method of grey incidence analysis and grey clustering. J. Clean. Prod. **168**, 1348–1357 (2017)
B. Liu, D. Chen, W. Zhou, N. Nasr, T. Wang, S. Hu, B. Zhu, The effect of remanufacturing and direct reuse on resource productivity of China's automotive production. J. Clean. Prod. **194**, 309–317 (2018)
V. Lofthouse, Ecodesign tools for designers: defining the requirements. J. Clean. Prod. **14**, 1386–1395 (2006)
J. Loomis, G. Helfand, *Environmental policy analysis for decision making* (Kluwer Academic Publishers, Dordrecht, 2001). ISBN 0-7923-6500-3
B. Lu, J. Liu, J. Yang, B. Li, The environmental impact of technology innovation on WEEE management by multi-life cycle assessment. J. Clean. Prod. **89**, 148–158 (2015)
B. Lu, J. Yang, W. Ijomah, W. Wu, G. Zlamparet, Perspectives on reuse of WEEE in China: Lessons from the EU. Resour. Conserv. Recycl. **135**, 83–92 (2018)
E.B. Machin, D.T. Pedroso, J.A. Carvalho Jr., Energetic valorization of waste tires. Renew. Sustain. Energy Rev. **68**, 306–315 (2017)
A. Mahvi, Sequencing batch reactor: a promising technology in wastewater treatment. J. Environ. Health Sci. Eng. **5**, 79–90 (2008)
A. Mahvi, A. Rajabizadeh, N. Yousefi, H. Hosseini, M. Ahmadian, Survey wastewater treatment condition and effluent quality of Kerman province hospitals. World Appl. Sci. J. **7**, 1521–1525 (2009)
G. Malcolm Richard, M. Mario, T. Javier, T. Susana, Optimization of the recovery of plastics for recycling by density media separation cyclones. Resour. Conserv. Recycl. **55**, 472–482 (2011)
O. Malik, *Primary vs. Secondary: Types of Wastewater Treatment* (2014). http://archive.epi.yale.edu/case-study/primary-vs-secondary-types-wastewater-treatment. Retrieved on 09 Aug 2018
P. Manomaivibool, Extended producer responsibility in a non-OECD context: the management of waste electrical and electronic equipment in India. Resour. Conserv. Recycl. **53**, 136–144 (2009)
E. Maranon, I. Vazquez, J. Rodriguez, L. Castrillon, Y. Fernandez, H. Lopez, Treatment of coke wastewater in a sequential batch reactor (SBR) at pilot plant scale. Bioresour. Technol. **99**, 4192–4198 (2008)
G.A. Marques, J.A.S. Tenório, Use of froth flotation to separate PVC/PET mixtures. Waste Manag. **20**, 265–269 (2000)
N. Martin, J. Skvarla, M. Sisol, *A possibility of using the flotation process to separate plastics*. Ann. Fac. Eng. Hunedoara Int. J. Eng. **3**, 275–278 (2011)
J.D. Martínez, N.P.R. Murillo, T. García, M.V. Navarro, A.M. Mastral, Waste tyre pyrolysis—a review. Renew. Sustain. Energy Rev. **23**, 179–213 (2013)
A.M. Mastral, R. Murillo, M.S. Callén, T. Garcia, C.E. Snape, Influence of process variables on oils from tire pyrolysis and hydropyrolysis in a swept fixed bed reactor. Energy Fuel **14**, 739–744 (2000)
A. Mayer, P. Kauppi, P. Angelstam, Y. Zhang, P. Tikka, Importing timber, exporting ecological impact. Science **308**, 359–360 (2005)
T. McAloone, N. Bey, *Environmental Improvement Through Product Development: A Guide* (2009). https://mst.dk/media/90192/environmental_improvement_through_product_development.pdf. Retrieved on 22 Aug 2018
T. McAloone, *Industry Experiences of Environmentally Conscious Design Integration: An Exploratory Study* (Cranfield University, UK, 1998). https://core.ac.uk/download/pdf/139063.pdf. Retrieved on 23 Aug 2018

R. McKenna, S. Reith, S. Cail, A. Kessler, W. Fichtner, Energy savings through direct secondary reuse: an exemplary analysis of the German automotive sector. J. Clean. Prod. **52**, 103–112 (2013)

MED-EUWI, *Mediterranean component of the EU Water Initiative (MED EUWI)*. Strategic Partnership on Water for Sustainable Development (2007)

B.A. Miller-Chou, J.L. Koenig, *A review of polymer dissolution*. Prog. Polym. Sci. **28**, 1223–1270 (2003)

M. Mirata, T. Emtairah, Industrial symbiosis networks and the contribution to environmental innovation: the case of the Landskrona industrial symbiosis programme. J. Clean. Prod. **13**(10–11), 993–1002 (2005)

E. Mokrzycki, A. Uliasz-Bochenczyk, Alternative fuels for the cement industry. Appl. Energy **74**, 95–100 (2003)

D. Molden, *Trends in water and agricultural development*, in *Water for Food, Water for Life: A Comprehensive Assessment of Water Management in Agriculture*, ed. by D. Molden (Earthscan/IWMI, 2007). ISBN 978-1-84407-397-9

L. Möller, Nachhaltige Entwicklunge Wege zur ökologischen, ökonomischen und sozialen Zukunftsfähigkeit (in German), in *Integratives Umweltmanagement e Systemorientierte Zusammenhänge zwischen Politik, Recht, Management und Technik*, ed. by M. Kramer (GWV Fachverlag, Wiesbaden, 2010), pp. 41–61. ISBN 978-3-8349-1947-2

O. Mont, A. Plepys, Sustainable consumption progress: should we be proud or alarmed? J. Clean. Prod. **16**(4), 531–537 (2008)

N. Morley, D. Eatherley, *Material Security: Ensuring Resource Availability for the UK Economy* (C-Tech Innovation, Ltd., 2008), pp. 1–36. https://static1.squarespace.com/static/5a60c3cc9f07f58443081f58/t/5ab3e4ba03ce64d747ce1df8/1521738947024/material_security.pdf. Retrieved on 23 Aug 2018

F. Murakami, A. Sulzbach, G.M. Pereira, M. Borchardt, M.A. Sellitto, How the Brazilian government can use public policies to induce recycling and still save money? J. Clean. Prod. **96**(1), 94–101 (2015)

H.S. Naess-Schmidt, L. Jensen, *Environmental Policy Analysis—Dealing with Economic Distortions, Nordic Council of Ministers* (2015). ISBN 978-92-893-4391-6. https://www.copenhageneconomics.com/dyn/resources/Publication/publicationPDF/1/331/1447412478/environmental-policy-analysis-copenhagen-economics.pdf. Retrieved on 14 May 2018

Y. Nakajime, M. Matsuyuki, Utilization of waste tires as fuel forcement production. Conserv. Recycl. **4**(3), 145–151 (1981)

J. Namias, *The Future of Electronic Waste Recycling in the United States: Obstacles and Domestic Solutions* (Columbia University, New York, 2013). http://www.seas.columbia.edu/earth/wtert/sofos/Namias_Thesis_07-08-13.pdf. Retrieved on 27 May 2018

B. Narasimhan, S.K. Mallapragada, Dissolution of amorphous and semicrystalline polymers: mechanisms and novel applications. Recent Results Dev. Macromol. Res. **3**(2), 311–324 (1998)

J. Nash, C. Bosso, Extended producer responsibility in the United States. J. Ind. Ecol. **17**(2), 175–185 (2013)

A.K. Naskar, A.K. Bhowmick, S.K. De, Thermoplastic elastomeric composition based on ground rubber tire. Polym. Eng. Sci. **41**, 1087–1098 (2001)

A.K. Naskar, S.K. De, A.K. Bhowmick, Thermoplastic elastomeric composition based on maleic anhydride–grafted ground rubber tire. J. Appl. Polym. Sci. **84**, 370–378 (2002)

National Development and Reform Commission of China, *Opinions on Promoting the Development of Remanufacturing Industry* (in Chinese). www.gov.cn/zwgk/2010-05/31/content_1617310.htm. Retrieved on 10 Aug 2018

National Research Council, *Water Reuse: Potential for Expanding the Nation's Water Supply through Reuse of Municipal Wastewater* (National Academy Press, Washington D.C., 2012). http://dels.nas.edu/Report/Water-Reuse-Potential-Expanding/13303. Retrieved on 09 Aug 2018

E.B. Nauman, J.C. Lynch. US Patent 5 198 471 (1993)

E.B. Nauman, J.C. Lynch. US Patent 5 278 282 (1994)

I. Naz, N. Khatoon, M.I. Ali, D.P. Saroj, S. A-u, N. Batool, S.Ahmed Ali, Appraisal of the tire derived rubber (TDR) medium for wastewater treatment under aerobic and anaerobic conditions. J. Chem. Technol. Biotechnol. **89**, 587–596 (2014)

W.P.Q. Ng, H.L. Lam, F.Y. Ng, M. Kamal, J.H.E. Lim, Waste-to-wealth: green potential from palm biomass in Malaysia. J. Clean. Prod. **34**, 57–65 (2012)

R. Ng, Z. Yeo, J.S.C. Low, B. Song, A method for relative eco-efficiency analysis and improvement: case study of bonding technologies. J. Clean. Prod. **99**, 320–332 (2015)

Nikkakyo.org, *How to Know If and When it's Time to Commission a Life Cycle Assessment* (2018). https://www.nikkakyo.org/sites/default/files/ICCA_LCA_Executive_Guide_en.pdf. Retrieved on 24 Aug 2018

S. Niza, E. Santos, I. Costa, P. Ribeiro, P. Ferrão, Extended producer responsibility policy in Portugal: a strategy towards improving waste management performance. J. Clean. Prod. **64**, 277–287 (2014)

D.C. Nugent, *Process for the separation and recovery of plastics* (1991). US Patent 5022985

J. Nygren, R. Antikainen, *Use of Life Cycle Assessment (LCA) in Global Companies* (2010). Reports of the Finnish Environment Institute, 16, 2010. https://helda.helsinki.fi/bitstream/handle/10138/39723/SYKEre_16_2010.pdf?sequence = 1. Retrieved on 26 Aug 2018

Organization for Economic Co-operation and Development (OECD), *Extended producer responsibility: a guidance manual for governments*. Paris, France (2001)

M. Oguchi, H. Sakanakura, A. Terazono, H. Takigami, Fate of metals contained in waste electrical and electronic equipment in a municipal waste treatment process. Waste Manag. **32**(1), 96–103 (2012)

Y. Ogushi, M. Kandlikar, Assessing extended producer responsibility laws in Japan. Environ. Sci. Technol. **41**, 4502–4508 (2007)

T. Okuda, K. Kurose, W. Nishijima, M. Okada, Separation of polyvinyl chloride from plastic mixture by froth flotation after surface modification with ozone. Ozone: Science and Engineering **29**, 373–377 (2007)

A. Oláh, H. Hillborg, G.J. Vancso, Hydrophobic recovery of UV/ozone treated poly (dimethylsiloxane): adhesion studies by contact mechanics and mechanism of surface modification. Appl. Surf. Sci. **239**, 410–423 (2005)

E.L. Olson, Green innovation value chain analysis of PV solar power. J. Clean. Prod. **64**, 73–80 (2014)

F.O. Ongondo, I.D. Williams, Greening academia: use and disposal of mobile phones among university students. Waste Manag. **31**, 1617–1634 (2011)

M. Ozdemir, C.U. Yurteri, H. Sadikoglu, Physical polymer surface modification methods and applications in food packaging polymers. Crit. Rev. Food Sci. Nutr. **39**, 457–477 (1999)

H. Pakdel, D.M. Pantea, C. Roy, Production of dl-limonene by vacuum pyrolysis of used tires. J. Anal. Appl. Pyrol. **57**, 91–107 (2001)

K. Palmer, W.E. Oates, P.R. Portney, Tightening environmental standards: the benefit-cost or the no-cost paradigm? J. Econ. Perspect. **9**, 119–132 (1995)

K. Palmer, H. Sigman, M. Walls, The cost of reducing municipal solid waste. J. Environ. Econ. Manag. **33**(2), 128–150 (1997)

G. Pappa, C. Boukouvalas, C. Giannaris, N. Ntaras, V. Zografos, K. Magoulas, A. Lygeros, D. Tassios, The selective dissolution/precipitation technique for polymer recycling: a pilot unit application. Resour. Conserv. Recycl. **34**, 33–44 (2001)

K. Parajuly, H. Wenzel, Potential for circular economy in household WEEE management. J. Clean. Prod. **151**, 272–285 (2017)

C.-H. Park, H.-S. Jeon, J.-K. Park, PVC removal from mixed plastics by triboelectrostatic separation. J. Hazard. Mater. **144**, 470–476 (2007)

R. Pascoe, B. O'Connell, Development of a method for separation of PVC and PET using flame treatment and flotation. Miner. Eng. **16**, 1205–1212 (2003a)

R. Pascoe, B. O'Connell, Flame treatment for the selective wetting and separation of PVC and PET. Waste Manag. **23**, 845–850 (2003b)

M. Patel, N. von Thienen, E. Jochem, E. Worrell, Recycling of plastics in Germany. Resour. Conserv. Recycl. **29**, 65–90 (2000)

A.W. Patrick, S. Mathias, M. Esther, G. Rolf, RoHS regulated substances in mixed plastics from waste electrical and electronic equipment. Environ. Sci. Technol. **46**, 628–635 (2012)

B. Peng, Tu Yu, E. Elahi, G. Wei, Extended producer responsibility and corporate performance: effects of environmental regulation and environmental strategy. J. Environ. Manage. **218**, 181–189 (2018)

PEW Research Center Report, Spring (2015). http://mo.techweb.com.cn/2016-05-11/2330764.shtml. Retrieved on 27 May 2018

D.C.A. Pigosso, *Ecodesign Maturity Model: A Framework to Support Companies in the Selection and Implementation of Ecodesign Practices*. Biblioteca Digital da USP (Universidade de São Paulo, Brasil, 2012). http://orbit.dtu.dk/files/117267549/ECODESIGN_MATURITY_MODEL.pdf. Retrieved on 23 Aug 2018

D.C.A. Pigosso, H. Rozenfeld, T.C. McAloone, Ecodesign maturity model: a management framework to support ecodesign implementation into manufacturing companies. J. Clean. Prod. **59**, 160–173 (2013)

D.C. Pigosso, T.C. McAloone, H. Rozenfeld, *Systematization of best practices for ecodesign implementation*, in International Design Conference—Design 2014. Dubrovinik, Croatia (2014), pp. 1651–1662

P. Pipilikaki, M. Katsioti, D. Papageorgiou, D. Fragoulis, E. Chaniotakis, Use of tire derived fuel in clinker burning. Cement Concr. Compos. **27**, 843–847 (2005)

Ł. Piszczyk, A. Hejna, K. Formela, M. Danowska, M. Strankowski, Effect of ground tire rubber on structural, mechanical and thermal properties of flexible polyurethane foams. Iran. Polym. J. **24**, 75–84 (2015)

D. Popp, International innovation and diffusion of air pollution control technologies: the effects of NO_X and SO_2 regulation in the US, Japan, and Germany. J. Environ. Econ. Manag. **51**(1), 46–71 (2006)

M.E. Porter, M.R. Kramer, Creating shared value. Harvard Bus. Rev. **89**, 62–77 (2011)

D.L. Presti, Recycled tyre rubber modified bitumens for road asphalt mixtures: a literature review. Constr. Build. Mater. **49**, 863–881 (2013)

Y. Qi, X. Chen, W. Wang, Improved life cycle assessment of recycling organic wastes for practice. J. Clean. Prod. **195**, 1558–1571 (2018)

B. Qian, Progress summarization non recycling, treatment & utilization for electronic garbage at home and abroad. China Environ. Prot. Ind. **8**, 18–23 (2010)

A. Quek, R. Balasubramanian, Liquefaction of waste tires by pyrolysis for oil and chemicals—a review. J. Anal. Appl. Pyrol. **101**, 1–16 (2013)

M. Rahmani, R. Nabizadeh, K. Yaghmaeian, A.H. Mahvi, M. Yunesian, Estimation of waste from computers and mobile phones in Iran. Resour. Conserv. Recycl. **87**, 21–29 (2014)

K.P. Raman, W.P. Walawender, T. Fan, Gasification of waste tires in a fluid bed reactor. Conserv. Recycl. **4**(2), 79–88 (1981)

B. Rånby, Surface modification of polymers by photoinitiated graft polymerization, in *Makromolekulare Chemie. Macromolecular Symposia* (Wiley Online Library, 1992)

A. Rashid, M.A. Farazee, P. Krajnik, C. Nicolescu, Resource conservative manufacturing. J. Clean. Prod. **57**, 166–177 (2013)

K.S. Rauwald, C.F. Moore, Environmental attitudes as predictors of policy support across three countries. Environ. Behav. **34**(6), 709–739 (2002)

M.S. Reddy, K. Kurose, T. Okuda, W. Nishijima, M. Okada, Separation of polyvinyl chloride (PVC) from automobile shredder residue (ASR) by froth flotation with ozonation. J. Hazard. Mater. **147**, 1051–1055 (2007)

M.S. Reddy, K. Kurose, T. Okuda, W. Nishijima, M. Okada, Selective recovery of PVC-free polymers from ASR polymers by ozonation and froth flotation. Resour. Conserv. Recycl. **52**, 941–946 (2008)

M.S. Reddy, T. Yamaguchi, T. Okuda, T.Y. Tsai, S. Nakai, W. Nishijima, M. Okada, Feasibility study of the separation of chlorinated films from plastic packaging wastes. Waste Manag. **30**, 597–601 (2010)

J. Rezaiyan, N.P. Cheremisinoff, *Gasification technologies. A primer for engineers and scientists* (CRC Press, Taylor & Francis, Boca Raton, 2005). ISBN 978-1420028140

F.D.M. Ribeiro, I. Kruglianskas, Improving environmental permitting through performance-based regulation: a case study of Sao Paulo State, Brazil. J. Clean. Prod. **46**, 15–26 (2013)

J.L. Richter, R. Koppejan, Extended producer responsibility for lamps in Nordic countries: best practices and challenges in closing material loops. J. Clean. Prod. **123**, 167–179 (2016)

Robert Bosch GmbH, *Stark wachsender Markt für Austauschteile - Bosch ist der führende Anbierter von wiederaufgearbeiteten Fahrzeugteilen - Größtes Programm für die zeitwertgerechte Reparatur* (Robert Bosch GmbH, Karlsruhe, 2011). (in German)

J.F.D. Rodrigues, A. Lorena, I. Costa, P. Ribeiro, P. Ferrão, An input-output model of extended producer responsibility. J. Ind. Technol. **20**(6), 1273–1283 (2016)

V.P. Rodrigues, D.C.A. Pigosso, T.C. McAloone, Measuring the implementation of ecodesign management practices: a review and consolidation of process-oriented performance indicators. J. Clean. Prod. **156**, 293–309 (2017)

D.C. Rodríguez, N. Pino, G. Peñuela, Monitoring the removal of nitrogen by applying a nitrification–denitrification process in a sequencing batch reactor (SBR). Bioresour. Technol. **102**, 2316–2321 (2011)

H. Rozenfeld, F.A. Forcellini, D.C. Amaral, J.C. Toledo, S.L. Silva, D.H. Alliprandini, R.K. Scalice, *Gestão do desenvolvimento de produtos: uma referência para a melhoria do processo*, 1st edn. (Editora Saraiva, Brazil, 2006, in Portuguese). ISBN 978-8502112094

H. Rozenfeld, *Reference model for managing product development*, in *Sustainability in Manufacturing: Recovery of Resources in Product and Material Cycles* (Springer, Berlin, 2007), pp. 193–206. ISBN 978-3-540-49870-4

Rubber Manufacturers Association, *2015 US scrap tire management summary*. Washington DC, USA (2016)

K. Ruediger, C. Fitzpatrick, L. Claudia, *E-Waste Challenges: Re-Use Practices Principles and Standards* (2011). https://unu.edu/publications/articles/e-waste-challenges-re-use-practices-principles-and-standards.html. Retrieved on 11 Aug 2018

J. Rybicka, A. Tiwari, G. Leeke, Technology readiness level assessment of composites recycling technologies. J. Clean. Prod. **112**, 1001–1012 (2016)

Y.M.B. Saavedra, A.P.B. Barquet, H. Rozenfeld, F.A. Forcellini, A.R. Ometto, Remanufacturing in Brazil: case studies on the automotive sector. J. Clean. Prod. **53**, 267–276 (2013)

M. Salgot, M. Folch, S.S. Unit, Wastewater treatment and water reuse. Curr. Opin. Environ. Sci. Health **2**, 64–74 (2018)

K. Sander, S. Schiling, N. Tojo, C. van Rossem, J. Vernon, C. George, *The producer responsibility principle of the WEEE Directive* (2007). DG ENV, Study Contract No. 07010401/2006/449269/MAR/G4, 2007

S. Saisinchai, Separation of PVC from PET/PVC mixtures using flotation by calcium lignosulfonate depressant. Eng. J. **18**, 45–54 (2013)

S.C. Santos, R.A. Boaventura, Treatment of a simulated textile wastewater in a sequencing batch reactor (SBR) with addition of a low-cost adsorbent. J. Hazard. Mater. **291**, 74–82 (2015)

E. Saouter, G. van Hoof, A database for the life-cycle assessment of procter & gamble laundry detergents. Int. J. Life Cycle Assess. **7**, 103–114 (2002)

F. Scala, R. Chirone, P. Salatino, Fluidized bed combustion of tyre derived fuel. Exp. Thermal Fluid Sci. **27**, 465–471 (2003)

J.W. Scheijgrond, Extending producer responsibility up and down the supply chain, challenges and limitation. Waste Manage. Res. **29**(9), 911–918 (2011)

J. Scheirs, *Overview of commercial pyrolysis processes for waste plastics*, in Feedstock Recycling and Pyrolysis of Waste Plastics: Converting Waste Plastics into Diesel and Other Fuels, ed. by J. Scheirs, W. Kaminsky (Wiley, New Jersey, 2006), pp. 383–433. ISBN 978-0470021521

M. Schluep, C. Hagelueken, R. Kuehr, F. Magalini, C. Maurer, E. Meskers, C. Mueller, F. Wang, *Recycling from E-waste to Resources. United Nations Environment Programme (UNEP)*. Sustainable Innovation and Technology Transfer Industrial Sector Studies DTI/1192/PA (2009). http://www.unep.org/pdf/pressreleases/E-waste_publication_screen_finalversion-sml.pdf. Retrieved on 15 Sept 2016

J.-P. Schöggl, R.J. Baumgartner, D. Hofer, *Improving sustainability performance in early phases of product design: a checklist for sustainable product development tested in the automotive industry*. J. Clean. Prod. **140**, 1602–1617 (2017)

S. Schwarzer, A.D. Bono, G. Giuliani, S. Kluser, P. Peduzzi, *E-waste—The Hidden Side of IT Equipment's Manufacturing And use. United Nations Environment Programme (UNEP)*. Early warning on emerging environmental threats no. 5. UNEP/GRID Europe (2005). http://www.grid.unep.ch/products/3_Reports/ew_ewaste.en.pdf. Retrieved on 26 May 2018

S. Sciacca, G.O. Conti, Mutagens and carcinogens in drinking water, Mediterranean. J. Nutr. Metab. **2**, 157–162 (2009)

C.E. Scruggs, N. Nimpuno, R.B.B. Moore, Improving information flow on chemicals in electronic products and E-waste to minimize negative consequences for health and the environment. Resour. Conserv. Recycl. **113**, 149–164 (2016)

M. Selbes, O. Yilmaz, A.A. Khan, T. Karanfil, Leaching of DOC, DN, and inorganic constituents from scrap tires. Chemosphere **139**, 617–623 (2015)

N. Seyring, M. Klinge, J. Weißenbacher, M. Hestin, L. Lecerf, F. Magalini, D.S. Khetriwal, R. Kuehr, *Study on WEEE Recovery Targets, Preparation for Re-use Targets and on the Method for Calculation of the Recovery Targets* (2015). http://ec.europa.eu/environment/waste/weee/pdf/16.%20Final%20report_approved.pdf. Retrieved on 12 Aug 2018

H. Shen, E. Forssberg, R. Pugh, Selective flotation separation of plastics by particle control. Resour. Conserv. Recycl. **33**, 37–50 (2001)

H. Shen, R. Pugh, E. Forssberg, Floatability, selectivity and flotation separation of plastics by using a surfactant. Colloids Surf. A **196**, 63–70 (2002a)

H. Shen, E. Forssberg, R. Pugh, Selective flotation separation of plastics by chemical conditioning with methyl cellulose. Resour. Conserv. Recycl. **35**, 229–241 (2002b)

H. Shent, R. Pugh, E. Forssberg, A review of plastics waste recycling and the flotation of plastics. Resour. Conserv. Recycl. **25**, 85–109 (1999)

C. Sherwin, *Innovative Ecodesign—An Exploratory and Descriptive Study of Industrial Design Practice*. Doctoral Thesis, Cranfield University, UK (2000)

J. Shibata, S. Matsumoto, H. Yamamoto, E. Kusaka, Flotation separation of plastics using selective depressants. Int. J. Miner. Process. **48**, 127–134 (1996)

V.L. Shulman, Tyre recycling. *Rapra review reports*, vol. 15, No. 7, UK (2004)

M. Sienkiewicz, J. Kucinska-Lipka, H. Janik, A. Balas, Progress in used tyres management in the European Union: a review. Waste Manag. **32**, 1742–1751 (2012)

W. Sierzchula, S. Bakker, K. Maat, B. van Wee, Technological diversity of emerging eco-innovations: a case study of the automobile industry. J. Clean. Prod. **37**, 211–220 (2012)

D.A.L. Silva, G.W.S. Renó, G. Sevegnani, T.B. Sevegnani, O.M.S. Truzzi, Comparison of disposable and returnable packaging: a case study of reverse logistics in Brazil. J. Clean. Prod. **47**, 377–387 (2013)

K.R. Skene, Circles, spirals, pyramids and cubes: why the circular economy cannot work. Sustain. Sci. **13**(2), 479–492 (2017)

J.P. Snudden, C. Ward, K. Potter, Reusing automotive composites production waste. Reinf. Plast. **2014**, 20–27 (2014)

P. Song, X. Zhao, X. Cheng, S. Li, S. Wang, Recycling the nanostructured carbon from waste tires. Compos. Commun. **7**, 12–15 (2018)

J.H. Spangenberg, A. Fuad-Luke, K. Blincoe, Design for Sustainability (DfS): the interface of sustainable production and consumption. J. Clean. Prod. **18**, 1485–1493 (2010)

A.P. Srivastava, B.E. Nauman. Kinetics of solvent induced melting of semicrystalline polymers. Polym. Mater. Sci. Eng., in *Proceedings of the ACS Division of Polymeric Material*, vol. 69 (1993), pp. 307–308

W.R. Stahel, The circular economy. Nature **531**(7595), 435–438 (2016)

L. Steg, C. Vlek, Encouraging pro-environmental behaviour: an integrative review and research agenda. J. Environ. Psychol. **29**(3), 309–317 (2009)

B. Stuckrad, K. Lohr, V. Vogt, *Sorting of waste plastic mixtures by flotation*, in *Proceedings of the XX International Mineral Processing Congress* (GMDB, Gesellschaft für Bergbau, Metallurgie, Rohstoff- und Umwelttechnik, 1997). ISBN 978-3980486040

B.T.O. Stückrad, *Sortierung von Kunststoffgemischen durch Flotation* (1996). Ph.D. Thesis Clausthal University of Technology, Germany

P.M. Subramanian, Plastics recycling and waste management in the US. Resour. Conserv. Recycl. **28**, 253–263 (2000)

A.A.M. Sultan, E. Lou, P.T. Mativenga, What should be recycled: an integrated model for product recycling desirability. J. Clean. Prod. **154**, 61–70 (2017)

J. Sun, W. Wang, Z. Liu, C. Ma, Recycling of waste printed circuit boards by microwave-induced pyrolysis and featured mechanical processing. Ind. Eng. Chem. Res. **50**, 11763–11769 (2011)

P. Sutanto, F. Picchioni, L.P.B.M. Janssen, Modelling a continuous devulcanization in an extruder. Chem. Eng. Sci. **61**, 7077–7086 (2006)

A. Takdastan, N. Mehrdadi, A.A. Azimi, A. Torabian, G. NabiBidhendi, Investigation of the excess sludge reduction in SBR by oxidizing some sludge by ozone. Iranian J. Chem. Chem. Eng. (IJCCE) **28**, 95–104 (2009)

B. Tansel, From electronic consumer products to e-wastes: global outlook, waste quantities, recycling challenges. Environ. Int. **98**, 35–45 (2017)

G. Tao, Q. He, Y. Xia, G. Jia, H. Yang, W. Ma, The effect of devulcanization level on mechanical properties of reclaimed rubber by thermal-mechanical shearing devulcanization. J. Appl. Polym. Sci. **129**(5), 2598–2605 (2013)

F. Testa, D. Styles, F. Iraldo, Case study evidence that direct regulation remains the main driver of industrial pollution avoidance and may benefit operational efficiency. J. Clean. Prod. **21**(1), 1–10 (2012)

E.T. Thostenson, T.W. Chou, Microwave processing: fundamentals and applications. Compos. A **30**(9), 1055–1071 (1999)

U. Tischner, M. Charter, Sustainable product design, in *Sustainable solutions: developing products and services for the future*, ed. by M. Charter, U. Tischner (Greenleaf Publishing Ltd., Sheffield, UK, 2001), pp. 118–138. ISBN 978-1351282468

V. Torretta, M. Ragazzi, A. Istratel, E.C. Rada, Management of waste electrical and electronic equipment in two EU countries: a comparison. Waste Manag. **33**, 117–122 (2013)

P. Trowbridge, A case study of green supply-chain management at advanced micro devices. Greener Manag. Int. **35**, 121–135 (2001)

W.-T. Tsai, Analysis of the sustainability of reusing industrial wastes as energy source in the industrial sector of Taiwan. J. Clean. Prod. **18**(14), 1440–1445 (2010)

UN-Water, UN-Water Annual Report 2017 (2017). https://www.unwater.org/app/uploads/2018/10/UN_Water_Annual_report_2017.pdf. [ONLINE]. Retrieved on 14 Aug 2018

UN-Water, *The United Nations World Water Development Report 4: Managing Water Under Uncertainty and Risk*, World Water Assessment Programme (WWAP), Paris, France (2012)

United Nations Environment Programme (UNEP), *E-waste volume II: E-waste management manual*. Division of Technology, Industry and Economics. International Environmental Technology Centre, Osaka, Japan (2007)

L.H.M. Vefago, J. Avellaneda, Recycling concepts and the index of recyclability for building materials. Resour. Conserv. Recycl. **72**, 127–135 (2013)

I. Vermeir, W. Verbeke, Sustainable food consumption: exploring the consumer "attitude–behavioral intention" gap. J. Agric. Environ. Ethics **19**(2), 169–194 (2006)

C. Vieira, J. Alves, M. Roque, *Manual Prático de Ecodesign*, AEP, Porto, Portugal. Ecodesign.pdf (2013, in Portuguese). ISBN 978-972-8702-83-0. http://certif-ambiental.aeportugal.pt/Documentation/Manual%20Pr%C3%A1tico%20de%20. Retrieved on 18 Aug 2018

V. Vogt, *Vorgänge an Grenzflächen bei der Flotation von Kunststoffen (in German)* (1981). Ph.D. Thesis, Clausthal University of Technology, Germany

N. Voulvoulis, Water reuse from a circular economy perspective and potential risks from an unregulated approach. Curr. Opin. Environ. Sci. Health **2**, 32–45 (2018)

H. Walker, L. Preuss, Fostering sustainability through sourcing from small businesses: public sector perspectives. J. Clean. Prod. **16**(15), 1600–1609 (2008)

K. Walker, N. Ni, B. Dyck, Recipes for successful sustainability: empirical organizational configurations for strong corporate environmental performance. Bus. Strategy Environ. **24**(1), 40–57 (2015)

S. Walker, Experiments in sustainable product design. J. Sustain. Prod. Des. **7**, 41–50 (1998)

G. Walther, J. Steinborn, T.S. Spengler, T. Luger, C. Herrmann, Implementation of the WEEE-directive-economic effects and improvement potentials for reuse and recycling in Germany. Int. J. Adv. Manuf. Technol. **47**, 461–474 (2010)

T.P. Wampler, *Applied Pyrolysis Handbook*, 2nd edn. (CRC Press, Boca Raton, 2006). ISBN 978-1574446418

C. Wan, G.Q. Shen, Perceived policy effectiveness and recycling behaviour: the missing link. Waste Manag. **33**(4), 783–784 (2013)

C. Wan, G.Q. Shen, A. Yu, Key determinants of willingness to support policy measures on recycling: a case study in Hong Kong. Environ. Sci. Policy **54**, 409–418 (2015)

L. Wang, M. Chen, Policies and perspective on end-of-life vehicles in China. J. Clean. Prod. **44**, 168–176 (2013)

H. Wang, C. Guo, J.G. Fu, Z.X. He, W. Liang, X.L. Chen, C.H. Zhuang, Adsorption behavior of weak hydrophilic substances on low-energy surface in aqueous medium. Appl. Surf. Sci. **257**, 7959–7967 (2011)

H. Wang, X.L. Chen, Y. Bai, C. Guo, L. Zhang, Application of dissolved air flotation on separation of waste plastics ABS and PS. Waste Manag. **32**, 1297–1305 (2012)

H. Wang, C. Wang, J. Fu, G. Gu, Wetting behavior and mechanism of wetting agents on low-energy surface. Colloids Surf. A **424**, 10–17 (2013a)

H. Wang, C.Q. Wang, J.G. Fu, Floatability of polymer materials modulated by frothers. Waste Manag. **33**, 2623–2631 (2013b)

P.C. Wang, Y.M. Lee, C.Y. Chen, Estimation of resource productivity and efficiency: an extended evaluation of sustainability related to material flow. Sustainability **6**, 6070–6087 (2014a)

H. Wang, C.Q. Wang, J.G. Fu, G.H. Gu, Flotability and flotation separation of polymer materials modulated by wetting agents. Waste Manag. **34**, 309–315 (2014b)

C.-Q. Wang, H. Wang, J.-G. Fu, Y.-N. Liu, Flotation separation of waste plastics for recycling—a review. Waste Manag. **41**, 28–38 (2015)

Q. Wang, J. Dou, S. Jia, A meta-analytic review of corporate social responsibility and corporate financial performance: the moderating effect of contextual factors. Bus. Soc. **55**, 1083–1121 (2016)

H. Wang, Y. Gu, L. Li, T. Liu, Y. Wu, T. Zuo, Operating models and development trends in the extended producer responsibility system for waste electrical and electronic equipment. Resour. Conserv. Recycl. **127**, 159–167 (2017)

WDD, Addressing the challenge of water scarcity in Cyprus, *Presentation by Charis Omorphos at Global Water Efficiency 2008. International Conference and Exhibition, 27–28, November 2008*, Limassol, Cyprus (2008).

M.J. Welfens, J. Nordmann, A. Seibt, Drivers and barriers to return and recycling of mobile phones. Case studies of communication and collection campaigns. J. Clean. Prod. **132**, 108–121 (2016)

S. Wibowo, H. Deng, Multi-criteria group decision making for evaluating the performance of e-waste recycling programs under uncertainty. Waste Manag. **40**, 127–135 (2015)

A. Wijkman, K. Skanberg, *The Circular Economy and Benefits for Society Jobs and Climate Clear Winners in an Economy Based on Renewable Energy and Resource Efficiency, A study report at the request of the Club of Rome with support from the MAVA Foundation* (2016). https://www.clubofrome.org/wp-content/uploads/2016/03/The-Circular-Economy-and-Benefits-for-Society.pdf. Retrieved on 16 Aug 2018

P.T. Williams, S. Besler, D.T. Taylor, The pyrolysis of scrap automotive tyres. Fuel **69**, 1474–1482 (1990)

P.T. Williams, Pyrolysis of waste tyres: a review. Waste Manag. **33**, 1714–1728 (2013)

E. Worrell, M.A. Reuter, Definitions and terminology, in *Handbook of Recycling, State-of-the-art for Practitioners, Analysts and Scientists* (Elsevier Inc., Amsterdam, 2014), pp. 9–16. ISBN 978-0123965066

X. Xu, X. Chen, J. Zhang, P. Guo, T. Fu, Y. Dai, S. Lin, X. Huo, Decreased blood hepatitis B surface anti body level slinked to e-waste lead exposure in pre-school children. J. Hazard. Mater. **298**, 122–128 (2015)

H. Yabar, M. Uwasu, K. Hara, Tracking environmental innovations and policy regulations in Japan: case studies on dioxin emissions and electric home appliances recycling. J. Clean. Prod. **44**, 152–158 (2013)

S. Yagneswaran, N. Tomar, D. W. Smith, Thermal and morphological of mechanochemical devulcanized ground tire rubber (GTR) filed epoxy composites. *Presented at the Fall 174th Technical Meeting of the Rubber Division of the American Chemical Society, Inc., Louisville, KY*, 14–16 Oct 2008, ISSN: 1547–1950

S. Yagneswaran, N. Tomar, D.W. Smith, Non-isothermal curing kinetics of epoxy/mechanochemical devulcanized ground rubber tire (GRT) composites. Polym. Bull. **70**, 1337–1351 (2013)

K.L. Yam, *Encyclopedia of Packaging Technology*, 3rd edn. (Wiley, U.S.A, 2009). ISBN 978-0470087046

K. Yamaguchi, K. Takeuchi, Consumer preferences for reduced packaging under economic instruments and recycling policy. Waste Manag. **48**, 540–547 (2016)

Z. Yang, X. Zhou, L. Xu, Eco-efficiency optimization for municipal solid waste management. J. Clean. Prod. **104**, 242–249 (2015)

C. Yao, S. Zhao, Y. Wang, B. Wang, M. Wei, M. Hua, Microbial desulfurization of waste latex rubber with *Alicyclobacillus* sp. Polym. Degrad. Stab. **98**, 1724–1730 (2013)

E.Y. Yazici, H. Deveci, Cupric chloride leaching (HCl-$CuCl_2$-NaCl) of metals from waste printed circuit boards (WPCBs). Int. J. Miner. Process. **134**, 89–96 (2015)

J. Yin, Y. Gao, H. Xu, Survey and analysis of consumers' behaviour of waste mobile phone recycling in China. J. Clean. Prod. **65**, 517–525 (2014)

J. Yu, E. Williams, M. Ju, Y. Yang, Forecasting global generation of obsolete personal computers. Environ. Sci. Technol. **44**, 3232–3237 (2010)

X. Zeng, J. Li, Measuring the recyclability of e-waste: an innovative method and its implications. J. Clean. Prod. **131**, 156–162 (2016)

X. Zeng, J. Li, N. Singh, Recycling of spent lithium-ion battery: a critical review. Crit. Rev. Environ. Sci. Technol. **44**, 1129–1165 (2014)

X. Zeng, J. Li, B. Shen, Novel approach to recover cobalt and lithium from spent lithium-ion battery using oxalic acid. J. Hazard. Mater. **295**, 112–118 (2015a)

X. Zeng, Q. Song, J. Li, W. Yuan, H. Duan, L. Liu, Solving e-waste problem using an Integrated mobile recycling plant. J. Clean. Prod. **90**, 55–59 (2015b)

X. Zeng, R. Gong, W.Q. Chen, J. Li, Uncovering the recycling potential of "New" WEEE in China. Environ. Sci. Technol. **50**, 1347–1358 (2016)

S.L. Zhang, Z.X. Zhang, Z.X. Xin, K. Pal, J.K. Kim, Prediction of mechanical properties of polypropylene/waste ground rubber tire powder treated by bitumen composites via uniform design and artificial neural networks. Mater. Des. **31**, 1900–1905 (2010a)

S.L. Zhang, Z.X. Zhang, K. Pal, Z.X. Xin, J. Suh, J.K. Kim, Prediction of mechanical properties of waste polypropylene/waste ground rubber tire powder blends using artificial neural networks. Mater. Des. **31**, 3624–3629 (2010b)

D.Q. Zhang, S.K. Tan, R.M. Gersberg, Municipal solid waste management in China: status, problems and challenges. J. Environ. Manage. **91**, 1623–1633 (2010c)

L.-W. Zhao, W. Zhang, A study on an integrative structure of business intelligence and ERP system based on multi-agent, in *2008 4th International Conference on Wireless Communications, Networking and Mobile Computing*, Dalian, China, 12–17 October 2008

Y.-B. Zhao, X.-D. Lv, H.-G. Ni, Solvent-based separation and recycling of waste plastics: a review. Chemosphere **209**, 707–720 (2018)

W. Zhou, Y. Zheng, W. Huang, Competitive advantage of qualified WEEE recyclers through EPR legislation. Eur. J. Oper. Res. **257**(2), 641–655 (2017)

Y. Zhu, D. Schwam, J.F. Wallace, S. Birceanu, Evaluation of soldering, washout and thermal fatigue resistance of advanced metal materials for aluminum die-casting dies. Mater. Sci. Eng. A **379**(1), 420–431 (2004)

W.A. Zisman, Relation of the equilibrium contact angle to liquid and solid constitution. Adv. Chem. **43**(1), 1–51 (1964)

G.I. Zlamparet, W. Ijomah, Y. Miao, A.K. Awasthi, X. Zeng, J. Li, Remanufacturing strategies: a solution for WEEE problem. J. Clean. Prod. **149**, 126–136 (2017)

Chapter 7
Practices on Cleaner Production and Sustainability

Abstract There are many factors that can influence the adoption of Cleaner Production practices, being rapid development, and economic gains the most influential. As CP practices call for a deep analysis and monitoring of processes, problematic areas quickly become visible, paving the road for improvement and gains. However, there is no universal solution as every company, industry and country has specific obstacles that need to be overcome. As previously stated, good CP practices start in the design phase of any product/process, leading to gains throughout the entire chain when an eco-design is implemented. Even though there many positive examples, still many factors hinder the application of CP policies such as disbelief of economic gains, lack of workforce training, lack of governmental investment, laws and enforcement, lack of foresight by top management and stakeholders, among others. For these reasons, the application of CP practices is scattered and has been avoided. In this chapter, the factors that influence the adoption of CP policies are discussed, why companies choose to pursue these policies and how SME's and specific countries attempt to apply these policies.

Keywords Cleaner Production practices · Cleaner Production in SMEs · Regional Cleaner Production practices · Factors affecting Cleaner Production

7.1 Factors Affecting Cleaner Production Practices

Rampant industrialization and the development of new materials, corresponding to the growing market demands, led the environment in general, and even the health of the populations in particular, to be relegated to the background. Products with characteristics which are now well recognized as generating serious human health complications and/or severe environmental problems, with effects that humanity is now trying to reverse (such as climate change or the melting of glaciers), were currently and abundantly used. Among these products, and only regarding industrial applications, are heavy metals, benzene, asbestos, vinyl monochloride, phthalates, polybrominated diphenyl ethers, polychlorinated biphenyls, dioxins, among many others. In agriculture, there are other similarly dangerous products such as

F. J. Gomes da Silva and R. M. Gouveia, *Cleaner Production*,
https://doi.org/10.1007/978-3-030-23165-1_7

organophosphate pesticides. Unfortunately, many of the materials and products are developed and applied, and only at a later time does science correlate the effect of exposure or consumption of a given product with chronic diseases that have been developing in the meantime. In fact, the current problems that the world population is facing today are the result of very rapid industrialization, generating economic development but also causing excessive consumption and depletion of resources. Also, the growth of the number of companies in each sector has led to an increase in competitiveness, which has translated to the development of materials that, although price competitive, are not always environmentally friendly.

Taking into account the study elaborated by Nilsson et al. (2007), it can be considered that there are five factors of particular relevance in the implementation of CP practices. These factors can be summarized as follows:

- Material replacement: Select materials which are less detrimental to the environment and/or that allow a significant improvement in the product life cycle. This change must be considered for both the product's raw materials and the auxiliary or operational materials used in its manufacture.
- Internal organization: Rationalize the use of materials and increase energy efficiency in manufacturing processes, avoiding waste of all nature.
- Internal recycling: Design and promote reuse cycles for materials, energy, water, solvents, and so on, creating new product lines that can take advantage of recycled materials, which would otherwise be considered unsuited for primary applications, in order to increase overall efficiency and drastically reduce waste, while gaining economic advantages.
- Technological update: Be attentive and integrate new technologies to improve processes, using technologies that minimize waste, produce less gas and effluent emissions and are generally more environmentally friendly and less dangerous for the overall operation of a company.
- Product improvement: Emphasize extended product life cycles, increase ease of repair through modular construction and easy access to component replacement, prepare the product so that it can be easily refabricated or recycled, avoiding the use of hazardous materials.

In any case, the implementation of CP practices has to take on other equally important aspects, such as:

- Scrutinize and compare the current use of company resources relative to its competitors, through an exhaustive analysis of everything that is incorporated directly or indirectly in the product, and everything that comes out in the form of product or waste.
- Observe and identify opportunities to reduce, reuse and recycle resources, quantify the savings that can be achieved, and set Cleaner Production targets.
- Carry out a thorough evaluation of the opportunities created by the implementation of CP practices.
- Cultivate an execution plan based on previous CP assessment.
- Carry out the thought plan, through consistent actions.

- Identify new CP opportunities through the new performance benchmarking.
- Returning to the first step, proceed to the identification of new opportunities for improvement in the environmental aspect, observing the good practices carried out by similar companies, and once again consider reviewing CP internal policy, improving it and reaching new levels of demand.

In order to achieve the best practices in terms of CP, it is necessary that the focus on the environment be present from the planning and design stages of the product, to the steps of recycling or disposal of the product, necessarily going through the phases of manufacturing and distribution (Yusup et al. 2013). The implementation of CP practices implies that the entire organization is predisposed to make this happen, especially from the top management, but also from all the teams that will be involved, from the management of the processes to its operationalization. Creating the necessary motivation in the organization is an extremely challenging task, and if it is not successfully overcome, it will hardly allow for a careful implementation of CP practices. However, creating this motivation often passes by the change of the company's internal culture, which is an extremely difficult task, but it can be more easily achieved if the implementation of CP practices is associated with a CSR (Corporate Social Responsibility) program, as discussed previously, because it requires greater involvement of the whole organization. However, companies also need to properly control all input and output variables in their system. In fact, the ecological footprint of its products can be significantly affected due to by-products or components that the company incorporates into its product, manufactured externally and without due environmental care, which will undermine the environmental performance of the whole set. Indeed, the pioneering companies to adopt CP practices within each of the clusters in which they are integrated will experience increased difficulties, as they will have to impose new rules for the degree of requirement and business approach, triggering chain reactions in the remaining supply companies of the cluster. For CP practices to be successful, these practices need to be replicated throughout the supply chain. In this sense, the company that leads the process of implementing CP practices will have to prepare internally to communicate assertively with its supply chain, obtaining from them the necessary commitment and adequate response to the requests of the leading company. Obviously, the supply chain will have to realize what advantages it will draw from the additional effort it will need to make to meet the new set requirements (World Bank 2012).

Analysts and technicians are unanimous in pointing out that there is no single way to successfully implement CP practices because each company has its own culture and internal organization. In any case, it is necessary that the process involves top management and the main departments of the organization, so that there are adequate involvement and knowledge of the process. It is also highly recommended that teams that will operationalize the CP implementation process be cross-functional, multilevel and match a matrix structure. The strategic and operational thinking of the company will have to be properly aligned so that there is no gap between the changes stipulated by the project, the implicit investment needs and the operationalization of the ideas in the manufacturing of products. Usually, the implementation of CP practices does not require large investments and can bring significant returns if the communication and sales processes are properly conducted, also bringing significant advantages to the environment. The implementation of a CP program in a company could be the opportunity to implement longtime employee suggestions, promoting the improvement of the overall performance of the process, often with very low investment needs. A key factor in ensuring that the CP implementation process is successfully achieved is needed for it to be led by a manager who is respected by all, and who is able to convey to all stakeholders the necessary confidence that the process will be successful. The manager must be able to explain at every stage to the stakeholders the steps that are being taken, and those that will be necessary to give in the future. This is the main prerequisite for success. Within this managerial stance, it is essential to establish a program to recognize the effort of different team members, to reward the emergence of innovative solutions and inspiration for change, in order to encourage the collaboration of all. Once again, the idea that the communication of the benefits to each of the parties involved is very important is reinforced, noting what each part of the stakeholders has to gain from this process, from the simple collaborator on the shop floor to the supplier, society in general and the environment.

Citing the Pew Center report essentially devoted to energy efficiency, Prindle (2010) refers to seven core practices and principles to implement CP successfully, as follows:

- The best strategy is efficiency: increasing the company's efficiency and implementing an effective strategy to achieve it are key factors.
- There are real leadership and organizational support, which are sustainable: leadership is noticed and respected by stakeholders, who feel that the process evolves in a progressive and sustained way over time, with results that are perceived by the stakeholders.
- The company has well-defined goals for SMART (Sustainable, Measurable, Responsible, Replicable and Time-Sensitive) efficiency: the CP implementation project is properly divided into phases, with a well-established schedule and well-defined process milestones, so as to allow adequate monitoring of implementation, a communication and incentive program and a real perception of the process's progress.
- The strategy for CP implementation is based on a robust tracking and measurement system: process management has been able to identify and monitor the indicators needed to be able, on the basis of reliable data collection, to have an accurate picture of the procedure evolution, allowing milestones compliance verification and taking action if there is any deviation from the initially set objectives. The indicators should make it possible to easily establish a percentage relationship between the starting data and the target results, so that it is possible to communicate the evolution graphically. It is not enough for the project leader to have generic data. It is necessary that the data be provided to you in detail, so that it may be possible to act in a timely manner on the process that is causing any problems toward the objectives.

- The organization provides sufficient resources in quantity and quality: in an initial phase, mainly, there will certainly be a need to reinforce the team, both in quantity, to meet the needs of greater definition in the projects, more abundant communication and focus on the changes that are necessary to implement, as in quality, because the level of demand becomes higher, due to the factors with which the processes need to be managed.
- The strategy of implementation of CP principles shows favorable results: based on the indicators established for process monitoring, it is possible to analyze the evolution in the desired direction, which must be communicated to stakeholders, never losing sight of the initially set objectives. Each critical factor should be measured and analyzed individually, such as raw material consumption, energy expenditure and water consumption, among others. The collection and processing of data are essential so that the process can be closely monitored. The units of measurement to be used shall be standardized and for easier comparison of results. The data may be expressed in standard units, e.g., liters of water per kg of raw material processed.
- The company effectively communicates the outcomes of the implementation of CP practices: taking into account all stakeholders, the company diffuses the results of the program as a stimulus to its own employees, customers and its supply chain, making stakeholders evolve jointly with the company.

Prindle (2010) found that 60% of surveyed companies noticed that their reputation had improved, 50% confirmed that their competitive position had been favored, 37% reported that the degree of employee satisfaction had improved, and 29% noted an increase in employee productivity. These results meet the general objectives of any CP practice plan, revealing very encouraging results, and certainly can be further improved.

7.1.1 Reasons Why Companies Adopt CP Practices

Kabongo (2004) argues that decisions about the choice of Cleaner Production essentially focus on two vital factors: top management vision and financial availability. A study, carried out through survey, taking into account French managers of SMEs companies with less than 250 employees showed that, despite there being a large widespread concern with the environment (82% of the respondents), about 72% of these managers had taken no effective action in their companies, and also had not planned to do so in the near future (Gault 2009). The same author claims that this is a result of companies' ignorance of how CP tools can be applied and of which management strategies and decisions to take to effectively pursue this path. The implementation of CP programs is even more difficult when it is known that there is a very wide diversity of types of organization and that there is a huge variety of sets of manufacturing processes, which require adaptation of the general procedures to each specific case. On the other hand, managers are also aware of the specificity required by each case, making decisions about adopting and implementing Cleaner Production even more difficult to take. Decision making through cleaner manufacturing processes requires in-depth knowledge about the performance of each process, the processes as a whole and the terms of comparison already existing with other technologies. Making a decision in this direction also means being fully aware of the potential of your employees to adapt to new technologies, what technical requirements are required, what the expected economic return on investment is, and what social impact that this decision can bring. Only in the face of credible data of these factors will it be possible to establish a strategy that will change paradigms in terms of posture toward the environment. Given the diversity of factors to be taken into account in decision-making, some authors argue that multi-criteria analysis is the best way to make the best decision (Laforest et al. 2013). The implementation of CP practices is much more complicated to perform in SMEs due to factors such as the lack of adequate training of intermediate managers, the exacerbated power of the main manager or main managers due to the concentration of power in very few people, gaps in terms of data recording, limited technical capacities, lack of employee involvement, deficient capacity for adequate monitoring, management too much focused on production, limited human resource capabilities, poor equipment maintenance, and difficulties in financing itself or in markets (Cooray 1999).

Greater population awareness regarding environmental concerns and a higher perception of how humanity is contributing negatively to the undesirable changes that are occurring in the climate has forced companies to adopt new strategies to be in tune with the concerns of their customers. On the other hand, competitiveness between companies has led to a global trend toward a reduction in the consumption of natural resources and also with a particular focus on the reduction of fossil energy consumption (Klemeš et al. 2012). To this end, the tools described above have been of great usefulness. The adoption of policies that have penalized polluters has also contributed consistently to the reduction of waste, since the rates to be paid by companies reduce their competitiveness, which also diminishes their field of action in the global market, and can compromise their survival. In order to avoid paying these fees, most companies started their CP process with "end-of-pipe" technologies. However, with the natural evolution over time and adoption of technologies that start a waste reduction process early in the design and/or manufacture of the product, it changed the strategy of companies that had adopted the end-of-pipe solutions, since the different studies developed clearly show that any CP strategy should begin with the design of the product, as described above.

7.1.2 Adoption of CP Practices in SMEs

International reports have pointed out the need for urgency in the implementation of CP practices in SMEs because, while mega-companies emit large amounts of pollution, the vast majority of companies are SMEs, where environmental care is not obviously safeguarded and the dimension of the organization does not allow the implementation of CP practices so easily, leading to the generation of a greater amount of environmentally harmful products per unit of manufactured product. Due to the inexistence of reliable data issued by these companies, it is not even possible to have a precise notion of the pollution generated by SMEs. In the case of a large part of the industrialized countries of Asia, as well as of Latin American countries, SMEs predominate, and there is no control over the release of effluents and other environmentally hazardous products that they emit (World Bank 2000). In the particular case of Mexico, 98.5% of companies are classified as SMEs (Reynoso et al. 2014), which are responsible for 66.2% of jobs.

Taking into account the implementation of CP practices in specific countries, Malaysia was studied as a case of a developed country that has undergone strong economic transformations in the last decades. The implementation of CP in Malaysia has been strongly encouraged by government bodies. As a country where services have played the leading role in the economy, the strong growth of the industrial sector in recent years has led government entities to set up specific programs for SMEs (since 2002) in order to avoid the serious environmental problems that the growth of industrial sectors entail. However, the information and awareness of entrepreneurs still do not seem to have reached the desired maturity due to the fact that there is still a very low level of adoption of CP practices in relation to what is desirable and necessary. After initially establishing some guidelines for the industry, the Malaysian government decided to establish the National Green Technology Policy as a priority, in order to encourage the industrial sector to adopt CP practices as soon as possible. Surveys carried out in Malaysia have led to the conclusion that the country's industrial managers are strongly aware of some factors that determine adherence to CP practices, such as the need to produce durable products, implement programs to minimize the consumption of raw materials, and generate waste, and the integration of principles of respect for the environment and safety at work (Yusup et al. 2016).

Given the difficulties experienced by companies in implementing certain CP tools, some techniques have emerged that aim to combine the best practices already developed, showing them as an example to companies that want to follow this same path. The best available technique (BTA) principle has been used in the same sense and was defined by the Integrated Pollution Prevention and Control Directive (IPPC), which was in use from 1996 until the end of 2013 and was subsequently replaced by the Green Action Plan for SMEs, as 99.8% of privately owned companies in the EU are SMEs.

Following the IPPC Directive (IPPC Directive 2008), the BAT principle was defined as "the most effective and advanced stage in the development of activities and their methods of operation which indicate the practical suitability of particular techniques for providing in principle the basis for emission limit values designed to prevent and, where that is not practicable, generally to reduce emissions and the impact on the environment as a whole". Each term of the BAT abbreviation has a very own meaning, which Laforest et al. (2013) described in a quite detailed form, as follows:

- 'Techniques' mean both the technique used and the way in which the set of processes is designed, interconnected, operated, maintained, and deactivated.
- 'Available techniques' mean those which having already been developed in such a way that their implementation in the industry can be considered technically feasible, can be easily conducted and bring relevant economic results, taking into account cost/benefit ratios. The selection and adoption of these techniques are not confined to those developed within the European Community.
- 'Best' is concerned with the effectiveness of the solution to achieve the highest level of environmental protection.

The main difference between the BAT principle and the application of CP lies in the state in which environmental protection begins to be applied. BATs are applicable mainly when it is not possible to act on the reduction of resources consumed, early in the process, being used as an end-of-pipe solution (Laforest 2008; Polders et al. 2012).

7.1.3 Adoption of CP Practices in Some Specific Countries

With regards to the analysis of the implementation of CP practices in different countries and geographic areas, Asia and particularly China, are among the most studied regions. This is due to the strong economic development witnessed in the last decades, generating concerns among the population in general, and in the scientific community in particular. However, other areas of the world are also concerning, particularly in some regions of Africa and South America, where some countries have also shown record developments, and where regulation is not always possible to lead to a capable and sustained implementation of CP practices. This subchapter will review some situations of CP practices in different countries or regions, allowing to understand how CP practices are spreading, especially in countries where industrialization is growing more strongly.

China

The pollution problem in China is known to everyone and represents a global threat (Zhang et al. 2013), as China may become the largest economic power in the world within a few years (Edward et al. 2014). In addition to the contribution, it brings to global warming and climate change, and it also causes serious health problems to neighboring countries. Effectively, Jiao et al. (2019) states clearly that economic growth in Dongguan (China), for example, is based on low labor cost and land availability but, on the other hand, the region is copiously invaded with dust, air pollutants, wastewater and solid waste, generated by the industrial plants of toys and electronic devices. In 2014, the industrial activity in the region accounted for 73% of the total energy consumed, accounting also for 80% of the pollutants generated (Jiao et al. 2019). Some Chinese researchers point out as the main cause of this situation that legislation does not require CP practices to be effectively applied by companies thus not encouraging the increase of sustained CP practices in China (Peng and Liu 2016). In this way, companies do not feel that the implementation of these practices is urgent, delaying their application and dissemination. Because the implementation of CP practices does not lead to immediate profits and, in the first instance, may even cause some cost penalties, causing some distortion in competitive conditions, it is more difficult for companies to voluntarily adopt these practices, penalizing the environment. Indeed, some Chinese companies deal with economic difficulties that prevent them from investing more heavily in more advanced production processes. As there is no sustained and muscular policy of economic support for companies pursuing CP, the implementation of CP practices ends up being relegated to second place, in an effort of satisfying the market and generate competitive and innovative products based on technologies that normally do not respect the environment. It is necessary that the Chinese government provide funds for companies willing to initiate a process of sustainable CP practices. (Zhang et al. 2016). As the disclosure of CP practices is not adequately promoted by the Chinese authorities, most managers do not understand the need for urgency of these practices, neglecting the necessity to acquire more efficient technologically and more proficient equipment (Bai et al. 2015). In fact, since the beginning of this century, more industrialized Chinese regions have opted for pollution control measures based more on end-of-pipe techniques than on the development of sustainable policies and decrease of material and energy consumption (Jiao et al. 2019). Since China is the largest developing country in the world, it has been the focus of numerous studies, some of which focused on how policies for CP practices are addressed by the various stakeholders, noting that incentives to CP practices are divided into four different strands: compulsory policies, audit-based policies, motivation-based policies, and punitive policies (Luken et al. 2016). Chang et al. (2015) stated that the development and adoption of CP practices in China can be divided into four distinct phases, which are directly correlated with the stages of knowledge acquisition of concepts, elaboration of legislation, transfer of knowledge to institutions, and improvement and deepening of practices. China was also heavily influenced and supported by the State Environmental Protection Administration (SEPA), as well as by the United Nations Industrial Development Organization (UNIDO) program, culminating in the merger of the Chinese National Cleaner Production Center (CNCPC) with the National Cleaner Production Centers (NCPCs), with a view to promoting the delivery of information to stakeholders, encouraging and supporting the training of company employees in CP practices, supporting consultancy and promoting the corresponding technical support to companies that need and want to evolve in adopting practices that take care of the environment (Luken et al. 2016). As far as training is concerned, 400 courses were held until 2011, involving about 20,000 CP technicians, trained to develop CP implementation programs in 95% of China's industrialized territory (where the main manufacturing companies are located), covering industries in a wide range of areas, such as chemical, metallurgical, civil construction, manufacturing of industrial equipment, electronics and lighting, among many others (Hong and Li 2013). Considering that the implementation of CP practices mainly involves the establishment of legislation, promotion of incentives to stakeholders and assistance in guiding the implementation of the practices, China has been the first country in the world to legislate on the matter, in 2002, through of the Cleaner Production Promotion Law, with a view to forcing companies and organizations to implement CP practices. This regulation, given the nature of the market for which it was designed, contained punitive aspects for those who did not comply with the terms of the legislation. In addition, it has also established a mandatory auditing system for more efficient use of material and energy resources (Bai et al. 2015), thus letting a more precise control of waste generation in terms of the manufactured products and processes used, which permits to have more reliable data on the resources consumption and the generation of pollutants, either through qualitative or/and quantitative information (Rahim and Raman 2015). Given that all other measures to adopt CP practices are voluntary, mandatory audits of CP practices as described above are the most assertive path toward positive developments toward a Cleaner Production in China, which have contributed in a crucial way to improve CP practices in China (Song et al. 2011). Compulsory audits consist essentially of five distinct stages, as described below (Bai et al. 2015):

- Local government authorities release the list of companies that will be audited.
- Companies under audit process are required to disclose information regarding their activity in relation to polluting emissions and energy consumption.
- The company is then audited.
- The company discloses the audit results.
- Finally, local government authorities take cognizance of the audit results, analyze and record them.

It is worthy of note that auditing is mandatory only for companies that exceed the emission limit values established by the Chinese government or incorporate into their manufacturing processes toxic or environmentally harmful products (Mol and Liu 2005). In 2010, Geng et al. (2010) carried out a retrospective of the results achieved through the implementation of CP practices in the Liaoning region, and the results can be considered extremely encouraging, although it is recognized that there is still much to be done. Based on the 2009 Liaoning Environmental Protection Bureau report, 693 companies had been audited, resulting in the implementation of 12,411 CP-related actions, which led to a total investment of about USD815 million in sectors such as coal extraction and treatment, coal-based power generation, petrochemical, and food processing industries. This investment resulted in operating gains of USD536 million, through savings in water consumed, in electric power and, above all, in a significant reduction of energy produced using coal. At the same time, as a consequence of these savings, there was a significant reduction in the emission of pollutants, such as SO_2, wastewater, reduced industrial ash, and reduced COD emissions. However, the scientific principles which have been at the basis of the audits are poor, being based on the comparison of the previous and current values recorded and using weighted scoring methods to compute the analysis, taking into account the level of pollutants emitted, the energy consumption, the consumption of material resources and the maximum potential to achieve Cleaner Production, as well as the toxicity of emissions, type of resources consumed, market potential, etc. (Hong and Li 2013). Furthermore, in China, companies are encouraged to adopt more sustainable policies for resource and energy consumption, essentially through regulation and policy constraint (Mol and Liu 2005; Geng et al. 2007; Peng and Liu 2016). However, the best results in terms of environmental protection are usually achieved through mixed government policies that reconcile compulsory and voluntary approaches in order to keep pace with market needs and their own dynamics, always accompanied by programs to support and encourage practices of CP (Iraldo et al. 2009). In addition, and despite some inefficiency in the measures given the size of China, financial incentives have been established to promote CP practices, especially to SMEs, by deducting expenses related to the training of their employees in CP practices, reimbursement of expenses related to environmental audits, exemption or reduction of taxes for companies generating new products from waste, among other encouraging actions. However, it should be safeguarded that measures to encourage CP practices in China are mostly optional, not mandatory (Hicks and Dietmar 2007). In the last decade, the Chinese government has also developed policies aimed at guiding companies and institutions to adopt more advanced technologies for the production and distribution of goods and services, as well as adopting advanced technologies for environmental protection (Kong and White 2010). Several authors argue that only a legislative system that obliges companies to comply with environmental protection standards can make CP practices more positive. In China, the implementation of mandatory industrial audits has been a very positive impetus for compliance with environmental standards, leading China's industrial sector in the right direction through greater respect for the environment (Bai et al. 2015; Peng and Liu 2016). However, Geng et al. (2010) clearly state that there is a lot to do in China in terms of CP, because there is no effective pressure for legislation to be applied and CP practices effectively implemented, which is ultimately the result of a fragmented organization framework, an uncooperative and resisting population and overall lack of awareness regarding environmental problems.

Regarding the Chinese SMEs, they did not spend time analyzing alternatives to overcome constraints to CP implementation. On the contrary, they overcame the limitations that existed, both in organizational and cultural terms, regarding the difficulties of access to CP information, diluting the resistance to change and lack of awareness about CP in the top management of the SMEs (Shi et al. 2008). This problem was previously identified when the SMEs used essentially end-of-pipe technology, without a real increase of technological knowledge which would have allowed to update production systems, making them more effective, more competitive and environmentally friendly (Wang 1999).

India

In India, it is customary for companies to be divided into clusters of business sectors, and very well-defined exportation flows of products, thus benefiting from economies of scale and synergies. In order to increase productivity as much as possible, CP strategies must also be properly aligned. This may allow resource sharing, which has no effect on the performance of each company, as this does not in any way contribute to companies losing their independence in terms of innovation, nor does it create any dependence on resources on an individual basis, but can significantly improve its profitability and competitiveness in the market. As is well known, the textile industry is a high consumer in the water sector. After a long study, some practices to be implemented in the textile cluster of the Tirupur region were established, which are exclusively CP practices and consist of the following: (1) replacement of chemical-based dyes by natural cotton colors; (2) technological improvement of existent processes; (3) reuse or reconstitution of the dye baths already in use; (4) recovery of used water, filtration, and elimination of some chemicals used in the process, through membrane filters (Narayanaswamy and Scott 2001). This joint effort to implement CP practices was due to the initiative and support of some Indian and international entities, such as India's Environment Division of National Productivity Council (NPC), India's National Cleaner Production Center (NCPC), and Institute for Communication and Analysis of Science and Technology (ICAST), Geneva, Switzerland, among others with less relevance in the undertaken process.

In India, the levels of pollution recorded in the mid-2010s do not seem to be as alarming as those detected in China, but the economic and industrial growth in India in recent years could also reverse negatively the existing data by 2014. From taking into account the Report of the Intergovernmental Panel on Climate Change (IPCC 2014), India presents a rate of 1.5 MT CO_2 per capita, which is within what is considered internationally permissible. As regards to total carbon emissions, around 8% relate to particulate matters and emissions from motor vehicles. On the basis of the foregoing, it is necessary that the GDP growth is not strongly accompanied by the growth of carbon emissions, so it is imperative that the growth of industrial activity be strongly controlled. According to news released on CNN in July 2014 (CNN 2014), the Indian Prime Minister allocated USD1.2 billion to increase the implementation of CP practices and development of smart cities, with a period of implementation between 2014 and 2016. These initiatives were also economically and technically supported by various countries, such as Japan, Germany, England, France, Canada, and Singapore. In 2015, the Indian government launched the "Make in India" program, which was based on a commitment by the industrial community, called "Zero Effect, Zero Defect" (ZED), which aimed to control numerous small and medium-sized enterprises, with the purpose of identifying and reducing situations of marked air pollution, wastewater production, excessive energy consumption and exaggerated overall resource consumption. This campaign, in addition to reducing the pollution generated by existing companies, also aimed at attracting international companies to India, in order to increase foreign investment, and to improve cooperation between companies within India. Another program launched by the Indian government in the same period called "Solar Missions" aimed at reducing energy consumption from polluting sources, by generating energy through renewable sources like solar, reducing by this way pollutant emissions. This program was supported by the US Import–Export Bank, which granted subsidized loans with very appealing rates, thus increasing the incentive of the Indian people to join these initiatives. More recently, India has also joined the 2030 Agenda for Sustainable Development program, which has involved many of the UN member states willing to pursue the goals outlined in the Sustainable Development Goals, which include a set of targets perfectly defined to be achieved by 2030. In this decade, there have been quite positive effects in India on the evolution of CP practices and sustainable development policies, mainly concerning recycling projects and environmentally friendly practices (Mahadevia 2011). To this end, much financial support has been provided by some members of the United Nations, as well as by the European Union, which contributed USD2 million to each Sustainable Cleaner Production project and having financed a total of 11 projects (Castro-Hallgren 2016). On a governmental level, for the period 2012–2017, India explicitly stated the need to pay attention to sustainability and environmental protection, through initiatives that are perfectly directed to the development of CP practices and to an efficient consumption of resources. According to Adapa (2018), the Indian government has shown a very positive attitude toward environmental protection, decentralizing decision making and involving citizens in the evolution processes for CP practices, creating the necessary conditions among the various organs of power and between the various ethnicities, so that the evolution toward environmental sustainability can be carried out more effectively. Through this process of decentralization, the government has created the technical conditions necessary for human resources to be properly trained in environmental management by public bodies and local municipal entities, facilitating the flow of information and training in companies in general, into the practice of CP. This policy allows, in a gentle way, to create the necessary roots for the culture of these practices to extend to a greater number of companies, improving environmental conditions (Hoelscher 2016).

A study performed by Govindan et al. (2016) using analytical network process (ANP) and interpretive structural modeling (ISM) methodologies has identified that the factor that most affects the implementation of CP practices in Indian SMEs is their high cost of implementation.

Vietnam

In addition to other sectors linked to agriculture and industry, Vietnam is the fourth largest motorcycle producer in the world, soon after China, India and Indonesia, as some international manufacturers have decided to deploy part of their production in this country (DucTho 2011). Like any other product, motorcycles are also under great pressure from the authorities and the market, requiring that this type of product reflects the environmental concerns that must be present in all sectors. Thus, according to Lin et al. (2013), studies have been undertaken to promote a more sustainable product and environmentally friendly production. In fact, consumers' environmental awareness tends to be reflected on the selection of the product and how it is designed and manufactured, which is also currently happening in this market.

Mitchell (2006) pointed out that the development of some projects of CP practices were not implemented in Vietnam because they did not consider top management a priority as well as the non-allocation of working time for employees to conveniently implement these projects, even when prior studies concluded positive feasibility and ensured an increase in profitability and competitiveness in the medium and long term. Thus, this behavior revealed lack of availability and sensitivity by part of top management, therefore missing the opportunity to establish strategies with the objective of improving global market competitiveness while complying with environmental requirements. As a result of this work, Mitchell (2006) proposed several lines of action, which can be listed as follows:

- To intensify the learning of CP practices in universities;
- To increase the training of government officials in this field;
- To stimulate the implementation of these practices in organizations;
- To provide training specifically for top managers with the objective of quickly changing their environmental paradigms;
- To intensify the qualification of employees to more quickly conduct internal audits and create control systems within each company.

South Korea

South Korea has experienced problems in excessive release of GHGs due to its powerful steel and petrochemical industries. For the reduction of GHG emissions through the restructuring of technological processes, three different strategies have been adopted: all-around; in-process-focused, and wait and see. However, given the types of industry involved in this situation, the strategy most commonly used is process-focused. In this case, there are limitations that do not exist in simpler processes, since investments in this type of industry are carried out with a view of relatively long paybacks, in the order of ten years, so that adapting to the new needs concerning the environment is not easy to accomplish. The focus on the consumption of other types of energy will inevitably draw some pressure from the petrochemical industry, but South Korea has made significant efforts to mitigate the emission of GHGs by adopting new technologies and CP practices that will reverse the initial situation (Lee 2013).

Egypt

The Middle East and North Africa region face titanic environmental challenges in the face of water and power shortages. In this region, the first country to set up CP centers was Tunisia, in 1996, being followed by other countries such as Morocco, Lebanon, and Syria (Arab SCP 2009), while other countries of that region are now developing their CP strategies, installing the first National Cleaner Production Centers. Contrary to what has happened in other countries, there has been no lack of governmental support in this region for implementations and dissemination of CP practices. However, the industrial sector has shown great apathy in the adoption of CP practices, which has translated into an implementation rate that is much lower than expected (Sakr and Sena 2017). Although Law 4 in Egypt dates back to 1994 and aims to increase the country's environmental protection, drawing the attention of stakeholders to this need and establishing guidelines for new industries to settle in the country, also establishing penalties for non-compliance with the law, the establishment of effective CP practices does not seem to have worked out as expected (Elwan et al. 2018). According to 2009 data published by AFED (2009), roughly 45% of the electricity in this area is consumed by the industrial sector in the MENA region, while in Egypt the industrial sector consumed about 37% of the country's total electric power consumption (Sakr and Sena 2017). In industrial terms, energy in Egypt is essentially spent by industries that require high levels of energy in their process, such as steelmaking, cement production, the petrochemical industry and the production of fertilizers, among others. In Egypt, electricity is generated mostly using fossil fuels (90%), with natural gas being the main fuel used for this purpose (EgyptERA 2015). The Egyptian government has developed a guide that allows companies to compare the industrial cost of their products by choosing CP practices, comparing them with traditional practices, using the industrial costing template according to the accounting procedures legally established in the country (Hamed and ElMahgary 2004); however, the guide is often ignored. Furthermore, based on a study carried out by the UNEP in 2011, it was possible to verify that there is a vast potential for an increase in energy efficiency of several industrial sectors in Egypt, with a potential saving of 32% in the textile industry, 30% in the manufacture of steel, 25% in the production of glass, and 21% in the manufacture of fertilizers (UNEP 2011). However, in the decade of 2010, energy consumption has risen at an average rate of 3% per year. This was reflected in a transformation in the status of the country, which changed from exporter of natural gas to importer of oil and its by-products. In fact, this excess demand and external dependence have led to successive cuts in the supply of energy to some industrial sectors, such as cement production. This situation has attracted the attention of government entities, which have been working to strengthen the energy supply system, but on the other hand, there has not been enough investment toward improving energy consumption efficiency (through adequate information, people awareness and national legislation in order to make mandatory the adoption of CP practices and energy savings). Thus, this region is singled out as one of the least energy-efficient areas in the world. Indeed, as reported by the Oxford Business Group, Egypt spends about 20% more energy than countries that have adhered to good energy sector practices and this holds true for both business consumption and household consumption (Oxford Business Group 2016). The main reason for this is the constant sharing of energy expenditure by governments, which discourages entrepreneurs from adopting CP practices, among others, by reducing energy consumption. Indeed, in 2013 the Egyptian government spent 7% of GDP on subsidies granted to the oil sector, making the fuel price sufficiently attractive for the complete discouragement of stakeholders to formulate energy-saving strategies (World Bank 2009). Finally, in 2014 the Egyptian government drew up a five-year plan to reduce subsidies to electric power for both domestic and industrial consumption with an aim for eliminating those same subsidies in 2019 and update the price of oil-based fuels (RCREEE 2015). Thus, it is not surprising that expectations indicate that only about 10% of companies are willing to adopt CP practices, even if they can count on government support for this purpose. In terms of legislation that encourages the efficient use of energy in the industry, only five countries in the region have worked toward this purpose (Morocco, Tunisia, Algeria, Jordan and Syria). There is a strong heterogeneity in the way each country in this region views CP practices and energy efficiency, as there are also very marked differences in the economic structure of these different countries, as some of them have strong energy reserves and are even strong exporters of hydrocarbons (Qatar and Saudi Arabia), while others (Morocco, Lebanon, and Jordan) are dependent on energy imports, as they do not have enough of their own resources. In this sense, Sakr and Sena 2017) make a very accurate diagnosis about the state of implementation of CP practices in the countries of this region, with a particular emphasis on Egypt.

Recently, more concrete steps have been taken toward Egypt actually implementing CP practices in a more sustained way, and this has been achieved in part because of external aid, in particular the IEE (Industrial Energy Efficiency) program, stimulated by UNIDO, and Egypt Climate Innovation Collaborative program, promoted by UKaid. The first program, financially supported by the GEF (Global Environmental Facility) and the Egyptian entities, involving a total amount that exceeds USD25 million, was implemented mainly by UNIDO in collaboration with local institutions (Egyptian Environmental Affairs Agency, Industrial Modernization Center, Egyptian Organization for Standardization and the Federation of Egyptian Industries), and aimed at creating a truly integrated approach to energy management, involving business managers, engineering, commercial entities linked to industrial equipment and technical entities with the capacity to transfer the necessary technology to fields such as energy efficiency in industrial processes (IEE 2018). On the other hand, the Egypt Climate Innovation Collaborative program (UKaid 2018), promoted by UKaid which, through intense effective contact with the Egyptian industrial reality, allowed to identify and classify numerous situations of lack of energy efficiency and bad industrial practices, as well as the creation of numerous business plans based on sustainable business principles. It was possible to identify the creation of more than 24 start-ups linked to sustainable CP practices, from the production of efficient energy systems to waste management or renewable energy production. This is the real breath of fresh air over a negative situation that lasted for a long time and did not do anything in favor of the Egyptian National bodies, local people, their health, and the environment in general. Despite these incentives and programs, companies continue to show some inertia in terms of proactivity, being only reactive to the stimuli created. It is also worth noting that even with beneficial effects in companies that are audited and intervened, other companies are slow to react and adopt the most efficient techniques. Despite recent progress, CP practices still have a huge implementation potential in Egypt, having the need for managers to realize the real benefits of implementing these practices. Research on CP should also be stepped up, helping to spread the need for Egypt to be more effectively involved in these practices and to be less dependent on external aid in this area.

Despite the above-mentioned 1994 Law 4 approved by the Egyptian Government, prohibiting the burning of agricultural crop residues, the well-known "black-cloud" which affects Cairo every year after harvesting rice represents a breach of this law. The source is the burning of around 3.1 million tons of rice straw per year (Said et al. 2013), also employed by farmers as disease control process (Afifi 2010). However, this biomass has a very high-energy potential which, if properly exploited, could reach around 2.5 GWh/year, thus avoiding the emission of about 1.2 million tons of CO_2 (Abdelhady et al. 2014). It should be noted that the carbon released during the burning of this biomass is equivalent to the carbon absorbed by photosynthesis during plant growth, so the environment is not harmed by this burning process and would avoid the need to generate energy using fossil fuels, which would generate a very negative balance in the generation of carbon (Weldemicheal and Assefa 2016). This could be one way, along with many others in Egypt, which could allow the industry to become more sustainable, lowering the need for energy production via the traditional and more pollutant methods (Abdulrahman and Huisingh 2018).

Ethiopia

In the African continent, the scenario is unfortunately not very encouraging either. In an effort to boost economic prosperity in the region, the depletion of many of the existing natural resources has been virtually ignored, as well as the generation of waste and pollutant emissions engendered by industrial activities. The first efforts to adopt CP practices in Ethiopia were made as early as 1996 through the creation of the National Cleaner Industrial Production Project of Ethiopia (NACIPPE), but soon there were criticisms due to the lack of a solid base of concepts related to CP practices, lack of a solid industrial sector in the country, lack of specialized technicians in this field and the economic context lived in the country at the time (Retta 1999). The objectives of the creation of NACIPPE were the same as many other CP Centers: to create knowledge in the area, to disseminate CP principles, to train people based on CP principles adapted to the needs of Ethiopia, to induce availability by top management of companies to join CP programs, increase the profitability of the industrial sector in Ethiopia by increasing sustainability, and develop capacity for the management of CP programs within the country. Due to the small size of the industrial sector in Ethiopia, the NACIPPE 1 and 2 programs have produced surprisingly positive results, as the proximity of program managers to top management of the main companies has become much easier, leading to the collection of data on the generation of waste and emissions (which was also easy). The design of appropriate solutions to achieve the greatest economic benefit in each case was also facilitated, as well as the creation of sectorial associations to continue the implementation of CP actions as a process of continuous improvement and with the help of UNIDO, the first CP Center in Ethiopia was set up for industrial activities (Retta 1999). In all of the two programs, 59 industrial audits were also developed with the main objective of reducing waste. These audits have led to the identification of wastes, formulation of ideas for their elimination and notable economic gains in industries such as textiles, tires, refining of oil-based products, and in the plastic products industry, with a notable national economic impact. The programs also allowed the formation of 61 specialized auditors and 156 assistant auditors, who were able to carry out the actions recommended by the first ones, based on the audits carried out and the detected waste situations and corresponding improvements.

Burkina Faso

Burkina Faso is another African country where increased energy supply difficulties are felt, as the supply chain has serious weaknesses and about 90% of the population does not have access to energy for kitchens or modern equipment. In addition, the energy produced to supply SMEs is essentially generated through biomass, with the main source being wood burning, which implies high levels of deforestation (roughly

250,000 ha of forest are cut each year for this end) (IOB 2013). In 2010, UNEP programs were implemented in Burkina Faso to implement sustainability practices for production and consumption, through reduction natural resource consumption and waste generation, which should be managed in an integrated manner. Based on this program and the experience acquired through it, strategies have been established for the dissemination of the knowledge to be acquired by the population in general, with a view to broadening and involving as many citizens as possible in adopting practices that save energy and other natural resources. Residuals from agricultural crops have also been used to produce briquettes which are subsequently used to generate energy. The briquettes are a great alternative to the charcoal use and wood burning for energy generation, since they produce a much smaller environmental impact, starting from the use of recycled resources and allows for the dispose of sludge (Maninder et al. 2012; Oliveira et al. 2017).

Turkey

In Turkey, the concept of CP began to be developed in 1999, through The Scientific and Technological Research Council of Turkey (TUBITAK) and Technology Development Foundation of Turkey (TTGV), a public body that defines the interconnection that should exist between industry, technology, and science. In this context, the concept of CP has been assumed as one of the priority areas of action of the Supreme Council for Science and Technology (BTYK) which is responsible for the definition of national policies for technology and science. These initial efforts were complemented by others carried out in 2003 through the Environment and Sustainable Development Panel which in the following years reinforced the need for the Turkish industry to follow CP practices. Moreover, this situation was also included and emphasized in the application for access to the European Community (Ulutas et al. 2012). Although initial efforts began early enough, in the mid-2000s, the CP concept was mainly applied to issues related to energy efficiency, largely due to the lack of internal capacity to implement technically and operationally these practices in Turkey. In order to overcome this situation, two projects in this area, the UNIDO Eco-efficiency Program, created and started in 2008 as a subprogram of a larger program entitled "Enhancing the Capacity of Turkey to Adapt to Climate Change—The United Nations Joint Program", under the support of the UN, and another called "Determination of the Framework Conditions and Research-Development Needs for the Dissemination of Cleaner (Sustainable) Production in Turkey", under the support of the Ministry of Environment and Forestry were undertaken. In the conclusions of a study carried out by Ulutas et al. (2012), Turkey presents several weaknesses regarding CP implementation, common to many other countries, and can be summarized as follows:

- Turkey's scientific capacity in terms of CP is still limited in terms of needs, because although there are several universities teaching and researching the subject, the quantitative results in terms of indexed scientific production are restricted to a much less significant number of universities, showing a clear weakness in terms of the local scientific community being prepared to face all the needs to which the country seems to be exposed.

- The implementation of CP practices in Turkey is very low yet and is still heavily dependent on entities based outside of the country, so that the lack of internal capacity for the development of projects of this nature among Turkish companies is reinforced. In fact, more in-depth knowledge of CP is concentrated in a very limited number of institutions within Turkey.
- The work performed up to 2012 in Turkey in terms of promoting CP practices had been carried out exclusively through external funding. Researchers demand financial support from public bodies in order to create the necessary conditions, in terms of stimulus and willingness, so that the number of actions to implement CP practices can be significantly increased, also showing a tougher commitment on the part of governmental entities and allowing the situation in the country can drastically change at this level.
- Stakeholders also show a lack of in-depth knowledge regarding the main differences between end-of-pipe techniques and the development of sustained CP policies and practices, producing waste and spending money on its treatment instead of avoiding it from the get-go. This shows a clear need for training and the creation of internal teams capable of informing and training company labor.
- Through an inquiry, it was possible to verify that there are entities with a solid background in environment issues, but that they do not have training and specialized knowledge in CP. This corroborates some of the above points, making it clear that Turkey needs external support to train internal teams with a strong specialized CP training. However, the skills accumulated by some public bodies should not be forgotten and should be strengthened.
- In 2012, there was a clear lack of coordination among stakeholders in the process of implementing CP practices, particularly among researchers, government agencies, and SME's. In order to enhance that coordination, more effective actions were taken such as increasing profit for implementing companies, improvements in terms of acquired knowledge, potentiation of results and adequate disclosure of the same, in order to potentiate new implementations.
- Increasing entrepreneur awareness in Turkey is extremely relevant, as the predisposition to implement CP practices is apparent among managers, in-depth knowledge on the subject is precarious and the concerns of these entrepreneurs are essentially focused on meeting the requirements imposed by the EU in matters related to occupational safety and health (so that the country can become part of this community) and not with an integrated vision that allows to adopt CP practices consistently.
- The eco-efficiency center, created through the UN-stimulated Eco-Efficiency program, should be able to provide accredited consultancy services both directly and through universities and research centers, thereby enhancing the response capability for the implementation needs of CP practices.
- The most internationally accredited institutions which promote CP activities (UNIDO, UNEP and OECD, for example) have a very limited activity in Turkey, a problem that needs to be reversed.

- Compared with other countries where CP practices have already reached very interesting levels, Turkey still has a long way to go.

A study carried out by Yuksel (2008) in Turkey found that most of the companies surveyed considered that the implementation of CP practices entailed a significant increase in costs and therefore showed strong resistance in allocating resources to these practices. The companies that applied these practices stated that they felt no advantage in this and that the perceived benefits were equivalent to the costs incurred in the implementation of these programs, so the responses were in line with the perceived passivity of joining CP programs. On the contrary, companies prefer to present a reactive attitude by looking for "end-of-pipe" technology to comply with current legislation, that is, they try to treat waste instead of avoiding its production. In the same study, some ideas deserve to be highlighted:

- Companies put great emphasis on the project, but the use of the tools to integrate environmental aspects in the project is still very limited.
- The logistical operations added to the manufacture and distribution of the product are not valued enough.
- The integration of environmental aspects into production is still very limited.
- There are not enough dedicated recycling programs to make this practice more common.
- There are numerous opportunities for companies to work in reverse logistics, but there is practically no activity in this sector.
- The use of renewable energy is far from reaching desirable levels.

This study also reveals that there is a lack of awareness by employees and the society in general regarding environmental issues. The idea that the government should promote more programs aimed essentially at avoiding the production of waste rather than requiring to treat it and punish offenders in environmental terms was also highlighted.

North America

In Canada, for example, the non-allocation of working time and the lack of involvement by workers in SMEs two decades ago made it impossible to implement projects to adopt CP practices, which were fully guaranteed as economic viability and would lead to significant economic gains in the medium term (Hilson 2000). A study carried out in 2006 by Côté et al. (2006) concluded that public bodies were attempting to develop environmental management tools that were truly useful for SMEs. A checklist was developed that proved to be appropriate for use in Nova Scotia for the pursuit and implementation of eco-design policies. The study also revealed that there is a huge margin of progression in the application of these and other CP practice tools.

The USA, being one of the most industrialized countries and the most important in foreign trade, was also one of the pioneers in the development of legislation corresponding to CP practices, due also to the strong influence of the UN and its programs. From 1970 to 1984, the USA had a period known as "Compliance Era", in which numerous environmental protection laws were drafted and promulgated.

During this period, there was also a deep reflection on the harm inflicted to the environment, a reversal of these damages by cleaning contaminated lands, reducing the release of pollutants into the waterways, and a greater control over polluting gaseous emissions released to the atmosphere. This period was essentially aimed at legislating and making legislation known, forcing companies to comply with regulations, preventing delayed implementation. This period was also essential for stakeholders to realize that in order to comply with regulations, wastewater treatment, the level of emissions released into the atmosphere and the creation of new operating rules for incineration plants and landfills were necessary. The payment of fees in response to the pollution burden forced companies to adhere to new control technologies and reduction of products dangerous to health and environmentally harmful. In the mid-1980s, the Comprehensive Environmental Response, Compensation and Liability Act had a strong impact on the US industry as it allowed managers to become aware that paying such fees would only contribute to liabilities that would be prolonged over time and that the transfer of pollution burden to other companies or their treatment were not the best way to solve the problem. Thus, a new paradigm has been created: generating waste represents a lack of efficiency in the use of resources. Through this new paradigm, the age of strategic environmental management began, from 1985 to 1994, where the focus was to avoid waste generation and atmosphere releases. This evolution in corporate thinking is certainly not dissociated from some of the accidents that have had a major impact in terms of the human life loss, which occurred at the beginning of this period and which has already been reported in this work (Union Carbide accident, among others). Beginning in 1986, state governments began to launch programs to minimize hazardous wastes, as well as programs to provide technical assistance for companies, with the aim of promoting management and the minimization of generated waste by these companies, without fear of penalties. Since the early 1990s, there has been an avalanche of laws and some funding programs for compliance with these same laws by the business community. However, these laws and funding sources in the USA continue to be affected by the category of effluent/polluting by-product; i.e., there are laws and funding programs to combat the release of hazardous products to the atmosphere, reduction of effluent dump in watercourses and their cleaning, and reduction in solid waste generation and corresponding treatment. However, as pollution burdens are not generated in isolation, this treatment makes it difficult to access support programs and understanding the needs as a whole. This difficulty can still be better perceived when it is observed that the Environmental Protection Agency (EPA) itself is organized in the same way. Given that, in addition to the fines imposed for non-compliance with the law, the implementation of programs and CP practices is optionally followed by companies. Only companies with delegations in other countries or companies that are essentially exporters to countries where those practices are required or valued tend to adhere to such practices. Fortunately, the use of chemicals in electronic products has been reduced, improving the possibility that these can more easily be recycled. However, this policy has largely been followed by the companies themselves and is not supported by any specific program to that end, following an integrated approach for the reduction of pollutants. Efforts carried out under government funds, including through the Government Performance and Results Act (1993), need to quantify the environmental benefits achieved through these funds, using the indicators established for that purpose by the Federal Government, communicating these results and corresponding performance. Pollution prevention efforts in the USA continue to be carried out and valued, as can be seen from the information released in 2018 by Pollution Prevention (P2) Organization (P2.org 2018a). Data previously published by the P2 Organization and reported in the work of Miller et al. (2008) show that pollution control programs in the USA achieved very positive results between 1990 and 200, with a reduction of about 78.5 billion tons of pollutants produced and about 182 million m^3 of water saved, in addition to the economic benefits achieved with this and a cost/benefit ratio of 1/5. In any case, since the mid-1990s, federal investment in policies related to pollution control has declined, and it is obvious at the end of the 2010's decade that there is no clear desire to conduct policies that openly protect the environment. The latest results released by the Organization P2 refer to the 2010–2012 period, and show that 90 programs were successfully conducted during this period, saving US $5.4 billion, reducing pollutant production by about 4 million tons, saving about 33 million m^3 of water and reducing electricity consumption by 1.4 billion of kW. The fact that no report has been issued since 2012 reveals a decrease in the importance given to these programs by the federal government (P2.org 2018). Although there are some successful initiatives promoted by some US states, most industrial associations offer great resistance to the switch from end-of-pipe control systems to prevention policies in the generation of pollutants (Armenti et al. 2011).

However, the pollution and public health hazards of some types of industry have dictated that these companies settle in other countries where legislation on environmental matters is not so restrictive. It is rather strange that the literature on cases studies carried out in the USA is rather scarce, and almost always come in the form of marketing advertising, such as the case of Levi's, where they started to use much less water in the manufacturing process of jeans, requiring fewer washes along its useful life. Of course, many other examples of success could be described, but there is no literary support for the progress of these developments in the USA. At the time of writing, the USA faces some political adversity in meeting the environmental goals that the most industrialized countries are looking for and want to settle in, stating a resistance that intends to defend the industrial sector of the country, but that will affect the whole world in the near future. Unfortunately, Brazil is also showing the first signs of adversity to internationally agreed environmental protection policies, which could be a setback in the tenuous progress achieved globally so far. The combination of these countries' adverse efforts could be an unprecedented step backwards toward a better environment through greater dissemination of CP practices.

Although not one of the best region examples of CP and environmental protection practices, North America has been conducting programs aimed at alleviating the harmful effect of emissions from large companies, particularly in the chemical sector. Thus, in the late 1980s, a program was launched and implemented by the US Environmental Protection Agency (US EPA), designated 33/50, to promote the reduction of the use, release and disposal of seventeen chemical substances known as harmful to the environment. This reduction was set to be 33% by 1992, increasing to 50% by 1995, based on the values initially documented in 1988 (Borkey et al. 1999). On the other hand, voluntary programs developed in Canada since 1998, at both the federal and local levels, have focused primarily on the substantial reduction of chemicals and by-products harmful to the environment in the manufacturing processes hitherto used (Chittock and Hughey 2011).

Mexico

In general, studies conducted in Latin American countries regarding the implementation of CP practices is somewhat old (Ehrenfeld et al. 2002; Grutter and Egler 2004; Jimenez 2007; Van Hoof and Herrera 2007; Blackman et al. 2009). More recently, a study was developed based on SMEs in Mexico (Van Hoof and Lyon 2013), where it has been proven that public–private partnerships can be successfully developed for the implementation of CP practices, which can generate considerable benefits for the environment. In this case, in spite of the initial action of external assistance teams and specialized consultants, the companies covered by the implementation program of CP practices later preferred the methodology of learning-by-doing, which allowed the solutions to be developed by the workforce, involving them in the process. However, the authors made clear idea that it will be necessary to analyze over time how the process evolves, given the barriers that normally arise in terms of passing on knowledge and performance evolution. It should be noted, however, that the return on investment period calculated for most of the different projects covered by this program was in the order of twelve months, which shows how attractive investment in CP practices can be. It should also be noted that the program has, as usual, reduced energy consumption, saved water, reduced waste and reduced the consumption of raw materials, acting in companies of different sectors, in a total of 1934 projects (Van Hoof and Lyon 2013).

Brazil

As in most countries, the reality of the Brazilian economy is strongly based on SMEs, showing the same problems of implementation of CP practices that other developing countries also present. Indeed, the main reason given for the mass non-adoption of CP practices by SMEs lies essentially in the lack of equity for the investment that needs to be made. According to several authors (Hamed and ElMahgary 2004; Mitchell 2006; Silvestre and Silva Neto 2014), the lack of cash flow leads to several gaps in opening up to the implementation of CP practices, which capacity for identifying government funding sources for the development and implementation of these practices, for a greater capacity to develop and manage change processes for sustainable policies, to identify cleaner technologies that can successfully replace the technologies currently in use, and by the predisposition of employees to integrate programs of this nature. First of all, companies will have to be able to identify the difficulties in implementing CP practices so that it is possible to quantify the necessary investment and government aid they can obtain, and then calculate payback time, taking into account the benefits achieved through the implementation of these practices. To this end, SMEs need to be aware of which processes are least economically and environmentally sustainable and to be able to identify alternatives or rely on specialized support in the field in order to perceive what needs to change and the corresponding investment needed. It is precisely in the identification of these foci of problems and corresponding solutions that SMEs feel more difficult (Oliveira Neto et al. 2017). From the perspective of developing countries, obtaining relatively short periods of return on investment may be a very good indicator that will serve to unlock the constraints commonly encountered in the adoption of CP practices by SMEs (Abdulrahman et al. 2015). As top management usually makes the decision of a given company to move to a CP adoption process, the availability of managers to move forward is usually proportional to the financial gains that are made through that implementation. In the case of Brazilian SMEs, five main difficulties were identified in the adoption of CP practices (Oliveira Neto et al. 2017): cultural issues, financial constraints, technical limitations, organizational deficiencies, and lack of government policies. These constraints are explained below.

- Cultural issues: a shortage of cultural capacity has been identified in terms of workers of Brazilian SMEs, as well as the managers of these companies, in order to adequately overcome the challenges to which these same companies are exposed in the global market.
- Financial constraints: the small size of Brazilian SMEs, as well as their weak structure, creates successive problems of access to financing with appealing yield spreads capable of being supported by those companies. Indeed, the credit rates practiced for companies in Brazil are among the highest in the world. Moreover, credit with reimbursed rates by the government is only accessible to large companies, since the high bureaucratic burden required to obtain this subsidy necessitates an organizational structure that most Brazilian SMEs do not have. In addition, fairly tight conditions are usually imposed, which prevent SMEs from accessing such state aid.
- Technical limitations: Brazilian SMEs normally employ workers with low academic and professional qualifications, also using technologically outdated manufacturing equipment, which hinders competitiveness and jeopardizes the sustainability of companies in the medium and long term.
- Organizational deficiencies: the lack of adequate skills by part of owners and managers of SMEs is a serious obstacle to their development, creating problems due to lack of adequate knowledge regarding the context in which the company is inserted. The small size of companies also limits the possibility of hiring technically gifted people with the necessary knowledge to make companies move toward a sustainable path. This represents a serious risk factor for corporate survival, as well as the implementation of more environmentally friendly techniques.
- Government policies: the existence of a high tax burden, as well as the lack of a more accessible means of communication for companies toward the government, hinder communication and prevents greater competitiveness.

However, other barriers have been mentioned in addition to those cited above, such as the rooted idea that complying with environmental regulations is an expensive burden which has no return, the usual resistance to change typical of some cultural environments, the preference for end-of-pipe technologies, some lack of awareness and knowledge about environmental issues and scarcity of time granted to employees to analyze and exploit information on how to implement CP practices. Moreover, this shortage of time granted to workers, referred previously by Mitchell (2006) regarding a study carried out with companies from Vietnam, was also corroborated by Silvestre and Silva Neto (2014) in Brazil. On the other hand, another study performed taking into account a SME Brazilian company revealed something even more serious: top management considered that it had no time to lose in investigating new production technologies and implementing CP practices, as the absolute priority of the company was to maintain the production pace, thus preventing employees from acquiring new knowledge about other alternative production technologies, and preventing their involvement in a future technological update of their manufacturing processes (D'Aquino et al. 2014). In fact, it seems to be clear from already published research that most workers in SMEs do not have adequate training in CP practices, and the opportunities given to them to evolve in this field are very small (Koefoed and Buckley 2008). According to D'Aquino et al. (2014), Brazil suffers from the same problem, where the labor force normally allocated to SMEs acting in the field of metalworking industry is very poorly trained and does not allow a solid implementation of CP practices. Taking this analysis into account in Brazil, similar constraints were noted in developing countries, such as Kenya, among others (Frijns and Van Vliet 1999). Although some of these problems may be considered across most of the world's SMEs, mainly in developing countries, others may be considered typical of the Brazilian reality. In fact, reports from 1999 (Gombault and Versteege 1999) show that in the Netherlands, companies had a relatively short period of time to identify which production flows generated the largest amount of effluents and undesirable emissions into the environment, as well as reducing the consumption of resources, thus allowing the prioritization of actions to be implemented with a view to making companies more sustainable in the short term.

In fact, it is not only in Brazil that SMEs give particular relevance to the production itself, relegating to the background learning and the elaboration of strategies that allow them to be viable, and even larger in the medium or long term. The same is true and referred to in several studies in various parts of the world, especially in developing countries. The fact that they do not have sufficient size to possess their own resources for research and development, prevent these companies from escaping from this situation. Thus, it is common for SMEs to contract external entities to carry out the processes of implementation of CP practices, but to do so, they must have capital to carry out this investment. One way that has shown good results in Brazil is to use continuous improvement (Kaizen) projects implemented using the Six Sigma methodology, allowing to implement CP practices. This path normally circumvents the lack of awareness and motivation of managers or owners of SMEs for the practices of CP, leading so that the gains obtained as result of those implementations can be used as a motivation for new actions (Calia and Guerrini 2006). In fact, Silva et al. (2013) identified in a study that quality tools as simple as the Ishikawa Diagram or 5W2H (Who, What, When, Where, Why and How Much) have proven to be powerful tools in identifying constraints to the implementation of CP practices. Lean manufacturing tools have also shown to be effective in addressing problems of analysis and improvement of corporate sustainability as CP practices, in addition to making companies more sustainable and competitive, also aim at improving environmental performance (Garza-Reyes 2015a). In 1995, Nedenes (1995) stated that CP practices were based on a process similar to the PDCA cycle, as the necessary steps perfectly fit that continuous improvement tool. In 2007, the EPA (EPA 2007) launched a brochure called "The Lean and Environment ToolKit" aimed at promoting a coordination between the application of lean methodologies and CP practices, briefly describing some case studies, in order to encourage increased productivity while also helping to reduce harmful environmental actions. The association between lean and CP practices has been the subject of a number of studies (Chiarini 2014; Garza-Reyes 2015a, b; Verrier et al. 2016; Cherrafi et al. 2016, 2017; Nadeem et al. 2017), due precisely to the synergies that can be obtained through this link, since managers, seduced by the savings that can be obtained through Lean methodologies, accept more easily the introduction of CP practices. In fact, there are points that can be considered common between the two practices, since Lean methodologies are essentially aimed at eliminating waste in processes, and CP aims to reduce the consumption of resources, so the two processes can occur simultaneously, and when an organization is implementing one of the methodologies, it is mandatory to contribute favorably to the other. In SMEs, where the main focus is on producing more, making the acceptance of Lean methodologies very favorable. Due to the increase in production, it is considered the good practice to take advantage of the application of these methodologies to introduce some improvements in CP practices, because it may be the only way to convince managers or owners of companies to improve their environmental performance, even in a tenuous and indirect way.

With regard to Brazilian SMEs, and in summary, it may be noted that these companies have significant limitations regarding investment capacity, which are accompanied by a lack of knowledge of the opportunities offered by governmental entities to finance programs for implementing CP practices, the difficulty in overcoming the barriers of lack of technological knowledge, the existence of resistance to change by managers and owners of SMEs, and lack of motivation and commitment on the part of these same elements of the organization in establishing Cleaner Production practices. In addition, most SMEs are not able to generate a hedging plan capable of analyzing the time needed for return on investment, which does not help managers feel motivated to invest in CP (Hilson 2000; Hamed and ElMahgary 2004; Mitchell 2006; Silvestre and Silva Neto 2014).

In a recent study carried out by Ramos et al. (2018) in Southern Brazil, 74 companies were asked to respond to a questionnaire on their willingness to adopt more sustainable practices with regard to resource saving and emission reduction for the environment, of which only 16 showed their available to participate in the study. In this study, it was identified that the main condition that determines the adoption of CP practices is the cost factor, and the second most important factor is to ensure the quality of the products sent to the market. Although these companies are all certified by ISO 9001 and 10 of them are certified by ISO 14001, companies have shown a greater availability to implement actions that promote the flexibility of their production system, than actions that aim to reduce their environmental impact. This is a situation typical of a developing country and replicated in many other developing countries, clearly showing that there is a great deal of work to develop in terms of raising managerial awareness regarding environmental concerns. Among the sixteen companies that actively participated in the study, only six had programs for continued CP practices, and five of these six companies were certified by ISO 14001. In this study, it was also clear that the size of the company is not the main factor affecting the willingness to adopt CP practices. In fact, two of the medium-sized companies that participated in the study presented excellent results in terms of environmental performance, while four large companies presented very poor results in this field. It remains again clear that having a greater budget availability and a larger organization are not factors that contribute most decisively to their willingness to improve their environmental performance in a sustained way. It is worthy to note that the same study revealed a better performance in practices of CP for companies that did not have any program established for the continued practices of CP, reason why it is confirmed that to have an established program is not a vital factor for the company to achieve an optimal environmental performance. Once again, it becomes clear that what makes the difference in terms of environmental behavior of a company is the predisposition of its managers to comply with established environmental goals, to have a strong environmental awareness and internally promote a culture of respect for the environment. In that study, it was also possible to verify that companies with better environmental performance also applied Lean tools to improve their production process.

Another study carried out in Brazil (Wasserman et al. 2017) focusing on 36 CP projects carried out in company related to the printing industry, revealed that in addition to the environmental goals pursued, very positive social and economic objectives were also achieved. Of the 32 projects under study, 24 were aimed at reducing the generation of waste, four intended to reduce water consumption and another four had as their main objective the reduction of energy consumption. Taking into account the philosophy of 3R (reduce, reuse, recycle), it was possible to verify that 51.52% of the projects focused mainly on reducing the resources used, 39.39% were able to establish procedures that allowed reuse, and 6.25% led to recycling of the produced waste. Based on these results, a procedure was proposed to be applied in the graphic industry, which consists essentially of: (a) training; (b) adequacy; and, level relevance. Regarding training, it was identified that engineering schools have a very restricted number of credits allocated to environmental and sustainability-related subjects. As for adequacy, it is necessary to analyze the complexity of the processes involved and the technological upgrades in the meantime, which effectively increase the environmental performance. Regarding the level relevance, it will be necessary to analyze which projects are actually feasible, taking into account the necessary investment and corresponding financial return, establishing priorities according to the effectiveness index of each action. Considering a list of steps to be followed in implementing CP practices in this type of industry, the degree of adherence of the organization to each stage of the process was also identified, also taking into account the four phases of the PDCA cycle, since any of the steps seem to be work perfectly regarding the four phases stipulated by this tool.

Another study developed by Severo et al. (2017) based on a survey carried out in collaboration with 762 companies in the metalworking sector and considering companies of different sizes, allowed to conclude that companies that have sustainable development processes in their products have a better financial performance. On the other hand, sustainable product development is invariably based on good CP practices.

Australia and New Zealand

In Australia, the cost of energy is considered relatively low and there are no demand control mechanisms, which represent a threat to the rationality of consumption. Moreover, the energy produced in Australia between 2000 and 2010 was predominantly obtained from fossil fuels, a factor that becomes unsustainable in the long term. A study by Pagan and Prasad (2007) on the food industry on the island of Queensland in 2007 identified a number of factors that posed threats to its long-term sustainability and are therefore a source of concern and desirable targets for the application of CP. The starting data for the study can be summarized as follows:

- In the early 2000s, Queensland's food industry accounted for 23% of the island's exports, employing more than 35,000 people and including meat-producing, bread-making, sugar processing, beverage processing, fruit and vegetables processing, among others.
- Some of the major threats of the food industry in Queensland were water consumption and scarcity (56 GL/year), non-renewable energy consumption (84 PJ), and the generation of greenhouse gases (2.2 Million tons of CO_2/year).
- In order to encourage the reduction of water consumption, and in order to amortize the water collection and distribution network, the authorities raised the water price by around 90% between 1997 and 2002.
- Wastewater discharges represent more significant costs for companies, and the amount charged is a function of the volume and type of wastewater discharged (organic loads, nutrient loads, etc.). In the same period considered in some regions of the island, rates doubled in order to encourage companies to avoid the discharge of wastewater.
- Given that the produced products need to be packaged, some pressure was also put on the food industry to reduce the consumption of raw materials used in packaging, promoting a double saving: reduction of consumed resources and lesser need for waste treatment.
- Concerning the solid waste produced by the food industry sector, about 80% can be recycled or reused, while 16% is organic waste and the remaining 4% will have to be landfilled. Solid waste costs are not the main sustainability concern for this type of industry, although this situation may also be improved.

Given the assumptions described above, the development of projects regarding CP practices is perfectly viable and can have very positive effects in the short term.

In order to be able to encompass the entire food industry of the island, it was essential to have a coordinating committee made up of the representatives of the sector, the various stakeholders, the funding bodies, and the eco-efficiency project implementation commission, promoting the involvement of companies and sharing of problems, information and solutions. After initial brainstorming sessions and sharing of information, concrete projects were developed for each company, taking into account the specificities of the problems presented by each one. The initial data sets were compiled in order to allow a comparison between the initial situation and the situation after the implementation of CP practices. Each company divided the savings achieved by the following four aspects: water savings, energy savings, reduction of waste production, and use of chemicals reduction. The application of the project has shown that the implementation of CP practices helps to greatly improve the competitiveness of enterprises, although it requires an initial economic effort. However, given the short period of return on investment required, it is unnecessary for governments to intervene through cost-effective financing because the investment is easily recoverable. Given that it is a very attractive investment, the biggest problem once again lies in the motivation of the management teams of companies.

References

S. Abdelhady, D. Borello, A. Shaban, F. Rispoli, Viability study of biomass power plant fired with rice straw in Egypt. Energy Proc. **61**, 211–215 (2014)

A.O. Abdulrahman, D. Huisingh, The role of biomass as a cleaner energy source in Egypt's energy mix. J. Clean. Prod. **172**, 3918–3930 (2018)

A.O. Abdulrahman, D. Huisingh, W. Hafkamp, Sustainability improvements in Egypt's oil and gas industry by implementation of flare gas recovery. J. Clean. Prod. **98**, 116–122 (2015)

S. Adapa, Indian smart cities and cleaner production initiatives – Integrated framework and recommendations. J. Clean. Prod. **172**, 3351–3366 (2018)

M. Afifi, *Rapid survey of agro residue based biomass sector*. 1st Interim report. Egyptian Environmental Affairs Agency (Cairo, Egypt, 2010)

Arab SCP, *Arab Regional Strategy for Sustainable Consumption and Production*. (The League of Arab States (LAS), Cairo, Egypt, 2018). http://www.unep.fr/scp/marrakech/publications/pdf/Final%20Draft%20Arab%20Strategy%20on%20SCP%20-%2006-10-09.pdf. Retrieved on 14 Dec 2018

K.R. Armenti, R. Moure-Eraso, C. Slatin, K. Geiser, Primary prevention for worker health and safety: cleaner production and toxics use reduction in Massachusetts. J. Clean. Prod. **19**, 488–497 (2011)

Y.J. Bai, J. Yin, Y. Yuan, Y.J. Guo, D.N. Song, An innovative system for promoting cleaner production: mandatory cleaner production audits in China. J. Clean. Prod. **108**, 883–890 (2015)

A. Blackman, E. Uribe, B. van Hoof, T. Lyon, *Voluntary Environmental Regulation in Colombia: Lessons for Capacity Building*, Resources for the future (RFF). RFF Report (2009)

P. Borkey, M. Glachant, F. Leveque, *Voluntary Approaches for Environmental Policy in OECD Countries, an Assessment* (Organization for Economic Cooperation and Development, Paris, France, 1999)

R.C. Calia, F.M. Guerrini, Estrutura organizacional para a difusão da produção mais limpa: uma contribuição da metodologia seis sigma na constituição de redes intra-organizacionais (in Portuguese). Gestão e Produção **13**(3), 531–543 (2006)

S. Castro-Hallgren, *Current trends on sustainable consumption and production and resource efficiency in India*. EU SWITCH-Asia Programme Regional Policy Support Component United Nations Environment Programme (UNEP) Background Paper (2016). https://www.switch-asia.eu/fileadmin/user_upload/RPSC/country-level-work/India/160803_India_Background_SCP_Paper_SCjuly_2016-RRG_edit.pdf. Retrieved on 11 Dec 2018

I.S. Chang, J. Wu, H.M. Qiao, Z.L. Zhang, The spatio-temporal approach to regional analysis on cleaner production in China. Renew. Sust. Energy Rev. **52**, 1491–1503 (2015)

A. Cherrafi, S. Elfezazi, A. Chiarini, A. Mokhlis, K. Benhida, The integration of lean manufacturing, six sigma and sustainability: a literature review and future research directions for developing a specific model. J. Clean. Prod. **139**, 828–846 (2016)

A. Cherrafi, S. Elfezazi, J.A. Garza-Reyes, K. Benhida, A. Mokhlis, Barriers in green lean implementation: a combined systematic literature review and interpretive structural modelling approach. Prod. Plann. Control **28**(10), 829–842 (2017)

A. Chiarini, Sustainable manufacturing-greening processes using specific Lean Production tools: an empirical observation from European motorcycle component manufacturers. J. Clean. Prod. **85**, 226–233 (2014)

D.G. Chittock, K.F.D. Hughey, A review of international practice in the design of voluntary pollution prevention programs. J. Clean. Prod. **19**, 542–551 (2011)

CNN, Cities of the future? Indian PM pushes plan for 100 'smart cities' (2014). http://edition.cnn.com/2014/07/18/world/asia/india-modi-smart-cities/index.html. Retrieved on 10 Dec 2018

N. Cooray, Cleaner Production Assessment in Small and Medium Industries of Sri Lanka, in *Global Competitiveness through Cleaner Production: Proceedings of the 2nd Asia Pacific Cleaner Production Roundtable and trade Expo* (Australia, 21–24 Apr 1999), pp. 108–114

R. Côté, A. Booth, B. Louis, Eco-efficiency and SMEs in Nova Scotia, Canada. J. Clean. Prod. **14**, 542–550 (2006)

S.F.D'Aquino, F.D.B.Prá, M.C. Goulart, L.M. Campos, P.A.C. Miguel, *Uma análise da aplicação empírica da produção mais limpa na manufatura no Journal of Cleaner Production (in Portuguese),* BASE **11**(3), 246–258 (2014). http://revistas.unisinos.br/index.php/base/article/view/base.2014.113.06/4316. Retrieved on 02 Jan 2019

DucTho, *Two Faces of the Motorcycle Market* (2011). http://vneconomy.vn/201102070659192P0C23/hai-mat-cua-thi-truongxe-may.htm. Retrieved on 08 Dec 2018

L. Edward, W. Martin, C. Zeng, Do Chinese Government subsidies affect firm value? Acc. Organ. Soc. **39**(3), 149–169 (2014)

AFED, *2009 Report of the Arab forum for environment and development*, Arab environment climate change, impact of climate change on Arab countries, in *Arab Forum for Environment and Development*, ed. by M.K. Tolba, N.W. Saab (Beirut, Lebanon, 2009). http://www.afedonline.org/afedreport09/Full%20English%20Report.pdf. Retrieved on 14 Dec 2018

EgyptERA, *Egyptian Electric Utility and Consumer Protection Regulatory Agency, 2015*. ERA Reports. http://egyptera.org/ar/elec_report.aspx. Retrieved on 15 Dec 2018

J.R. Ehrenfeld, W. Ashton, A. Luque, *Mejores practicas para la produccion más limpia su fomento e implementacion en la pequeña empresa (Best practices for the dissimination and implementation of cleaner production in small firms)*, Interamerican Development Bank (IADB), Multilateral Investment Fund (Washington, DC, U.S.A, 2002)

Y. Elwan, T. Ahmed, M. Hamed, *Cleaner production as a tool for pollution prevention in lake Nasser*. Documentation of the UN (2018). http://unpan1.un.org/intradoc/groups/public/documents/arado/unpan020870.pdf. Retrieved on 15 Dec 2018

EPA, *The Lean and Environment Toolkit* (2007). https://www.epa.gov/sites/production/files/2013-10/documents/leanenvirotoolkit.pdf. Retrieved on 03 Jan 2019

J. Frijns, B. Van Vliet, Small-scale industry and cleaner production strategies. World Dev. **27**(6), 967–983 (1999)

J.A. Garza-Reyes, Green lean and the need for six-sigma. Int. J. Lean Six Sigma **6**(3), 226–248 (2015a)

J.A. Garza-Reyes, Lean and green e a systematic review of the state of the art literature. J. Clean. Prod. **102**, 18–29 (2015b)

G. Gault, *Quelle perception de l'environnement chez les dirigeants de PME? - Résultats d'un nouveau sondage ADEME/TNS SOFRES* (in French), Colloque PM4E: Environnement et Maitrise de l'énergie (Paris, 17–18 Mar 2009)

Y. Geng, M. Haight, Q.H. Zhu, Empirical analysis of eco-industrial development in China. Sust. Dev. **15**(2), 121–133 (2007)

Y. Geng, W. Xinbei, Z. Qinghua, Z. Hengxin, Regional initiatives on promoting cleaner production in China: a case of Liaoning. J. Clean. Prod. **18**, 1502–1508 (2010)

M. Gombault, S. Versteege, Cleaner production in SMEs through a partnership with (local) authorities: successes from the Netherlands. J. Clean. Prod. **7**(4), 249–261 (1999)

K. Govindan, K.M. Shankar, D. Kannan, Application of fuzzy analytic network process for barrier evaluation in automotive parts remanufacturing towards cleaner production–a study in an Indian scenario. J. Clean. Prod. **114**, 199–213 (2016)

J. Grutter, H.P. Egler, From cleaner production to sustainable industrial production modes. J. Clean. Prod. **12**, 249–256 (2004)

M.M.E. Hamed, Y. ElMahgary, Outline of a national strategy for cleaner production: the case of Egypt. J. Clean. Prod. **12**(4), 327–336 (2004)

C. Hicks, R. Dietmar, Improving cleaner production through the application of environmental management tools in China. J. Clean. Prod. **15**, 395–408 (2007)

G. Hilson, Barriers to implementing cleaner technologies and cleaner production (CP) practices in the mining industry: a case study of the Americas. Miner. Eng. **13**(7), 699–717 (2000)

K. Hoelscher, The evolution of smart cities agenda in India. Int. Area Stud. Rev. **19**(1), 28–44 (2016)

J. Hong, Y. Li, Speeding up cleaner production in China through the improvement of cleaner production audit. J. Clean. Prod. **40**, 129–135 (2013)

Industrial Energy Efficiency Project, *What is IEE?* (2018) http://ieeegypt.org/iee-2/what-is-iee/. Retrieved on 15 Dec 2018

IOB Evaluation, *Impact Evaluation of Improved Cooking Stoves in Burkina Faso - The impact of two activities supported by the Promoting Renewable Energy Programme* (Nov 2013). https://www.government.nl/documents/reports/2013/11/18/iob-impact-evaluation-of-improved-cooking-stoves-in-burkina-faso. Retrieved on 16 Dec 2018

IPCC, *Climate Change 2014 – Synthesis Report*, Intergovernmental Panel of Climate Change (2014). https://www.ipcc.ch/site/assets/uploads/2018/05/SYR_AR5_FINAL_full_wcover.pdf. Retrieved on 10 Dec 2018

European Council, *Integrated Pollution Protection and Control (IPPC)*. Directive 2008/1/EC of the European Parliament and of the Council (2018). https://eur-lex.europa.eu/legal-content/EN/TXT/PDF/?uri=CELEX:32008L0001&from=EN. Retrieved on 28 Aug 2018

F. Iraldo, F. Testa, M. Frey, Is an environmental management system able to influence environmental and competitive performance? The case of an ecomanagement and audit scheme (EMAS) in the European Union. J. Clean. Prod. **17**, 1444–1452 (2009)

Y. Jiao, M. Su, C. Ji, S. Yang, P. Zhang, C. Li, L. Liu, Cleaner production instruments assisting sustainable transition at urban scale: a case study of Dongguan, a typical manufacturing city in China. J. Clean. Prod. **210**, 1449–1461 (2019)

O. Jimenez, Voluntary agreements in environmental policy: an empirical evaluation for the Chilean case. J. Clean. Prod. **15**, 620–637 (2007)

J. Kabongo, Intégrer économie et écologie: le cas de l'industrie canadienne (in French). Vertigo **5**, 1–12 (2004)

J.J. Klemeš, P.S. Varbanov, D. Huisingh, Recent cleaner production advances in process monitoring and optimization. J. Clean. Prod. **34**, 1–8 (2012)

M. Koefoed, C. Buckley, Clean technology transfer: a case study from the South African metal finishing industry, 2000–2005. J. Clean. Prod. **16**(1), S78–S84 (2008)

G. Kong, R. White, Toward cleaner production of hot dip galvanizing industry in China. J. Clean. Prod. **18**, 1092–1099 (2010)

V. Laforest, G. Raymond, É. Piatyszek, Choosing cleaner and safer production practices through a multi-criteria approach. J. Clean. Prod. **47**, 490–503 (2013)
V. Laforest, Applying Best Available Technologies in Environmental Management Accounting? in *From the Definition to an Assessment Methodology, 5th EMAN Book "Environmental Management Accounting for Cleaner Production", EMAN* (Springer, 2008), pp. 29–48. ISBN 978-1-4020-8912-1
S.-Y. Lee, Existing and anticipated technology strategies for reducing greenhouse gas emissions in Korea's petrochemical and steel industries. J. Clean. Prod. **40**, 83–92 (2013)
R.-J. Lin, K.-H. Tan, Y. Geng, Market demand, green product innovation, and firm performance: evidence from Vietnam motorcycle industry. J. Clean. Prod. **40**, 101–107 (2013)
R.A. Luken, R. Van Berkel, H. Leuenberger, P. Schwager, A 20-year retrospective of the National Cleaner Production Centres programme. J. Clean. Prod. **112**, 1165–1174 (2016)
D. Mahadevia, Branded and renewed? policies, politics and processes of urban development in the reform era. Econ Polit. Wkly **46**(31), 56–64 (2011)
R.S. Maninder, S.Grover Kathuria, S. Grover, Using agricultural residues as a biomass briquetting: an alternative source of energy. J. Electr. Electr. Eng. **1**, 11–15 (2012)
G. Miller, J. Burke, C. McComas, K. Dick, Advancing pollution prevention and cleaner production e USA's contribution. J. Clean. Prod. **16**, 665–672 (2008)
C.L. Mitchell, Beyond barriers: examining root causes behind commonly cited Cleaner Production barriers in Vietnam. J. Clean. Prod. **14**(18), 1576–1585 (2006)
A. Mol, Y. Liu, Institutionalising cleaner production in China, the cleaner production promotion law. Int. J. Environ. Sust. Dev. **4**(3), 227–245 (2005)
S.P. Nadeem, J.A. Garza-Reyes, S.C. Leung, A. Cherrafi, T. Anosike, M.K. Lim, Lean manufacturing and environmental performance e exploring the impact and relationship. in *IFIP International Conference on Advances in Production Management Systems (APMS 2017): Advances in Production Management Systems. The Path to Intelligent, Collaborative and Sustainable Manufacturing* (Hamburg, Germany, Springer, 3–7 Sept 2017), pp. 331–340
V. Narayanaswamy, J.A. Scott, Lessons from cleaner production experiences in Indian hosiery clusters. J. Clean. Prod. **9**, 325–340 (2001)
O. Nedenes, *Best practices guide for cleaner production programmes in Central and Eastern Europe,* OECD working paper **3**(66), 39. Organisation for Economic Co-Operation and Development (Paris, France, 1995)
L. Nilsson, P.O. Persson, L. Rydén, S. Darozhka, A. Zaliauskiene, *Cleaner Production Technologies and Tools For Resource Efficient Production* (The Baltic University Press, Uppsala, Sweden, 2007). ISBN 978-91-975526-1-5
R.S. de Oliveira, S.M. Palácio, E. Antonio, Briquettes production for use as power source for combustion using charcoal thin waste and sanitary sewage sludge. Environ. Sci. Pollut. Res. Int. **24**(11), 10778–10785 (2017)
G.C. Oliveira Neto, R.R. Leite, F.Y. Shibao, W.C. Lucato, Framework to overcome barriers in the implementation of cleaner production in small and medium-sized enterprises: multiple case studies in Brazil. J. Clean. Prod. **142**, 50–62 (2017)
Oxford Business Group, E*gypt's growing population increases demand for energy production* (2016). https://oxfordbusinessgroup.com/overview/opening-new-channels-growing-population-means-looking-energy-production-traditional-and-non. Retrieved on 15 Dec 2018
P2.org, *Most Valuable Pollution Prevention Efforts* (2018). http://www.p2.org/most-valuable-pollution. Retrieved on 05 Jan 2019
P2.org, *P2 Results* (2018)*, 2010–2012*. http://www.p2.org/p2-results/. Retrieved on 05 Jan 2019
B. Pagan, P. Prasad, The Queensland food eco-efficiency project: reducing risk and improving competitiveness. J. Clean. Prod. **15**, 764–771 (2007)
H. Peng, Y. Liu, A comprehensive analysis of cleaner production policies in China. J. Clean. Prod. **135**, 1138–1149 (2016)

C. Polders, L. Van den Abeele, A. Derden, D. Huybrechts, Methodology for determining emission levels associated with the best available techniques for industrial wastewater. J. Clean. Prod. **29–30**, 113–121 (2012)

W.R. Prindle, *From Shop Floor to Top Floor: Best Business Practices in Energy Efficiency,* Pew Center on Global Climate Change (Fairfax, Virginia, 2010), https://www.c2es.org/site/assets/uploads/2010/03/PEW_EnergyEfficiency_FullReport.pdf. Retrieved on 1 Sept 2018

R. Rahim, A.A.A. Raman, Cleaner production implementation in a fruit juice production plant. J. Clean. Prod. **101**, 215–221 (2015)

A.R. Ramos, J.C.E. Ferreira, V. Kumar, J.A. Garza-Reyes, A. Cherrafi, A lean and cleaner production benchmarking method for sustainability assessment: a study of manufacturing companies in Brazil. J. Clean. Prod. **177**, 218–231 (2018)

Regional Center for Renewable Energy and Energy Efficiency (RCREEE), 2015. *Arab Future Energy Index™(AFEX) Energy Efficiency 2015*. http://www.rcreee.org/sites/default/files/afex_ee_2015_engish_web_0.pdf. Retrieved on 15 Dec 2018

N. Retta, Cleaner industrial production practice in Ethiopia: problems and prospects. J. Clean. Prod. **7**, 409–412 (1999)

C.F. Reynoso, M.A.A. Osuna, L.E.O. Figeroa, Micro, small and medium business in Jalisco: their evolution, and strategic challenges. Rev. Bus. Finan. Stud. **5**(2), 27–43 (2014)

N. Said, S. El-Shatoury, L. Diaz, M. Zamorano, Viability study of biomass power plant fired with rice straw in Egypt. Renew. Sust. Energy Rev. **24**, 84–91 (2013)

D. Sakr, A.A. Sena, Cleaner production status in the Middle East and North Africa region with special focus on Egypt. J. Clean. Prod. **141**, 1074–1086 (2017)

E.A. Severo, J.C.F. de Guimarães, E.C.H. Dorion, Cleaner production and environmental management as sustainable product innovation antecedents: a survey in Brazilian industries. J. Clean. Prod. **142**, 87–97 (2017)

H. Shi, S.Z. Peng, Y. Liu, P. Zhong, Barriers to the implementation of cleaner production in Chinese SMEs: government, industry and expert stakeholders' perspectives. J. Clean. Prod. **16**(7), 842–852 (2008)

D.A.L. Silva, I. Delai, M.A.S. Castro, A.R. Ometto, Quality tools applied to cleaner production programs: a first approach toward a new methodology. J. Clean. Prod. **47**, 174–187 (2013)

B.S. Silvestre, R. Silva Neto, Are cleaner production innovations the solution for small mining operations in poor regions? the case of Padua in Brazil. J. Clean. Prod. **84**, 809–817 (2014)

D. Song, N. Duan, Y. Bai, J. Yin, Exploration and Practice on the Compulsory Cleaner Production Auditing of Key Enterprise in China (2011). in *3rd International Workshop Advances in Cleaner Production* (Sao Paulo, Brazil, 18–20 May 2011). http://www.advancesincleanerproduction.net/third/files/sessoes/4B/5/Song_Danna%20-%20Paper%20-%204B5.pdf. Retrieved on 10 Dec 2018

UKaid, Egypt Climate Innovation Collaborative, *Supporting Clean Tech Entrepreneurs in Egypt Through Partnerships and a Market Development Focus* (2018). http://documents.worldbank.org/curated/en/446291528869419204/pdf/Concept-Busines-Plan-Egypt-Clean-Tech-Startup-Support.pdf. Retrieved on 15 Dec 2018

F. Ulutas, E. Alkaya, M. Bogurcu, G.N. Demirer, The national capacity assessment on cleaner (sustainable) production in Turkey. J. Clean. Prod. **5**, 30–36 (2012)

UNEP, *Development of a System of Energy Intensity Indicators for the Egyptian Economy, Environics-ERC, October*. United Nations Environment Programme (UNEP) (Vienna, Austria, 2011)

B. Van Hoof, C. Herrera, *La Evolución y el futuro de la Producción más Limpia en Colombia 1992–2007 (Evolution and future of cleaner production in Colombia 1992–2007)*. Revista de ingeniería, Facultad de ingeniería, Universidad de Los Andes, (Bogotá, Colombia, 2007)

B. van Hoof, T.P. Lyon, Cleaner production in small firms taking part in Mexico's sustainable supplier program. J. Clean. Prod. **41**, 270–282 (2013)

B. Verrier, B. Rose, E. Caillaud, Lean and Green strategy: the lean and green house and maturity deployment model. J. Clean. Prod. **116**, 150–156 (2016)

J. Wang, China's national cleaner production strategy. Environ. Impact Assess. Rev. **19**(5–6), 437–456 (1999)

J.C. Wasserman, O.L.G. Quelhas, G.B.A. Lima, Analysis of cleaner production practices in a printing company in Brazil. Environ. Qual. Manage. **4**, 45–63 (2017)

Y. Weldemicheal, G. Assefa, Assessing the energy potential and GHG (greenhouse gas) emissions mitigation potential of biomass resources for Alberta. J. Clean. Prod. **112**(5), 4257–4264 (2016)

World Bank, *Greening Industry: New Roles for Communities, Markets and Governments*. (Oxford University Press, New York, NY, 2000), http://www.worldbank.org/research/greening. Retrieved on 15 July 2008

World Bank, *Tapping a Hidden Resource: Energy Efficiency in the Middle East and North Africa*. (World Bank, Washington, DC, 2009). https://openknowledge.worldbank.org/bitstream/handle/10986/17151/483290ESW0MNA01C0Disclosed061301091.pdf?sequence=1&isAllowed=y. Retrieved on 15 Dec 2018

World Bank, *Getting to green - A Sourcebook of Pollution Management Policy Tools for Growth and Competitiveness,* Report (2012). http://siteresources.worldbank.org/ENVIRONMENT/Resources/Getting_to_Green_web.pdf. Retrieved on 31 Aug 2018

H. Yuksel, *An empirical evaluation of cleaner production practices in Turkey*. J. Clean. Prod. **16S1**, S50–S57 (2008)

M.Z. Yusup, W.H. Wan Mahmood, M.R. Salleh, T. Rahayu, A review on optimistic impact of cleaner production on manufacturing sustainability. J. Adv. Manuf. Technol. **7**(2), 79–99 (2013)

M.Z. Yusup, W.H.W. Mahmood, M.R. Salleh, M.N.A. Rahman, *The implementation of cleaner production practices from Malaysian manufacturers' perspectives*, J. Clean. Prod. **108**, Part A, 659–672 (2016)

B. Zhang, S.C. Yang, J. Bi, Enterprises' willingness to adopt/develop cleaner production technologies, an empirical study in Changshu, China. J. Clean. Prod. **40**, 62–70 (2013)

J.Y. Zhang, J.H. Jiang, G.M. Tian, The potential of fertilizer management for reducing nitrous oxide emissions in the cleaner production of bamboo in China. J. Clean. Prod. **112**, 2536–2544 (2016)

Chapter 8
Sustainable Production Cases

Abstract In both the primary and secondary sectors, many efforts have been made to reduce the use of pollutants, transform effluents into reusable products, modify technologies, adapt manufacturing processes, etc., in an effort to minimize environmental impact. For example, conventional agriculture is responsible for approximately 10.3% of all greenhouse gases produced worldwide due to the use of pesticides and effluent of animal by-products. In fact, the abundant use of monocrops and chemicals in agriculture is turning soils barren, affecting crop production and yield. However, in this sector several new advances have been made which promote and incorporate the use of environmentally friendly biological fertilizers and pesticides such as composts, biochar, mycorrhizal fungi, rhizobacteria, and botanical pesticides. Also, the use of nanotechnology is starting to be notorious with the application of nanoclay minerals, nanoparticles, and smart fertilizers. The energy sector is also very worrisome as much of the world's energy is still produced from fossil-fuel-derived products such as coal, natural gas, and crude oil. Although there has been an uprising in renewable energy sources in developed countries, exploring solar, hydro, and wind sources, as well as through the incorporation and further development of alternative fuel and energy sources such as biofuels and hydrogen, it is still insufficient to offset the consumption made by largely industrialized and developing countries. Likewise, the chemical sector generates several emissions and effluents that are detrimental to humans and natural alike. The production of products such as lubricants and solvents from fossil fuels generates a significant negative eco-footprint, fact that has led researchers to explore the areas of biolubricants, biosolvents, water-based paints among other eco-friendly alternatives. For instances, it is probably unknown to the general public that the paper production industry and the paint production industry have a highly negative impact on Nature and human health, from the planting and colleting of raw materials (i.e., eucalyptus trees for paper production) which is changing the landscape of forests, to the incorporation of harsh chemicals used during bleaching processes, or even the evaporation of health jeopardizing solvents all through the simple action of paint drying. Indeed, although prohibited in several developed countries, the use of lead-based paints is still a dangerous reality, as their use is still practiced in large industrialized countries with laid back polices and low labor costs. These lead-based paints are utilized on products that are then exported all over the world. The production of cement products is also a heavy polluter as it is a

F. J. Gomes da Silva and R. M. Gouveia, *Cleaner Production*,
https://doi.org/10.1007/978-3-030-23165-1_8

high energy-consuming process, being dependent on the burning of fuels to generate the necessary process conditions. The incorporation of alternative energy sources to enhance thermal efficiency is a reality that is currently under development, through the inclusion of recycled/waste fuels (i.e., agricultural waste fuels such as rice straw). Furthermore, inclusion of natural fibers or recycled polymer fibers allows to improve overall mechanical properties of cement/mortar products, allowing the incorporation of materials that would otherwise end up in a landfill. Another very important industrial sector is the automotive/transportation industry. In the past decades, this sector has been constantly trying to improve its negative impact, from the incorporation of catalytic converter devices on gasoline engines, particle filter on diesel engines, non-leaded fuels, recycled and natural materials such as cork, alternative population systems (i.e., hybrid and electric motors), among many others. Manufacturers are using tools such as life cycle assessments, Design for X, lean manufacturing, simulation and finite element analysis, robotics and automation, among others, to generate lighter, more fuel-efficient and overall higher-quality and environmentally friendly products. Therefore, as the agricultural, energy production, chemical, construction, and automotive sectors are among the most environmentally damaging, they will be discussed in this chapter, laying out what is wrong and all that can be implemented, improved, or changed to help mitigate and invert the current situation.

Keywords Sustainable production examples · Sustainable production cases · Sustainable production by sector · Composts · Biochar · Mycorrhizal fungi · Bacilli rhizobacteria · Botanical pesticides · Nanoclay minerals · Nanoparticles · Smart fertilizers · Biomass · Alternative power generation · Biofuels · Hydrogen fuel · Sustainable lubricants · Pulp and paper sustainable production · Sustainable paint products · Chemical sustainability · Construction sustainability · LCA · Design for X · DfX · Design for Assembly · DfA · Design for Manufacturing · DfM · Vehicle weight minimization · Lean and green · Automotive sustainability

8.1 Agriculture Sector

Agriculture is an essential activity for human survival. However, with the push to increase productivity, soils have been overexploited, degrading their capability, showing signs of compactness, erosion, severe contamination, loss of biodiversity, and organic matter, among others (Turpin et al. 2017). Furthermore, despite the growing demand for food due to the continued growth of the world's population, agriculture has faced serious challenges to increase its productivity, due to a variety of factors such as the appearance of pests and diseases, the proliferation of weeds, and climate change, which has become even more serious in the twenty-first century. In fact, it is estimated that crops can be affected by some 67,000 different species of microorganisms, which, if not properly controlled, could endanger about 70% of world agricultural production (Campos et al. 2018). To control these harmful effects on agriculture, agrochemicals were developed to preserve crops, even increasing their yield. However, the agrochemicals initially developed did not include studies on the environmental impact that their use could cause in the medium and long term, with subsequent studies revealing harmful effects, namely high chemical soil concentration and consequent leaching to water resources through rainfall, toxicity-induced crops, effectiveness decrease over time, and the passage of toxic substances into the food chain. However, the negative impact of agrochemicals on the environment and on the arable land where they are used is also known as negative to the environment and may even interact with the food chain through bio-accumulation. In addition, the persistent use of this type of chemicals adversely affects the living fraction of the soil, the microbiota (Chaplain et al. 2011; Hartmann et al. 2015).

In order to circumvent the problems caused by pests and diseases in crops, genetically modified species were developed, becoming resistant to certain aggressions triggered by the surrounding environment. It was reported by Perry et al. (2014) that farmers who had maize and soybean sown genetically modified to resist insects required a much smaller amount of insecticide than conventional seedlings of the same products. However, the same authors also reported that the same farmers needed to increase the number of herbicides used when using genetically modified seeds relatively to farmers who continued to use non-engineered seeds, as a rise of glyphosate-resistant weeds was noted. This finding shows that the use of engineered seeds does not by itself eliminate the need for herbicides and may even increase their need.

Conventional agriculture and climate change have evolved negatively together, with clear reciprocal influences. Indeed, conventional agriculture is a strong generator of greenhouse gases (GHG), which are at the root of global warming and climate change that have been felt in the last decades. The production of GHG by the agricultural sector comes essentially from two aspects: from the agricultural production itself, due to the use of nitrogen-rich fertilizers, producing nitrous oxide emission, and also through animal production, contributing with methane gas generated by ruminant animals and grazing animals' dejections (Tilman et al. 2002). Regarding the European Union reports, agriculture contributes 10.3% of GHG generated by global economic activity (EU 2015). On the other hand, these climatic changes have also led to greater difficulties in the agricultural sector, particularly through prolonged periods of drought, torrential rains over relatively short periods of time, storms, and even disease infestation development affecting the quality and quantity of crops. The resilience of cultivation is also lower when farmers opt for monocultures, enabling the generation and spread of disease infestations much easier (Lin 2011). Negative influences on climate depend heavily on the area under consideration and on the type of crops considered. In fact, global warming tends to weaken warmer countries through droughts and a greater propensity to generate disease infestation, while colder countries will tend to benefit from warming and even from the carbon fertilizer effect (Stokes and Howden 2010). The productivity of a given soil is defined by its fertility, which relies on the soil's ability to provide sufficient water and the necessary nutrients that allow the growth of a given crop without the intervention of toxic elements that may impair the development of that crop (Havlin et al. 2014). This fertility is directly related to the physical, chemical, and biological characteristics of this soil (Igalavithana et al. 2015). In some regions where the soil is arid or semiarid, there is a greater difficulty in retention of water and nutrients needed for the plants, which limits the aptitude of the soil to respond to certain types of crops. In line with this situation are the tropical regions, where the main nutrients are taken from the more superficial layers of the soil to deeper ones by heavy rain, which, combined with higher temperatures and abundant content of decomposers, results in the favoring of the organic matter mineralization of the soils (Bruun et al. 2015; Khalifa and Yousef 2015). Considering the conventional farming system, inorganic fertilizers have been the mostly used way to promote an increase in soil productivity. Nitrogen and phosphorus are the most needed nutrients for plant growth. However, this is not a sustainable way to achieve this effect as soils become dependent on these fertilizers and their intensive use easily becomes a threat to their quality and fertility.

World population growth and improved access to essential goods have increased the consumption of food, thus requiring the agricultural sector to develop positively, in order to provide the market with more quantity and better quality (Patel and Minocheherhomji 2018). In order to meet this demand, conventional farming has adopted some very harmful practices for the environment, including the use of fertilizers and infestation attack inhibitors, with the aim of increasing quantity, as well as to ensure product quality and competitiveness (Kumar et al. 2010). Moreover, as evidenced by Pretty (2008), the growth of fertilizer use and intensive cultivation have been closely related. However, as previously mentioned in this paper, chemicals used as fertilizers may contain substances which are extremely harmful toward the environment, depending on their origin and composition. These substances tend to migrate to watercourses through land, also causing environmental problems. Linked to the agricultural sector is also the production of livestock and poultry, which can also cause severe environmental problems if care is not taken into account in their treatment. Fortunately, the agricultural sector has been heavily investing in modernization of the sector, which involves much more sustainable management models, especially in developed countries, and is increasing in developing countries. Many studies can be found related to this issue, creating solutions to transform conventional agriculture into a more sustainable sector, since legislation has also strongly contributed with stakeholder pressure, encouraging measures to be taken to make agriculture more sustainable. The term sustainable agriculture is defined by Pretty (2008) as "practices that meet current and future societal needs." The principle of sustainable agriculture argues that a smaller amount of fertilizers and pesticides should be applied in agriculture, regardless of the yield crop achieved. However, it should be noted that the non-use of inorganic chemicals is not a valid condition in itself to allow a given crop to be called as sustainable. A study carried out by Agovino et al. (2018) on the sustainability of agriculture and climate change in the EU-28 has led to the conclusion that:

- Climate change and crop yields are directly related; that is, when there is too much rain or too much sun, there is a drop in crop yields.
- Sustainable agriculture does not contribute to a decline in yield crops.
- The intensive conventional farming system adversely affects the development of the sustainable agriculture system.

8.1.1 Composts

Another simple, sustainable, and environmentally friendly option is the use of compost, which is characterized as a stabilized organic matter, with a low production cost and that has a negative environmental impact. It takes advantage of the waste generated during agricultural activities and avoids the use of harmful chemicals, such as inorganic fertilizers, being capable as well of controlling weed dissemination. Moreover, composting is still capable of sequestering carbon and preventing the spread of weed seeds and disease infestations (Vázquez and Soto 2017). Composting can be defined as "the controlled aerobic biological decomposition of organic matter into a stable, humus-like product called compost. It is essentially the same process as natural decomposition except it is enhanced and accelerated by mixing organic wastes with other ingredients to optimize microbial growth" (USDA 2000). In fact, composting is a process that aims to convert manure, agro-industrial processing residues, other organic wastes, and even unsold agricultural products into more environmentally stable compounds (Pergola et al. 2018). The result of the composting is not only destinated to the application in industrial crops. The European Union has encouraged this practice to be carried out even on small farms and that the compost obtained is used for the remediation of any type of land, including abandoned landfills, sports grounds, gardens, and public recreational areas. Thus, this practice also attempts to minimize the deposition of organic waste in landfill and puts into practice the principles of Circular Economy, with very low costs and with an extremely positive impact on the environment.

Composting is based on the principle of the decomposition of microorganisms into simple organic material in order to obtain energy and fertilizer material, which is vital for fertilization of the further crops. The final product, usually called compost, consists of biomass of dead and living microorganisms, as well as non-degradable parts of the raw material, such as stable composting by-products. In an integrated way, this process allows to properly manage the waste produced by a farm, taking advantage of it as organic fertilizer, with a process that can be relatively simple and environmentally sustainable. In addition, the resulting product can be perfectly transferred and/or shared by other farms, allowing a full use if management is properly carried out. In order for composting to take place, it is essential to allocate a specific site for it to occur and then necessary to have the corresponding recipe, i.e., the percentage of each waste type in order to obtain tailored and optimized compost formulations that can be employed in specific applications. However, in order for composting to take place, certain conditions must be ensured, namely the provision of nutrients, oxygen, and water. However, composting organisms require well-established pH and temperature conditions so that their growth can be at the highest possible rate. Operation and maintenance plans should also be conveniently planned. Economic and operational aspects should also be taken into account, allowing the composting process to use as much waste as possible while still providing the necessary quantities to ensure farming needs, or even allowing part of it to be sold or shared with other farms. The composting process is ensured by a population of mainly aerobic microorganisms which decompose organic components in order to grow and reproduce. This activity is promoted by managing the carbon–nitrogen ratio, the oxygen supply, the present moisture, the temperature, and pH of the biomass fed to the system. If these parameters are properly controlled, the decomposition is increased and the heat generated becomes sufficient to destroy weeds and pathogens, and the final product can also act as a pesticide.

After creating the mixture, the composting process consists essentially of two distinct phases. In a first phase, usually designated as the active phase or bio-oxidative phase, there is a greater activity of microbes, in which easily degradable materials disintegrate. The same happens with some difficult-to-decompose products, such as cellulose. The biomass used as raw material should be porous enough to enable oxygen access to feed the activity of these microorganisms (Pergola et al. 2018). During this phase, a wide range of temperatures occurs in the biomass stack, which provides developmental conditions for some microorganisms, while hindering the survival and development of others. The biomass pile may need two to three days to reach 40 °C. The temperature increases as a function of microbial activity until it reaches about 70 °C. It is this temperature level that allows the destruction of weed seeds and pathogens. At that time, all of the easiest degradable biomass will already be decomposed and the temperature begins to decrease, due to a lower microbial activity (the decrease in microbial activity results in less heat generation). After reaching the peak temperature at about 70 °C, the stack may take between 10 and 60 days to return to temperatures around 40 °C, depending on the operating conditions. The second phase, usually known as curing or maturing phase, consists of a period with less microbe activity, and subsequent decomposition of the products formed during the first phase of active composting. The curing phase begins when the biomass in compost reaches once again 40 °C. Compost stabilization is achieved at the end of this second phase. During the curing period, it is necessary to have proper moisture and oxygen management, in order to preserve microbial activity. The curing time required will be greater the lower the microbial activity and the more sensitive the culture where it will be used. In order to avoid recontamination of the compost, it may be necessary to cover the obtained product or relocate it to another suitably protected area. The total curing phase may extend for a period of time between 30 and 180 days (USDA 2000).

On-farm composting can also take on three different systems: (a) passive composting, (b) static composting with active aeration, and (c) composting in confined systems. In the case of passive composting (a), it consists of the proper mixing of different biomasses in the form of a stack, or using a windrow, and it is necessary to stir the mixture periodically so that aeration is carried out and the necessary porosity is restored, with a view to allow convenient access of oxygen to the deposited biomass. However, in this process, the oxygen acts passively, essentially through that porosity. This composting system is quite economical, being suitable for small farms or where space is not a concern. The static composting system with active aeration (b) is used essentially when there is a need to shorten the composting time or when there are problems of available space. In this case, blowers are used to ventilate the stack, either continuously or periodically, promoting forced aeration as well as temperature control. Therefore, higher care should be taken with stack composition as it is necessary to ensure that the bottom layers have sufficient porosity to allow adequate airflow and uniform distribution. Also, top layers need to ensure odor retention and avoid compost contamination from fly larvae or other insects and that the heat, moisture, and ammonia present in the stack be preserved. For these purposes, straw or wood chips may be used for the bottom layer, while finished compost or wood sawdust may be used for the top layer. Composting in confined systems (c) is a method that is rarely used in more industrialized agricultural production systems, essentially because it becomes more suitable for situations where it is necessary to deal with large amounts of compost biomass and, simultaneously, there are problems of available space. In this system, silos, previously used containers for other uses, wooden boxes, among others, are used. These systems may or may not be covered. It becomes a relatively expensive system because of the initial investment needed, as well as the time spent in the physical management of the biomass.

The way in which composting is carried out depends a lot on the conditions on the farm where the biomass is produced, the type of crop grown, the type of biomass produced, the existing machinery, and the space available for that purpose, given the necessary cycle time. It is a very cheap process, which allows, in an extremely sustainable and environmentally friendly way, to avoid the deposit of waste in landfills, also avoiding complex logistic systems while promoting the removal/sequestration of carbon generated by other agricultural activities. It encourages in a direct way the Circular Economy as the generated waste does not end up disturbing the environment, being used as a way to energize soil, allowing for increase of productivity and avoidance of disease infestation. This is a practice that has been strongly encouraged by the European Union, since it applies Cleaner Production practices in an extremely economical way.

8.1.2 Biochar

There is a growing demand for fertilizers and other products that can increase the fertility and productivity of soils, avoid the generation of GHG during their manufacture, and are environmentally friendly. The problems outlined above are already thoroughly dissected, and some techniques have already been developed to overcome these problems, which go through the so-called non-traditional agriculture, which uses science to circumvent the environmental problems that previous technologies could not ensure. In recent years, biochar has been developed, which is produced intentionally from biomass, and that can assume different compositions, being able to be tailored according to the soil where it is going to be applied and the crop that is intended to be improved (El-Naggar et al. 2019). Biochar can be characterized as a solid carbonaceous product with a high specific area due to its high porosity, strong resistance to decomposition, and high degree of aromatization, containing in its structure carbonyl, carboxyl, and hydroxyl, among other functional groups. It is an environmentally sound product suitable for the use as a reducing agent for organic or inorganic contaminants in wastewaters, such as pharmaceuticals, paints, and heavy metals, presenting as well other beneficial effects on the environment, which will be discussed below. In the case of highly concentrated wastewaters, activated biochar reveals a greater capacity for absorption of contaminants than pristine biochar (Tan et al. 2016). In order to improve the properties of biochar, leading it to have better characteristics for environmental protection, it is desirable to improve its physical, chemical, and biological properties, for example by modifying pH, increasing surface area and porosity, increasing cation exchange capacity, enhancement of adsorption capacity, etc. Modified biochar is usually called engineered biochar, as its characteristics can be selected in order to obtain tailor-made properties.

Biochar is obtained by thermochemical conversion of different types of biomass. Typically, such thermochemical processes consist of a pyrolysis process, but there are some alternatives, such as hydrothermal carbonization, since gasification and torrefaction are not considered as suitable processes to generate adequate biochar (Lee et al. 2018). In order for the produced biochar to have adequate characteristics, it is necessary that the parameters used in these processes be extensively studied and conjugated. Figure 8.1 shows the schematic diagram of two facilities that allow to produce biochar, in conjunction with biofuel, gas, or black carbon, the latter used in the manufacture of tires, among others.

In fact, biochar is mainly produced by thermal degradation of organic materials in a depleted oxygen environment, i.e., by pyrolysis. Depending on the biomass used for the production, as well as the pyrolysis process, the biochar will assume different properties in terms of water absorption capacity, pore size distribution, surface area, composition, pH, and electrical conductivity. In the course of the pyrolysis process, the organic components present in the biomass are thermally decomposed, with a subdivision into the released gases and generated biochar. The amount of biochar produced will depend on the heating rate imposed on the pyrolysis process. If the option is to generate as much biochar as possible, the "slow" pyrolysis should be chosen, with a heating rate in the order of 5–10 °C/min, and temperatures between 300 and 800 °C, whereas if it the bio-oil production is preferable, the option should fall into "fast" pyrolysis, with heating rates of 300 °C/min and temperatures in the range of 400–600 °C (Onay and Kockar 2003; Mohan et al. 2006). The types of pyrolysis previously focused also promote biochar with different characteristics. When biochar is produced at higher temperatures, it reduces the release of carbon-rich compounds, while lower-molecular-weight gases continue to release, giving rise to the more carbon-rich biochar (Ronsee et al. 2013). These high pyrolysis temperatures also lead to the release of volatile matter, which contributes to a higher porosity and surface area of the produced biochar. On the other hand, when lower pyrolysis temperatures are used with low heating rates, the amount of biochar formed is higher (Onay 2007) but, even so, about 25–30% of gas (syngas) and bio-oil can be obtained from this type of pyrolysis. It is worthy to note that both syngas and bio-oil are considered renewable energy sources. Depending on the feedstock being used as biomass, pretreatments may be necessary, which essentially pass through the heating and drying of this biomass, before being fed to the reactor responsible for pyrolysis (Lee et al. 2018).

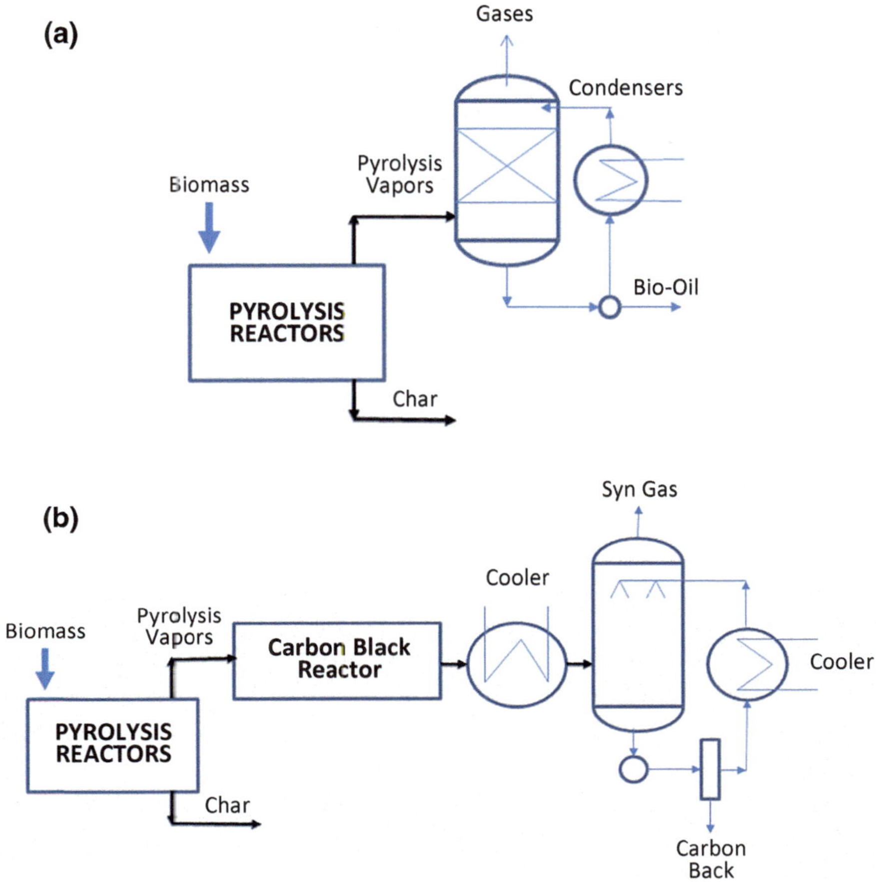

Fig. 8.1 Schematic diagrams of some biochar production methods, which include gas or biofuel production. Adapted from Garcia-Perez et al. (2010)

Another process that can be used to produce biochar is hydrothermal carbonization, which has some advantages over pyrolysis. It presents high efficiency rates and is performed in water at a temperature between 180 and 260 °C, requiring cycle times between 5 min and 6 h (Hoekman et al. 2013). With the water almost in the liquid state under pressure, the process is carried out under water vapor saturation pressure. Under these conditions, some organic parts of the biomass become soluble in this saturated water, making it possible to eliminate some hazardous waste, such as polychlorinated dibenzofurans and polychlorinated biphenyls (Weber et al. 2002; Lee et al. 2018). This process is designed to be fed mainly with wet biomass (wood waste), thus waving the need to perform a pyrolysis preheating, making it a more economical process as it avoids an energy-consuming step (Benavente et al. 2015; Lee et al. 2018). However, the consumption of water is considerable. Through this process, and in the same way as pyrolysis, biochar, bio-oil, and even small amounts of CO_2 are obtained. The percentage of each of these products resulting from the process depends heavily on the parameters selected for the process, with temperature being the most influential process in the quantities and final properties of the products obtained (Kambo and Dutta 2014; Lee et al. 2018), and the content of biochar obtained through this process may vary between 40 and 70%. In any case, the biochar obtained is still composed of liquid and solid phases and therefore requires mechanical or thermal treatment if it is to be used as fuel.

In order to improve the surface characteristics of biochar, different techniques may be used, as described in Fig. 8.2.

The sources of biomass in agriculture are essentially: manure, sewage slurry, grass, or agroforestry waste. Depending on the source of biomass used and pyrolysis parameters, biochar properties will differ, as can be observed in the database provided by the University of California, and designated by Davis Biochar Database (University of California 2015), through hundreds of case studies. When higher temperatures are used in pyrolysis, a higher porosity is created in the biochar, promoting a greater internal area, increasing its adsorption capacity and cation exchange capacity, as well as its pH (Park et al. 2015; Singh et al. 2015a). Also, properties of biochar can still be altered due to their soil aging process (Mukherjee et al. 2014). Typically, biochar is composed of condensed and volatile aromatic organic substances, as well as some inorganic substances (Brewer et al. 2011; Spokas et al. 2012). Thus, depending on the soil and biochar characteristics, the action of biochar in the soil may be more

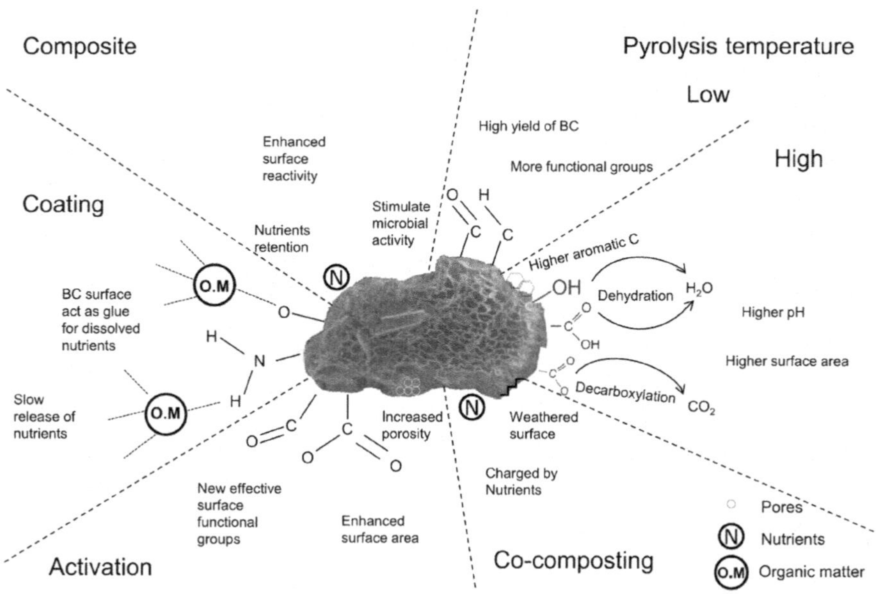

Fig. 8.2 Different techniques used to improve the surface characteristics of the biochar. Adapted from El-Naggar et al. (2019)

or less effective. However, it is proven that its characteristics advise its use for the mitigation of climate change and allow a more adequate management of agricultural sector waste, when adequately applied, promoting all the previously reported benefits seen with fertile soils. Indeed, through several studies carried out in recent years, it has been proven that biochar is extremely useful in improving quality and quantity of crops and boosting productivity (Lehmann and Joseph 2015; Kuppusamy et al. 2016). Furthermore, biochar can help in a wide range of extremely important environmental factors, such as (a) carbon sequestration (due to its high chemical stability); (b) allowing the proper use and management of waste (El-Naggar et al. 2018); (c) improvement of soil quality through an increase in the bioavailability of soil nutrients, as well as higher water retention (Omondi et al. 2016; Dai et al. 2017), and (d) remediation of contaminated soils and water reserves, through the immobilization of pollutants present in soils (Beiyuan et al. 2017; Igalavithana et al. 2017). These benefits also include an increase in soil fertility, soil nutrient enrichment, CO_2 emissions, C_4H and N_2O, absorption of organic and inorganic contaminants, among others (Zheng et al. 2017; Dai et al. 2017; Niazi et al. 2018). In fact, biochar has the ability to provide nutrients to the soil due to its ability to act as an organic fertilizer, conceding to the soil the nutrients it maintained from the biomass that gave rise to it (Randolph et al. 2017; Zhang et al. 2017). In addition, it also has the ability to retain other micronutrients, such as N (Gul and Whalen 2016). However, the positive action of biochar is not limited to this, as it also allows to increase soil nutrient retention and enrich its efficiency, reducing the action of nutrient leaching caused by heavy rains (Randolph et al. 2017). The higher the soil's degree of nutrient deficiency, the more effective the application of biochar will be. At the limit, in soils with high fertility, the action of biochar in terms of productivity increase may be practically insignificant, although all the other advantages of carbon sequestration mentioned above remain, which bring very significant environmental benefits. The increase of fertility achieved through the addition of biochar to arid, degraded, or acidic soils has been the subject of several studies, which have reported benefits that lead to the doubling of soil fertility (Kimetu et al. 2008; Laghari et al. 2015), for example, significantly increasing the dry weight of sorghum grown in arid land in China. Regarding acidic soil, far better results were obtained in the cultivation of common beans and maize, considering two sowing/harvesting actions per year, over a period of six months (Raboin et al. 2016). These favorable results were attributed to a higher soil pH, as well as to a greater fixation of aluminum in the soil. However, no significant improvements were noted in a rice plantation under the same conditions.

In addition to the active effect that biochar can have on soil productivity, it should be noted here that in many cases, the use of biochar can perfectly substitute inorganic fertilizers commonly used in intensive crops, minimizing by this way the dissemination of harmful products on the environment. Also, biochar promotes two other important effects: It uses biomass usually generated by the agricultural sector and sequesters carbon produced in the same activity. In environmental terms, these facts are extremely important. It should also be noted that biochar has a longer effect in terms of soil remediation because the carbon component present in biochar exhibits greater stability and higher recalcitrant effects. Due to its lower degradation rate, when compared to manure and compost, biochar has a longer lift time. However, despite all that has been previously mentioned, biochar cannot be considered a fertilizer. In fact, biochar acts in many other ways, which favor the productivity of a particular soil, especially those that, by nature or by weakness, have fertility problems. Biochar improves fertility by properly adjusting soil pH, promotes water retention, avoids leaching of desired nutrients, and improves cation exchange capacity. It should be noted that not all types of biomass can produce biochar with the best characteristics for certain types of soil, which may present problems in nutrient retention (Ogbonnaya and Semple 2013).

Reports from several researchers have also conveyed some failures in the application of biochar (Major et al. 2010; Deenik et al. 2011; Spokas et al. 2012; Borchard et al. 2014; Kloss et al. 2014; David 2015; Jeffery et al. 2015; Haider et al. 2017). There seems to be a consensual opinion that the fertilizing effect of biochar stands out mainly when applied in soils typically with low fertility, while in fertile soils its fertilizing effect can be null or even adverse. However, as noted above, the beneficial effect on the environment remains, as it assists with waste management and sequesters carbon, in addition to other minor effects.

For different purposes, mycorrhiza has also been studied in order to promote plant growth and root development through a more favorable absorption of nutrients without the use of inorganic fertilizers. Mycorrhiza comes from a link between the root of plants and biotrophic mycorrhizal fungi, which promotes a network of filaments linked to this root, giving rise to the beneficial effects described above (Basu et al. 2018). Among the five existing variants of Mycorrhiza, the most common is arbuscular mycorrhiza, which can be found in about 80% of the vascular plants, contributing through the previously described factors, and allowing soil remedy and a greater circulation of nutrient-rich compounds, thus contributing to the improvement of the ecosystem and to a more sustained growth of all plants in which this effect of association between fungi and root can be achieved.

8.1.3 Mycorrhizal Fungi

In view of the aforementioned problems with the application of agrochemicals, it is necessary to rethink the strategies to be implemented in agriculture to improve their yield, without harming their own arable land and the environment. These strategies should include greater crop rotation in each area, as well as the use of biofertilizers and biopesticides, which are environmentally friendly, and which do not saturate the land where they are applied or cause food chain risks. However, research linked to the genetic manipulation of certain species may allow the production of appreciable quantities of natural compounds independently (Daviet and Schalk 2010). Hence, in line with current legislation in most parts of the world, biopesticides and biofertilizers are treated in the same way as synthetic fertilizers and pesticides, so their application is still dependent on long approval processes, as well as fairly accurate toxicological analyses (Pavela and Benelli 2016). Fortunately, the European Food

Safety Authority has recently promoted the classification of some botanical products as "low risk," through the 1107/2009 Regulation (EU 2009), which has made requirements to approve the market entry for some of these products (Marchand 2015).

In fact, mycorrhizal fungi act on the rhizosphere, the fine area of the soil around the roots, where bacteria and other fungi also usually act. It is in this area that there is a competition between the pathogenic microorganisms that attack the plants and those who defend them. Therefore, it is here that the action of biological products must act to avoid the use of environmentally harmful inorganic chemicals. In the last decade, efforts have been expended to understand how microbial volatile organic compounds (MVOCs) can interact with the rhizosphere to promote plant root defense and assist in proper plant nutrition (Piechulla and Degenhardt 2013). Indeed, MVOCs are naturally derived environmentally safe compounds. A more in-depth study carried out in 2008 made it clear that only 400 of the 10,000 microbial species have the true capacity of producing MVOCs (Ortiz-Castro et al. 2008). For some decades now, in order to avoid the use of pesticides, fungicides, and fertilizers, the cultivation of genetically modified plants emerged. However, the effects of this genetic modification are far from being completely dissected and, in this scenario of uncertainty, governments in various countries have been pushing for legislation and regulatory systems that have delayed the spread of these crops. This opens a window of opportunity for the development of MVOCs, capable of being an environmentally friendly alternative through a sustainable, cost-effective, efficient, and effective solution as MVOCs can act as biofertilizers and/or biopesticides. MVOCs were extensively studied in the laboratory, showing extremely promising results. However, when studies were started in the field, the results fell short of expectations. Perhaps because of this, and despite strong and growing demand for the natural fertilizer and pesticide in the global market, MVOCs still account for only about 4% of the market share (Wilson et al. 2013). In fact, there may still be a long way to go in understanding the interactions between MVOCs and microorganisms so that their effectiveness can be improved, regardless of farmers' and market expectations (Kanchiswamy et al. 2015).

8.1.4 *Bacilli Rhizobacteria*

It is already known that some bacteria can bring significant benefits to crops through nitrogen fixation, as is the case of diazotrophic bacteria (Andrade et al. 2013). However, it is also known that some of these bacteria, if disseminated in the soil without adequate conditions of temperature, pH, and humidity, will not produce the desired results, since the direct inoculation of these bacteria will lead to their loss of efficiency due to the competitive effect promoted by other native soil microorganisms (Calabi-Floody et al. 2018). Plant rhizospheres are border areas between the roots and the surrounding soil, being also known as the microbe storehouse (Gouda et al. 2018). It is the place where there is a strong appetite for the interaction of microbes between them. These interactions have the great advantage of promoting the attraction of beneficial bacteria from the soil to the roots, and including other actions such as the formation and control of phytohormones, regulation of nutrient release, and control of phytopathogens (Egamberdieva and Adesemoye 2016). In this way, plant rhizospheres congregate a wide range of commonly known bacteria such as plant growth-promoting rhizobacteria (PGPR) (Aloo et al. 2019). Several studies carried out since 2012 have shown that rhizobacteria have a very positive effect in different aspects, namely in the aid of plant growth, nutrient solubilization, N_2 fixation, enzyme biosynthesis, and increasing the immunity of plants to pathogenic organisms (Beneduzi et al. 2012; García-Fraile et al. 2015; Gupta et al. 2015a; Aloo et al. 2019). Thus, rhizobacteria can be considered as a biofertilizer and a biopesticide as it, respectively, contributes significantly to plant nutrient collection and fights the action of certain pathogenic organisms. Due to the way rhizobacteria facilitates relationships between the roots of plants and the soil around them, it is considered as one of the most growth-promoting biofertilizers (Castanheira et al. 2017), and successful studies have recently been carried out to analyze the viability of their application as biofertilizers, replacing inorganic fertilizers (Patel and Minocheherhomji 2018). Some of the most researched rhizobacteria for promoting crop growth are, for example, Acinetobacter, Alcaligenes, Arthrobacter, Azospirillum, Azotobacter, Bacillus, Burkholderia, Enterobacter, Klebsiella, Pseudomonas, and Serratia, having their effectiveness been tested successfully in the cultivation of several species, such as bean (Stefan et al. 2013), cucumber (Islam et al. 2016), maize (Krey et al. 2013), potato (Dawwam et al. 2013), and wheat (Govindasamy et al. 2014), among others. Moreover, a huge list of uses, effects, and target application can be found in a recent review carried out by Gouda et al. (2018). These studies have made it clear that different rhizobacteria species have the capacity to improve seed germination, increase root size, and even increase leaf area, but it is still necessary to know in depth the mechanisms that are the basis of experimentally proven phenomena. This knowledge will also allow the study of a possible optimization of its effect, as well as ensuring that food security is guaranteed (Wang et al. 2016; Rosier et al. 2018). However, rhizobacteria also have bioremediation properties, as it contributes to the degradation of pollutants and reduction of toxicity in soils contaminated with heavy metals, through reduction process (Goswami et al. 2016). Further studies are also needed to select which rhizosphere bacteria are best suited to each crop/root type, as well as to promote the production of microbial communities that can be used in sustainable agriculture as a substitute for inorganic fertilizers. To achieve this, multi-disciplinary studies involving biotechnology, nanotechnology, and other related sciences are needed to test new formulations and verify their effectiveness.

Bacilli rhizobacteria is one of the most abundant species in plant rhizospheres (Sivasakthi et al. 2014), being also one of the most studied (Souza et al. 2015). This rhizobacteria has shown a particularly close affinity with plant roots, being able to be contained in the root or in its periphery/border and able to generate extremely intimate connections with the majority of plants. This is due to its ability to lodge itself on to the tissues of the hosting plants, thereby showing greater effectiveness in controlling pathogenic microorganisms and in the control of plant roots. This bioprotection can be carried out in different means, i.e., through the production of antibiotics, enzymes, siderophores, as well as volatile organic compounds, as recently described in a paper presented by Patel and Minocheherhomji (2018). A synthesis of the mechanisms of production of these substances can be found in a review paper recently compiled by Aloo et al. (2019). From the vast work carried out by numerous researchers mainly between 2008 and 2019, it can be said that Bacilli rhizobacteria, through its action of bioprotection, hormone production, and nutrient solubilization, already represents a strong alternative to conventional inorganic fertilizers and pesticides, constituting an environmentally sustainable solution. There is still a lot of work to be done with regard to optimizing it and ensuring that it is a safe alternative within the food chain. It should also be noted that its formulation and commercialization has been an extremely easy step to overcome, but close cooperation with the biotechnology field can also help to further improve and disseminate this biofertilizer and bioprotector, helping to may still have a greater impact in the near future.

8.1.5 Botanical Pesticides

Plants can use self-generated biopesticides as a form of self-defense against diseases, weeds, or insect attacks. Botanical pesticides are characterized by exhibiting a broad-spectrum antimicrobial activity. However, its action cannot be directed only to some species and may therefore affect the soil microbiota population (Conti et al. 2014). These biopesticides can be made up of various chemical constituents such as citrus oil, nicotine, pyretrins, or urushiol, and, as they are natural, their toxic effect will be minimal or null to the environment; however, they have an extremely effective action against insects, further bypassing the problem of insects becoming more resistant to inorganic pesticides (Campos et al. 2018). Indeed, it is these oils that are primarily responsible for the aromas and flavors of most aromatic plants (Nagegowda 2010), and can act on several fronts, that is, as repellents, insecticides, bactericides, fungicides, and so on. However, depending on the medium that is attacking plants, these oils may have different forms of reaction. Thus, when they are intended to combat fungi and bacteria, such oils promote disruption of cell walls and cytoplasmic membranes of organisms, while in the fight against insects and mammals which can lead to pests, these oils act essentially on the central nervous system (Campos et al. 2018). It is also known that, despite its repellent, insecticidal and antipathogenic effect, its action is not equally effective with all kinds of pests plaguing the agricultural sector (Mossa 2016). Although botanical pesticides are already relatively well positioned on the market, there are still some problems to be solved in their application as a means of pest management, namely high volatility, low solubility in aqueous media, and relatively easy oxidation (Pavela and Benelli 2016). Following are some examples of biopesticides in market use and their specific application:

- Insecticides: nicotine, essential garlic oil, oil of neem, and rotenone;
- Insecticides/Repellents: citrus oil and essential oil of thyme;
- Insecticide and Animal Feed Supplement: essential oil of oregano.

The management of pests can be improved by combining various combat techniques, thus widening the range of species covered. In addition, if botanical pesticides are used in conjunction with microorganism-based bio-insecticides, it appears to be a very promising route for more effective control of pests that can affect crops (Ezhil Vendan 2016).

8.1.6 Nanoclay Minerals

In the absence of solutions capable of sustainably replacing fertilizers and pesticides usually used in traditional agriculture, one can choose to intervene in the way they act. In fact, there is a time lag between the moment fertilizers release nutrients (such as the nitrogen) and the moment in which plants uptake it. This promotes a loss of efficiency, requiring larger doses and greater soil exposure without any added benefits, as the plant cannot take advantage of the fertilizer's properties efficiently. In this case, nitrogen is retained as NH_4^+, resulting from the hydrolysis of urea. The same applies to pesticides which, if released in a smaller quantity, will promote less pollution of soil, air, and hydric resources. Thus, if the release of pesticides and fertilizers is controlled, damage to the environment can be lessened and the efficacy of harmful products in crops will be minimized (Bhardwaj et al. 2012). In order to control the release of pesticides, herbicides, and fertilizers, substances such as nanocomposites, clay minerals, or zeolites (hydrated aluminosilicates composed of alkali and alkaline earth metals) may be added (Binner et al. 2017). These minerals control the release of substances used in soils, improve seed germination, and increase the water absorption and retention. Nanoabsorbents can be used, being composed of kaolin nanoclays in a hydrophilic three-dimensionally cross-linked polymer network, resulting in a composite polymer (Sarkar et al. 2013; Manjaiah et al. 2018). Nanoclays, as well as zeolites, due to their low cost of production and easy availability, have been increasingly used in multiple applications, including in environmental cleanup operations (Kumararaja et al. 2017; Mukhopadhyay et al. 2017; Manjaiah et al. 2018). Furthermore, nanoclays can also act as enzyme stabilizer, improving their catalytic action (Calabi-Floody et al. 2018). Moreover, the zeolites present a high affinity for nutrients, which favors the use of these minerals in crops in which one intends to increase their productivity. Very interesting synergies can be created between clay minerals, zeolite, and fertilizers, through an adequate mixture of these components, as already confirmed by de Campos Bernardi et al. (2011) in tomato and lettuce plantations. However, the action of these nanocomposites also extends to the retention of carbon in irrigated or rain-hit soils. Thus, they can be used as remedial of less fertile soils, promoting a better absorption of nutrients, thus improving crop yields. Besides not harming the environment, they promote a better use of soil in terms of cultivation, without saturation problems (Manjaiah et al. 2018). In fact, the soil's sequestration capacity of organic carbon ends up being controlled by the type and quantity of clay used in the nanocomposite, showing that the greater the bond between clay particles and the soil's carbon humus, the better the carbon sequestration capacity (Singh et al. 2017). Moreover, these minerals can still help the environment through the removal of nitrates from wastewater, while simultaneously solving difficult problems, such as manure management.

8.1.7 Nanoparticles and Smart Fertilizers

Weeds are a concrete problem that agriculture has to struggle with as they feed on the nutrients intended for crops, have rapid development, and consequently block sun exposure for sown/planted species. Herbicides can act according to two principles: They can inhibit photosynthesis, inhibiting cellular growth, or they can inhibit the absorption of enzymes of weeds, not letting them develop. Among the most commonly used herbicides is atrazine, which acts as a photosynthesis inhibitor, and belongs to the family of herbicides known as triazine (Trebst 2008). However, this herbicide corresponds to the traditional class of herbicides that are entrenched in the soil and damages yields in the medium and long terms, as well as contaminating hydric resources (Graymore et al. 2001). In fact, herbicides have attracted the attention of several researchers who have studied how to detect and remove them, either from reserves and watercourses, or from arable land (Tabani et al. 2015; Shah et al. 2018; Sharma and Krishnan 2018; Otero et al. 2019).

As previously mentioned, nanoparticles have been developed that aim to reduce or eliminate the need for these herbicides, with added benefits to the environment. These nanoparticles promote the biological activity of some chemicals, improving the way they are transported to the roots of plants, reducing the amount of chemicals needed, and thus avoiding a greater contamination (Liu et al. 2008; Wanyika et al. 2012). The poly (epsilon-caprolactone) polymer, due to its excellent physicochemical properties, high biocompatibility, and easy biodegradability, has been used to obtain nanoparticles carrying biologically active compounds, due to its physicochemical properties (Sinha et al. 2004). In a study developed by Pereira et al. (2014), it has been proven that nanocapsules or nanospheres of poly (epsilon-caprolactone) can be successfully used in atrazine encapsulation, allowing it to be released in a slower and more controlled way, giving atrazine a more targeted action against weeds, without disturbing the normal development of the plants to protect. In addition, the soil mobility of atrazine was improved, so its effectiveness against weeds was increased. It was also clear that the conjugated effect of poly (epsilon-caprolactone) with atrazine decreases the harmful effects of this substance when in the free state, reducing the harmful environmental effects.

A study recently published by Calabi-Floody et al. (2018) details a list of recently developed studies on smart fertilizers, which can take on a number of functions, among which the following can be highlighted:

- Controlled release of fertilizers or biofertilizers;
- Retention of the nitrogen;

- Microbial carrier in bioformulations;
- Encapsulation of microorganisms.

When we talk about nanotechnology and the ability to deal with materials on a nanometric scale, it is possible to obtain other physical, chemical, and biological properties, which allows us to face the challenges of agriculture in a different way. Particles with dimensions between 1 and 100 nm can be engineered in another way, allowing to perceive some types of physical, chemical, and biological phenomena that have not been accessed in a higher dimension, allowing to design and synthesize other types of compounds that can meet the needs of the agricultural sector. Nanotechnology in this sector is in full development and is seen as a solution to provide nutrients in exact proportions to satisfy plant needs (Bley et al. 2017), increasing the efficiency of the application of fertilizers, herbicides, and pesticides, and avoiding the saturation of soils with these products.

Nanoclays have been previously treated, now highlighting biodegradable polymers, such as natural polymeric carbohydrates. Although they still need to undergo development, due to their hydrophilic properties, these natural polymers attracted attention because of their low cost, high degradability, and low environmental accumulation tendency (Naz and Sulaiman 2016). This coating is provided with fine pores which will help the coated product to release in a slow manner, as required by the crops. However, non-degradable polymers used as a coating or capsule of fertilizers have shown a greater efficacy in retarding the release of these encapsulated fertilizers compared to biodegradable polymers based on cellulose acetate (Jarosiewicz and Tomaszewska 2003; Calabi-Floody et al. 2018). Non-degradable polymers that have been used as nanocomposites in fertilizer encapsulation are polystyrene, polyvinyl chloride, polysulfone, polyurethane, among others (Lü et al. 2016). On the other hand, polylactide acids, biopols, polycaprolactone, alginates, starches, and so on, can be used as biodegradable polymers (Calabi-Floody et al. 2018).

As seen, nanotechnology is not only used in this sector as a coating for bacteria or microorganisms. Therefore, it is possible to state that there are three classes of nanofertilizers: synthesized nanoparticles that constitute a nanofertilizers, raw products that are added using nanoconstituents, or nanoparticles coated or filled products (Mastronardi et al. 2015).

8.1.8 Energy Generation from Agriculture Biomass

The generation of energy through biomass is also a way of making food production and agricultural activity cleaner. Given that many countries are still largely dependent on fossil fuels, energy production through the use of biomass generated in the agricultural sector is a very significant step toward cleaner and more sustainable production. For this, it is important that each country maps the generation of waste, allowing the identification of the areas where biomass is generated and strategically placing power plants near them with the intent of facilitating the reception of waste and energy generation. Recently, Morato et al. (2019) mapped the generation of waste from agriculture in Bolivia, concluding that approximately 98% of the country's agricultural waste was generated in a single province (Santa Cruz), which greatly facilitates waste management and energy production, being able to reach 4.5 TWh/year, taking into account the amount of biomass providing from the agricultural activity in that area annually (3.3 M dry ton/year).

8.1.9 Other Cases of the Implementation Cleaner Production Practices

As mentioned above, the main sources of environmental problems generated by agricultural activity are relatively well identified and are mainly based on the consumption of energy, selected fuel type and water consumed on the one hand, and solid and wastewater generated by other. A study based on a juice producing industry located in Kuala Lumpur, Malaysia, which produced approximately 3456 m^3 of natural juice each year, distributed in 16 different flavors and following three different ranges (concentrated, flavored, and cordial juices), initially generated 0.07 g of CO_2/L of juice produced, being that 88% of this CO_2 generation was due to energy consumption. After an intensive survey of the problems that gave rise to such a high CO_2 generation, six measures were considered vital to lower these values: (a) development of standard procedures for different operations across the company, (b) waste classification and segregation, (c) implementation of first-in–first-out system of inventories, (d) optimization of production, reducing as much as possible changes in flavor and type of juice in equipment, (e) planning maintenance actions, and (f) training of operators. The first item (a) was aimed at saving energy and water consumption. The second action (b) was aimed at improving the aptitude and sensitivity for the recycling and reuse of products. The third measure (c) was intended to reduce or even eliminate the existence of stored expired products, which was another source of waste or residues. The fourth action (d) was to save water consumption in cleaning operations. The fifth measure (e) aimed to avoid spills and corresponding wastes. Finally, action (f) was aimed at preventing the creation of waste due to improperly performed product packaging tasks. With the accomplishment of this work and the application of the aforementioned practices, it was expected to obtain savings of around 20% in the volume of CO_2 produced, reducing the amount of CO_2 generated as a function of production rate to values of 0.042 g CO_2/L of juice produced, and it was also estimated that the return on investment of these implementations would be made in a six-month period (Rahim and Raman 2015).

8.1.10 Remarks

Agricultural and agroforestry occupy about 40–50% of the Earth's land surface (Smith 2007), which represents a significant portion. The accelerated population growth rate brings new challenges to the agricultural sector in order to produce sufficient food to guarantee human well-being. The existence of chemical fertilizers to increase the productivity of arable land does not seem to be a big problem as nitrogen soil needs can be met by using the Haber–Bosch method, with affordable rates and prices (Dawson and Hilton 2011). Phosphorus however is more critical as soil productivity depends heavily on this element, and due to the finite reserves of phosphate rocks, its market price might vary. However, the use of phosphorus has other concerns: phosphorus-rich fertilizers, in certain less fertile lands, can be completely absorbed by the soil without even reaching the plants. This is a current concern which needs to be resolved (Cordell and White 2011). Increasing the efficiency of phosphorus application would lessen future supply concerns and consequently the depletion of natural reserves. Furthermore, it would also eliminate an environmental problem as the effects of phosphorus leaching are even more disturbing. In recent years, a number of much greener solutions have emerged which need to be conveniently tested in terms of food chain safety and environmental impact and may be strong alternatives to conventional fertilizers, herbicides, and pesticides as they are capable of producing similar effects without degrading the environment.

8.2 Cleaner Energy Production

There has been a steady trend toward an increase in the world population, which has been accompanied by a sustained increase in energy consumption, which is due to an increase in urban areas and industrialization, increasingly supported by the need for electricity (Ebrahimi and Rahmani 2019). By 2015, and taking the 1950s as a reference, the world population had grown 2.5 times, while electricity consumption in the same period increased seven times. Until the last decades of the twentieth century, electricity was essentially produced from fossil fuels, nuclear power plants in more developed countries, and renewable water resources. The successive oil crises, and the constant threats of early finite reserve depletion due to high consumption, were alerting the world population and the scientific community that it was necessary to find other resources to ensure the generation of electricity. On the other hand, the environmental concerns promoted the discussion and development of alternative techniques for electricity production, passing by a decarburization process, which mainly went through the generation of electric energy based on renewable resources, such as solar energy and wind energy. These strands have been heavily exploited, and efficiency has grown considerably. However, a relatively recent report issued by the UNECE (2015) found that more than two-thirds of world-produced electricity is still fossil-based products and that the generation of electric power is strategically distributed by places with easy access to fossil fuels and water for corresponding refrigeration needs. It is expected that electricity generation in the near future will still remain strongly based on fossil fuels, mainly because this fuel is expected to continue to be abundant and supplied at relatively low prices. Moreover, the least developed countries already have generation centers based on this technology and find it difficult to promote the generation of energy based on renewable resources without external aid. Indeed, there are still a large number of countries where power generation is mainly based on thermal power plants, such as the USA, Russia, Israel, Poland, Estonia, Bosnia–Herzegovina, Ukraine, and Serbia, among others. There are even some countries, such as Tajikistan and Kyrgyzstan where, although electricity is generated mainly through renewable sources, namely water resources, they are planning to build thermal power plants, and are also building coal reservoirs (UNECE 2015).

Energy production is one of the major contributors to CO_2 emissions, which reached global figures of 34.6 Gton in 2011 (Oliver et al. 2012) and from that year to 2013 rose by 2%, reaching in 2013 about 35.2 Gton (Oliver et al. 2014). If CO_2 emissions continue to grow at this rate, by the end of this century the temperature may have risen by about 6 °C (Martinez-Duart et al. 2015). The use of renewable energy for the production of electricity has undergone a notable increase, but, in world terms, it represents only about 10% of consumption. According to the European Union's expectations, in 2050 it is expected that the electricity generation based on renewable energy will account for about 55% of the electricity needs at that time (EC 2012). However, these scenarios are also based on the prediction that energy storage systems will also make significant progress, as otherwise there will be some barriers related to the adequacy of production and consumption over time. While thermal power plants have the ability to respond quickly to network requests, the energy generated from renewable resources is dependent on Nature, especially from rain, wind, or the sun. This dependence, i.e., the need to respond in situations of peak consumption and the absence of natural factors that allow the generation of energy required to cover the needs, as long as it exists, will always prevent thermal power plants from being put out of order. Thus, the solution can be to increase the efficiency of these thermal plants, as a reduction of 2–3% in CO_2 emissions is achieved for each percentage point that can be gained in its efficiency.

Even in light of the rising energy consumption and growing awareness of the need to generate electricity through the use of renewable energy, many countries still have a very small percentage of generated electricity based on renewable sources (Alipour et al. 2017). Moreover, the more intensive use of renewable energy sources poses new challenges to research, as it becomes necessary to create systems capable of storing energy (Martinez-Duart et al. 2015), since the capacity to generate electricity through renewable sources depends heavily on Nature, while consumption follows a law with some randomness and does not coincide in time with the capacity to produce energy by renewable means. The use of fossil fuels as an energy source for electricity production is still the most widely used method, as can be seen in the graph of Fig. 8.3, and the growth trend seems to remain solid until 2030, according to the studies carried out.

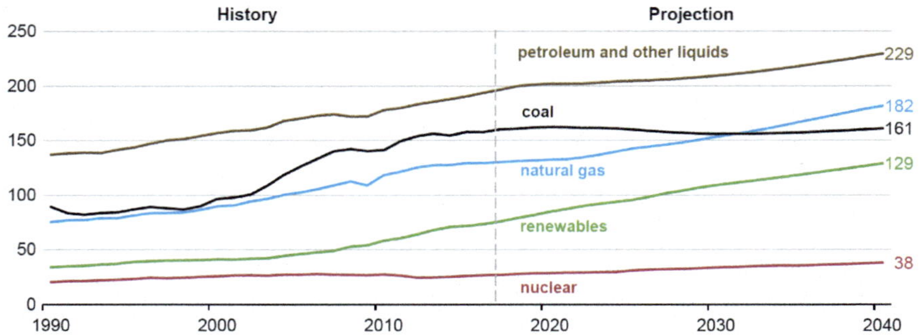

Fig. 8.3 Worldwide history and perspective of electricity consumption from 1970 to 2040, taking into account the different sources of energy used in its production. Consumption expressed in quadrillion of BTU (EIA 2018)

Figure 8.4 shows the relatively recent European reality (2016) compared to 1995, with decreasing fossil fuel use by −11.5%, although renewable sources grew from 5.1 to 13.2% in total production (EC 2018), i.e., an increase of about 160% in the same period. From this, it can be concluded that the various global warnings for climate change have had some effect on the bodies responsible for the sector, but the effort made to date is still far short of what is needed.

The use and mainly the production of electrical energy through renewable resources are argued by Ebrahimi and Rahmani (2019), also with a view to the development of rural areas, since solar and wind power can be generated closer to the places of consumption, but in less developed areas and in a perfectly decentralized way, generating some jobs in areas farthest from urban centers, thus contributing to lower migration of people to large cities and desertification of some undeveloped areas.

However, other scenarios have been developed to overcome the difficulty of energy storage. One of the solutions is to transform the power generated by wind and solar means in energy, i.e., a gas, which can be stored and used as an energy source later, using appropriate technology. In fact, this unused energy can be used to generate hydrogen, which can later be used to generate other compounds, namely methane or methane, or even to be used in the processing of raw materials such as hydrocarbons of long molecular chains (Blanco et al. 2018). The yield of the pyrolysis process usually used to convert the generated and unused electrical energy into methane is usually 70%, while the yield to convert the hydrogen to methane is around 75% (Lehner et al. 2014; Götz et al. 2016). This technique has many advantages, since it does not need other infrastructures than the existing ones, allows the use of already well-studied vessels as storage means, if stored in the form of methane has a high specific energy capacity, can be stored for long periods of time without suffering degradation, and reduces carbon emissions. However, the technology used in these transformations still needs to be greatly improved, and the most economically profitable ways to exploit the gas generated by this process have yet to be explored.

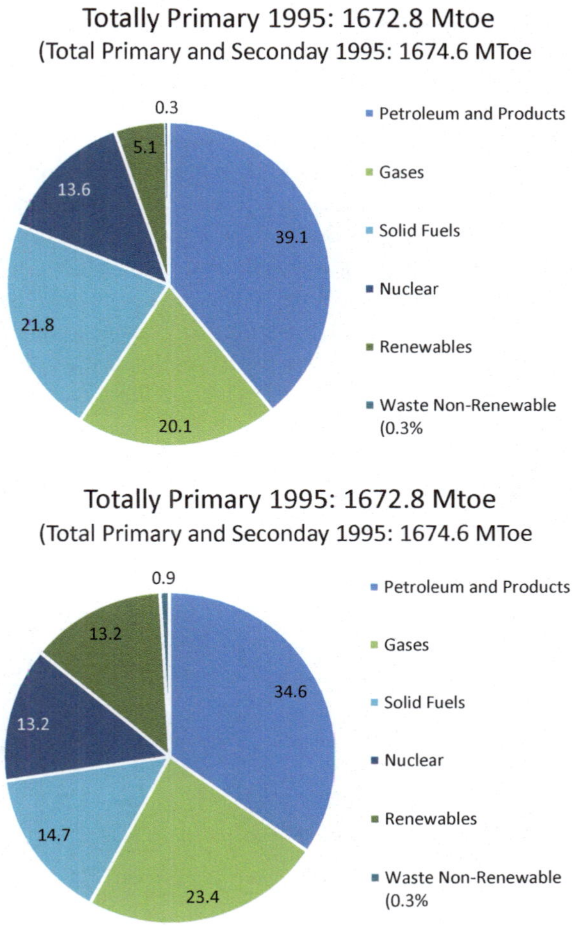

Fig. 8.4 European mix of electricity production sources: comparative situation between 1995 and 2016. Adapted from EC (2018)

Global biomass-based electricity production, mainly through the transformation of biofuels, has grown significantly from a production of 227 TWh in 2004 to 646 TWh in 2016, which is accompanied by the production of biodiesel, which also rose from 2.4 billion L in 2004 to 30.1 billion L in 2015 (Archer and Steinberger-Wilckens 2018). However, there needs to be an adequate balance between the portion of cultivation that is allocated to biofuels and that which is necessary for food production, without any conflicts of interest. The production of biofuels cannot be sustained by severe deforestation. Otherwise, the imbalance of the ecosystem could lead to serious consequences for mankind. The different pathways that can be used for biomass utilization and biofuel extraction were synthesized by Archer and Steinberger-Wilckens (2018), and the corresponding diagram can be seen in Fig. 8.5. In that work, an exhaustive analysis of the various processes that can be used for an adequate use of the different types of biomass as well as extracted biofuel and corresponding by-products are carried out.

Although CO_2 is a problem when emitted freely into the atmosphere, it may however be useful as a raw material. In fact, by reacting CO_2 with hydrogen, methane is obtained, which can be used as automotive fuel, releasing much lower carbon emissions than current fuels. If CO_2 is conveniently captured and stored, when properly combined with hydrogen, it will give way to methane, which can be stored and distributed through the existing natural gas distribution structure. In order to implement this form of energy generation, Audi built a plant in Werlte (Germany) to produce hydrogen based on the water's electrolysis, having this plant the capacity to generate 6 MW. By subsequently reacting the hydrogen with the CO_2, methane is produced, obtaining energy and water as the results of the process. In this way, Audi took the opportunity to prove that cars that use this type of fuel will have very low carbon emissions.

Because renewable resources such as solar energy and wind are already widely dissected, although it is still possible to optimize many of the systems already implemented, this is not the main scope of this work, so more attention will be given to other alternative ways of generating energy, which will be dealt with below.

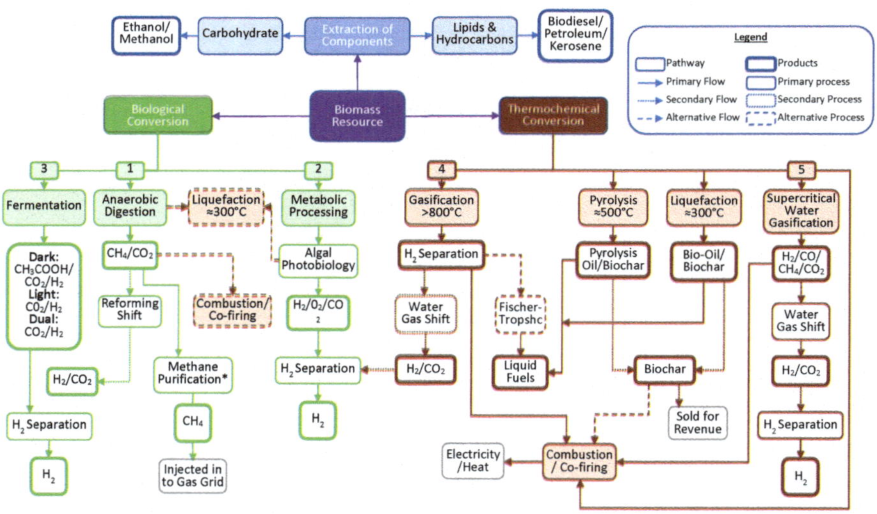

Fig. 8.5 Different pathways for biomass conversion and biofuel extraction. Adapted from Archer and Steinberger-Wilckens (2018)

8.2.1 Biofuels

Biofuels can be defined as products enriched in chemical energy through biological techniques, or from biomass containing microorganisms, such as bacteria, microalgae, or plants (Rodionova et al. 2016). Although some biofuels continue to have a fossil base, namely the biofuels from the first generation, the name bio essentially comes from the fact that they release less GHG and the life cycle time of these gases is much lower than traditional fossil fuels (Yong et al. 2016). Biofuels can also be produced on the basis of photosynthetic microorganisms processing, such as photosynthetic bacteria, micro- and macro-algae, as discussed later in this chapter. In a first stage, biofuels can be presented in the gaseous, liquid, or solid form, being later converted through biochemical, physical, and thermochemical methods. These biofuels can still be grouped into two main classes: primary and secondary biofuels, the primary ones being produced directly from the burning of forest waste or crop residues and sewage (Enagi et al. 2018). Secondary biofuels can still be subdivided into three generations that are generated indirectly based on plant residues and animal residues.

The production of biofuels has undergone a remarkable evolution, mainly in the selection of the products that are the origin of the process of biofuels generation, both to be used in static production of electric energy, and for the locomotion of motor vehicles. Ethanol represents the first generation of biofuels, being produced from starch-rich plants such as sugarcane, sunflower oil, and so on, or biodiesel produced from fats from slaughter animals, as well as fats from cooking oils and food wastes (Rodionova et al. 2016). However, the use of some of these species posed serious obstacles to the cultivation of these species for food, as it has become more profitable for farmers to produce these species for biofuels than for feeding populations. The evolution reported in the meantime has given rise to a second generation, whose production has been mainly synthetized from biomass residues and has been focused essentially on bio-ethanol resulting from non-food cellulosic biomass and biodiesel extracted from seeds of oil-rich plants such as soy, jatropha, wheat straw, corn cob, cassava, among other (Leong et al. 2018). This generation of biofuels no longer entailed the previously stated problems of conflict of interest between the feeding of populations and the crop of species for biofuels (Trabelsi et al. 2018). However, despite the strong development that this second generation of biofuels has suffered in recent years, there are still well-established barriers in its evolution, as it is still necessary to pretreat the lignocellulosic biomass in order to ferment the sugars so that they can generate the type of the desired biofuel. All processes around the production of these biofuels tend to lower their rate of production and profitability, such as parallel spontaneous reactions, or due to other factors that may affect the process, such as the action of inhibitors of the necessary reactions, which prevents successful metabolic processes (Jönsson et al. 2013; Ghimire et al. 2015). The third generation is based on cyanobacteria, microalgae, and other microbes, which give rise to biofuels capable of generating electricity or being used in motor vehicles. In fact, a few years ago, biofuel was extracted essentially from plants, because it was the well-known and most effective route for this purpose (Alaswad et al. 2015). However, the study of the metabolic process of microalgae and the possibility of carrying out specific cultures with a view to their use for energy production has known significant developments in recent years, and microalgae culture in the form of biofilm is now under development. With the recent study of microalgae behavior, it was realized that they are able to transform atmospheric CO_2 into sugars, through the process of photosynthesis using sunlight as a source of energy, being these sugars the raw material necessary for the production of bio-ethanol and biomethanol (Dias et al. 2009), such as the microalgae species *Botryococcus braunii* and *Chlorella protothecoides* (Razaghifard 2013), *or Escherichia coli* and *Bacillus subtilis* when genetically engineered (Rodionova et al. 2016). Obviously, the costs involved are very low, since CO_2 is available in the atmosphere, and sunlight only depends on Nature. Furthermore, biofuel production using microalgae is a few hundred times higher than the production rate achieved with the best plants for this purpose. The microalgae biomass is obtained through the microalgae fermentation and the generation of microorganisms, giving rise to the biofuel. At present, several microorganisms that can give rise to biofuels are identified, as well as some techniques and drivers to improve biofuel generation, as follows:

- Improvement of the rate of hydrogen production through genetic engineering and metabolic engineering of cyanobacteria and similar;
- Conversion of carbohydrates into biofuels and biohydrogen through dark fermentation of bacteria;
- Production of biohydrogen through the biological photosynthesis of microalgae;
- Increased ethanol production rate, admitting a higher alcohol concentration, through genetic engineering of yeast;
- Increased production of bio-ethanol and biobutanol through genetic manipulation of microorganisms resulting from the fermentation of carbohydrates;
- Identification and classification of microalgae species that present greater capacity for the production of oil that results in biodiesel;
- Increased production of biofuels that result from fermentation of carbohydrates from plants with cell walls.

The correct knowledge of the composition of the different species of microalgae allows determining their content in the substances that produce the highest amount of biomass, such as lipids, proteins, and carbohydrates. Some microalgae species clearly allow better utilization of its biomass, being able to generate more energy, such as the *Schizochytrium* species, which contain between 50 and 77% w/w of lipids, *Spirulina maxima*, which has between 60 and 71% w/w of proteins, or *Porphyridium cruentum*, which presents carbohydrates in a percentage between 40 and 57% of its weight (Rodionova et al. 2016). Microalgae should also be selected according to the type of fuel to be obtained, since some species have a higher yield in the production of certain fuels, to the detriment of others. Genetic engineering has shown that it is possible to produce more hydrogen at the expense of ethanol production through the species Caldicellulosiruptor, Pyrococcus, Thermococcus, and Thermotoga. This finding reveals the importance of genetic engineering in the biochemical production of bacterial cells, as well as the direct link between the culture form of these microorganisms and the production of biofuels (Carere et al. 2012). It should also be noted that there are also species suitable for the production of raw materials, such as Bacillus and *E. coli*, which provide lactic and glutamic acids (Hasunuma et al. 2013).

8.2.2 Hydrogen

Hydrogen is regarded as an excellent alternative to fossil fuels in a number of situations, from which it can highlight the production of electrical energy, heat generation, or mechanical work, such as in car locomotion. These sectors, together with the transport sector, account for around 65% of global CO_2 emissions (Staffell et al. 2018). Obviously, it will be as a means of producing electricity that hydrogen will be explored in this section. Moreover, hydrogen has a high efficiency as an energy source and is harmless to the environment, because its combustion generates only water as the resultant by-product (Marone et al. 2014). It is noteworthy stating that hydrogen has a specific energy of about 120–142 kJ/g (purity degree dependent), which when compared to the specific energy presented of diesel (42.9 kJ/g), gasoline (41.2 kJ/g), or even coal (25–35 kJ/g), shows greater potential to become one of the energy sources of the future (Das et al. 2008).

Hydrogen fits perfectly into what is defined as renewable energy, and more than that, it also fits perfectly into the concept of Circular Economy because it can be regenerated by Nature. Since hydrogen has a lower density than air, it is not a freely formed gas, as it tends to rise in the atmosphere. Thus, for current use, hydrogen needs to be fabricated as it is normally combined with other elements. However, it is one of the most abundant elements in the Earth's crust. Due to its low density, hydrogen presents simultaneously the highest energy per unit weight and the lowest amount of energy per unit volume. Like electricity, hydrogen can also be considered an energy carrier because it can be stored in different ways and transported through pipelines, just as electricity is transported through electrical conductors. In certain cases, it is more economical to transport hydrogen through pipelines, than to carry energy through electrical cables, mainly regarding long distances. The sustainability of hydrogen is further strengthened by the fact that it can sequester carbon which represents an extremely current need, can aid in waste management as it can be produced from different types of biomass, including organic effluents, or can be produced from several bacteria that can be found in several environments or be grown exclusively for this purpose, and be able to be produced at ambient temperature and atmospheric pressure, requiring only renewable resources for their synthesis, such as sunlight as a source of energy (Das et al. 2008).

Since it is not possible to find hydrogen in the form of gas, it needs to be obtained industrially by electrolysis of water or by steam reforming, the latter being the most economical process. This process can be carried out from methane gas, one of the ways of storing hydrogen with high energetic potential, but the steam reforming process will give rise to harmfully GHG. Storing hydrogen in the form of methane is not only disadvantageous as it may result from a CO_2 sequestration process, as already mentioned above. On the other hand, electrolysis is a rather expensive process which requires electrical energy, and which gives rise to high purity hydrogen, being produced by the dissociation of water in oxygen, which will accumulate in the positively charged anode and hydrogen, which will be collected at the negatively charged cathode. Hydrogen generation processes are not limited to these main methods, as it is still possible to obtain hydrogen from biomass, by photo-electrolysis and also through algae and bacteria, with the help of solar radiation, as will be described later.

Hydrogen production can still be generated on the basis of renewable energy, making the entire production process even cleaner. Nevertheless, there are also other less environmentally friendly hydrogen production techniques, usually known as traditional, when it is produced from fossil fuels such as coal gasification, partial oxidation of hydrocarbons, or even natural gas reforming (Kothari et al. 2008). On the other hand, the production of hydrogen from biomass does not have any negative impact on the environment and assists in the management of waste (Khan et al. 2017). Indeed, in countries that have surpluses in terms of electricity production based on hydroelectric resources, and in times when demand is clearly below the supplying capacity, water can be stored in reservoirs, but there will be times when it needs to be discharged without being turbinated. In these cases, the energy produced in excess can be used to promote the production of hydrogen by electrolysis, which is easily storable, unlike electrical energy. Moreover, hydrogen can be used for other purposes, such as heating or the production of raw materials. In addition, hydrogen can be used as a source of energy in the event of a loss of water flow, lack of wind, sun, or any other source of renewable energy that depends on the conditions of Nature.

However, water and electrolysis are not the only means of generating hydrogen. Given the increasing crowding of people in urban centers, generation of organic waste is also increasing. This organic waste, as well as properly treated wastewaters, can also give rise to hydrogen or hydrogen-rich gases, allowing the problem of excess waste to be converted into a source of energy. Recent studies have shown that hydrogen can also be produced through a method called anaerobic digestion, which was initially used to produce methane from waste (Shen et al. 2015; Mao et al. 2015). Although early studies date back to the early twenty-first century, it is also well established that anaerobic digestion can produce biohydrogen from waste (Hwang et al. 2004; Cui and Shen 2012; Cheng and Zhu 2013; Angeriz-Campoy et al. 2015; Intanoo et al. 2016). Anaerobic digestion consists of a series of biochemical reactions, through which organic materials are transformed by microorganisms into a mixture of methane and carbon dioxide in the absence of oxygen (Adekunle and Okolie 2015). Although feasibility has been described as perfectly proven, there remain numerous difficulties surrounding the production and storage of hydrogen, factors which have prevented its manufacture by this process to evolve in the desired way and be running in the current market. The difficulties inherent to production are related to the optimization of process parameters and the identification of the conditions that lead to the inhibition of the transformation process, besides making the process able to transform different types of substrate, consisting essentially of four distinct stages: hydrolysis, acidogenesis, acetogenesis and methanogenesis (Khan et al. 2017). Methane cannot be obtained directly from the resulting acids of the second stage therefore, a third stage is necessary. This third stage promotes the formation of hydrogen and CO_2 (which are present in almost irrelevant quantities at the end of the second phase) and acetate. It is only after this third stage that the last phase of conversion of anaerobic digestion occurs, with the production of methane and CO_2 through methanogenic Achaea, in which two groups take action: The acetotrophic group of methanogen microorganisms guzzle the acetates and convert them into methane and CO_2, and on the other hand, the hydrogenotrophic group also converts the hydrogen and CO_2 formed in the previous steps also into methane (Sarmiento et al. 2011; Andre et al. 2016). In order to proceed with the production of biohydrogen by anaerobic digestion and the corresponding reactions required at each step, several reactors have been developed and tested, of which a summary can be found in the work of Khan et al. (2017). In 2009, the studies focused mainly on dark fermentation were developed, but the performance of the process was considered at that time as not adequate to the desired end, reason why it was improved over time, having evolved to the designated bio-electrochemical systems. More recently, in 2016, works based on microbial electrolysis cell have begun to be published, a subject that is being studied with some intensity at the moment in order to find the most profitable way to produce biohydrogen, essentially trying to minimize the energy required to produce electrolysis. This evolution resulted in a productivity increase of over 100%, since it was possible to go from about 4.0 mol H_2/mol glucose in the dark fermentation process to about 8.55 mol H_2/mol glucose in the microbial electrolysis cell process and go from energy consumption values in conventional electrolysis of 4.5–50.6 kWh/m^3 H_2 to 0.6–1.0 kWh/m^3 H_2 with the newly developed process (Premier et al. 2013; Rahman et al. 2016; Kadier et al. 2016). It should be noted, however, that all of the above reported values strongly depend on the type of biomass/organic material that is used in the process. The quality of the hydrogen produced and the yield of the process will be greatly affected by the lack of quality of the organic material from which the biohydrogen will be obtained. However, the inhibition mechanisms cannot be forgotten, because they can undermine the process. Some of the factors that may promote the inhibition of the required anaerobic digestion reactions are the presence of furan, alcohol, phenol derivates, bacteria capable of consuming hydrogen, metal ions, among others.

Basically, dark fermentation consists of the acidogenic fermentation of carbohydrate-rich materials in a special environment (exempt of light and oxygen), producing biohydrogen, also resulting as effluent a mix of acetic acid, butyric acid, propionic acid, and alcohols such as ethanol (Bundhoo 2017). Regarding dark fermentation, recent studies have revealed that it is possible and desirable that the process be improved and even optimized, as there are several residues of various types of crop that can be harnessed for this purpose (Boodhun et al. 2017). Overall, it is estimated that about 58 Mm^3 of rice straw, as well as about 35 Mm^3 of wheat straw, is generated annually, which can be made available for the production of bio-energy, which is equivalent to a generation potential of about 620 and 370 PJ, respectively. However, if such waste is pretreated, its energy potential is still substantially higher (Bundhoo 2018). Although studies initially developed in 2009 are not as promising as expected, dark fermentation is currently considered a suitable process for the treatment of large amounts of biomass in a systematic way, being pointed out as a process that consumes relatively little amounts of energy and allows the treatment of a wide range of substrates, such as crop residues, industrial, domestic, and municipal solid wastes, and can be coupled to other types of technologies in order to maximize yield (Lukajtis et al. 2018). If sugarcane residues normally condition the amount of biohydrogen generated at the above-mentioned values, other substrates such as lignocellulosic biomass further condition these values and may jeopardize the economic viability of the process. As such, it is convenient to carry out a pretreatment which consists of destroying the lignin seal which surrounds the cellulose molecules, facilitating their release in the form of solution, subsequently destroying the crystalline structure of cellulose and proceeding to a depolymerization process, with the aim at increasing the enzymatic digestibility and acidogenic fermentation, or in other words, the production of biohydrogen (Sun et al. 2016). Considering the various physical, chemical, physicochemical, and biological methods that can be used to perform pretreatment, it has been reported that the most effective are acid or thermal pretreatments (Bundhoo et al. 2015). Pretreatments may also be applied to microbial cultures, which may contain H_2-consuming species which need to be destroyed in order to improve biohydrogen yield. These pretreatments unsure that the hydrogen-promoting bacteria survive, while others do not (Wang and Yin 2017; Bundhoo et al. 2015).

The biological methodology of biomass transformation by metabolic processing has the drawback of consuming a little more biomass feedstock for the same amount of energy produced, but is an alternative process to consider, since it can be used in conjunction with another technology, which allows the improvement of their performance. In the process of metabolism processing, algae may assume a prominent position as they are the most abundant form of photosynthetic life in the terrestrial crust, playing a key role in the energy flow dynamics of ecosystems. Moreover, algae also have a high ecological and financial importance because they are a source of numerous natural substances useful for various purposes, namely in food and pharmaceutical products (Khetkorn et al. 2017). In addition, green algae can still be used as a source of biodiesel, bio-oil, biogas, and biohydrogen (Khetkorn et al. 2013; Skjånes et al. 2013). Because microalgae can be produced on non-arable land using seawater or wastewater effluent streams, their culture does not conflict with the usual concerns of arable land, where a competition between the production of species for energy production and for human or animal consumption exists, as seen with the first generation of biofuels. Furthermore, microalgae are able to achieve higher yields per unit of cultivated area than traditional crops due to short life cycles and greater production efficiency (Schenk et al. 2008).

In fact, it is possible to produce biohydrogen from photosynthetic and non-photosynthetic microorganisms, such as green algae, photosynthetic bacteria, cyanobacteria, among others, using distinct metabolism pathways (Khetkorn et al.

2017). Thus, microalgae can be used as a renewable source of biohydrogen generation, through the photosynthetic microorganisms of which they are constituted (Nagarajan et al. 2017). Usually a subdivision is made between the algae used for this purpose, with green algae being included in the eukaryotic group, while algae with green–blue coloration are from the cyanobacteria group (Wetzel 2001). Moreover, this biohydrogen generation process is conducted at room temperature and at atmospheric pressure, without the need for the presence of noble metallic elements that promote the catalytic reaction, making this process truly environmentally friendly (Eroglu and Melis 2016). Algae are in general a group of highly phototrophic organisms, which may assume single-celled forms, and are the predominant group regarding the photobiological hydrogen production or multi-cellular configurations. The potential of microalgae in environmental terms is enormous, since in addition to being able to produce biohydrogen only using solar energy, it can also reduce the molecular nitrogen present in the atmosphere to ammonia and/or capture CO_2 from the environment. The production of biohydrogen is facilitated by the fact that microalgae have, by nature, very efficient photosynthesis systems for this purpose, using just water electrons and solar light as energy source (He et al. 2017). This effect is achieved by separating H_2O into oxygen (O_2) and hydrogen (H^+) protons on the one hand, and simultaneously converting solar energy into chemical energy, promoted by the pigment molecules contained in microalgae under the sunlight effect (Yialmaz et al. 2016). This process of photobiological production of hydrogen from green microalgae is considered one of the most efficient because the energy it needs is minimal and supplied naturally: sunlight. The biological production of biohydrogen can be carried out through several mechanisms already well known in the scientific community, such as direct biophotolysis of water, possible to perform based on green microalgae and cyanobacteria, photofermentative fixation of nitrogen with resources to photosynthetic bacteria and heterocysts of filamentous cyanobacteria, non-photosynthetic methods from the fermentation of organic compounds that require the use of anaerobic bacteria, and nitrogen fixation when the biomass used is based on fermentative bacteria (Eroglu and Melis 2011). In addition to contributing to the natural creation of biohydrogen, algae also have other environmentally friendly functions, such as carbon capture, both in organic and inorganic forms. Algae that have the ability to uptake organic carbon have already been described previously, while others are able to absorb inorganic carbon in the form of CO_2 or $-HCO_3$. In these cases, reducing energy and ATP (adenosine triphosphate) from the photosynthesis is required so that the absorption process can be carried out, also converting the inorganic carbon to organic substrate (Eroglu et al. 2013). Thereafter, biohydrogen can be used as fuel in the fuel cells, which also present high efficiency in electricity production process, as previously mentioned. Khetkorn et al. (2017) highlight the high cost of biohydrogen produced from microalgae, as well as the energy produced by this way, which makes the process too expensive to be economically viable. However, the most recent advances in terms of metabolic and genetic engineering can make this form of production viable in the near future.

With a view to designing a photo-bioreactor capable of operating with microalgae as biomass and capable of generating and collecting the desired gases, designers need to consider some fundamental requirements, such as (a) there must be a sufficient and generous sunlight input to illuminate the largest surface possible. Flat-plate type ensures the best results due to a better surface-to-volume ratio; (b) the blend may be suitably carried out; (c) there may be temperature and pH control in the culture within the reactor, in order to maintain values between 20 and 30 °C, and between 7.0 and 8.5, respectively; (d) there is a correct supply of CO_2 to the biomass; (e) there is an efficient removal of O_2 from the reactor; (f) there is efficient extraction and storage of biohydrogen; (g) ensure that the H_2 pressure near the reactor head is sufficiently low; (h) ensure that there is efficient delivery of essential nutrients to cells; (i) have mechanisms to prevent invading microorganisms from contaminating crops; (j) ensure stability of operating conditions over time; (k) be easy to build and carry out the corresponding maintenance operations; and (l) ensure that the facility does not compete with food crops (Burgess et al. 2011; Adessi and de Philippis 2014). The rate of biohydrogen production is one of the key factors in assessing the feasibility of a given process. In the case of the biological generation of hydrogen, this rate is dependent on several factors, such as (a) geometry of the photo-bioreactor and its dimensions; (b) material used in the construction of the photo-bioreactor; (c) electron source; (d) type of algae used; (e) initial cell concentration; (f) conditions imposed on the mixing and cultivation of algae; and (g) conditions used in lighting (Eroglu and Melis 2016). Given that these processes require water, it is now evident that the proper use of wastewaters for the generation of free-carbon energy will be the next step to be taken, through the integration of processes, generating the necessary energy through the production of biohydrogen and being able to contribute for the reduction of contaminants present in wastewaters. The challenges in the development of microbial fuel cells are their reduction in size and viability of modularization, in order to make these systems more portable and to avoid some associated losses, instead of increasing the size of each power production unit. The materials used in bioreactors and fuel cells should also be carefully selected to increase their life span. The control and storage of the generated energy is also a part that must be explored in the short term, in order to obtain the maximum yield of the generation of energy (Gajda et al. 2018).

Fuel cells are one of the means used to directly convert the energy present in hydrogen into electricity, releasing pure water and heat as by-products. Basically, a fuel cell consists of an anode (oxidation) and a cathode (oxygen reduction). All cells are provided with an electrolyte, which has the function of transporting the ions produced at the anode, or at the cathode, to the counter electrode, and a catalyst, which accelerates the electrochemical reactions at the electrodes (Archer and Steinberger-Wilckens 2018). At the beginning of the twenty-first century, strong expectations were created around fuel cells, expectations that have come to fruition, in part due to the cost of the systems not having decreased as expected, and on the other hand the performance of the systems did not evolve both when desired. However, since then, and especially in the present decade, the situation has evolved in a much more favorable way, and some government support for this type of energy generation has started to decarbonize the energy production sector (Staffell et al. 2018). In the automotive sector, and taking into account the launch of Honda's Clarity fuel cell vehicle model, Cackett (2019) expressed disappointment at the lack of evolution in the fuel cell sector, noting that they have promised a lot but have brought the automobile market very little.

Compared to traditional combustion technologies, hydrogen fuel cells are two to three times more efficient and have the great advantage of not generating any kind of pollution. Important research steps have been taken to optimize fuel cells so that they become a safe and economically viable alternative to other means of electric power generation, as well as locomotion for cars or aircraft propulsion. A fairly direct relationship is established between fuel cells and hydrogen, but this need not necessarily be so because (Staffell et al. 2018):

- Hydrogen can be produced on the basis of a varied set of feedstocks, biomasses, and greenhouse gases, requiring different processes, investment costs, and corresponding support structures.
- Hydrogen may be used for the production of energy in various forms, including or not fuel cells. On the other hand, fuel cells also do not rely solely on hydrogen to generate energy.
- The use of hydrogen and fuel cells, either together or separately, is able to cover all energy needs that are currently known.
- Although the cost of installing infrastructure is expensive, there are alternatives that have not yet been properly explored.

Fuel cells are the most efficient way of converting the energy of a fuel into electrical energy and heat through electrochemistry, practically without losses and without harmful consequences for the environment, since the released product is only water (de Miranda 2019). As mentioned by de Miranda (2019), it is curious to note that some materials commonly used as fuel have more complex chemical structures, such as wood, but this is not directly translated into specific energy, which is 20.6 MJ/kg. On the other hand, natural gas has a less complex chemical structure than petroleum and coal, having a distinctly higher specific energy, which is about 52.2 MJ/kg. Curiously, the hydrogen content also increases systematically from wood to coal, from coal to oil, and from oil to natural gas, reaching its peak with pure hydrogen, which presents a specific energy of 142.2 MJ/kg (Mazloomi and Gomes 2012). On the contrary, and taking into account the same materials, the carbon content also decreases systematically from wood to hydrogen.

Fuel cells have several very significant advantages, which can change the paradigm of energy production and distribution. In fact, the centralized production on which the current model of electric power generation is based could be decentralized, making local production possible, thereby circumventing some obstacles in the distribution of electricity. Another important factor that could be countered by fuel cells is that electricity production and distribution grids have several inefficiencies, inducing losses, which are reflected in the quality and cost of market energy supply, especially when long distances are involved. The introduction of harmonics in the electrical grid is also a problem that could be solved with the model of local energy production through fuel cells, as these harmonics overload the network, generating disturbances in the supply and overheating of transformers.

Another vitally important aspect of hydrogen is the economic activity currently taking place around fuels. Given that oil, coal and natural gas are concentrated in some specific geographic areas, giving economic and social benefits to some countries in detriment of others. This possession has generated numerous global conflicts, due to the greed for the possession of areas where the extraction of these natural goods can be done. These conflicts have consequent economic benefits for some and much sacrifice to others. Furthermore, political convulsions have been a constant in the last century, mostly due to quarrels caused by the eagerness to control fossil-fuel-rich areas. The production of energy or heat through fuel cells would allow a much more distributed management of energy generation, and it would be possible for each country or region to manage its natural resources in order to produce hydrogen to fulfill its needs, without depending on third parties and avoiding the concentration of economic power in certain countries, regions, or regimes. This is probably not a particularly favorable situation for those who currently have the power to have coal, oil, or natural gas reserves, but could be an extremely important factor in world stability, creating a new economic organization that could balance different economies. This system of production would also allow for a business model identical to the one currently in place, since international fuel trade could take place in the same way as now, but with greater flexibility, since hydrogen can be stored in a variety of ways, from pure form to combined with other elements in the form of gases or carbon-rich compounds. Any need for a country or region can easily be suppressed by an external supply, in the same way that is practiced today with natural gas or oil.

Although hydrogen production has already been extensively studied and exploited, with several ways of delivering the gas in a sufficiently pure state for the intended applications, there are also consistent studies on the possibility of exploitation of pure hydrogen from the Earth's crust. de Miranda (2019) indicates a series of conditions that can be identified as propitious to the exploration of pure hydrogen. Indeed, several geological conditions seem to be identified which lead to a relatively easy identification of the locations that could allow the exploitation of natural hydrogen, and these conditions can be found in countries such as the USA, Russia, Turkey, Brazil, the Philippines, Mali, among others. In fact, it is in Mali that the first natural hydrogen exploration was carried out by Petroma Inc. in the Bourakebougou region. The discovery was achieved when, carrying out a water prospection, natural hydrogen was found which, after being analyzed, showed that it contained 98% pure natural hydrogen, mixed with 1% methane and 1% nitrogen (de Miranda 2019). Given the characteristics of that exploration, the reserves appear to be relatively close to the Earth's surface, i.e., about 100 m deep, and the gas appears to be housed in shallow wells, which makes the extraction operation relatively easy and economical, even more economical than the previously mentioned production processes. Taking advantage of the similarities between the characteristic of Mali and Brazilian subsoils, where there is natural hydrogen, it is possible to draw some inferences about which areas in the Earth's crust should be explored, in order to easily identify the places where these explorations may be more likely to succeed.

Due to a properly designed development, installation, and training program, Japan has evolved notably in distributed power generation, through two main types of fuel cells: polymeric electrolyte membrane fuel cells (PEMFC) and solid oxide fuel cells (SOFC). The implementation of the PEMFCs started in 2009, with about 30,000 systems installed that year, having evolved to about 2 million systems by 2017. The implementation of the SOFCs started later in 2011, having evolved from around 20,000 systems in that year to about 400,000 by 2017. As reported by Archer and Steinberger-Wilckens (2018), the SOFC represents the best option in technological terms, as they enable the use of a wide range of biofuels and enable installations that can assume relatively modest power values, but which can be scaled up to 100 kW while maintaining high efficiency. Market demand for these systems has led to new, more scaled-up production systems, which has led to a sustained lowering of market prices, with PEMFCs costing in 2017 about one-third of what they cost in 2009. SOFCs also saw their price fall by half from 2011 to 2017 due to large-scale production and optimization. Staffell et al. (2018) also refers that the global installed capacity of fuel cells for energy generation in 2015 reached 1 GW, but the installed capacity of the stationary fuel cell has grown at a rate of about 25% per year, which makes it possible to predict that by 2025 a capacity of around 10 GW should be installed, expecting to reach 30 GW by 2030.

Nanotechnology, using material particles on a scale usually between 1 and 100 nm, allows to obtain very particular characteristics of these same materials and to interfere positively in the improvement of certain properties, impossible to reach until a few years ago. These nanoparticles have some very interesting properties, such as high crystallinity, large surface area relatively to its volume, chemical stability, catalytic activity, and strong adsorption capacity. In this way, many sectors have been exploring this new aspect of materials, with nanoparticles already used in sectors such as electronics, dermocosmetics, pharmaceuticals, and even in agriculture (Tyagi et al. 2018). Given the innovative characteristics provided by nanoparticles for various purposes, it became clear that those nanoparticles could be useful in the development of biofuels, acting on the factors identified as limiting a broader and more cost-effective application of this energy generation process. One of the above limiting factors is the inhibitory effect of the reaction required to generate biofuel. Improvement of electron transfer capacity and anaerobic activity are also critical factors in the generation of biofuels, which are believed to be greatly improved by the incorporation of certain nanoparticles (Sekoai et al. 2019). It has already been studied and proven that nanoparticles can act in the metabolic process involving microorganisms, increasing the electron transfer under anaerobic conditions. An improvement in the dynamics of biohydrogen generation process is thus achieved (Ali et al. 2017). Nanoparticles have also already been used in the process of generating energy by dark fermentation, but it has been found that the number of nanoparticles to be added is critical to the process, since it can generate an inhibition of the growth of microorganisms. The process of generating biohydrogen in anaerobic batch reactor was improved by the inclusion of silver nanoparticles, which increased the conversion to glucose by 62%, allowing to achieve biohydrogen production rates of 2.48 mol H_2/mol glucose (Zhao et al. 2013). However, slightly higher yields have been achieved using 13.64 nm nickel nanoparticles in the same biohydrogen dark fermentation process (Mullai et al. 2013). Also nanoparticles of Fe^0 were added in the production of biohydrogen by the dark fermentation process, using grass under mesophilic conditions as biomass, significantly improving the activity of microbes responsible for the generation of biohydrogen, improving biohydrogen generation in about 100% (Yang and Wang 2018). Fe^0 nanoparticles were also used to encapsulate microorganisms used in the production of biohydrogen by dark fermentation, leading to improvements in the production rate of around 58% compared to cultures in which these nanoparticles were not used (Elreedy et al. 2017). Nanoparticles synthesized from Fe_3O_4 and SiO_2 have also been shown to have good ability to improve the biohydrogen production process via dark fermentation, as they improve the catalytic process, yet exhibit great stability and flexibility in terms of pH in the medium in which they are applied. In addition, it has the advantage of being rerecovered after the fermentation process (Mohan et al. 2008; Abbas et al. 2014). The addition of nanoparticles in the biogas production through the anaerobic digestion process has also shown quite positive effects, improving the hydrolysis of organic materials (Sekoai et al. 2019). In this case, the most widely used nanoparticles in studies to date have been those of Fe^0, although experiments with Fe_2O_3, CuO and γ-Al_2O_3 have already been performed. The review recently presented by Sekoai et al. (2019) summarizes in a very good way the knowledge in this area. As in the above cases, the use of nanoparticles essentially aims to improve the production rate of biofuel. In most cases, they improve the catalytic effect and provide elements that promote easier exchange of electrons in the process. As mentioned by Sekoai et al. (2019), there is still a long way to go in the successful application of nanoparticles in bio-energy generation. Thus, it will always be necessary to ensure that the nanoparticles used are not toxic to the culture of the microorganisms where they will be applied, that the cost/benefit ratio of the use of nanoparticles is indeed advantageous to the process, and that the nanoparticles used do not cause any damage to the environment.

8.2.3 Remarks

Some of the countries with the largest oil and natural gas reserves promote policies to subsidize tariffs for electricity and some fuels, with the aim of making their economies more attractive to foreign industrial investment. These policies tend to distort the market and mask reality by preventing further research and awareness of alternatives based on less polluting resources and technologies that are closer to what standards of electricity production are supposed to be in the next decades.

Nuclear energy is clearly a commitment of the European Community as an alternative to traditional fossil fuel sources. Although it is perfectly clear in the Energy Roadmap 2050 (EC 2012) document, due to the negative public opinion that nuclear energy represents, it is a very little discussed subject. However, in the scientific and business circles, nuclear power generation has gained increasing importance, and there is a great deal of ongoing research to optimize reactors and render them less vulnerable to serious damage, even though radioactive waste is a matter that is hardly discussed in public terms. Moreover, globally, 5.8% of the energy consumed is generated in nuclear power stations (Boodhun et al. 2017).

The world's population is periodically bombarded with news of tragedies related to breakdowns or accidents at nuclear power plants, such as the recent case of Fukushima, usually with very serious consequences for the surrounding populations, so that the establishment of a nuclear power station is never welcomed by the surrounding population. Only ignorance about the true risks associated with these power plants allows populations to accept their installation. According to the World Nuclear Association (WNA 2019), nuclear power is the basis of about 11% of the electricity generated, according to data from 2016, from about 450 reactors spread around 30 countries. Some countries, such as Italy and Denmark, even depend on the supply of nuclear power from other countries. At the end of 2010, more than 60 reactors are under construction, which will contribute to an increase of the electric power generation capacity by about 16% relatively to the current installed capacity, and regarding just the nuclear generation process. In the last years, the energy supplied worldwide has increased very slightly (about 0.4% between 2016 and 2017), with a noticeable decrease in capacity in Western Europe, while Eastern Europe has increased installed capacity. The USA also has increased installed power during the last five decades, but in a more moderate way in the last three decades, while in Asia there is a strong growth in energy production through nuclear power plants. Although this pathway can bring environmental benefits in terms of reducing carbon emissions, it also generates waste that is extremely difficult to manage and represents a danger to public health. The security risks to neighboring populations are also an unavoidable factor, which politicians are ignoring in the absence of truly sound alternatives to power generation via thermal power stations. These plants also represent very high investments, which are not available to a large number of countries, so the spread was initially made by the developed countries and is currently being done through some developing countries (Khor and Lalchand 2014). The USA is the country with the largest capacity to generate electricity through nuclear power plants, with about 805 TWh installed in 2017, which accounts for about 20% of its global electricity needs. However, France has about 379 TWh of installed capacity, which covers about 75% of its electricity needs, with the country having its electricity consumption more reliant on own production in nuclear power plants (WNA 2019).

8.3 Chemical Sector

Some of the biggest environmental disasters are linked to the chemical sector, as described above. The development of the chemical industry with a view to solving some of humanity's problems has also raised other concerns. The lack of studies on the life cycle of some compounds, as well as their detrimental effect on the environment, has led to a considerable degradation of the environment, which now needs to be reversed. This section will describe some of the advances made in the development of new materials, compounds, or processes, in order to replace those that were manifestly harmful to the environment. Some of the products are based on hydrocarbons because, in addition to fuels, they have been abundantly used for other purposes, namely in the design of raw materials as a large part of polymers, or as lubricants, among many others.

8.3.1 Lubricants

Lubricants are an extremely demanded product, since their use avoids considerable losses of energy, being therefore an element that also contributes in a clear way to sustainability, avoiding wasted energy. Indeed, lubricants are an essential product in all situations where there is slipping contact between moving parts in mechanical systems, and the materials involved do not exhibit self-lubricating characteristics. Over a decade ago, Mang and Dresel (2007) estimated annual lubricant consumption to be between 30 and 40 million ton, with about 55% of this amount contaminating the environment for many years.

Most of the lubricants used are based on petroleum products (Mang and Dresel 2017), which entails increased care to avoid spills and evaporation, with the problem of having a very long life cycle (Panchal et al. 2017). However, growing concerns about the environment have put a negative note on the use of synthetic lubricants based on petroleum products. Thus, synthetic lubricants have gradually begun to be replaced by biolubricants, which are named primarily because they are produced on the basis of live species, in particular plant oil and animal fats, presenting good biodegradability properties and not presenting any type of toxicity for humans. In the case of lubricants based on vegetable oils, they may be used in their pure form or chemically modified with a view to improve properties and fields of application. An estimate recently issued by the US Department of Agriculture (USDA) points to a production of vegetable-based oils in the 2016–2017 biennium of about 185 million Mton, the largest ever produced (USDA 2016). Development work carried out in the last decades, when the environmental issues began to be placed with greater acuity, allowed to verify that lubricants based on vegetable oils can have a high flash point, a perfect suitable viscosity index, elevated lubricity, low evaporation losses, good shear resistance, and very good adhesion to metal when compared to mineral oils. This is due to the hydrocarbon long molecules with 4–32 carbon atoms with a carboxyl group associated on one end, which are able to interact with metal surfaces, forming a metallic soap layer with low shear stress able to improve the movement between metallic moving parts (Soni and Agarwal 2014). In addition, these long molecular chains make vegetable oil an amphiphilic surfactant par excellence, advising that it can be used in boundary lubrication (Jain and Suhane 2012). Nevertheless, the properties of vegetable oils are conditioned by the presence of fatty acids in esters, which will govern some of its main characteristics, such as viscosity and oxidation resistance. More recently, studies have been carried out to provide market alternatives that are biodegradable and renewable, based essentially on natural products, using sustainable processes and allowing performances identical to those the market was accustomed.

Transesterification is a reaction process capable of transforming an ester into another ester by changing the alkyl group due to an excess of alcohols. However, the amount of alcohol is a critical factor in this reaction because if the alcohol ratio is not adequate and is in excess, recovery of glycerol will be more difficult (Schuchardt et al. 1998). When applied to vegetable oils, transesterification gives rise to the synthesis of various alkyl esters of fatty acids by the reaction of these oils with alcohols of different molecular chain lengths (Asadauskas and Erhan 1999). The final results will be fatty acid alkyl esters which derived from vegetable oils due to base-catalyzed reactions, and can be used as lubricants in several applications while being environmentally friendly. The base-catalyzed reactions are most commonly used commercially because their reaction velocity is about 4000 times faster than acid-catalyzed reactions, making the latter less economically profitable. It is further noted that acids used in acid-catalyzed reactions, such as sulfuric acids, hydrochloric acids, and phosphoric acids, can promote corrosion within the reactors, which makes their use even less desirable.

Another reaction process used to produce lubricants from vegetable oil is epoxidation. An epoxy is a cyclic ether with three atoms forming a ring. This ring roughly defines an equilateral triangle that makes it extremely tough. This tight ring makes epoxy more reactive than the other ethers. The synthesis of epoxide groups is processed through a reaction known as epoxidation, wherein an alkene is reacted with an organic peroxy acid. This reaction is normally used in unsaturated fatty acids, taking advantage of their double bonds to provide the necessary reactions. Unsaturated fatty acids have as main limitation the fact that they are not suitable for high-temperature applications. However, this situation has been improved in recent years by the inclusion of some additives also from vegetable oils, such as antioxidants and pour point depressant, which has allowed the use of these lubricants in applications with higher working temperatures (Saurabh et al. 2011). However, chemical modification seems to be the most promising technique to increase physical properties and oxidation resistance of vegetable oil-based lubricants, allowing them to extend the range of temperatures at which they can operate (Syahir et al. 2017).

In order to overcome the frequent use of solvents in the production of lubricants, a new methodology has been used: enzyme-catalyzed synthesis. Indeed, the enzymes have also assumed a prominent role in the development of biolubricants. In the last decade, various formulations of lubricants based on natural vegetable oils have been developed (Panchal et al. 2017; Syahir et al. 2017). Based on the above, the study of biolubricants has gained an increasing number of researchers interested in its development. Within the group of enzymes, lipases can be found which present as one of the main characteristics the possibility of catalyzing various types of reactions, such as ester synthesis, carbohydrate modification, fatty acid enhancement, and even biodiesel production. Since, in the case of liquid enzymes, the rate of ester production increases with the chain length of the reactants, the production of long chain biolubricants becomes more profitable. Thus, the soluble lipase is clearly more advantageous than the immobilized lipase enzyme or chemical catalysts, commonly used in the production of lubricants (Trivedi et al. 2015).

Some two-decade-old studies had already shown that *Candida rugosa* lipase could be successfully used, because transesterification of trimethylolpropane lipase with rapeseed fatty acid methyl ester to triester was confirmed showing very interesting results: 64% of the trimethylolpropane was transformed to the biodegradable triester trimethylolpropane in 24 h under very simple conditions: 5.3 kPa, 47 °C, and 15% of added water (Uosukainen et al. 1998). However, using a higher temperature (in the order of 120 °C) and sodium methylate as the catalyst, the production rate is significantly improved, rising from 64% in 24 h to 99% in just 10 h. It is thus clear that enzymatic synthesis has very considerable environmental advantages over chemical synthesis methods, since they consume much less energy, have a less expensive waste treatment, have moderate production conditions, do not require use of environmentally harmful reagents harmful, and can be based on relatively inexpensive raw materials (Kleinaitė et al. 2014).

Some studies have focused on the biodegradable and renewable lubricants monoricinolein and diricinolein (Guo et al. 2017; Saboya et al. 2017). However, being monoricinolein and diricinolein natural elements, they do not abound in nature, so its preparation from castor oils needs to be prepared synthetically. These lubricant products can be prepared from vegetable oils through chemical processes (lipase-catalyzed esterification), which involve ricinoleic acid preparation from castor oil, using methanol as the reaction substrate. As an alternative to the preparation of these same lubricant products, glycerolysis of castor oil may also be used. However, glycerolysis as an industrial process incorporates the use of inorganic catalysts and relatively high temperatures, contrary to the principles of sustainability that are intended to be achieved. Then, also based on castor oil, the enzymatic glycerolysis process was developed to obtain the most sustainable monoricinolein and diricinolein, avoiding the use of inorganic solvents. Using this new technology, it was possible to obtain the biodegradable lubricant through a process that is much more environmentally friendly, giving rise to lubricating products with high functional performance (Sun et al. 2018a).

The work presented by McNutt and He (2016) presents a comprehensive description of different methods that can be used for the modification and improvement of different vegetable oils. When used as a pure lubricant, chemically modified vegetable oils are more advantageous than vegetable oils in their original state, mainly in terms of reducing friction and wear due to higher viscosity, compared to chemically unmodified vegetable oils, which leads to the formation of a lubricating fluid film that exhibits greater efficiency during movement. Chemically modified vegetable oil also has a higher resistance to oxidation, which significantly reduces the formation of oxidation products, thereby reducing the amount of contact wear.

Ionic liquids, also known as molten salts in liquid state, have also been studied as lubricants or as additives to other lubricants. These substances were initially studied by Ye et al. (2001), with a view to be applied as high-performance lubricants, and since then have been heavily investigated, originating a significant number of articles and patents. Ionic liquids usually come from amino acids, carboxylic acids, or other organic sources, usually exhibiting better friction-lowering properties than mineral-based lubricants. Ionic liquids are typically composed of organic cations and organic or inorganic anions (Zhou and Qu 2017). The cations have larger dimensions, due to the long alkyl chains, and are asymmetric in structure, while the anions are smaller. In ionic liquids, one or both ions present delocalized charge, promoting diffusion of charge between ions. The electrostatic force between the oppositely charged ions is small, making it impossible to form a stable crystalline structure. Thus, the ionic liquids can remain stable in the liquid state at temperatures below 100 °C. This very singular kind of structure gives them many very interesting properties in the field of lubrication such as (Xiao 2017; Zhou and Qu 2017):

- They can be used as lubricants or additives in a wide range of applications, as there is a wide variety of organic and inorganic materials that are soluble in ionic liquids.
- The volatility shown by ionic liquids is very low, which facilitates the application of this lubricant in more demanding situations, such as high-vacuum systems, in addition to reducing the chances of contamination that normally characterize synthetic lubricants.
- Due to their ionic nature, these lubricants have a greater ability to be easily adsorbed by the surfaces they are supposed to protect.
- Ionic liquids benefit from the fact that there is currently a wide variety of cations and anions available, which facilitates the diversification and customization of lubricants as required by each application or environment.
- Lubricants based on ionic liquids can be applied in environments where the working temperature can reach 400 °C, which represents an added advantage over synthetic lubricants, which tend to decompose at this temperature range. Even in this temperature range, they present high thermal stability and low flammability.

In contact between moving parts, they have the ability to form a thin layer which exhibits very good stability even under limit lubrication conditions, due to their polar nature which allows them to be easily absorbed by the metal surfaces, decreasing by this way the contact friction coefficient (Xiao 2017). Lubricants based on ionic liquids have already been tribologically tested on the contact between metal/metal, metal/ceramic, and ceramic/ceramic pairs, presenting very good results in any of the situations. Regarding metal/metal contact, it has already been proven that ionic liquids based lubricants have a higher load capacity. Compared to perfluoropolyether- or phosphazene-based lubricants, they have a 50% higher loading capacity (Ye et al. 2001), essentially because a compact lubricant layer is formed which prevents contact between the moving parts (due in large part to the ease of adsorption of the ionic liquids by the metallic surfaces). Based on the same pair of metallic surfaces, ionic liquids also showed a coefficient of friction generated between the moving parts of about 20–30% lower than that obtained when poly α-olefin (PAO) was used under the same contact conditions (García et al. 2014). Invariably, several researchers have tested different pairs of metal alloys in contact, from cast iron/cast iron to steel/lightweight alloys, and it has been found that the coefficient of friction and wear resistance conditions are always favored when ionic liquids are applied, compared to other commercially used lubricants, for the same contact conditions (Qu et al. 2009a, b; Somers et al. 2013; Murdokhovic et al. 2013). Ionic liquids have also been tested in dissimilar contacts between ceramic and metallic materials, and the general decrease in both the coefficient of friction and wear volume has also been confirmed in comparison with conventional lubricants normally used in these situations (Liu et al. 2002; Zeng et al. 2007; González et al. 2010; Battez et al. 2013; Cai et al. 2014). These lubricants were also successfully tested in polymer/metal pairs, considering external use as well as inclusion in the matrix of the polymer utilized as counterpart in the contact, and excellent results were obtained with regard to the reduction of the coefficient of friction between epoxy resin and AISI 316 stainless steel. However, especially when used in an external way, ionic liquids have not always allowed a substantial reduction in the coefficient of friction with other polymers when in contact with metallic counterparts (Sanes et al. 2006, 2007). It should also be noted that the above-mentioned behaviors are valid for the use of ionic liquids either as neat lubricants or acting as additives to other lubricants. Ionic liquids have the great advantage of being completely neutral to the environment, a factor that has been crucial in its increasing application and constant development, also presenting as disadvantages the fact that they are an expensive option and anions present an increased chemical activity, which causes corrosion problems especially when applied to corrosion-sensitive metal surfaces. This problem is attempting to be overcome by developing the synthesis of halogen-free ionic liquids, as well as the study of other types of ionic liquids which do not contain elements that facilitate the advent and development of corrosive processes (Xiao 2017).

8.3.2 Pulp and Paper Sustainable Production

The paper industry is a sector that integrates an appreciable amount of chemicals, in the sense of turning natural wood into pulp and sheets of white paper. The impact of paper pulp production on the environment is easy to observe, because obtaining the necessary pulp implies the planting and cutting of trees, which have a relatively rapid growth, due to the strong absorption of water from lands. The pollution caused by each of these companies is directly related to the technology used, the degree of treated water reuse, the quantity and type of additives it uses in its process, and even the degree of purity of the supplied raw material (Tiku et al. 2007). Indeed, to supply cellulose, a huge area of agroforestry based essentially on *eucalyptus* is necessary, which could be used for other purposes. This reduces biodiversity, induces an increased need for irrigation, rises the risk of fire in these areas with the consequences in terms of atmospheric emissions, and promotes soil compaction through the need to use heavy machinery. In certain cases, fertilization and pesticide or herbicide may also contribute to soil and water contamination. In cellulose plants, it is common to adopt biomass boilers (wood residues), where the material considered as secondary to the production process is burned, which includes twigs, barks, knots, sticks, and others. Thus, these wastes are a lesser problem, since they are very useful as biomass. However, residues resulting from the burning of such waste in boilers must be properly controlled and treated.

However, the production of pulp also has a strong impact on the environment due to the consumption of water that the process requires, as well as the chemical agents that are used in its bleaching and the energy needed to carry out the process. In Europe, the European Union Environmental Guideline for Pulp and Paper Production obliged producers to reduce freshwater consumption through the installation of closed-loop water recirculation systems (Suhr et al. 2015). However, fully closed-loop production is a utopia, since a production system that does not produce any discharge is technically impossible in this sector because there is a permanent need to purge dissolved solids and more conventional wastes, such as ash, dregs, and slaker grits (Bajpai 2018).

On the other hand, the papermaking process can be considered as a strong consumer of energy, due to the needs that the processes require, through the boilers that generate the necessary steam to the manufacturing process. A recently developed study reports that the factor that most influences the high GHG emission is the energy consumed in the process (Sun et al. 2018b). In fact, this same study points to an emission of around 950 kg CO_2 eq/ton of paper produced, which clearly indicates that saving energy in this sector is urgent. It is noteworthy that about 50% of the environmental impact caused by this type of industry is centered on pulp production and the other 50% can be attributed to the production of paper and paperboard as final products. However, techniques have been applied to reuse resources. Thus, an important portion of the energy consumed ends up coming from internal resources, through biomass from tree branches, as from the black liquor that is produced by the process itself, as it is referred forward.

The consumption of water in the pulp industry is another major problem to be solved. It is possible to find water consumption values up to 60 m^3/ton (Thomson et al. 2001). This value can be reduced as the internal recirculation and efficiency of washing equipment and processes of recycling of alkali filtrates increase, with direct effects on effluent discharge. The minimization of liquid effluents is intended to be solved through the reduction of effluents at the source where they are generated, through recovery systems, with the concept that each effluent generating area is also responsible for its appropriate management. The use of secondary treatment of water effluents is already a common practice in the pulp and paper mills and is sometimes conceived as outsourcing. Regarding atmospheric emissions, new technologies have been used, which use the so-called black liquor burning, through the use of low-odor recovery boilers, which require the burning of black liquor. The old cascade-type evaporators, which were significant sources of odor emission, were practically banned, having been replaced by falling film evaporators, which generate highly concentrated black liquor (75–80%), also generating less contaminated condensates, which can later be used to wash the pulp or lime mud. The actions to be carried out at mill shop floor should include the transition to a closed-loop water consumption system, as well as the reuse of sludge in the preparation of the raw material. These actions will save water and fiber and generate a smaller amount of effluents, reducing the energy required for the operation of the wastewater treatment plant (Žarković et al. 2011). In the case of integrated pulp mills, where the pulp is produced and directly transformed into a final product, significant savings can be achieved in the consumption of water and energy, if the installation of vacuum extraction tanks is carried out, water used in pump sealing is reduced, and the use of freshwater in felt showers is reduced (Avsar and Demirer 2008). These actions, if properly implemented, could lead to a reduction in freshwater consumption by about 50%, an increased degree of fiber recovery and filling (about 28%), and more efficient operation of wastewater treatment plants (Žarković et al. 2011). When more efficient recycling of filtrates within the bleaching plant is promoted, the results are a lower freshwater consumption and lower effluent flows from the bleaching plant and lead as well to a mass reduction of specific emissions of elements such as adsorbable organic halogens (AOX) and chemical oxygen demand (COD) (Bajpai 2018). Ashrafi et al. (2014) developed a study about different wastewater treatments to minimize the effect of harmful products in the effluents providing from pulp and paper mills, concluding that both aerobic and anaerobic methods are perfectly suitable for this treatment. Both treatments produce good results in terms of COD removal, but anaerobic treatment generally showed better results in terms of removal of lignin and AOX. In order to increase the process efficiency of contaminant removal, it has been proven that in order to obtain better results, it is ideal to use a combination of systems, either by joining physicochemical techniques with biological techniques, or by using a combination of two biological techniques. In this way, it is possible to achieve better bleaching and lower environmental impact. On the other hand, the anaerobic or mixed/hybrid system induces a greater generation of GHG due to the methane formed in the anaerobic bioreactor, a situation that does not happen in the aerobic one, despite both emitting N_2O. However, if methane is conveniently collected, it can be used as a source of energy, reducing energy costs of the whole process and minimizing the previously reported negative effects. In terms of operation stability, the hybrid system was the one that presented better results, when compared to the individualized operation of each of the other systems (Ashrafi et al. 2014). Wastewater treatment systems may utilize aerobic and/or anaerobic biological processes to remove organic contaminants from effluents. Pulp and paper mills typically prefer aerobic processes because of their ease and cost of operation, as well as lower initial investment (Mulligan 2002). Among aerobic technologies, activated sludge and aerated lagoons are probably the most widely used systems in the pulp and paper industry (Pokhrel and Viraraghavan 2004). On the other hand, although the option for anaerobic processes in the pulp and paper industry is less common, there are some anaerobic technologies implemented because they generate a smaller amount of sludge, allow the production of biogas, occupy a smaller land area, and promote a more pronounced degradation of pollutants (Habets and Driessen 2007). Both aerobic and anaerobic processes present some limitations, which include the high amount of sludge generated by aerobic processes and the high sensitivity of some anaerobic bacteria to certain toxic materials (Ashrafi et al. 2014).

It is noteworthy to remember that paper accounts for about 40% of domestic waste generated in developed countries, much of which is tied to the packaging sector. Almost all pulp used in papermaking is produced from eucalyptus wood (short fiber), as seen in the alkaline kraft process, with significant impacts on biota and the health of the people involved. This has served as the main incentive for the adoption of new production methods with a view to the conservation of raw materials, energy, and water, as well as the use of more sophisticated controls. The strong paper consumption and associated environmental problems also led to recycling of paper, where recycled fibers function as an alternative source of raw material and are incorporated in the pulp after several cleaning steps (intended to remove paints, adhesives, and other contaminants reminiscent from the previous uses) (Žarković et al. 2011).

In an accurate analysis of the gross atmospheric emissions of cellulose pulp plants, i.e., without any treatment, it is possible to find the existence of products such as carbon monoxide, carbonyl sulfide, chlorine/chlorine dioxide, chloroform, dioxins and furans, hydrochloric acid, nitrogen oxides, particulate matter, phenols, sulfur oxides, reduced sulfur compounds, acid resins, terpene alcohols, acetaldehyde, nitrates, fungi (*Aspergillus fumigatus* and *Aspergillus versicolor*), bio-aerosols (endotoxins), chlorinated aromatic compounds, and other volatile organic compounds (including dichloroacetic acid, methyl ester, 2,5-dichlorothiophene, styrene, benzene, toluene, and xylene), among others. The effects of these products on the environment are not well-known and however are thought to be extremely harmful.

Legislation has emerged in several countries to limit the emissions of these effluents and to regulate this industry, which has had considerable effects on effluents (end-of-pipe solutions), but it is urgent to modify the processes and products used, in order to minimize negative impacts on the environment. More than a thousand chemicals have already been identified in the paper production process, which include organochlorine compounds, heavy metals, acids, and resins.

Recycling of sludge and solid waste has been a common practice in pulp and paper pulp producers and has been one of the sector's recent bets. Segregation of waste by type and generating source, for separate treatment, is already common practice in industry. Recently, Singh et al. (2018) studied the incorporation of deinking paper mill sludge in the manufacture of bricks for the construction of buildings. In such sludge, various fluxing agents, such as Fe_2O_3, TiO_2, CaO, MgO, and K_2O, have been found which are useful in the brick firing process, carried out at about 950 °C. In order to maintain the compressive strength properties of bricks specified in the standards, the maximum quantity of sludge of this nature admissible in the mixture is 15%. It was also reported that water uptake increased with sludge content as well as the porosity of the set. In addition, the thermal conductivity of the bricks decreased as the sludge content from the deinking process was increased. This incorporation does not drastically reduce clay consumption, but it prevents sludge from being deposited in landfills, giving this type of industry a cleaner and environmentally friendly character.

As mentioned above, the process of manufacturing cellulose involves the use of various chemicals, which increase risks to the environment and population health. Until the end of the twentieth century, bleaching of paper pulps, regardless of their origin in soft or hardwood, employed large amounts of chlorine and chlorine-based chemicals. Currently, most pulp and paper mills worldwide use chlorine dioxide (ClO_2) as the elemental chlorine-free (ECF) bleaching agent for the production of high-quality white paper (Bajpai 2012; Singh et al. 2019). Chlorine gas, for example, is used in the bleaching of cellulose pulp in older units, usually in liquid form (dissolved in water, in the form of sodium hypochlorite or hypochlorous acid). Given the danger it presents, it has been progressively abandoned as it poses risks to business assets, the environment, and workers' health. These effects are proportional to the concentration of the product and the time of exposure. Chlorine dioxide is widely used in bleaching processes of ECF kraft pulp. It is usually used in the form of an aqueous solution, being a powerful biocide. Among the many harmful effects that it has, some are: the ability to become an explosive gas in concentrations greater than 12% (in air), corrosiveness, high ingestion toxicity and harmfulness towards plants and animals (with aquatic life being particularly sensitive to its effects when in concentrations higher than 3%). Given the high organic content of the wood pulp that is mixed with chlorine dioxide used in the bleaching process, organochlorine compounds are formed, which subsequently need to be discharged as effluents. It has been reported that these organochlorine compounds, classified and measured as AOX, can cause genetic and reproductive impairment in living beings, both in the aquatic and terrestrial environments, and even in humans (Sharma et al. 2014). According to the digestion process adopted, the production of cellulose involves the use of large amounts of sulfur salts (sodium sulfate, in the case of the kraft process), which during the digestion process undergo chemical reactions, transforming into reduced compounds of sulfur, responsible for the characteristic odor of pulp mills, which is due to the presence of substances such as methyl mercaptan, perceptible to human smell even at low concentrations. In addition to potentiating problems with the community surrounding the plant, certain sulfides have the property of drastically smoothing the smell when above certain concentrations, which can lead to problems of intoxication or suffocation in situation of high concentration. Sodium hydroxide is one of the strongest alkaline substances and is added to the digester for the extraction of lignin in the kraft process. It is another highly aggressive agent to the human health, which when in direct skin contact can cause severe burns, with deep ulceration. Eye contact avoidance is also imperative as it may cause severe irritation, leading to light scars or even permanent blindness, while inhalation might cause pulmonary edema. Cases of severe exposure can even lead to death. There is also a wide variety of additives used to improve paper properties or process efficiency, which include detergents, bleaches, anionic and cationic polymers, dyes, antifoams, resins, and many other compounds. Several of these substances may have toxic and/or irritant effects, which makes it essential to know their potential effects on human health and on the environment, as well as the emergency procedures in case of accidental spills, contamination, or intoxication.

Various alternatives have been considered to ban chlorine from the process, eliminating the above problems. For this, the use of extended cooking or the delignification based on the introduction of oxygen, hydrogen peroxide, or ozone was considered. However, any of these alternatives has very high costs to be implemented on a large scale, as entails significant changes in the process. Thus, although they have been explored since 1992, enzymes emerge as an economically more viable and technically adequate alternative to reduce the use of chlorine dioxides and other chemicals in the bleaching process, allowing higher paper glossiness levels. Enzymes are usually produced by mesophilic organisms, such as fungi, molds, yeasts, and various types of bacteria. This type of mechanism of enzyme production confers a lower thermal and pH stability, as well as a lower consistency in the presence of salts. In addition, there is an added difficulty in customizing enzymes depending on the substrate used. There is still open field for an improvement in the catalytic performance of enzymes through the bioengineering of proteins (Singh et al. 2019). The first report of the application of enzymes in pulp bleaching date back to 1994, when Bajpai et al. (1994) found that treatment of eucalyptus pulp with Novozyme 473 xylanases and Cartazyme HS-10, already commercially available on the market at that time, allowed a reduction in chlorine consumption by 31% and considerably increased final paper brightness. The process does not involve large investments in altering the process, and the xylanase enzyme as prebleaching agent is already being used industrially in the USA, Europe, and Asia (Bajpai 2012; Singh et al. 2019). Prebleaching is a more environmentally friendly process that is responsible for the removal of lignin, saving 10–15% of chlorine and other chemicals (Garg et al. 2011), while the bleaching process is then used for the removal step of 5–10% residual lignin which still remains in the pulp after the above-mentioned prebleaching process. The use of xylanase in the process of paper bleaching, increasing its brightness, and attempting to reduce the consumption of chemicals in the process has been extensively studied (Beg et al. 2000; Sindhu et al. 2006; Ayyachamy and Vatsala 2007; Senthilkumar et al. 2008; Nagar et al. 2013; Gupta et al. 2015b; Kumar et al. 2016; Kaur et al. 2017). After the xylanases were extensively explored, researchers' attention was focused on the laccases, which have been the most exploited enzymes for biobleaching pulp in recent years. These enzymes are oxidative biocatalysts that have attracted the attention of researchers due to their exceptional properties as bleaching enzymes, and have also motivated special attention from several researchers (Singh et al. 2008, 2010, 2015b, 2019; Sharma et al. 2014). Sharma et al. (2014) also performed experiments with this enzyme, noting that laccase, when used alone, was able to reduce the use of ClO_2 by 25%, whereas xylanase had only been able to reduce the use of ClO_2 by only 15%. Subsequent experiments using these two enzymes together allowed to see that the effect of reducing the use of ClO_2 was further increased, achieving a reduction of 32%. The bleaching process is still followed by several steps of paper whitening operations that normally use chlorine dioxide as a bleaching agent, which is an elemental chlorine-free (ECF) element, but this practice is still not universally implemented (Singh et al. 2019). Normally, the use of elemental chlorine-free agents is preceded by the use of oxygen delignification (ODL), in order to improve the efficiency of the bleaching process and increase paper brightness. Some researchers reported that biobleaching of pulps by laccases in the absence of a mediator component is not feasible due to the lower redox potential (E0) of laccases relative to non-phenolic lignin structures (Camarero et al. 2007; Singh et al. 2015b). It is necessary to realize that the use of enzymes depends on the type of raw material that is being used to produce the pulp. This restriction may prevent enzymes from being cultivated in a more universal way, making necessary to customize the enzymes according to the type of wood with which the process is being fed (Sharma et al. 2015). The investigations proceed in the sense that the paper bleaching process may be completely chlorine-free. To this end, different enzyme recipes need to be studied in order to achieve this goal in an economical way, which would clearly benefit the environment. The production processes of the enzymes mentioned above should also merit the attention of researchers in the coming years as only an expeditious and economical process will allow the use of enzymes to be widely applied, effectively reducing environmental concerns in this sector.

Recently, rice straw has also been exploited as a raw material for the production of pulp in order to increase environmentally friendliness and avoid deforestation problems, taking advantage of agricultural waste (Elhelece 2018). The successful application of this reuse makes the pulp industry more sustainable. However, a study recently published by Sun et al. (2018b) clearly states that the production of straw-based pulp has a very high environmental impact, mainly because of the energy involved in the process. Pineapple leaf fiber was also used as raw material for pulp and paper production, along with newspaper waste, in order to produce recycled paper that could present a quality level acceptable by market standards, requiring less forest harvesting. It should be noted that about 10% of the paper produced in world terms is no longer dependent on the harvesting of trees, and the alternative raw material depends on the most appropriate resources available in each region or country. In the case of China, the alternative raw materials to wood pass through wheat straw and other non-wood fibers (Holik 2006). A more rigorous analysis of its properties has led to the conclusion that this combination of raw materials gives rise to good quality board which can be perfectly used in the manufacture of packaging. However, it is not suitable for the manufacture of relatively thin sheets of paper. The process also proved that the product obtained is feasible, since the costs and ease of production are quite advantageous (Yusof et al. 2012).

8.3.3 Paint Products

The manufacture of paints is another large sector of the chemical industry. Paints enrich the surfaces on which they are applied, enhancing the aesthetic quality of the world we live in, as well as they extend the life span of many facilities or goods which are part of our daily lives. However, in addition to the more basic functions, other equally important functions should be added such as the possibility of providing illumination, promoting waterproofing of surfaces, contributing to the safety of fires or similar accidents, assisting in visual communication, assisting in hygiene, providing a decorative environment, and preventing or delaying the wear of a substrate (Sichieri and Salado 2007).

Two or three decades ago, in developed countries, environmental concerns and the health of populations regarding this type of industry were summarized. In the meantime, legislation in more developed and developing countries has begun to restrict the application of certain products, minimizing the risks to public health and avoiding effluents that could contaminate water reserves and streams. Presently, there are two types of paints that are the subject of environmental concerns: paints and varnishes used in construction and household goods, and inks used in paper and cardboard printing. In this section, priority will be given to paints used in construction and household goods, as they represent a greater volume of consumption. However, paper printing inks will also be reviewed. Products such as decorative or protective paints and printing inks are produced from the combination of various components that may have fossil, mineral, biological, or synthetic origin. By themselves, these components already promote a certain environmental impact, even before they are delivered to the paint manufacturers. To this initial environmental impact, the impact caused by the transportation, application, and drying of this paint must also be added. On the other hand, paint applied to buildings and consumer goods helps to protect and extend the useful life of the products, so this positive impact must also be taken into account and needs to be subtracted to the negative impact mentioned above. In order to meet customer expectations, there are certain properties that must be present in the paint film, such as low absorption and good water resistance, water vapor transfer resistance, resistance to ion passage, resistance to chemical agents, high dry and wet film adhesion, resistance to abrasion, good elasticity and elongation capacity, and finally, resistance to bacteria and fungi. These properties are extremely important and condition paint composition, sometimes leading to the addition of harmful elements, jeopardizing the environment and the health of populations. In order to make a fair assessment of the sustainability of applied techniques or products, it is necessary to consider the entire life cycle of the coated or printed product, in order to assess the overall environmental impact. The concept of life cycle thinking is critical to understanding the upstream and downstream implications of any change, anywhere in the supply chain.

Most of the manufacturing processes of paints for buildings, maintenance, and household appliances, as well as the automotive industry, occur according to the following stages: (a) separation and weighing of raw materials; (b) premixing of raw materials; (c) milling; (d) adjusting the composition of the raw material; (e) color fine-tuning with pigment; (f) quality control; and (g) filtration and packaging. Taking into account the toxicity, organic solvent-based paints have a much higher intensity than aqueous-based paint, even though both paints present some toxic portion in their content, some organic solvents have some components that are highly harmful to humans, and regarding solvent-based paints, varnishes used in woods and metals are also included. Even though there are currently several types of paints based on organic solvents, aqueous base, among others, they are produced with basic raw materials. The basic polymeric compounds of paints are made with elements such as solvents, additives, resin, and pigments. The most important part of the fixed (or solid) of a paint transfer vehicle is composed of the oils and resins which make it up, and their characteristics are determined by these same resins. Hence, it is common to use the name of the resins that make up the fixed vehicle when characterizing the paint systems. Synthetic resins are the most used in the production of paints. Examples of that are alkyd resins, which consist of an oil-modified polyester. Phenolic resins, chlorinated rubber, epoxides, acrylics, nitrocellulose, polyurethane, amines, and polyester, among others, are also used in the manufacture of synthetic paints. Solvents are organic compounds and have an enormous capacity to dissolve some materials, without interfering or altering chemical properties. The name given to this interaction is solubilization. Through this line of study, plasticizers can be mentioned, which also belong to this class of products. However, their high boiling point classifies them as nonvolatile materials with regard to environmental requirements, unlike the so-called volatile solvents. The use of the solvents in paints has two totally different purposes: to give the correct viscosity for a good and easy application and to promote resin solubilization, promoting film formation. Some properties that characterize solvents are: surface tension, toxicity, odor, evaporation rate (biodegradation), flash point, among others. The solvents used in paint manufacture are classified into two types: hydrocarbons and oxygenates. The first are constituted of elements such as hydrogen and carbon, while the second has oxygen atoms in a molecule that directly assists in increasing the polarity of the compound.

The most worrying paint pollutants are lead and volatile organic compounds, in addition to others that are added as additives, but are therefore present in lesser amounts. Although lead appears to have been banned from the paint industry, recent studies show that this is true only in developed countries (Kessle 2014; O'Connor et al. 2018). Indeed, a study developed in 2017 by Wu et al. (2017) has shown that, despite the limitations imposed on lead use since 1986, Chinese paint products still provide large quantities of lead-based paint to the market, ignoring international recommendations. The consequences of lead exposure, especially of children, are well known, including influence of intellectual development, and may lead to death through poisoning. Although it is internationally established that the range of use of lead-based products should be between 90 and 600 mg/kg, 90 mg/kg being the most conservative and most advisable value, many companies in developing countries are producing paints with several tens of thousands of mg/kg of lead concentration. The problem now arises over non-compliance with international specifications, but there is a further risk of exposure during paint removal operations, which could also cause major public health constraints. Despite the efforts made at the International Conference on Chemicals Management (ICCM) in Geneva in 2009 for lead-based products to be irradiated from paints in the year 2020, only 71 of the 193 UN member states actually set limits on such use, regarding paints (UNEP 2018). O'Connor et al. (2018) report that about 50% of countries still exceed 600 mg/kg of lead-based products in the manufacture of paints, so it is urgent that legislation be more restrictive in these countries because, particularly in West Europe, North America, and Oceania, the recommended range of values is perfectly complied with, ensuring good paint quality. Lead has represented an essential component of paints for many years, essentially because of the durability characteristics it imparted to them, as well as very good paint/substrate adherence (Gilbert and Weiss 2006; Lin et al. 2009). Lead has been used in various forms, namely as pigment for white or yellow paints, using as pigments lead carbonate ($PbCO_3$) and lead chromate ($PbCrO_4$), respectively. Lead carbonate can also be used as a coating facilitator, avoiding the appearance of cracks by helping the coating remain compact and flexible (Crow 2007).

Volatile Organic Compounds (VOCs) which are included in paint producrs are another problem for the environment. Indeed, paints and varnishes contribute consistently to the worsening of air quality, both in the areas surrounding their manufacturing plants as well as where they are applied. As stated by the World Health Organization (WHO), VOCs are all organic compounds that exhibit a boiling point in the range of 50 and 260 °C (Liu et al. 2015). In order to obtain more environmentally friendly paints that do not pose health problems, low VOC paints should be manufactured and selected, which are characterized by VOC levels below 75 g/L. When a wall is painted, the solvent evaporates, promoting the drying of the paint. If a paint contains VOCs, as it dries it continues to release toxins, which can happen for a long time. Even the continued exposure to low levels of VOCs has the potential to cause eye, nose, and throat irritation, headaches, loss of coordination, and nausea, among other more serious diseases. In solvent-based paints of hydrocarbons, VOC levels can reach 350–450 g/L, which may include benzene, toluene, xylene, glycol ether, phenol, formaldehyde, and methylene chloride. All these components are carcinogenic. Water-based paints typically have VOC levels between 30 and 80 g/L. Even water may act as a solvent, and paints may continue to contain potentially hazardous components. On the other hand, low VOC paints typically contain VOC levels between 1 and 75 g/L. Several types of VOCs are photochemically sensitive, which promotes the formation of ozone and other products when they are exposed to nitrogen oxides and sunlight (You et al. 2017). Natural paints are paints with low content or even VOC-free and without any petroleum or hydrocarbon by-products, replacing these ingredients with natural oils such as safflower oil or linseed oil, and isoalifate, a less harmful organic petrochemical solvent. Natural paints also have the great advantage of letting walls breathe, due to a greater permeability to the passage of gases. Although water-based paints cost more than conventional synthetic paints, prices have been converging, so there are less and less reasons to choose health harming paints.

8.3.4 Other Products in the Chemical Industry

Catalysis is a widely used process in the manufacture of chemicals that are produced on a large scale, as it favors reactions, making them incomparably faster, allowing products to become sufficiently competitive to be used in a standardized way. In fact, the production of inorganic chemicals such as sulfuric acid (H_2SO_4), with a production of around 200 million ton worldwide annually, is carried out through the aerobic oxidation of sulfur dioxide (SO_2), originating sulfur trioxide (SO_3). This process uses a temperature generally comprised between 400 and 500 °C and a pressure slightly above atmospheric pressure and usually uses as homogeneous catalyst a silica-supported solution of vanadium oxides in molten salts. On the other hand, ammonia is also produced and consumed in large quantities, in the order of 130 million ton/year, worldwide, being manufactured on the basis of the reaction of hydrogen (H_2) with nitrogen (N_2) in the gaseous form, using a heterogeneous iron oxide-based catalyst under relatively high pressure and temperature conditions (150–250 bar, 400–500 °C, respectively). In addition to a number of industrial uses, a considerable part of the production of these inorganic chemicals is linked to the manufacturing of commonly used inorganic fertilizers in agriculture and is the basis of the growing food production that fuels the equally growing world population, although with the corresponding environmental damage previously mentioned. However, many other inorganic chemicals are also produced by the same route, namely pharmaceuticals and pesticides. A large proportion of the most commonly used polymers, such as polyethylene and polypropylene, also come from organic chemicals, and their production is also carried out on the basis of homogeneous or heterogeneous catalysis. They are used in products as common as ordinary plastic bags, numerous packaging, labeling, textiles, and plastic films, among many other applications, each year consuming between 50 and 80 million ton of each of these plastics. For the production of these polymers, catalysts such as titanium (III) chloride or chromium (IV) oxide on silica are used. In fact, since its initial development, which was developed by J. J. Berzilius in 1935, catalysts have experienced a less positive phase for the environment by aiding in the production of inorganic chemicals which, in spite of aiding in the development of agricultural production, also greatly contributed to the pollution of soils and waterways. However, in the last decades, catalytic reactions of reduction have been developed that have greatly aided the environment. In fact, the three-way catalyst (TWC) can be considered as something that has considerably impeded the increase of the degradation of the environment by motor vehicles. This catalytic converter, which is mandatory for petrol-driven car engines in developed countries, converts through well-known catalytic reactions, three gases widely known to be extremely harmful to the environment: carbon monoxide (CO), nitrogen oxides (NO_x), and non-combustion fuels (HC), in less harmful or even environmentally friendly products such as carbon dioxide (CO_2), dinitrogen (N_2), and water (H_2O). The materials used as catalysts in this process may be small particles of some metals, such as palladium, rhodium, or platinum, to which rare earths are added as stabilizers, supported on aluminum oxide with high surface area (Al_2O_3), allowing the whole function as a heterogeneous catalyst. This material is extruded in the form of monolithic blocks with extremely thin parallel holes, where exhaust gases will have to pass through, being exhausted to the atmosphere after the catalytic conversion reaction in a more environmentally friendly way. The inclusion of these catalysts in automobile vehicles resulted from smog problems in large urban centers as a result of emissions from the burning of hydrocarbons and from the formation of low-altitude ozone as a result of excessive NO_x emissions. However, the emission of oxides of nitrogen is not exclusive to motor vehicles, as about half of the emissions of this gas into the atmosphere result from other processes, such as the generation of electricity, the production of cement, or the incineration of waste. Thus, it was imperative that conditions be created to minimize these emissions, which was achieved through selective catalytic reactions (SCR) of NO_x using for this purpose ammonia as a reductant. Through this converter, the nitrogen oxides are converted into dinitrogen (N_2) using high temperatures and the injection of ammonia. VOCs are emitted by a wide variety of industrial sectors, from coffee roasting to waste incineration, which may include dangerous dioxins. With the injection of oxygen, these gaseous emissions can be efficiently transformed into products with little or no harmful effects on the environment, such as water or CO_2, through the action of a Pt-based catalyst installed in the industrial pipelines through which combusted gases need to cross. Lastly, catalysts designated as environmentally friendly should be mentioned by the fact that they are capable of removing potential pollutants before they are used. As a striking example of this type of catalyst can be mentioned, which is capable of removing sulfur in the petroleum refining process, this catalyst consists of a mixture of cobalt molybdenum oxide in aluminum oxide supports ($CoO-MoO_3/g-Al_2O_3$), which is capable of removing 85–99% of the sulfur during the refining process. Typically, the conditions required for the catalytic action go through 30–150 bar H_2 pressure at 300–400 °C. This process avoids the environmental emission of millions of ton of sulfur gases per year globally. Furthermore, this catalytic action is also responsible for a 90% reduction of NO_x emissions after the combustion process, having a double positive effect on the environment.

However, the area of catalytic reactions has seen significant improvements and may still become more environmentally friendly, notably through energy savings that can still be achieved. Currently, one of the main research routes is the use of solar energy to trigger catalytic reactions through photocatalysts. A good example of how this process that can be further explored in the future is the possibility of decomposition of water, allowing the formation of hydrogen in a perfectly sustainable way. Titanium dioxide (TiO_2) also has very interesting properties in this respect, as it has been shown to be a sun-sensitive material, allowing the fundamental redox process to promote the split-up of hydrogen and oxygen from water, by the excitation of its photons to the conductive band under the action of sunlight. This process may tend to replace the electrolysis process of water based on the supply of electrical energy, since it is a more sustainable process. The heterogenization of homogeneous catalysts through the distribution of catalytic solutions on the surface of a porous inert material has shown promising results in catalytic reactions, giving rise to a fixed-bed process and continuous flow, without the traditional problems of homogeneous industrial catalysis, which require the separation of the catalysts and their convenient recycling. Heat transfer from exothermic reactions can also be made more efficient through homogeneous catalytic systems. It should be noted however that few liquid-phase catalysts have proved to be successful in this aspect due to the evaporation of the solvent from the catalyst or the condensation of vapors, flooding the catalyst and drastically increasing the resistance to mass transport. More recently, a new category of solvents, usually named ionic liquids, have been developed which are based on salts with melting temperature which may reach values below 100 °C or even room temperature, and wherein said salts are constituted by an organic cation and an organic or inorganic anion. The hydrophobic or hydrophilic character of the solvent may be adjusted over a fairly wide range by appropriate selection of the ion pairs, thereby allowing to calibrate the solubility of the catalyst, binder, reagent, and product metals (Fehrmann and Riisager 2017).

Yates et al. (2017) studied three different pathways of obtaining value-added oxygen-containing substances from limonene, a by-product of the orange juice manufacturing process, seeking to obtain products as sustainable as possible, through an adequate selection of catalysts, reactants, and solvents with low toxicity, as well as to find procedures that have the least negative effect on the environment. The study was developed with a view to finding the most suitable catalysts for different activation means such as considering conventional heating, microwave, and solar energy. The use of citrus peel oils (Limonene) is intended to produce terephthalic acid, carvone, carveol, carvacrol, or other limonene oxides, products considered as renewable raw materials (RRM). The catalysts used were iron or manganese, titania, and sepiolite, all of which presented low toxicity. Conversion rates normally achieved in this type of production are around 35% for any of the activating media used. However, the types of catalysts and their quantity may influence the conversion rate and the sustainability of the process. An increase in activity was observed when calcined catalysts were used at lower temperatures. Lower conversion rates may be associated with greater selectivity for epoxides, being understood that these are intermediates of other products which exhibit a much higher level of oxidation. Solar-fueled heating allows conversion rates similar to those achieved through conventional heating or microwave heating and is in fact the most sustainable and cost-effective option for activation.

The inclusion of some industries' goods in consumer products manufactured by other types of industries has been the way many researchers have found to minimize the environmental impacts caused by the excessive generation of waste. Kairytė et al. (2018) studied the influence of the inclusion of paper production waste sludge in the manufacture of rigid polyurethane foams (RPUFs), showing several benefits in this inclusion. In fact, despite the density of RPUFs growing between 35 and 63%, which may be problematic for some applications, all other factors are positive; namely, flammability characteristics have been improved, decreasing from 373 to 352 kW/m^2 when 20% of PPWS was added, the release of toxic gases such as CO_2 or CO was also reduced, while mechanical strength and thermal insulation properties were improved. It is also worth mentioning that the addition of PPWS to RPUFs helps to reduce the environmental impact of this product without increasing the cost, making it a perfectly sustainable solution to recycle PPWS.

There is also an enormous diversity of small studies that aim to solve numerous specific problems, reducing the energy spent on processes, minimizing the use of toxic products, integrating waste into other products, improving their performance, etc. The edition of scientific articles does not stop growing in this area, showing solutions to problems with which we lived peacefully until now. Particularly in this type of industry, it is necessary to consider some tools that may allow relatively easy access to the way products are produced, and what their environmental impact is throughout their life cycle. The more frequent and more conscious use of these tools is a must for a future that we hope will be better for all of us.

The manufacture of a wide variety of products offers different ways of achieving the same end through the use of different raw materials, diverse manufacturing processes, and dissimilar recipes. In order to adequately address environmental concerns, one of several techniques developed to assess the impact of a given product in terms of the environment can be selected. Life cycle assessment (LCA) analysis is perhaps the most comprehensive because it focuses not only on the materials and manufacturing processes used, but also on how the product interacts with the environment throughout its life cycle. This analysis is important in all sectors, but it is even more important in the chemical industry because, as we have seen in this chapter, there are products that can release pollutants during a good part of their useful life (solvents, for example) or be difficult to recycle (like some polymers), making their degradation problematic. Pistikopoulos et al. (1994) developed another technique based on the ACL, called the minimum environmental impact methodology (MEIM), which also includes the global and local environmental impact assessments, establishing indices that take into account the environment harm and waste created by the end-of-life of the product to be discharged in either water, air, soil, or other.

8.4 Sustainability in Construction

In construction, cement is one of the most frequently used products, and in considerable quantities. It assumes particular importance in the formation of concrete, where it acts as a binder, providing the mixture with high strength. Cement production normally accompanies the development of economic activity in local and global terms. In 2016, according to data released by Cembureau (2017), around 4.65 billion ton of cement was produced worldwide. Roughly 52% of this cement was produced in China. It is known that cement production is an intensive energy-consuming industry. Some strategies have been adopted to minimize the typical energy consumption of this production process, namely by replacing the raw materials currently used by other materials that require less energy, improving the thermal efficiency in the production process and reduction of CO_2 emissions, minimizing the environmental impact of this industrial sector. In Europe, significant efforts have been made in the last two decades to reduce this sector's energy consumption, with reductions achieving around 30%. It should be noted that a significant part of the energy consumption recorded in cement production is carried out in the preparation of the raw materials for the process, in particular in particle grinding, so that they are sufficiently small. This consumption may vary widely, depending on the type of process installed, but is usually between 16.5 and 63.5 kWh/ton, where the average value is 36 kWh/ton (Tokyay 1999). The grinding process is very important because the granulometry of the raw material should not exceed a specific limit, creating better conditions for the hardening process (Duda 1977). It is also reported in the literature that the cement industry accounts for about 5% of the global anthropogenic emissions of carbon dioxide emitting about 900 kg of CO_2/ton of cement produced (Ba-Shammakh et al. 2008; Adebiy et al. 2015). On the other hand, the level of dust emitted was also drastically reduced, now presenting about 10% of the values that were typical of this industry twenty years ago. This is due to a strong investment that this industry has made in the redefinition of processes and in technology capable of minimizing harmful emissions (IPPC Bureau 2013).

However, although still be highly harmful to the environment, the cement industry remains influential in most countries as a crucial economic development activity, avoiding the creation of external dependence in vital sectors such as construction. The cement manufacturing process includes three main steps: preparation of the raw material, clinker production, and cement production. In the preparation stage of the raw material, the extraction is initially carried out, which is followed by the prehomogenization stage and then the raw materials are ground. The second stage takes place in the cement kiln, where the clinker is produced by a chemical reaction caused by heat supplied from burning fuel. The clinker is then produced thanks to the chemical reaction between the raw materials introduced into the furnace after previous homogenization and the ashes of the fuel used in the cement kiln. This clinker is then stored in silos suitable for this purpose. Only after is the production of the cement itself carried out, where the clinker is mixed with grinding aids to make the final fine-tuning of the composition. The cement is then ready to be shipped. About 50% of the CO_2 emissions produced in a cement plant come from the decarbonation reaction of raw materials. Limestone is usually used as raw material, which is converted into lime. Other raw materials, such as materials rich in calcium carbonates ($CaCO_3$), may also be used. The combustion of the fuels used in the cement kiln contributes 40% to the total CO_2 emissions produced by this process (Benhelal et al. 2013). In terms of the electrical energy consumed in the process, about two-thirds are absorbed by the raw material grinding process (used to reduce the grain size of the particles) (Tsakalakis and Stamboltzis 2008). Since this type of industry consumes a lot of energy, it is fundamental to study and improve its thermal efficiency. Based on this objective, a number of studies have been carried out to minimize energy expenditure, meeting two other goals: reducing greenhouse gas emissions (GHGs) and reducing cement production costs. In order to optimize the cost of operating cement production units, Kookos et al. (2011) developed a mixed integer linear programming model that could be adapted to various circumstances. This model allows to enter different important variables in the process, such as the raw materials to be used, the incorporation of fossil fuels, and waste derived fuels and to analyze the best combination to be used in order to reduce operating costs. The model developed was based on the need to optimize a company in the sector, and five different types of raw materials were selected, as well as six different types of energy sources. The model proved to be able to be applied under these conditions, providing positive results in terms of operating costs, bringing as well environmental benefits. Indeed, the application of the model to the real case above-mentioned allowed the introduction of a new factor: The hypothesis of using waste derived fuel as co-firing in cement kiln was considered. Several fossil fuel substitution rates have been studied, starting at a replacement rate of 10% and ending at a replacement rate of 32%. This was considered the maximum feasible fuel substitution rate in the company-installed process used as a case study. Given that the company considered had an installed production capacity of 2 Mton clinker/year, using a fossil fuel substitution rate of 17%, it would make possible to save about M€1.67 annually. More importantly, the environment would receive 112 kton less of CO_2 emissions per year from this company, which is a noticeable gain in environmental terms.

Carpio et al. (2008) performed similar work using other mathematical principles, which included the quality of the cement produced and environmental concerns. Regarding these goals, they used a methodology called particle swarm optimization (PSO), which is based on a Cauchy and Gauss distribution, taking into account several constraints of the process. Through the developed model, it was verified that it is possible to add wastes to the fuel used in the production of clinker, reducing the cost of cement production without creating environmental problems. In this case, the quality of the produced clinker was quantified through the analysis of its chemical composition, in order to study the influence of the inclusion of wastes as co-fuel, while the environmental impact induced by this inclusion was measured through CO_2 and SO_2 emissions. Later, Oyepata and Obodeh (2015) also used the PSO methodology to elaborate a model dedicated to the optimization of the working conditions of a cement production unit, comparing the results through a genetic algorithm (GA) and pattern search. The model was validated through a case study where mineral coal, pet-coke, and heavy oil were used as main fuels. As alternative fuels, some agricultural waste, such as sugar waste, rice husk, and ground shell, were considered. The study aimed to optimize the process by analyzing three main variables: quality of the cement produced, environmental benefits achieved, and cost/benefit ratio for companies wishing to use alternative fuels. The application of the model to the case study allowed to discover that the use of an alternative fuel could induce reduction in the cost of cement production between 30 and 70%, without the quality of the final product being significantly impaired. The authors also noted that these results can be obtained together with a benefit to the environment, which is not quantified in that work. The study performed by Adebiy et al. (2015) focused on a significant reduction in CO_2 emissions. Considering the facilities chosen to validate the developed algorithm, this savings represents about 100 kton of emissions per year. Indeed, the total amount of emissions released annually by the company considered in that study is 549.471 kton. Four hypotheses have been studied to mitigate emissions: to increase thermal and electrical efficiency of installations, to use alternative fuels, to replace carbon-intensive clinker with lower carbon materials and good cementitious properties, and finally carbon capture and storage. Bearing in mind all considered options, it is possible to achieve an 18% reduction in CO_2 emissions, but this leads to a 9.1% increase in overall production costs and 14.3% in total operating costs. The results of this study could help cement plant managers significantly reduce carbon emission levels to the atmosphere by implementing various retrofitting options to achieve desired results, significantly improving their environmental performance. Obviously, obtaining these results implies a considerable technological update in the company's production system. A similar study was also developed by Ba-Shammakh et al. (2008), which aimed to establish a mathematical programming model that would be able to establish the most appropriate strategy to mitigate CO_2 emissions, at the lowest possible cost. The objective was twofold, trying to minimize operating costs, as well as the costs of investing in new technologies. For this, the model considered essentially three variables: the CO_2 generating sources, the efficiency of control options, and cost. Research has led to the conclusion that the use of different technologies to improve emissions presents good solutions for emission reduction rates of up to 10%. In addition, the model also led to understand the actions that need to be implemented, according to the level of ambition in the reduction of emissions. If the targets are more ambitious, such as a 20% reduction, then a change of fuel will have to be considered. For objectives above 20%, it is necessary to adopt carbon capture technologies, since the technologies considered for efficiency improvement were not sufficient to achieve such levels of mitigation. The study also showed that cement production costs increase significantly when emission reduction targets exceed 20% because carbon capture technologies are the costliest. In fact, an analysis of the cost impact on cement produced with such ambitious low emission levels generated a 57% price increase. However, in a more recent study developed by Ishak et al. (2016) using mixed integer linear programming (MILP), much more interesting results were obtained. In this case, the variables considered were the use of co-fuels, kiln improvement, and the use of carbon capturing technologies, the latter as in the previous case, but considering the technological developments that have taken place in this field. The study showed that the highest rate of CO_2 reduction possible to achieve, taking into account all the combinations considered in the investigation, is 79%. For this emission reduction rate, the cost of clinker will be USD 136.45/ton, taking into account the necessary investments, and the CO_2 emitted to the atmosphere will be 179.90 kg/ton of clinker. In the study, other scenarios were considered that allowed a final clinker cost of USD 90.21, but the reduction of emissions did not meet the desired minimums. The optimum composition for cement production was found to be 1309.32 kg/ton clinker limestone, 37.24 kg/ton clinker clay, 128.19 kg/ton clinker sand, and 15.03 kg/ton of clinker iron source. Fuel consumption is 42.47 kg/ton clinker coal and 17.06 kg/ton tire-derived fuel clinker.

Municipal solid waste (MSW) is a difficult problem to solve, especially in large urban centers. However, these wastes have very interesting energy levels, which can be converted into heat. It should be noted however that incineration generates by-products, such as ash and carbon emissions. Thus, it is necessary to evaluate if it is feasible to use the energy contained in MSW, as well as its ash and carbon dioxide in cement manufacturing, adapting the manufacturing process through the insertion of new steps and procedures. In a paper recently published by Ghouleh and Shao (2018), the cement was synthesized using more than 85% ash and a temperature of 1000 °C, conditions that meet what is typical in an incinerator. It was observed that cement produced in this way presents a superior ecological performance when comparing to traditional Portland cement as it does not need such an intensive use of traditional resources, consumes less energy, and emits less carbon to the atmosphere. Although it does not show any hydraulicity, it can be easily activated using carbon dioxide. Some samples were tested, allowing to realize that after two hours of activation through carbonation, the average compressive strength of the cement obtained by the described method was 53 MPa, while the sequestered CO_2 was 6.7%. This process represents a huge benefit because it gives use to wastes that would have to be incinerated anyway or landfilled. In addition, it is further demonstrated that the cement obtained mainly from ash originating from the incineration of MSW has a high reactivity in the presence of CO_2, also having a higher bond strength. It has further been found that this process requires fewer virgin additives and can be carried out within the typical temperature range of a traditional incinerator. It should also be mentioned that incineration can take place in the same place as the production of cement, allowing the reuse of heat generated in the process, thus saving a greater amount of energy.

A study developed by Hassan et al. (2019) aimed at promoting the valorization of wood biomass ash through its incorporation in one-part white geopolymer cement, using diatomite as its main precursor. In order to produce white geopolymer cement capable of hardening at 23 ± 2 °C after being mixed with water, a dry activator was prepared which has a white color and elevated calcium content, mixed with natural diatomite, also presenting a white color. This dry activator is prepared using NaOH-treated $CaCO_3$-rich wood biomass ash according to different $NaOH/CaCO_3$ molar ratios. The pretreated ash content resulting from the burning of the wood biomass added to the diatomite should increase as the $NaOH/CaCO_3$ molecular weight decreases to the equivalent of 3 wt% in all mixtures. Tests made possible to detect different properties of the cement prepared by this process, depending on the added ash content. Percentages of ash ranging from 8 to 43 wt% were tested. For those addition limits, the 28-day tests allowed to obtain compressive strength between 36 and 27 MPa, respectively. A more detailed study allowed to conclude that the percentage capable of maximizing the resistance of the obtained compound is 21.5 wt% of wood biomass ash, having obtained the compressive strength of 48 MPa for a same curing time.

Siddique (2010) also analyzed the inclusion of MSW incineration ash in the manufacture of cements and mortars. This study allowed us to conclude that MSW ashes can be added successfully in the manufacture of cement or mortar, within certain limits. Indeed, the production of clinker admits the inclusion of MSW ash of up to about 44%, properly aggregated with $CaCO_3$ and very small portions of SiO_2 and Fe_2O_3. However, these ashes may also be used in the composition of mortars to replace cement, but quantities greater than 10% may significantly increase setting times. It has been found that the inclusion of quantities of not more than 10% ash does not lead to a significant degradation of the compressive strength properties of the cement or mortar, so this value should be considered as the limit. It was also observed that the inclusion of these ashes in the composition of the cements and mortars does not affect the shrinkage or the loss of mass. Mortar and concrete have also been the subject of studies regarding the inclusion of waste from the pultrusion process (Ribeiro et al. 2011). The purpose of the pultrusion process is to produce constant cross-sectional profiles of polymer matrix composite material (Silva et al. 2017a, 2018a) and has deserved the attention of some researchers to improve their performance, both in technical and environmental terms (Silva et al. 2012a, 2014). Indeed, products made in this way have glass fibers embedded in thermosetting resin, which makes its recycling extremely difficult. Moreover, the typical cross-linked structure of thermosets renders recycling impossible or extremely difficult, a situation that is further aggravated by the presence of fiberglass reinforcements. In addition, the process typically generates waste, especially at the beginning of each production batch. The common destination of these wastes is landfill, which is a problem for the environment, since they are materials that are resistant to degradation. Furthermore, this also degrades the economic performance of the manufacturing process, since landfilling entails the payment of fees which worsen the product cost, implying a loss of competitiveness. Thus, some researchers have studied the possibility of grinding these polymeric matrix composites waste in concrete and mortars. Ribeiro et al.

(2011) investigated the inclusion of contents between 4 and 8 wt% of glass-fiber-reinforced plastics (GFRP) waste originated from pultrusion processes. The wastes were ground, giving rise to two different particles sizes. Samples of mortars with different GFRP contents and different particle sizes were then produced in order to study their mechanical properties. It was observed that the compressive strength of these mortars was higher the greater the weight of waste GFRP added, in any case presenting a higher compressive strength than the mortar without any addition of waste GFRP. The flexural strength measured in mortars with GFRP waste added was also higher than that presented by the original mortars. However, this property showed some dependence on the size of waste particles added. In fact, different particle size trends were observed, which were divided into coarse particles and fine particles. When fine particles are used, the flexural strength increases substantially and linearly when increasing the amount of added GFRP waste. In the case of the coarse particles, there is a reduction in the increase of the flexural strength for an addition of 8% of coarse particles. According to the authors, this difference in behavior may be related to the heterogeneity observed in the fiberglass content present in larger particles. It was also observed that for the same content of added GFRP waste, polymer mortars added with coarse particles usually present better mechanical behavior, compared to original mortars. It was also found that both types of waste particles improve ductility, clearly reducing the tendency for brittle fracture of mortars. As the main conclusion of this study, it can be noticed that the combination of waste GFRP to polymeric mortars improves its mechanical properties, eliminates the need to deposit GFRP in landfills, and allows to obtain new composite materials that may be of great interest in the construction sector. In a further study published later, Ribeiro et al. (2013) studied the incorporation of GFRP particles between 4 and 12 wt% in polyester-based mortars as sand aggregates and filler replacements, as well as the effect on mechanical properties of the addition silane as resin binder. Design of experiments (DOE) and data processing were performed using full factorial design and ANOVA. The experimental results obtained were quite promising, leaving to be expected that the recyclability of GFRP residues as partial replacement of aggregates and reinforcement of polymer mortar materials is perfectly possible and feasible, generating significant improvements in terms of mechanical properties. In this study, four different levels of GFRP waste addition were tested, as well as two different levels of silane, employed as binding agents. The addition of GFRP wastes aimed at replacing aggregate sands. Flexural strength and the compressive strength were investigated. The obtained results endorsed to conclude that both added products—GFRP wastes and silane—contribute positively to the increase of compressive resistance of the polyester-based mortar. However, it is the GFRP waste that contributes most to this increment, corresponding to 51% of the total variance for this property. In fact, the high amount of glass fiber present in GFRP wastes is believed to act as a bulking agent. On the other hand, it is verified that silane contributes with 92% of the global variance for flexural strength, showing that the amount used has a great influence on this property. The bonding effect of silane is important for the increased stiffness shown by mortars obtained with additions of GFRP wastes and silane, and this can be seen in both flexural strength and compressive strength behavior. It has also been observed that there is interaction between the additions of GFRP wastes and silane, which is reflected in the overall mechanical strength of the formulations obtained. This is particularly noticeable when amounts of GFRP greater than 8 wt% are added. Considering the different portions of GFRP wastes used in that work, it was possible to observe that the mechanical properties are particularly favored up to additions of 8 wt% of GFRP wastes. For higher values, the effect on mechanical properties no longer is interesting. Indeed, the most favorable mechanical properties increase in the polyester-based mortars was observed for 8 wt% of GFRP wastes and 1% of silane as resin binder. The results obtained in this work allowed to take into attention the addition of GFRP wastes as a desirable alternative for polyester-based mortars, which can be used as concrete–polymer-based products. In addition to improving the mechanical properties of mortars, the competitiveness in the production of profiles by pultrusion in polymer matrix composite materials is improved, and it is no longer necessary to discard the waste generated in the pultrusion process. Another similar study carried out later by the same team of authors allowed to analyze the influence of the particle size of the GFRP wastes added to the polymer mortars as sand aggregates and filler replacements (Ribeiro et al. 2015) and detail some effects on the mechanical properties. As in the previous work, an improvement was observed in both the compressive strength and the flexural strength of the polymer mortars subject to the addition of GFRP waste and silane binder. However, in this work it was possible to confirm the maximum quantities that have an effective influence on these properties, showing that the effects are not exactly the same. Indeed, as far as flexural strength is concerned, the most relevant positive effect is observed for a maximum addition of 4 wt%, while for compressive strength, the maximum improvement of this property is achieved with the addition of 8 wt% of GFRP waste. This study also showed that the addition of coarse particles leads to better results in improving mechanical strength, both in terms of bending and compression. This differentiation of effects between coarse particles and fine particles is due to several factors, namely particle geometry and intrinsic contact area. In fact, while smaller particles act as simple fillers in replacing sand aggregates, the larger particles act as true reinforcements, improving the mechanical properties more effectively. The research allowed corroborating the previously obtained results regarding the ideal proportions of additives: 8 wt% GFRP wastes and 1% silane are those which maximize compressive and flexural strength. As a main conclusion of this work, it is possible to realize that there are solutions that enhance the use of scrap from pultrusion, avoiding landfilling or incineration, and creating products with better mechanical properties that can be used in the construction sector. In addition to the economic benefits, this solution eliminates one more factor contributing to environmental degradation, making the process more sustainable. A similar study presented by Castro et al. (2014) using a different mathematical approach allowed to conclude that the best compromise for the maximization of compressive and flexural strength lies in the 6 wt% of waste GFRP added, with a 17% increase in the compressive strength and 12% in flexural strength, comparatively to nonadditive mortar samples. This study also concludes that this methodology of composite material recycling is not only valid for scrap resulting from the pultrusion process, but can easily be extended to all sectors that manufacture composite materials based on thermosetting resins, as well as in the recycling of end-of-life products, significantly improving the sustainability of the manufacturing parts in these composite materials sectors.

The stone processing sector, used for construction, also presents sustainability problems and increased environmental concerns. A study carried out by Bai et al. (2015) presented a CP practice evaluation system that allows decision makers to quantitatively assess the sustainability of their companies, allowing a critical analysis of what factors should be improved. To achieve this, a three-level structure was created, based on 6 first-level indicators and 24 subindicators, according to the context and characteristics of the stone processing. The assessment criteria were established based on the laws, procedures, regulations, and standards of pollutant discharges in China, as well as the level of technology for clean production to date. The developed model, based on fuzzy analytical hierarchy process (FAHP), was later validated through a case study. The evaluation of the sustainability situation in the stone processing sector carried out through this model showed that there was a clear evolution in this sector between 2011 and 2013. In fact, the implementation of CP strategies in this sector allowed clear advantages in terms of environmental and economic management. The use of indicators, previously developed by experts in this field of research, and the developed model allowed a deeper knowledge of the advantages obtained through the implementation of these CP strategies in this sector. Based on this knowledge, the top management of the companies will be able to make decisions and control, in a much more assertive way, the implementation of CP practices, realizing, through the indicators, their real effectiveness in the field. In addition, problem identification has also become much easier, through the analysis of the proposed new indicators and analysis methodology. In terms of management, the developed model also allows the rapid perception of deviations from the defined strategy, allowing the quick correction of procedures and the planning of improvement actions, in order to pursue the desired levels of sustainability. In any case, the authors assumed from the outset that the model can be improved because it is much more focused on the production system and production indexes than on the study and analysis of the life cycle of stone products. Another limitation found in the developed model is related to the ability to update the technologies related to CP practices. In this way, the model must have the necessary flexibility to allow technological updating and produce reliable results, not limited to the technology existing at the time of development.

The production of tiles using the traditional wet process is an industry known as highly energy-consuming, promoting considerable pollution to the surrounding environment. About 50% of the tiles produced worldwide are manufactured in China. Much of the development carried out through the work of Shu et al. (2010) was performed around the granulator. In this equipment, the ceramic suspension is directly sprayed into fine suspension droplets. However, dry ceramic powder is blasted onto the suspended droplets, enveloping them, isolating them, and promoting their absorption. In this way, coarser grains are formed, which are rolled, tamped, and dried, using for that purpose pressed powders. Laboratory experiments were implemented on the preparation of pressed powders using this new developed technology. The analysis of the obtained powders allowed to observe that they had a quasi-spherical shape, with grain size following a normal distribution. The measured bulk density was relatively high, and the flowability shown was quite favorable. These properties confirm that the granules obtained with this process are very similar to those obtained with the wet process. However, although the properties are very similar, the effects on the environment are far less harmful. In fact, the emission of pollutant dust is reduced by 98% through this new process, and energy and water savings of 24 and 69%, respectively, are achieved. The emission of sulfur oxides is reduced by about 65% and nitrogen oxides emitted also decrease by about 71%. Thus, the authors assure that this type of tile production process is much cleaner and more sustainable than the traditionally wet process, constituting a significant advance in the area of construction.

8.5 Sustainability in the Automotive and Truck Industry

The automotive industry has become globalized as for some decades, facing great challenges in terms of competitiveness. Thus, when discussing sustainability, there is a huge challenge, which is to provide equal opportunities for all companies involved in this sector. Given that the ground transportation in general, and the car in particular, is something that is part of our daily lives, serious challenges in several aspects are constantly present, such as

- Adopting a wrong strategy in terms of product development can have extremely harmful effects on the environment, which can proliferate in an unexpected manner and under difficult-to-control ways. Thus, the development of a product is of particular importance, not only in terms of the propellant system, but also in all the materials that are used in its manufacture, and which must be thoroughly studied in terms of life cycle.
- A car generates emissions that are harmful to the environment. Thus, the legislation of several countries has established a progressive reduction of emissions over time as an unambiguous target, which has been a constant challenge for the development of combustion engines. However, even the maintenance of a car can have an influence on this factor, so everything must be properly prepared so that emissions can be contained throughout the life cycle of the car. Many advances have been made in this regard, and it is possible to observe that manufacturers have developed technologies capable of reducing the level of emissions, which is particularly noticeable in the case of diesel engines. However, the recent example of excessive emissions from Volkswagen engines shows that it is still difficult to reconcile low emissions with low levels of fuel consumption and good vehicle performance as required by consumers.
- Emissions from combustion engines have triggered intense investigations into other forms of vehicle propulsion. Several currents of thought have been generated, from hydrogen propulsion, to the latest trend in electric propulsion, and there is still a lively trend: hybrid engine systems. None of the systems developed so far has

unequivocal advantages. Even electric propulsion raises serious environmental issues, namely in the methods of generating electricity generation feed electric vehicles and also in the problems that arise during the manufacturing and end-of-life of electrical storage systems, i.e., batteries.

- The management of production systems in the automotive industry entails high energy consumption, as well as the generation of solid, liquid and gaseous effluents which, due to their quantity and danger, require special care. Thus, although there is strong regulation with regard to the automotive industry, namely through ISO (International Organization for Standardization) standards and more recently, even more specific through the regulation issued by the International Automotive Task Force (IATF), business management needs to be cared for and supervised. Thus, this industrial sector promotes audits with a reduced periodicity, in order to control all processes, ensuring that market requirements and legislation are being duly fulfilled. The competitiveness of the mark*et al*so requires constant control and concern.

In addition to manufacturing itself, the study of the sustainability of the automotive industry, considering the final product, can be divided into four distinct strands: studies to evaluate the product life cycle, the analysis of the end-of-life of the product, the Design for X, and studies to minimize the weight of vehicles with a view to reducing energy consumption, and corresponding emissions. Some of these items will be developed below.

8.5.1 Life Cycle Assessment of Vehicles

It is usual to consider that the life cycle of a product begins with the process of extraction and processing of its raw material and ends with its end-of-life, i.e., reuse, recycling, incineration, or landfill. It is common to define life cycle assessment as the methodology used to describe the environmental impacts that are associated with a given product or service throughout its life cycle (Pennington et al. 2004; Sundin 2004). Generally, the life cycle can be divided into four main phases: raw material extraction, processing, product use, and disposal. However, this definition is neither consensual nor comprehensive for all products. As a matter of fact, Ashby (2009) stated that a phase related with the transport of raw materials and products also needs to be added to the four phases mentioned above. Considering the car as a product, the same author suggests that another phase should be added in the product life cycle, since it promotes an environmental load that is related to its function and that is inevitable. This methodology was refined in time and came to translate into an ISO family of standards, which assumed the designation of ISO 14,000, unfolding in several particular norms, according to its objective. The methodology should be applied in the early stage of the concept and project, in order to apply other concepts such as the reduction of material consumption, energy consumed in the extraction and processing of raw materials, as well as in the manufacturing processes of the product, prospect of reuse or recycling, etc.

In order to properly assess the environmental impact of a passenger car, it is necessary to take into account not only its use (vehicle emissions, for example), but also its entire life cycle, which covers three main stages: production (raw material processing and vehicle production), use (including maintenance and repair), and end-of-life vehicle (including reuse and recycling). Thus, the concept of LCA provides a very useful approach to the analysis of the environmental sustainability performance of the automotive industry. Especially in the case of battery electric vehicles (BEVs), three aspects can be identified that justify a critical evaluation of the strategy proposed for a "greener" automotive industry (Nordelöf et al. 2014):

- During the use phase of a BEV, the environmental burden promoted by the electric power consumption varies significantly, depending heavily on the mix of sources that are used to produce this electric energy. With regard to Hawkins et al. (2012) study, when compared to a conventional gasoline vehicle, the environmental burden induced by a BEV can vary from −31 to +27%, when considering a mix of the several electricity production methods in European countries, or if the electricity is produced mainly on the basis of coal, respectively. As global electricity production is still strongly based on the use of fossil fuels (70%), especially in the use of coal (40%) (IEA 2015), the environmental benefit of BEVs depends on factors other than just how they are produced. Thus, carbon capture and storage (CCS) is seen as a promising solution to this problem in the near future (Johnson et al. 2015; Bertram et al. 2015; Eom et al. 2015). However, as already mentioned above, the success of such technologies is still unknown.
- Theoretically, the BEVs transfer the environmental load from the phase of use to the stage of manufacture of the vehicle. This is due to the fact that more resources are needed for the manufacture of batteries, since the resources spent on the combustion engine may be considered to be equivalent to those consumed in the manufacture of the electric motors. It is tacitly assumed that about 20% of the environmental burden caused by a conventional vehicle is centered on its production phase (Carbon Trust 2011). This ratio is different in the case of BEVs because emissions during the use phase are considered to be zero (assuming only the use of renewable energy), and the manufacturing phase requires a higher environmental burden due to the production of batteries. Thus, the duration of the batteries is of particular relevance in assessing vehicle sustainability in the case of BEVs (Nieuwenhuis and Katsifou 2015). The longer the battery life, the lower the environmental impact of the car.
- The use and processing of specific materials and resources linked to the production of an electric power train leads to a change in the weight of the generated environmental burdens. Indeed, if some aspects related to the new power system have better environmental indicators, others will remain similar as they are at present with vehicles powered by internal combustion engines, and other environmental indicators will even be heavily penalized by the introduction of BEVs, since the materials used in the batteries have a considerable negative environmental impact.

Given the particular case of batteries used in BEVs, there is a strong uncertainty in how technology, materials, and manufacturing processes evolve around this component (Ritthoff and Schallaböck 2012). It is expected that the technological evolution will allow the specific capacity of the batteries to increase consistently and noticeably, the increase of the life cycle of batteries, and the development of more environmentally friendly batter chemistries. However, it is also necessary to consider that the environmental burden caused by these components can be considered reduced at the present time due to the youth of these vehicles and the reduced quantity still being manufactured and marketed but, given the current trend and the prospects announced by the main manufacturers of this type of vehicles, the world may be in the presence of a very serious environmental problem in the relatively near future, due to the difficulty of recycling these batteries, taking into account the current technology. In short, it can be concluded that the introduction of electric power trains will essentially promote a transfer of environmental burdens without addressing the problems that society is currently facing: reducing resource consumption and significantly decreasing the aggregate environmental burden regarding the mobility of people.

Schöggl et al. (2017) argued that the LCA cannot be effectively applied in the initial stages of product development because there is a strong lack of data to support this study. In this way, the authors developed a checklist, which was called the checklist for sustainable product development, intending to serve as a qualitative tool to support the decision of designers in the early stages of product development in the automotive area. This checklist is intended to encourage these designers to drive product development by respecting the best sustainable performance indices, in an iterative and structured way, by identifying tasks and operations that violate established sustainability principles. As usual, a case study based on a composite front hood was presented, which allowed to validate the developed tool successfully. In turn, Stoycheva et al. (2018) states that there are a large number of models based on qualitative principles, but there is a strong gap in terms of interconnection among such important factors as social behavior, economic situation, and respect for the environment, the "triple bottom line" on which the development of communities must be based. Sustainability has been widely dissected and discussed around automobile production, but different approaches have taken the crucial factors into account in an isolated way, making it impossible to obtain a truly effective problem-solving methodology (Ahmed Ali et al. 2015). In order to overcome these obstacles, Stoycheva et al. (2018) developed a model based on quantitative analysis for its application in the automotive industry. The model was based on a multi-criteria decision analysis, pondering the values presented by the top management of the companies linked to the automotive industry and the decision makers, with a view for making decisions regarding the materials to be applied in automobiles. The authors of the model have proven that it is possible to make decisions regarding the choice of more sustainable materials in the automobile industry, through quantifiable factors. In addition, a sensitivity analysis was also performed to evaluate the robustness of the obtained solutions through the application of the model. Through the use of this tool, decision makers can base their choices on perfectly measurable factors and on much more sustainable alternatives.

8.5.2 Design for X

The concept of sustainability can be conveniently explored in the design phase if the Design for X (DfX) methodology is conveniently used to design a product that is intended to be sustainable. Both the vehicle and its subsystems and components can be designed in accordance with DfX principles, in order to reduce the environmental impact of a product from specific design aspects such as ease of disposal and operational safety. The Design for Manufacturing (DfM) methodology must be adopted at the design stage of a motor vehicle, taking into account the productive means the company has or intends to have, the family of products in which it will be integrated, the structure of the product outlined, and the respective components (Yüksel 2010). It implies increased care with the processes used, and their equipment and tools, in order to minimize cycle time and, obviously, costs. The Design for Assembly (DfA) methodology emerges as a natural evolution of the DfM (Dieffenbach and Mascarin 1993), which aims to design the product in such a way that it can be easily assembled, which implies to think about all the necessary details in the design regarding the assembly operations. DfA's strategy may contain the number of parts reductions as target, through the agglutination of some parts, reducing assembly operations and corresponding adjustments (Boothroyd and Alting 1992; Boothroyd 1994). Concerns regarding the environment have also brought new methodologies to the design stage of vehicles, i.e., concerns about their end-of-life. Thus, a new methodology has emerged that has assumed several nomenclatures, such as Design for Recycling, Design for Disassembly, Design for Remanufacturing, and Design for Deconstruction. All these nomenclatures aim at the same: to conceive and design products that allow a greater ease of separation of its components divided by materials at the end-of-life of the product, allowing its components to be reused or easily rebuilt or recycled. This is also referred to by Luttropp and Lagerstedt (2006), who highlight that, rather than making the disassembly process more agile and expeditious, this methodology aims to allow more systems and components to be recovered from each vehicle, minimizing the negative environmental effect that this could have. However, the scope and definition of Design for Remanufacturing in the automotive industry is a little more complex, since it aims to conceive and design vehicles that allow the disassembly of systems and/or components that, through refurbishing operations, regain identical capacity to the one they had initially, making possible their reassembly without degrading the properties of the set or system where are to be inserted. This concept allows the reuse of systems or components with little or no wear or corrosion, avoiding the use of new resources (Coulter et al. 1996), avoiding consumption of energy or other resources in their recycling, or landfill. The target of these remanufactured components is obviously the aftermarket, allowing the introduction of parts and systems at very competitive prices (Palmer 2009). Obviously, prices will be more competitive the easier disassembling and recovery operations are therefore, this should be taken into account during the design stage. A very important rule suggested by Bulucea et al. (2010) is to avoid the joint use of different materials in assemblies. This drastically facilitates the process of disassembling, separating, collecting, organizing, and reusing materials. Reuter and van Schaik (2005) corroborated this idea, reiterating that the number of different plastic and non-plastic materials used in a given product should be minimized in order to increase the recyclability of the product.

In the case of motor vehicles, it makes perfect sense to also take into consideration the energy consumed by the vehicle during its useful life. In this particular case, as in other similar cases, this factor will depend on the use of the product. Thus, to fill this gap, a methodology called Design for Energy Efficiency emerged. In the case of vehicles, this concept is not restricted only to the efficiency demonstrated by the propulsion system. In fact, the reduction of the total weight of the vehicle also strongly influences its energy efficiency, as referred previously, so that the design of the vehicle must have several aspects in mind, such as to be able to combine them in the most beneficial way possible, since energy consumption throughout the life of a vehicle has a strong influence on the environmental impact it causes (Ungureanu 2007; Jawahir et al. 2007; Mayyas et al. 2012) (Fig. 8.6).

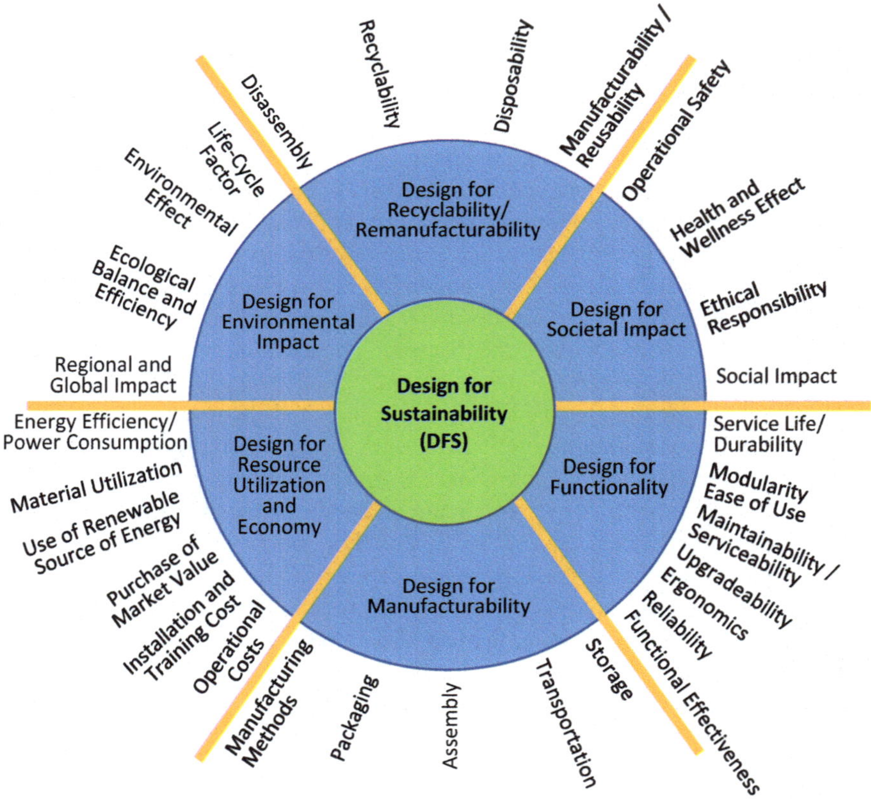

Fig. 8.6 Main factors contributing to Design for Sustainability. Adapted from Ungureanu (2007), Jawahir et al. (2007) and Mayyas et al. (2012)

In Europe, legislation has been stricter in this respect for two decades, and a directive was published in 2000 (Directive 2000/53/EC—the "ELV Directive") which aimed to establish the necessary guidelines to be respected by manufacturers. It was subsequently supplemented by Directive 2005/64/EC, which establishes new guidelines for the reuse, recyclability, and recovery of systems and components incorporated in motor vehicles. These directives set out very clear quantitative targets, which need to be achieved within the time limits laid down in those directives. Obviously, older vehicles do not meet these requirements, so some countries have developed controlled vehicle cutting programs, with only those that pose a greater risk to the environment being eligible (FAS 1992). As a result, a lot of cars have been eliminated from circulation, which are dismantled in a controlled way, avoiding more significant problems for the environment. These initiatives aim to ensure that, irrespective of the method of wrecking being carried out, there is no danger of dissemination of heavy metals or toxic substances.

8.5.3 Vehicle Weight Minimization

Another extremely effective methodology, as mentioned previously, is to reduce the amount of material used, both in manufacturing and throughout the life cycle of the product. This methodology significantly reduces the environmental impact of vehicles, which should be lighter, consume less material in their manufacture, and minimize consumption and corresponding emissions throughout their useful life stage. This stage represents about 87% of the total life cycle energy consumption (McAuley 2003). In automotive vehicles, this methodology has undergone a growing use of light alloys, such as aluminum alloys, and polymer matrix composite materials, both in vehicle bodies, interior and some powertrain components (Cheah and Heywood 2011; Santos et al. 2017). Indeed, a saving of 10% of vehicle weight represents the possibility of traveling 5% further with a given volume of fuel (Mayyas et al. 2011).

In order to minimize vehicle weight, material selection models have been developed to establish the necessary guidelines to achieve this specific objective. Santos et al. (2017) have developed a flowchart which aims to help decision makers make choices about replacing heavier materials with lighter ones. As a case study truck component was presented, which was manufactured in conventional cast iron, and for which was proposed the exchange for an identical part, but made out of a high-strength cast iron material, allowing the reduction of thickness/mass of the part, thus reducing the weight of that part and overall weight of the vehicle. If this procedure is carried out systematically with respect to other identical components, the overall weight may be reduced considerably, which will result in a reduction in the consumption and pollutant emissions of a vehicle. As in this study, other approaches have been developed, taking into consideration other factors, such as the risk management considered in the work developed by Cucuzzella (2016), design thinking using the framework for sustainable strategic development, which could lead to a new approach usually called sustainable design thinking (Shapira et al. 2017), concepts of sustainable values usually associated with other concepts such as risk management (Kim and Moon 2017) or the checklist for sustainable product development (Schöggl et al. 2017). In relation to the wide field of application of sustainable design, there are several types of approaches and fields of application, which can vary from civil construction, to the components that constitute an aircraft. These applications, having been the starting point of the investigations carried out, are generally aimed at validating the models developed in the meantime by researchers. Car producers have achieved very interesting values in terms of emission reduction following different strategies, which have gone through the improvement of the efficiency of combustion engines. Also, there has also been a strong collaboration in the selection of materials and the effect this has on the weight of vehicles (Mayyas et al. 2013; Raugei et al. 2015; Dhingra and Das 2014; Simões et al. 2016). Low-density alloys and composites are increasingly used in the automotive industry. Alloys of aluminum, magnesium, composites of polymeric matrix, among others, are widely used to reduce the overall weight of motor vehicles, replacing ferrous alloys (Mayyas et al. 2016).

Lightweight materials such as the A456 T6 aluminum alloy are commonly used in some applications due to their lower cost among alloys that have low density and good castability, allowing for thin-walled parts and complex geometries. However, light alloys have limited mechanical strength (about one-third of Young's modulus) and some limitations in terms of ductility (William et al. 2011). Structural components used in automobiles and trucks require high tensile strength and high elongation and need to include thin walls and complex geometries (Hardeman 2015). Thus, it is common that in these cases the selection of materials falls on ferrous alloys, which have higher specific mechanical strength, better Young's modulus, and higher ductility. In addition, the cost per unit of resistance is significantly lower when compared to light alloys (William et al. 2011; Gomes et al. 2017; Silva et al. 2018b). High-strength cast ferrous alloys have a high strength-to-weight ratio. As reported by Mohrbacher (2013), the increase in mechanical strength from 200 to 550 MPa allows reducing the weight of a given automobile component to 62.5%, by reducing its section. According to McAuley (2003), the option of using polymers in vehicle manufacturing can lead to significant energy savings over its useful life, saving 30 times more energy than the one spent during its manufacturing process.

Horton et al. (2019) recently developed a study to optimize the use of materials in the automotive industry. The main focus of this work was the metal chassis used in motor vehicles. To this end, a multi-disciplinary team was created to study five components of motor vehicles in detail. Through a structured design methodology, he sought a more efficient use of material. This research has identified opportunities for improvement in the order of 20% of material, which can be reflected in savings of £9 million and reduction of 5 kton of CO_2 emissions per year. It was also detected that the most productive phase to achieve this saving is in the product development cycle, before the production method is conditioned by the geometry of the component. The project was partially implemented, generating savings of £1.8 million and emission of less than 1.5 kton of CO_2 per year, with only a 3% reduction in the material applied.

In addition to the optimization of the weight of motor vehicles, a concern has arisen in the last decade. This concern has led to composites, which are already lighter in weight than the metal alloys traditionally applied in motor vehicles, to become more sustainable. In general, natural fibers are low cost, recyclable, and environmentally friendly. Studies carried out over the last few years have revealed that fiber adhesion influences the mechanical properties of polymer matrix composites. Chemical and physical modification methods have been used to improve the adhesion characteristics of the fiber to the matrix. In these studies, jute, flax, and coconut fibers have been considered to describe how processing techniques can be used to improve the general properties of composites reinforced with these natural fibers (Malkapuram et al. 2009). In general, it can be said that vegetable fibers are a composite material produced by Nature. These fibers consist of an amorphous lignin reinforced with rigid, crystalline cellulose and hemicellulose matrix microfibers. With the exception of cotton, most of the plant fibers are composed of cellulose, hemicellulose, lignin, waxes, and some water-soluble compounds. The characteristic properties of its constituents will contribute to the definition of the overall properties of natural fiber. Each constituent contributes certain characteristics. It is known that the biodegradation, micro-absorption, and thermal degradation of the fiber are conducted essentially due to the portion of hemicellulose, while the lignin content will condition some physical characteristics, since it is thermally stable, although susceptible to UV degradation. From fiber to fiber, the proportions of these constituents are different. In a coarse form, it can be said that fibers contain between 60 and 80% cellulose, 5–20% lignin, and up to 20% moisture. In order to assist lignocellulose to acquire an additional thermal degradability, the cell wall of fibers may be subjected to a chemical treatment known as pyrolysis, which consists of a process carried out at elevated temperatures, thus contributing to the formation of coal (Nguong et al. 2013; Bongarde and Shinde 2014; Kamath et al. 2017).

The investigations carried out in recent years have led to the application of natural fibers as reinforcements, which avoids the use of synthetic fibers, such as glass fiber and carbon fiber. The issue of mechanical strength is a limiting factor, but natural fibers have been studied, showing very interesting values, enabling to be applied as reinforcement in composite materials of polymer matrix. In this sense, Al-Oqla and Sapuan (2014) described as feasible the use of natural fibers in composite materials of polymer matrix, encountering difficulties in defining the criteria that must be followed to analyze this type of material. This gap is even more important because it entails two problems: (a) These fibers could be used in a common way in several industrial applications, so resources are being wasted; (b) when not being utilized, these fibers constitute a waste, being treated as such in environmental terms. In order to ensure the competitiveness and potential for future use of date palm fiber (DPF) in the development of sustainable products around the automotive industry, several comparative studies have been carried out between DPF and other types of fibers commonly used in this industry. The DPF showed to present the best set of properties among the different fibers tested, having stood out the ratio between specific Young's modulus and cost. Taking into account the technical properties of DPF and environmental, social and economic performance aspects, it can be seen that sustainability and productivity in the automotive industry can be improved by introducing this type of fiber in the production of components in polymer matrix composites. In addition, the use of these fibers avoids the need to treat them as waste, allowing a more sustainable waste management. Other studies have been developed to study various types of natural fibers from plants such as coir, jute, basalt, cotton, banana, bamboo, hemp, and so on (Selvaraj 2018).

Bessadok and Roudesli (2009) studied the alpha fibers after being subjected to various chemical treatments. Treatments carried out on fibers with chemicals such as maleic anhydride, styrene, or acrylic acid have been extensively studied in order to investigate their effect on fiber water absorption capacity. For this purpose, studies have been carried out by microscopy, surface energy measurement, and infrared spectroscopy, and it has been verified that these chemical products effectively reduce the amount of water absorbed by the natural fiber. Symington et al. (2009) characterized, performing tensile tests to various types of natural fibers, such as jute fiber, kenaf, flax, abaca, sisal, hemp, and coconut. For this, the fibers were initially exposed to severe conditions, beyond normal ambient conditions, such as 65% moisture content, 90% moisture content, and soaking. The effect of alkalinization was also studied through the use of a solution of 3% NaOH in fibers such as kenaf, sisal, kenaf, and fibers of abacá. The tests allowed to conclude that a saturation treatment can lead to a decrease of the properties initially presented by the tested natural fibers. In order to study the cellular structure of the Sansevieria Cylindrica fiber, Sreenivasan et al. (2011) used polarized light microscopy and scanning electron microscopy. This fiber revealed to have a hierarchical cellular structure, lumen of intermediate fibers, and lamellae. The porosity of the fiber structure was also analyzed, which was estimated at 37%.

Comparing to fiberglass, natural fibers consume less energy in their processing, presenting some advantages regarding application in light and non-heavily requested structures. It may be mentioned as an example that, comparatively to fiberglass, kenaf fiber is highly available and has characteristics which satisfactorily meet the requirements usually imposed by the automotive industry. However, their properties can still be optimized via the treatment of fibers, use of additives, and appropriate conjugation with the matrix through a suitable selection process and with appropriate selection of a processing method (Hassan et al. 2017).

8.5.4 Sustainable Automotive Production Through "Lean and Green"

Sustainability in the automotive industry goes without a doubt by eco-efficiency, which can translate into producing the same products but with more efficiency and less expense (Martinez-Jurado and Moyano-Fuentes 2014). However, in our busy day-to-day life, the question rarely raises whether products or services are really needed or not. When we look more closely at sustainable production and consumption, we need to think further: How will our needs and desires be met in a more sustainable way through more sustainable products and production systems? The automotive industry has been marked by a growing outsourcing of production activities, maintaining tight control over supply chain management (Araújo et al. 2016). The large original equipment manufacturers (OEM), which detain assembly lines, outsource the manufacturing of modular components, requiring the application of optimization techniques in production systems based on lean tools. OEMs maintain the control of the activities directly related to the customer, namely through innovation and development (Rosa et al. 2017a, b, 2018; Antoniolli et al. 2017; Martins et al. 2018). Taking into account what is referred by Nieuwenhuis and Katsifou (2015), OEMs that control assembly lines and product design remain agile, while requiring a strong rationalization of the component production from their suppliers. This strategy takes some pressure off top builders, allowing them not to overproduce, minimizing the effects this can have on sustainability.

Sustainability in the automotive industry has also seen a strong escalation through the modernization of processes, which through automation have made processes more competitive and reliable, improving quality indices and generating less waste. Magalhães et al. (2019) have developed an automatic system that, coupled with a wire CNC bending machine, allows to organize the bent wires that make up suspension mats used in car seats, facilitating the task and avoiding problems of safety and ergonomics. Moreira et al. (2017) have developed a new concept of integrated wire-rope production products used in door actuation assemblies in motor vehicles, minimizing the quantity of work in progress material during their manufacturing. This manufacturing system avoids production wastage generated with the mismatching of quantities to be produced between workstations. In addition, the developed equipment also made operations safer and more ergonomic, contributing to the well-being of workers. Costa et al. (2018) solved numerous quality problems related to the manufacture of transmission systems used in the operation of windscreen wipers. This new concept of equipment drastically reduced the possibility of manufacturing non-conforming parts, which greatly reduced the scrap levels generated in the process, thus contributing to greater sustainability. The cycle time was also improved, lowering the production cost of these components and improving competitiveness. Nunes and Silva (2013) developed a new concept of equipment capable of efficiently assembling twenty different types of drive systems for automotive widescreen wipers, composed of two aluminum alloy parts and one rubber. This system reduced the cycle time from 11 to 7 s, avoiding numerous errors which originated from operator fatigue as the previously process was manual and of intensive labor. Improving quality indices also minimized defective parts, which resulted in fewer parts needing to be recycled or landfilled. Costa et al. (2017) developed a fully automated system for the production of very similar components to those used by Nunes and Silva (2013); however, the main parts are polymeric material and rubber washers. The equipment is fully automated and fully flexible, the assembly of numerous models of components of the same family, with very short and completely automated setups, also allowing the control of assembled sets and separating OK sets from NOK sets. Again, the quality and reliability of the process allowed a strong reduction in the number of defective sets, which minimizes waste. In addition to flexibility, the cycle time was drastically reduced and human intervention was avoided, which significantly reduced the cost of the final automotive component, increasing its competitiveness.

The quest to minimize quality problems in the production of components for the automotive industry has motivated numerous investigations. Pinto and Silva (2017) used advanced simulation software to determine the most appropriate position of runners and venting channels in order to minimize the defects frequently observed in small high-pressure die-casting injected parts. By determining the most suitable positioning for those channels, it was possible to drastically reduce the appearance of pores inside parts, significantly reducing the generation of scrap and avoiding the recycling of large quantities of Zamak alloy. Gomes et al. (2017) and Silva et al. (2018b), using simulation processes, were able to identify the problems that led to the appearance of cracks and high spring-back in high-strength steel sheets, drastically reducing the amount of scrap generated. This was achieved by performing small modifications to the tools and blanks used in the production of stamped parts for the automotive industry. The knowledge added with these will certainly be useful so that other analogous situations can be corrected, avoiding high costs of non-quality and the generation of scrap that, even being recyclable, implies the consumption of energy and the displacement of resources so that the material may become once again a commercially usable product.

Nunes et al. (2017) studied the application of advanced coatings in molds used in the high-pressure die-casting process. Indeed, the wear of some components of these molds is very intense due to the high temperature and pressure to which they are subject. The application of modern technical coatings allowed to significantly increasing the useful life of these components, avoiding breakdowns during production and generation of defective parts. The coatings applied in this situation are sustainable since the effluents generated do not pose any health hazard or require any specific treatment. Indeed, coatings synthesized or deposited via PVD and CVD techniques are environmentally friendly processes, not generating hazardous effluents (Baptista et al. 2018). In addition, they can be applied in an enormous diversity of substrate materials and in numerous applications, both related to the automotive industry (Silva et al. 2011, 2012b, 2017b) and to other types of industries (Silva et al. 2002, 2004; Fernandes et al. 2017, 2018, 2019).

References

M. Abbas, B. Rao, M.N. Islam, S. Naga, M. Takahashi, C. Kim, Highly stable-silica encapsulating magnetite nanoparticles (Fe_3O_4/SiO_2) synthesized using single surfactantless-polyol process. Ceram. Int. **40**, 1379–1385 (2014)

K.A. Adebiy, O.O. Akinyemi, A.L. Akintan, A.O. Ajayeoba, Modelling the hazard control in a wet process plant—a case of cement plant. Braz. J. Oper. Prod. Manage. **12**, 338–348 (2015)

K.F. Adekunle, J.A. Okolie, A review of biochemical process of anaerobic digestion. Adv. Biosci. Biotechnol. **6**, 205–212 (2015)

A. Adessi, R. de Philippis, *Photobioreactor design and illumination systems for H_2 production with anoxygenic photosynthetic bacteria: a review*. Int. J. Hydrogen Energy **39**, 3127–3141 (2014)

M. Agovino, M. Casaccia, M. Ciommi, M. Ferrara, K. Marchesano, Agriculture, climate change and sustainability: the case of EU-28. Ecol. Indic. (2018 in press). Available online https://doi.org/10.1016/j.ecolind.2018.04.064

B.A. Ahmed Ali, S.M. Sapuan, E.S. Zainudin, M. Othman, Implementation of the expert decision system for environmental assessment in composite materials selection for automotive components. J. Clean. Prod. **107**, 557–567 (2015)

A. Alaswad, M. Dassisti, T. Prescott, A. Olabi, Technologies and developments of third generation biofuel production. Renew. Sustain. Energy Rev. **51**, 1446–1460 (2015)

A. Ali, R. Mahar, R. Soomro, S. Sherazi, Fe_3O_4 nanoparticles facilitated anaerobic digestion of organic fraction of municipal solid waste for enhancement of methane production. Energy Sour. Part A Recovery Util. Environ. Eff. **39**, 1815–1822 (2017)

M. Alipour, S. Alighale, R. Hafezi, M. Omranievardi, A new hybrid decision framework for prioritizing funding allocation to Iran's energy sector. Energy **121**, 388–402 (2017)

B.N. Aloo, B.A. Makumba, E.R. Mbega, The potential of Bacilli rhizobacteria for sustainable crop production and environmental sustainability. Microbiol. Res. **219**, 26–39 (2019)

F.M. Al-Oqla, S.M. Sapuan, Natural fiber reinforced polymer composites in industrial applications: feasibility of date palm fibers for sustainable automotive industry. J. Clean. Prod. **66**, 347–354 (2014)

M.M.M. Andrade, N.P. Stamford, C.E.R.S. Santos, A.D.S. Freitas, C.A. Sousa, J.M.A. Lira, Effects of biofertilizer with diazotrophic bacteria and mycorrhizal fungi in soil attribute, cowpea nodulation yield and nutrient uptake in field conditions. Sci. Hortic. **162**, 374–379 (2013)

L. Andre, M. Ndiaye, M. Pernier, O. Lespinard, A. Pauss, E. Lamy, T. Ribeiro, Methane production improvement by modulation of solid phase immersion in dry batch anaerobic digestion process: dynamic of methanogen populations. Bioresour. Technol. **207**, 353–360 (2016)

R. Angeriz-Campoy, C.J. Alvarez-Gallego, L.I. Romero-Garcia, Thermophilic anaerobic co-digestion of organic fraction of municipal solid waste (OFMSW) with food waste (FW): enhancement of bio-hydrogen production. Biores. Technol. **194**, 291–296 (2015)

I. Antoniolli, P. Guariente, T. Pereira, L.P. Ferreira, F.J.G. Silva, Standardization and optimization of an automotive components production line. Procedia Manuf. **13**, 1120–1127 (2017)

W.F.S. Araújo, F.J.G. Silva, R.D.S.G. Campilho, J.A. Matos, Manufacturing cushions and suspension mats for vehicle seats: a novel cell concept. Int. J. Adv. Manuf. Technol. **90**, 1539–1545 (2016)

S.A. Archer, R. Steinberger-Wilckens, Systematic analysis of biomass derived fuels for fuel cells. Int. J. Hydrogen Energy **43**, 23178–23192 (2018)

S. Asadauskas, S.Z. Erhan, Depression of pour points of vegetable oils by blending with diluents used for biodegradable lubricants. J. Am. Oil. Chem. Soc. **76**, 313–316 (1999)

M. Ashby, *Materials and the Environment*, 1st edn. (Butterworth-Heinemann, Oxford, UK, 2009). ISBN 978-0123859716

O. Ashrafi, L. Yerushalmi, F. Haghighat, Wastewater treatment in the pulp-and-paper industry: a review of treatment processes and the associated greenhouse gas emission. J. Environ. Manage. **158**, 146–157 (2014)

E. Avsar, G.N. Demirer, Cleaner production opportunity assessment study in SEKA Balikesir pulp and paper mill. J. Clean. Prod. **16**, 422–431 (2008)

M. Ayyachamy, T.M. Vatsala, Production and partial characterization of cellulose free xylanase by *Bacillus subtilis* C01 using agriresidues and its application in biobleaching of nonwoody plant pulps. Lett. Appl. Microbiol. **45**, 467–472 (2007)

M. Ba-Shammakh, H. Caruso, A. Elkamel, E. Croiset, P.L. Douglas, Analysis and optimization of carbon dioxide emission mitigation options in the cement industry. Am. J. Environ Sci. **4**(5), 482–490 (2008)

S. Bai, J. Zhang, Z. Wang, A methodology for evaluating cleaner production in the stone processing industry: case study of a Shandong stone processing firm. J. Clean. Prod. **102**, 461–476 (2015)

P. Bajpai, *Environmentally Benign Approaches for Pulp Bleaching*, 2nd edn. (Elsevier, USA, 2012). ISBN 978-0444594211

P. Bajpai, *Biermann's Handbook of Pulp and Paper: Raw Material and Pulp Making* (Elsevier, USA, 2018), pp. 527–546. ISBN 978-0128142417

P. Bajpai, N.K. Bhardwaj, P.K. Bajpai, M.B. Jauhari, The impact of xylanases in bleaching of eucalyptus kraft pulp. J. Biotechnol. **36**, 1–6 (1994)

A. Baptista, F. Silva, J. Porteiro, J. Míguez, G. Pinto, Sputtering physical vapour deposition (PVD) coatings: a critical review on process improvement and market trend demands. Coatings **8**, 402–424 (2018)

S. Basu, R. Rabara, S. Negi, AMF: the future prospect for sustainable agriculture. Physiol. Mol. Plant Pathol. **102**, 36–45 (2018)

A.H. Battez, R. González, J. Viesca, A. Fernández-González, M. Hadfield, Lubrication of PVD coatings with ethyl-dimethyl-2-methoxyethylammonium tris(pentafluoroethyl) trifluorophosphate. Tribol. Int. **58**, 71–78 (2013)

Q.K. Beg, B. Bhushan, M. Kapoor, G.S. Hoondal, Effect of amino acids on production of xylanase and pectinase from Streptomyces sp. QG-11-3. World J. Microbiol. Biotechnol. **16**, 211–213 (2000)

J. Beiyuan, Y.M. Awad, F. Beckers, D.C.W. Tsang, Y.S. Ok, J. Rinklebe, Mobility and phytoavailability of As and Pb in a contaminated soil using pine sawdust biochar under systematic change of redox conditions. Chemosphere **178**, 110–118 (2017)

V. Benavente, E. Calabuig, A. Fullana, Upgrading of moist agro-industrial wastes by hydrothermal carbonization. J. Anal. Appl. Pyrol. **113**, 89–98 (2015)

A. Beneduzi, A. Ambrosini, L.M.P. Passaglia, Plant growth-promoting rhizobacteria (PGPR): their potential as antagonists and biocontrol agents. Genet. Mol. Biol. **4**, 1044–1051 (2012)

E. Benhelal, G. Zahedi, E. Shamsaei, A. Bahadori, Global strategies and potentials to curb CO_2 emissions in cement industry. J. Clean. Prod. **51**, 142–161 (2013)

C. Bertram, N. Johnson, G. Luderer, K. Riahi, M. Isaac, J. Eom, Carbon lock-in through capital stock inertia associated with weak near-term climate policies. Technol. Forecast. Soc. Change **90**, 62–72 (2015)

A. Bessadok, S. Roudesli, Alfa fibers for unsaturated polyester composites reinforcement: effects of chemical treatments on mechanical and permeation properties. Compos. A Appl. Sci. Manuf. **40**(2), 184–195 (2009)

D. Bhardwaj, M. Sharma, P. Sharma, R. Tomar, Synthesis and surfactant modification of clinoptilolite and montmorillonite for the removal of nitrate and preparation of slow release nitrogen fertilizer. J. Hazard. Mater. **227–228**, 292–300 (2012)

I. Binner, S. Dultz, M. Schellhorn, M.K. Schenk, Potassium adsorption and release properties of clays in peat-based horticultural substrates for increasing the cultivation safety of plants. Appl. Clay Sci. **145**, 28–36 (2017)

H. Blanco, W. Nijs, J. Ruf, A. Faaij, Potential of power-to-methane in the EU energy transition to a low carbon system using cost optimization. Appl. Energy **232**, 323–340 (2018)

H. Bley, C. Gianello, J.S. Santos, P.L.R. Selau, Nutrient release, plant nutrition, and potassium leaching from polymer-coated fertilizer. Revista Brasileira de Ciência do Solo **41**, e0160142 (2017)

U.S. Bongarde, V.D. Shinde, Review on natural fiber reinforcement polymer composites. Int. J. Eng. Sci. Innov. Technol. **3**(2), 431–435 (2014)

B.S.F. Boodhun, A. Mudhoo, G. Kumar, S.-H. Kim, C.-Y. Lin, Research perspectives on constraints, prospects and opportunities in biohydrogen production. Int. J. Hydrogen Energy **42**(45), 27471–27481 (2017)

G. Boothroyd, L. Alting, Design for assembly and disassembly. CIRP Ann. Manuf. Technol. **41**(2), 625–636 (1992)

G. Boothroyd, Product design for manufacture and assembly. Comput. Aided Des. **26**(7), 505–520 (1994)

N. Borchard, J. Siemens, B. Ladd, A. Müller, W. Amelung, Application of biochars to sandy and silty soil failed to increase maize yield under common agricultural practice. Soil Tillage Res. **144**, 184–194 (2014)

C.E. Brewer, R. Unger, K. Schmidt-Rohr, R.C. Brown, Criteria to select biochars for field studies based on biochar chemical properties. Bioenergy Res. **4**, 312–323 (2011)

T.B. Bruun, B. Elberling, A.D. Neergaard, J. Magid, Organic carbon dynamics in different soil types after conversion of forest to agriculture. Land Degrad. Dev. **26**, 272–283 (2015)

C.A. Bulucea, D.A. Nicola, L.S. Lazarescu, A. Stinga, in *Building Awareness of Sustainable Automobile Manufacturing Within Industrial Ecology Framework*. Proceedings of the International Conference on Urban Sustainability, Cultural Sustainability, Green Development Green Structures and Clean Cars (USCUDAR '2010), Malta, 15–17 Sept 2010, pp. 148–153

Z.M.A. Bundhoo, Effects of microwave and ultrasound irradiations on dark fermentative biohydrogen production from food and yard wastes. Int. J. Hydrogen Energy **42**, 4040–4050 (2017)

Z.M.A. Bundhoo, Potential of bio-hydrogen production from dark fermentation of crop residues: a review. Int. J. Hydrogen Energy (2018 in press). Available online https://doi.org/10.1016/j.ijhydene.2018.11.098

M.A.Z. Bundhoo, R. Mohee, M.A. Hassan, Effects of pretreatment technologies on dark fermentative biohydrogen production: a review. J. Environ. Manage. **157**, 20–48 (2015)

S.J. Burgess, B. Tamburic, F. Zemichael, K. Hellgardt, P.J. Nixon, Solar-driven hydrogen production in green algae. Adv. Appl. Microbiol. **75**, 71–110 (2011)

N. Cackett, *Honda Clarity FCV Review* (2019), https://www.autocar.co.uk/car-review/honda/clarity-fcv. Retrieved on 19 Jan 2019

Z.B. Cai, H.M. Meyer, C. Ma, M.F. Chi, H.M. Luo, J. Qu, Comparison of the tribological behavior of steel-steel and Si_3N_4-steel contacts in lubricants with ZDDP or ionic liquid. Wear **319**, 172–183 (2014)

M. Calabi-Floody, J. Medina, C. Rumpel, L.M. Condron, M. Hernandez, M. Dumont, M. de la Luz Mor, Smart fertilizers as a strategy for sustainable agriculture. Adv. Agron. **147**, 119–157 (2018)

S. Camarero, D. Ibarra, A.T. Martinez, J. Romero, A. Gutierrez, J.C. del Rio, Paper pulp delignification using laccase and natural mediators. Enzym. Microb. Technol. **40**, 1264–1271 (2007)

E.V.R. Campos, P.L.F. Proença, J.L. Oliveira, M. Bakshi, P.C. Abhilash, L.F. Fraceto, Use of botanical insecticides for sustainable agriculture: future perspectives. Ecol. Indic. (2018 in press). Available online https://doi.org/10.1016/j.ecolind.2018.04.038

Carbon Trust, *International Carbon Flows—Automotive* (Carbon Trust, London, 2011), https://www.carbontrust.com/media/38401/ctc792-international-carbon-flows-automotive.pdf. Retrieved on 10 Mar 2019

C.R. Carere, T. Rydzak, T.J. Verbeke, N. Cice, D.B. Levin, R. Sparling, Linking genome content to biofuel production yields: a meta-analysis a major catabolic pathways among select H_2 and ethanol-producing bacteria. BMC Microbiol. **12**, 295–315 (2012)

R.C. Carpio, F.S. Júnior, L.S. Coelho, R.J. Silva, Alternative fuels mixture in cement industry kilns employing particle swarm optimization algorithm. J. Braz. Soc. Mech. Sci. Eng. **30**(4), 335–340 (2008)

N.L. Castanheira, A.C. Dourado, I. Pais, J. Samedo, P. Scotti-Campos, N. Borges, P. Fareleira, Colonization and beneficial effects on annual ryegrass by mixed inoculation with plant growth promoting bacteria. Microbiol. Res. **198**, 47–55 (2017)

A.C.M. Castro, J.P. Carvalho, M.C.S. Ribeiro, J.P. Meixedo, F.J.G. Silva, A. Fiúza, M.L. Dinis, An integrated recycling approach for GFRP pultrusion wastes: recycling and reuse assessment into new composite materials using Fuzzy Boolean Nets. J. Clean. Prod. **66**, 420–430 (2014)

Cembureau, *Activity Report 2017* (2017), https://cembureau.eu/media/1716/activity-report-2017.pdf. Retrieved on 06 Mar 2019

V. Chaplain, L. Mamy, L. Vieublé-Gonod, C. Mougin, P. Benoit, E. Barriuso, S. Nélieu, Fate of pesticides in soils: toward an integrated approach of influential factors, in *Pesticides in the Modern World-Risks and Benefits* (InTech, 2011), https://www.intechopen.com/books/pesticides-in-the-modern-world-risks-and-benefits/fate-of-pesticides-in-soils-toward-an-integrated-approach-of-influential-factors. Retrieved on 15 Jan 2019

L.W. Cheah, J. Heywood, Meeting U.S. passenger vehicle fuel economy standards in 2016 and beyond. Energy Policy **39**, 454–466 (2011)

J. Cheng, M. Zhu, A novel anaerobic co-culture system for bio-hydrogen production from sugarcane bagasse. Biores. Technol. **144**, 623–631 (2013)

B. Conti, G. Flamini, P.L. Cioni, L. Ceccarini, M. Macchia, G. Benelli, Mosquitocidal essential oils: are they safe against non-target aquatic organisms? Parasitol. Res. **113**, 251–259 (2014)

D. Cordell, S. White, Peak phosphorus: clarifying the key issues of a vigorous debate about long-term phosphorus security. Sustainability **3**, 2027–2049 (2011)

R.J.S. Costa, F.J.G. Silva, R.D.S.G. Campilho, A novel concept of agile assembly machine for sets applied in the automotive industry. Int. J. Adv. Manuf. Technol. **91**, 4043–4054 (2017)

M.J.R. Costa, R.M. Gouveia, F.J.G. Silva, R.D.S.G. Campilho, How to solve quality problems by advanced fully-automated manufacturing systems. Int J. Adv. Manuf. Technol. **94**, 3041–3063 (2018)

S. Coulter, B. Bras, G. Winslow, S. Yester, Designing for material separation: lessons from automotive recycling. J. Mech. Des. **120**(3), 501–509 (1996)

J.M. Crow, *Why Use Lead in Paint?* (2007), https://www.chemistryworld.com/news/why-use-lead-in-paint/3004319.article. Retrieved on 1 Feb 2019

C. Cucuzzella, Creativity, sustainable design and risk management. J. Clean. Prod. **135**, 1548–1558 (2016)

M. Cui, J. Shen, Effects of acid and alkaline pretreatments on the biohydrogen production from grass by anaerobic dark fermentation. Int. J. Hydrogen Energy **37**, 1120–1124 (2012)

Z. Dai, X. Zhang, C. Tang, N. Muhammad, J. Wu, P.C. Brookes, J. Xu, Potential role of biochars in decreasing soil acidification—a critical review. Sci. Total Environ. **581–582**, 601–611 (2017)

D. Das, N. Khanna, N.T. Veziroğlu, Recent developments in biological hydrogen production processes. Chem. Ind. Chem. Eng. Q. **14**, 57–67 (2008)

J.M. David, Biochar and compost increase crop yields but the effect is short term on sandplain soils of Western Australia. Pedosphere **25**, 720–728 (2015)

L. Daviet, M. Schalk, Biotechnology in plant essential oil production: progress and perspective in metabolic engineering of the terpene pathway. Flavour Fragr. J. **25**, 123–127 (2010)

C.J. Dawson, J. Hilton, Fertiliser availability in a resource-limited world: production and recycling of nitrogen and phosphorus. Food Policy **36**, S14–S22 (2011)

G.E. Dawwam, A. Elbeltagy, H.M. Emara, I.H. Abbas, M.M. Hassan, Beneficial effect of plant growth promoting bacteria isolated from the roots of potato plant. Ann. Agric. Sci. **58**, 195–201 (2013)

A.C. de Campos Bernardi, G.B. Souza, J.C. Polidoro, M.B. de Melo Monte, Yield, quality components, and nitrogen levels of silage corn fertilized with urea and zeolite. Commun. Soil Sci. Plant Anal. **42**, 1266–1275 (2011)

J.L. Deenik, A. Diarra, G. Uehara, S. Campbell, Y. Sumiyoshi, M.J. Antal, Charcoal ash and volatile matter effects on soil properties and plant growth in an acid Ultisol. J. Soil Sci. **176**, 336–345 (2011)

P.E.V. de Miranda, *Science and Engineering of Hydrogen-Based Energy Technologies* (Academic Press, Cambridge, MA, USA, 2019). ISBN 978-0128142516

R. Dhingra, S. Das, Life cycle energy and environmental evaluation of downsized vs. lightweight material automotive engines. J. Clean. Prod. **85**, 347–358 (2014)

M.O.S. Dias, A.V. Ensinas, S.A. Nebra, R.M. Filho, C.E.V. Rossell, M.R.W. Maciel, Production of bioethanol and other bio-based materials from sugarcane bagasse: integration to conventional bioethanol production process. Chem. Eng. Res. Des. **87**, 1206–1216 (2009)

J.R. Dieffenbach, A.E. Mascarin, Body-in-white material systems: a life-cycle cost comparison. J. Miner. Met. Mater. Soc. (JOM) **47**, 16–19 (1993)

W.H. Duda, *Manual Tecnológico del Cemento* (in Spanish) (Editores Técnicos Associados, Barcelona, Spain, 1977). ISBN 84-7146-195-5

M. Ebrahimi, D. Rahmani, A five-dimensional approach to sustainability for prioritizing energy production systems using a revised GRA method: a case study. Renew. Energy **135**, 345–354 (2019)

D. Egamberdieva, A.O. Adesemoye, Improvement of crop protection and yield in hostile agroecological conditions with PGPR-based biofertilizer formulations, in *Bioformulations: for Sustainable Agriculture*, ed. by N.K. Arora, S. Mehnaz, R. Balestrini (Springer, India, 2016), pp. 199–211. ISBN 978-81-322-2777-9

W.A. Elhelece, Rice straw as a raw material for pulp and paper production, in *Reference Module in Materials Science and Materials Engineering* (2018 in press). Available online https://doi.org/10.1016/B978-0-12-803581-8.10596-X

A. El-Naggar, S.M. Shaheen, Y.S. Ok, J. Rinklebe, Biochar affects the dissolved and colloidal concentrations of Cd, Cu, Ni, and Zn and their phytoavailability and potential mobility in a mining soil under dynamic redox-conditions. Sci. Total Environ. **624**, 1059–1071 (2018)

A. El-Naggar, S.S. Lee, J. Rinklebe, M. Farooq, H. Song, A.K. Sarmah, A.R. Zimmerman, M. Ahmad, S.M. Shaheend, Y.S. Ok, Biochar application to low fertility soils: a review of current status, and future prospects. Geoderma **337**, 536–554 (2019)

A. Elreedy, E. Ibrahim, N. Hassan, A. El-Dissouky, M. Fujii, C. Yoshimura, A. Tawfik, Nickel-graphene nanocomposite as a novel supplement for enhancement of biohydrogen production from industrial wastewater containing mono-ethylene glycol. Energy Convers. Manage. **140**, 133–144 (2017)

I.I. Enagi, K.A. Al-Attab, Z.A. Zainal, Liquid biofuels utilization for gas turbines: a review. Renew. Sustain. Energy Rev. **90**, 43–55 (2018)

Energy Information Administration, *International Energy Outlook* (2018), https://www.eia.gov/pressroom/presentations/capuano_07242018.pdf. Retrieved on 20th Jan 2019

J. Eom, J. Edmonds, V. Krey, N. Johnson, T. Longden, G. Luderer, K. Riahi, D.P. Van Vuuren, The impact of near-term climate policy choices on technology and emission transition pathways. Technol. Forecast. Soc. Change **90**, 73–88 (2015)

E. Eroglu, A. Melis, Photobiological hydrogen production: recent advances and state of the art. Biores. Technol. **102**, 8403–8413 (2011)

E. Eroglu, A. Melis, Microalgal hydrogen production research. Int. J. Hydrogen Energy **41**(30), 12772–12798 (2016)

E. Eroglu, M. Timmins, S.M. Smith, Green hydrogen: algal biohydrogen production, in *Natural and Artificial Photosynthesis: Solar Power as an Energy Source*, ed. by R. Razeghifard (Wiley Hoboken, NJ, USA, 2013), pp. 267–284. ISBN 978-1-118-16006-0

European Commission, *Regulation (EC) No 1107/2009 of the European Parliament and of the Council of 21th October 2009* (2009), https://eur-lex.europa.eu/legal-content/EN/TXT/PDF/?uri=CELEX:32009R1107&from=EN. Retrieved on 16 Jan 2019

European Commission, *Energy Roadmap 2050* (2012), https://ec.europa.eu/energy/sites/ener/files/documents/2012_energy_roadmap_2050_en_0.pdf. Retrieved on 20 Jan 2019

European Commission, *EU Agriculture and Climate Change* (2015), https://ec.europa.eu/agriculture/sites/agriculture/files/climate-change/factsheet_en.pdf. Retrieved on 07 Jan 2019

European Commission, *EU Energy in Figures, Statistical Pocketbook 2018* (2018), https://publications.europa.eu/en/publication-detail/-/publication/99fc30eb-c06d-11e8-9893-01aa75ed71a1/language-en. Retrieved on 20 Jan 2019

S. Ezhil Vendan, Current scenario of biopesticides and eco-friendly insect pest management in India. S. Indian J. Biol. Sci. **2**, 268–271 (2016)

FAS—Federation of American Scientists, *Retiring Old Cars Programs to Save Gasoline and Reduce Emissions* (1992), https://ota.fas.org/reports/9232.pdf. Retrieved on 12 Mar 2019

R. Fehrmann, A. Riisager, Catalysis for a sustainable chemicals production, environment, and the future, in *Encyclopedia of Sustainable Technologies*, ed. by M. Abraham (Elsevier, Amsterdam, The Netherlands, 2017), pp. 621–628. ISBN 978-0128046777

L. Fernandes, F.J.G. Silva, M.F. Andrade, R. Alexandre, A.P.M. Baptista, C. Rodrigues, Increasing the stamping tools lifespan by using Mo and B_4C PVD coatings. Surf. Coat. Technol. **325**, 107–119 (2017)

L. Fernandes, F.J.G. Silva, O.C. Paiva, A. Baptista, G. Pinto, Minimizing the adhesion effect in food packages forming by the use of advanced coatings. Procedia Manuf. **17**, 886–894 (2018)

L. Fernandes, F.J.G. Silva, R. Alexandre, Study of TiAlN PVD coating on stamping dies used in tinplate food package production. Micromachines **10**, 182–197 (2019)

I. Gajda, J. Greenman, I.A. Ieropoulos, Recent advancements in real-world microbial fuel cell applications. Curr. Opin. Electrochem. **11**, 78–83 (2018)

A. García, R. González, A.H. Battez, J. Viesca, R. Monge, A. Fernández-González, M. Hadfield, Ionic liquids as a neat lubricant applied to steel–steel contacts. Tribol. Int. **72**, 42–50 (2014)

P. García-Fraile, E. Menéndez, R. Rivas, Role of bacterial biofertilizers in agriculture and forestry. AIMS Bioeng. **2**, 108–205 (2015)

M. Garcia-Perez, T. Lewis, C.E. Kruger, *Methods for Producing Biochar and Advanced Biofuels in Washington State. Part 1: Literature Review of Pyrolysis Reactors. First Project Report.* (Department of Biological Systems Engineering and the Centre for Sustaining Agriculture and Natural Resources, Washington State University, Pullman, WA, 2010), 137 p. https://fortress.wa.gov/ecy/publications/documents/1107017.pdf. Retrieved on 09 Jan 2019

G. Garg, S.S. Dhiman, R. Mahajan, A. Kaur, J. Sharma, Bleach-boosting effect of crude xylanase from *Bacillus stearothermophilus* SDX on wheat straw pulp. New Biotechnol. **28**, 58–64 (2011)

A. Ghimire, L. Frunzo, F. Pirozzi, E. Trably, R. Escudie, P.N.L. Lens, G. Esposito, A review on dark fermentative biohydrogen production from organic biomass: process parameters and use of by-products. Appl. Energy **144**, 73–95 (2015)

Z. Ghouleh, Y. Shao, Turning municipal solid waste incineration into a cleaner cement production. J. Clean. Prod. **195**, 268–279 (2018)

S.G. Gilbert, B. Weiss, A rationale for lowering the blood lead action level from 10 to 2 μg/dL. Neurotoxicology **27**, 693–701 (2006)

T. Gomes, F.J.G. Silva, R.D.G.S. Campilho, Reducing the simulation cost on dual-phase steel stamping process. Procedia Manuf. **11**, 474–481 (2017)

R. González, A.H. Battez, D. Blanco, J. Viesca, A. Fernández-González, Lubrication of TiN, CrN and DLC PVD coatings with 1-butyl-1-methylpyrrolidinium tris(pentafluoroethyl) trifluorophosphate. Tribol. Lett. **40**(2), 269–277 (2010)

D. Goswami, J.N. Thakker, P.C. Dhandhukia, Portraying mechanics of plant growth promoting rhizobacteria (PGPR): a review. Cogent Food Agric. **2**, 1–9 (2016)

M. Götz, J. Lefebvre, F. Mörs, A. McDaniel Koch, F. Graf, S. Bajohr, R. Reimert, T. Kolb, Renewable power-to-gas: a technological and economic review. Renew. Energy **85**, 1371–1390 (2016)

S. Gouda, R.G. Kerry, G. Das, S. Paramithiotis, H.-S. Shine, J.K. Patra, Revitalization of plant growth promoting rhizobacteria for sustainable development in agriculture. Microbiol. Res. **206**, 131–140 (2018)

V. Govindasamy, M. Senthilkumar, K. Annapurna, Effect of mustard rhizobacteria on wheat growth promotion under cadmium stress: characterization of acdS gene coding ACC deaminase. Ann. Microbiol. **65**, 1679–1687 (2014)

M. Graymore, F. Stagnitti, G. Allinson, Impacts of atrazine in aquatic ecosystems. Environ. Int. **26**, 483–495 (2001)

S. Gul, J.K. Whalen, Biochemical cycling of nitrogen and phosphorus in biochar amended soils—review paper. Soil Biol. Biochem. **103**, 1–15 (2016)

S. Guo, C. Li, Y. Zhang, Y. Wang, B. Li, M. Yang, X. Zhang, G. Liu, Experimental evaluation of the lubrication performance of mixtures of castor oil with other vegetable oils in MQL grinding of nickel-based alloy. J. Clean. Prod. **140**, 1060–1076 (2017)

G. Gupta, S.S. Parihar, N.K. Ahirwar, S.K. Sneni, V. Singh, Plant growth promoting rhizobacteria (PGPR): current and future prospects for development of sustainable agriculture. J. Microb. Biochem. Technol. **7**, 96–102 (2015a)

V. Gupta, S. Garg, N. Capalash, N. Gupta, P. Sharma, Production of thermoalkali- stable laccase and xylanase by co-culturing of *Bacillus* sp. and *B. halodurans* for biobleaching of kraft pulp and deinking of waste paper. Bioprocess Biosyst. Eng. **38**, 947–956 (2015b)

L. Habets, W. Driessen, Anaerobic treatment of pulp and paper mill effluents e status quo and new developments. Water Sci. Technol. **55**(6), 223–230 (2007)

G. Haider, D. Steffens, G. Moser, C. Müller, C.I. Kammann, Biochar reduced nitrate leaching and improved soil moisture content without yield improvements in a four-year field study. Agr. Ecosyst. Environ. **237**, 80–94 (2017)

M. Hardeman, *Feasibility Study for a Cast Steel Guideline*, M.Sc. thesis, KTH Industrial Engineering and Management, Stockholm, Sweden, 2015

M. Hartmann, B. Frey, J. Mayer, P. Mäder, F. Widmer, Distinct soil microbial diversity under long-term organic and conventional farming. ISME J. **9**, 1177–1194 (2015)

F. Hassan, R. Zulkifli, M.J. Ghazali, C.H. Azhari, Kenaf fiber composite in automotive industry: an overview. Int. J. Adv. Sci. Eng. Inf. Technol. **7**(1), 315–321 (2017)

H.S. Hassan, H.A. Abdel-Gawwad, S.R. Vásquez-García, I. Israde-Alcántara, N. Flores-Ramirez, J.L. Rico, M.S. Mohammed, Cleaner production of one-part white geopolymer cement using pre-treated wood biomass ash and diatomite. J. Clean. Prod. **209**, 1420–1428 (2019)

T. Hasunuma, F. Okazaki, N. Okai, K.Y. Hara, J. Ishii, A. Kondo, A review of enzymes and microbes for lignocellulosic biorefinery and the possibility of their application to consolidated bioprocessing technology. Biores. Technol. **135**, 513–522 (2013)

J.L. Havlin, S.L. Tisdale, W.L. Nelson, J.D. Beaton, *Soil Fertility and Nutrient Management: An Introduction to Nutrient Management*, 8 edn. (Pearson, Upper Saddle River, NJ, USA, 2014). ISBN 978-0130278241

T.R. Hawkins, B. Singh, G. Majeau-Bettez, A.H. Strømman, Comparative environmental life cycle assessment of conventional and electric vehicles. J. Ind. Ecol. **17**(1), 53–64 (2012)

S. He, X. Fan, S. Luo, N.R. Katukuri, R. Guo, Enhanced the energy outcomes from microalgal biomass by the novel biopretreatment. Energy Convers. Manage **135**, 291–296 (2017)

S.K. Hoekman, A. Broch, C. Robbins, B. Zielinska, L. Felix, Hydrothermal carbonization (HTC) of selected woody and herbaceous biomass feedstocks. Biomass Convers. Biorefinery **3**, 113126 (2013)

H. Holik, *Handbook of Paper and Board* (Wiley-VCH Verlag GmbH & Co.KGaA, Weinheim, Germany, 2006). ISBN 3-527-30997-7

P.M. Horton, J.M. Allwood, C. Cleaver, Implementing material efficiency in practice: a case study to improve the material utilisation of automotive sheet metal components. Resour. Conserv. Recycl. **145**, 49–66 (2019)

M.H. Hwang, N.J. Jang, S.H. Hyun, I.S. Kim, Anaerobic bio-hydrogen production from ethanol fermentation: the role of pH. J. Biotechnol. **111**, 297–309 (2004)

IEA, *Key World Energy Statistics 2015* (OECD/IEA, Paris, 2015), https://www.connaissancedesenergies.org/sites/default/files/pdf-actualites/keyworld_statistics_2015.pdf. Retrieved on 10 Mar 2019

A.D. Igalavithana, Y.S. Ok, A.R. Usman, M.I. Al-Wabel, P. Oleszczuk, S.S. Lee, The effects of biochar amendment on soil fertility, in *Agricultural and Environmental Applications of Biochar: Advances and Barriers* (Soil Science Society of America, Inc., USA, 2015)

A.D. Igalavithana, S.-E. Lee, Y.H. Lee, D.C.W. Tsang, J. Rinklebe, E.E. Kwon, Y.S. Ok, Heavy metal immobilization and microbial community abundance by vegetable waste and pine cone biochar of agricultural soils. Chemosphere **174**, 593–603 (2017)

P. Intanoo, P. Chaimongkol, S. Chavadej, Hydrogen and methane production from cassava wastewater using two-stage upflow anaerobic sludge blanket reactors (UASB) with an emphasis on maximum hydrogen production. Int. J. Hydrogen Energy **41**, 6107–6114 (2016)

IPPC Bureau, *Integrated Pollution Prevention and Control: BET in the Cement, Lime and Magnesium Oxide Manufacturing Industries* (European Commission, May 2013), http://eippcb.jrc.ec.europa.eu/reference/BREF/CLM_BATC_Published.pdf. Retrieved on 26 Apr 2018

S.A. Ishak, H. Hashim, T.S. Ting, Eco innovation strategies for promoting cleaner cement manufacturing. J. Clean. Prod. **136**(Part B), 133–149 (2016)

S. Islam, A.M. Akanda, A. Prova, M.T. Islam, M. Hossain, Isolation and identification of plant growth promoting rhizobacteria from cucumber rhizosphere and their effect on plant growth promotion and disease suppression. Front. Microbiol. **6**, 1–12 (2016)

A. Jain, A. Suhane, Research approach & prospects of non-edible vegetable oil as a potential resource for biolubricant—a review. Adv. Eng. Appl. Sci. Int. J. **1**, 23–32 (2012)

A. Jarosiewicz, M. Tomaszewska, Controlled-release NPK fertilizer encapsulated by polymeric membranes. J. Agric. Food Chem. **51**(2), 413–417 (2003)

I.S. Jawahir, K.E. Rouch, O.W. Dillon, L. Holloway Jr., A. Hall, Design for sustainability (DFS): new challenges in developing and implementing a curriculum for next generation design and manufacturing engineers. Int. J. Eng. Educ. **23**(6), 1053–1064 (2007)

S. Jeffery, M.B. Meinders, C.R. Stoof, T.M. Bezemer, T.F. van de Voorde, L. Mommer, J.W. Groenigen, Biochar application does not improve the soil hydrological function of a sandy soil. Geoderma **251**, 47–54 (2015)

N. Johnson, V. Krey, D.L. McCollum, S. Rao, K. Riahi, J. Rogelj, Stranded on a low-carbon planet: implications of climate policy for the phase-out of coal-based power plants. Technol. Forecast. Soc. Change **90**, 89–102 (2015)

L. Jönsson, B. Alriksson, N. Nilvebrant, Bioconversion of lignocellulose: inhibitors and detoxification. Biotechnol. Biofuels **6**, 1–10 (2013)

A. Kadier, M.S. Kalil, P. Abdeshahian, K. Chandrasekhar, A. Mohamed, N.F. Azman, W. Logroño, Y. Simayi, A.A. Hamid, Recent advances and emerging challenges in microbial electrolysis cells (MECs) for microbial production of hydrogen and value added chemicals. Renew. Sustain. Energy Rev. **61**, 501–525 (2016)

A. Kairytė, M. Kirpluks, A. Ivdre, U. Cabulis, S. Vaitkus, I. Pundienė, Cleaner production of polyurethane foam: replacement of conventional raw materials, assessment of fire resistance and environmental impact. J. Clean. Prod. **183**, 760–771 (2018)

S.S. Kamath, D. Sampathkumar, B. Bennehalli, A review on natural areca fibre reinforced polymer composite materials. Ciência e Tecnologia dos Materiais **28**(3), 106–128 (2017)

H. Kambo, A. Dutta, in *Hydrothermal Carbonization (HTC): an Innovative Process for the Conversion of Low Quality Lignocellulosic Biomass to Hydrochar for Replacing Coal*. Proceedings of the 9th Annual Green Energy Conference (IGEC-IX), Tianjin, China, 25–28 May 2014, pp. 25–28

C.N. Kanchiswamy, M. Malnoy, M.E. Maffei, Bioprospecting bacterial and fungal volatiles for sustainable agriculture. Trends Plant Sci. **20**(4), 206–211 (2015)

K. Kaur, G. Singh, V. Gupta, N. Capalash, P. Sharma, Impact of phosphate and other medium components on physiological regulation of bacterial laccase production. Formul. Eng. Biomater. **33**(2), 541–548 (2017)

R. Kessler, Lead-based decorative paints: where are they still sold—and why? Environ. Health Perspect. **122**, A96–A103 (2014)

N. Khalifa, L.F. Yousef, A short report on changes of quality indicators for a sandy textured soil after treatment with biochar produced from fronds of date palm. Energy Procedia **74**, 960–965 (2015)

M.A. Khan, H.H. Ngo, W. Guo, Y. Liu, X. Zhang, J. Guo, S.W. Chang, D.D. Nguyen, J. Wang, Biohydrogen production from anaerobic digestion and its potential as renewable energy. Renew. Energy **129**(B), 754–768 (2017)

W. Khetkorn, N. Khanna, A. Incharoensakdi, P. Lindblad, Metabolic and genetic engineering of cyanobacteria for enhanced hydrogen production. Biofuels **4**, 535–561 (2013)

W. Khetkorn, R.P. Rastogi, A. Incharoensakdi, P. Lindblad, D. Madamwar, A. Pandey, C. Larroche, Microalgal hydrogen production—a review. Biores. Technol. **243**, 1194–1206 (2017)

C.S. Khor, G. Lalchand, A review on sustainable power generation in Malaysia to 2030: historical perspective, current assessment, and future strategies. Renew. Sustain. Energy Rev. **29**, 952–960 (2014)

S. Kim, S.K. Moon, Sustainable platform identification for product family design. J. Clean. Prod. **143**, 567–581 (2017)

J.M. Kimetu, J. Lehmann, S.O. Ngoze, D.N. Mugendi, J.M. Kinyangi, S. Riha, L. Verchot, J.W. Recha, A.N. Pell, Reversibility of soil productivity decline with organic matter of differing quality along a degradation gradient. Ecosystems **11**, 726–739 (2008)

E. Kleinaitė, V. Jaška, B. Tvaska, I. Matijošytė, A cleaner approach for biolubricant production using biodiesel as a starting material. J. Clean. Prod. **75**, 40–44 (2014)

S. Kloss, F. Zehetner, B. Wimmer, J. Buecker, F. Rempt, G. Soja, Biochar application to temperate soils: effects on soil fertility and crop growth under greenhouse conditions. J. Plant Nutr. Soil Sci. **177**, 3–15 (2014)

I.K. Kookos, Y. Pontikes, G.N. Angelopoulos, G. Lyberatos, Classical and alternative fuel mix optimization in cement production using mathematical programming. Fuel **90**, 1277–1284 (2011)

R. Kothari, D. Buddhi, R.L. Sawhney, Comparison of environmental and economic aspects of various hydrogen production methods. Renew. Sustain. Energy Rev. **12**, 553–563 (2008)

T. Krey, N. Vassilev, C. Baum, B. Eichler-Lobermann, Effects of long-term phosphorus application and plant-growth promoting rhizobacteria on maize phosphorus nutrition under field conditions. Eur. J. Soil Biol. **55**, 124–130 (2013)

J.I.N. Kumar, A. Bora, M.K. Amb, Chronic toxicity of the triazole fungicide tebuconazole on a heterocystous, nitrogen-fixing rice paddy field cyanobacterium, *Westiellopsis prolifica* Janet. J. Microbiol. Biotechnol. **20**, 1134–1139 (2010)

R. Kumar, J. Kaur, S. Jain, A. Kumar, Optimization of laccase production from *Aspergillus flavus* by design of experiment technique: partial purification and characterization. J. Genet. Eng. Biotechnol. **14**, 125–131 (2016)

P. Kumararaja, K.M. Manjaiah, S.C. Datta, B. Sarkar, Remediation of metal contaminated soil by aluminium pillared bentonite: synthesis, characterisation, equilibrium study and plant growth experiment. Appl. Clay Sci. **137**, 115–122 (2017)

S. Kuppusamy, P. Thavamani, M. Megharaj, K. Venkateswarlu, R. Naidu, Agronomic and remedial benefits and risks of applying biochar to soil: current knowledge and future research directions. Environ. Int. **87**, 1–12 (2016)

M. Laghari, M.S. Mirjat, Z. Hu, S. Fazal, B. Xiao, M. Hu, Z. Chen, D. Guo, Effects of biochar application rate on sandy desert soil properties and sorghum growth. Catena **135**, 313–320 (2015)

J. Lee, A.K. Sarmah, E.E. Kwon, Production and formation of biochar, in *Biochar from Biomass and Waste*, ed. by Y. Ok, D. Tsang, N. Bolan, J. Novak (Elsevier, Amsterdam, The Netherlands, 2018), pp. 3–18. ISBN 978-0128117293

J. Lehmann, S. Joseph, Biochar for environmental management: an introduction, in *Biochar for Environmental Management: Science, Technology and Implementation*, 2nd edn., ed. by J. Lehmann, S. Joseph (Earthscan from Routledge, London, 2015), pp. 1–1214. ISBN 978-0415704151

M. Lehner, R. Tichler, H. Steinmüller, M. Koppe, *Power-to-Gas: Technology and Business Models*, 1st edn. (Springer International Publishing, Switzerland, 2014). ISBN 978-3-319-03995-4

W.-H. Leong, J.-W. Lim, M.-K. Lam, Y. Uemura, Y.-C. Ho, Third generation biofuels: a nutritional perspective in enhancing microbial lipid production. Renew. Sustain. Energy Rev. **91**, 950–961 (2018)

B. Lin, Resilience in agriculture through crop diversification: adaptive management for environmental change. Bioscience **61**(3), 183–193 (2011)

G.Z. Lin, R.F. Peng, Q. Chen, Z.G. Wu, L. Du, Lead in housing paints: an exposure source still not taken seriously for children lead poisoning in China. Environ. Res. **109**, 1–5 (2009)

W. Liu, C. Ye, Y. Chen, Z. Ou, D. Sun, Tribological behavior of sialon ceramics sliding against steel lubricated by fluorine-containing oils. Tribol. Int. **35**(8), 503–509 (2002)

Y. Liu, Z. Tong, R.K. Prud'homme, Stabilized polymeric nanoparticles for controlled and efficient release of bifenthrin. Pest Manage. Sci. **64**, 808–812 (2008)

B.H. Liu, L. Chen, L.X. Huang, Y.S. Wang, Y.H. Li, Distribution of volatile organic compounds (VOCs) in surface water, soil, and groundwater within a chemical industry park in Eastern China. Water Sci. Technol. **71**, 259–267 (2015)

S. Lü, C. Feng, C. Gao, X. Wang, X. Xu, X. Bai, N. Gao, M. Liu, Multifunctional environmental smart fertilizer based on L-aspartic acid for sustained nutrient release. J. Agric. Food Chem. **64**(24), 4965–4974 (2016)

R. Łukajtis, I. Hołowacz, K. Kucharska, M. Glinka, P. Rybarczyk, A. Przyjazny, M. Kaminski, Hydrogen production from biomass using dark fermentation. Renew. Sustain. Energy Rev. **91**, 665–694 (2018)

C. Luttropp, J. Lagerstedt, EcoDesign and the ten golden rules: generic advice for merging environmental aspects into product development. J. Clean. Prod. **14**, 1396–1408 (2006)

A.J.A. Magalhães, F.J.G. Silva, R.D.S.G. Campilho, A novel concept of bent wires sorting operation between workstations in the production of automotive parts. J. Braz. Soc. Mech. Sci. Eng. **41**, 25–36 (2019)

J. Major, M. Rondon, D. Molina, S.J. Riha, J. Lehmann, Maize yield and nutrition during 4 years after biochar application to a Colombia savanna oxisol. Plant Soil **333**, 117–128 (2010)

R. Malkapuram, V. Kumar, Y.S. Negi, Recent development in natural fiber reinforced polypropylene composites. J. Reinf. Plast. Compos. **28**(10), 1169–1189 (2009)

T. Mang, W. Dresel, *Lubricants and Lubrication*, 1st edn. (Wiley-VCH, NJ, USA, 2007). ISBN 978-3-527-29536-4

T. Mang, W. Dresel, *Lubricants and Lubrication*, 3rd edn. (Wiley-VCH, NJ, USA, 2017). ISBN 978-3-527-32670-9

K.M. Manjaiah, R. Mukhopadhyay, R. Paul, S.C. Datta, P. Kumararaja, B. Sarkar, Clay minerals and zeolites for environmentally sustainable agriculture, in *Modified Clay and Zeolite Nanocomposite Materials—Environmental and Pharmaceutical Applications,* 1st edn., ed. by M. Mercurio, B. Sarkar, A. Langella (Elsevier, Amsterdam, The Netherlands, 2018), p. 362. ISBN 978-0128146170

C. Mao, Y. Feng, X. Wang, G. Ren, Review on research achievements of biogas from anaerobic digestion. Renew. Sustain. Energy Rev. **45**, 540–555 (2015)

P.A. Marchand, Basic substances: an opportunity for approval of low-concern substances under EU pesticide regulation. Pest Manage. Sci. **71**, 1197–1200 (2015)

A. Marone, G. Izzo, L. Mentuccia, G. Massini, P. Paganin, S. Rosa, C. Varrone, A. Signorini, Vegetable waste as substrate and source of suitable microflora for biohydrogen production. Renew. Energy **68**, 6–13 (2014)

J.M. Martinez-Duart, J. Hernandez-Moro, S. Serrano-Calle, R. Gomez-Calvet, M. Casanova-Molina, New frontiers in sustainable energy production and storage. Vacuum **122**(B), 369–375 (2015)

P. Martinez-Jurado, J. Moyano-Fuentes, Lean management, supply chain management and sustainability: a literature review. J. Clean. Prod. **85**, 134–150 (2014)

M. Martins, R. Godina, C. Pimentel, F.J.G. Silva, J.C.O. Matias, A practical study of the application of SMED to electron-beam machining in automotive industry. Procedia Manuf. **17**, 647–654 (2018)

E. Mastronardi, P. Tsae, X. Zhang, C.M. Monreal, M.C, DeRosa, Strategic role of nanotechnology in fertilizers: potential and limitations, in *Nanotechnologies in Food and Agriculture*, ed. by M. Rai, C. Ribeiro, L. Mattoso, N. Duran (Springer, Cham, Switzerland, 2015), pp. 25–67. ISBN 978-3-319-14023-0

A.R. Mayyas, Q. Shen, A.T. Mayyas, M. Abdelhamid, D. Shan, A. Qattawi, M. Omar, Using quality function deployment and analytical hierarchy process for material selection of body-in-white. Mater. Des. **32**, 2771–2778 (2011)

A. Mayyas, A. Qattawi, M. Omar, Dongri Shan, Design for sustainability in automotive industry: a comprehensive review. Renew. Sustain. Energy Rev. **16**, 1845–1862 (2012)

A.T. Mayyas, A. Qattawi, A.R. Mayyas, M. Omar, Quantifiable measures of sustainability: a case study of materials selection for eco-lightweight auto-bodies. J. Clean. Prod. **40**, 177–189 (2013)

A.T. Mayyas, A.R. Mayyas, M. Omar, *Sustainable lightweight vehicle design: a case study in eco-material selection for body-in-white, design, manufacturing, analysis and performance*, in *Lightweight Composite Structures in Transport* (Elsevier, Amsterdam, The Netherlands, 2016), pp. 267–302. ISBN 978-1782423256

K. Mazloomi, C. Gomes, Hydrogen as an energy carrier: prospects and challenges. Renew. Sustain. Energy Rev. **16**(5), 3024–3033 (2012)

J. McAuley, Global sustainability and key needs in future automotive design. Environ. Sci. Technol. **37**(23), 5414–5416 (2003)

J. McNutt, Q. He, Development of biolubricants from vegetable oils via chemical 1126 modification. J. Ind. Eng. Chem. **36**, 1–12 (2016)

D. Mohan, C.U. Pittman, P.H. Steele, Pyrolysis of wood/biomass for bio-oil: a critical review. Energy Fuels **20**, 848–889 (2006)

S.V. Mohan, G. Mohanakrishna, S.S. Reddy, B.D. Raju, K.S.R. Rao, P.N. Sarma, Self-immobilization of acidogenic mixed consortia on mesoporous material (SBA-15) and activated carbon to enhance fermentative hydrogen production. Int. J. Hydrogen Energy **33**, 6133–6142 (2008)

H. Mohrbacher, Reverse metallurgical engineering towards sustainable manufacturing of vehicles using Nb and Mo alloyed high performance steels. Adv. Manuf. **1**, 28–41 (2013)

T. Morato, M. Vaezi, A. Kumar, Assessment of energy production potential from agricultural residues in Bolivia. Renew. Sustain. Energy Rev. **102**, 14–23 (2019)

G. Mordukhovich, J. Qu, J.Y. Howe, S. Bair, B. Yu, H. Luo, D.J. Smolenski, P.J. Blau, B.G. Bunting, S. Dai, A low-viscosity ionic liquid demonstrating superior lubricating performance from mixed to boundary lubrication. Wear **301**(1), 740–746 (2013)

B.M.D.N. Moreira, R.M. Gouveia, F.J.G. Silva, R.D.S.G. Campilho, A novel concept of production and assembly processes integration. Procedia Manuf. **11**, 1385–1395 (2017)

A.-T.H. Mossa, Green pesticides: essential oils as biopesticides in insect-pest management. J. Environ. Sci. Technol. **9**, 354–378 (2016)

A. Mukherjee, A.R. Zimmerman, R. Hamdan, W.T. Cooper, Physicochemical changes in pyrogenic organic matter (biochar) after 15 months of field aging. Solid Earth **5**(2), 693–704 (2014)

R. Mukhopadhyay, K.M. Manjaiah, S.C. Datta, R.K. Yadav, B. Sarkar, Inorganically modified clay minerals: preparation, characterization, and arsenic adsorption in contaminated water and soil. Appl. Clay Sci. **147**, 1–10 (2017)

P. Mullai, M.K. Yogeswari, K. Sridevi, Optimisation and enhancement of biohydrogen production using nickel nanoparticles—a novel approach. Biores. Technol. **141**, 212–219 (2013)

C.N. Mulligan, *Environmental Biotreatment: Technologies for Air, Water, Soil, and Waste* (Government Institutes, Maryland, USA, 2002)

S. Nagar, R.K. Jain, V.V. Thakur, V.K. Gupta, Biobleaching application of cellulose poor and alkali stable xylanase from *Bacillus pumilus* SV-85S. 3 Biotech **3**(4), 277–285 (2013)

D. Nagarajan, D.-J. Lee, A. Kondo, J.-S. Chang, Recent insights into biohydrogen production by microalgae—from biophotolysis to dark fermentation. Biores. Technol. **227**, 373–387 (2017)

D.A. Nagegowda, Plant volatile terpenoid metabolism: biosynthetic genes, transcriptional regulation and subcellular compartmentation. FEBS Lett. **584**, 2965–2973 (2010)

M.Y. Naz, S.A. Sulaiman, Slow release coating remedy for nitrogen loss from conventional urea: a review. J. Controlled Release **225**(10), 109–120 (2016)

C. Nguong, S. Lee, S. Debnath, A review on natural fibre reinforced polymer composites. Int. J. Chem. Biomol. Metall. Mater. Sci. Eng. **7**(1), 33–40 (2013)

N.K. Niazi, I. Bibi, M. Shahid, Y.S. Ok, E.D. Burton, H. Wang, S.M. Shaheen, J. Rinklebe, A. Lüttge, Arsenic removal by perilla leaf biochar in aqueous solutions and groundwater: an integrated spectroscopic and microscopic examination. Environ. Pollut. **232**, 31–42 (2018)

P. Nieuwenhuis, E. Katsifou, More sustainable automotive production through understanding decoupling points in leagile manufacturing. J. Clean. Prod. **95**, 232–241 (2015)

A. Nordelöf, M. Messagie, A.M. Tillman, M.L. Söderman, J. Van Mierlo, Environmental impacts of hybrid, plug-in hybrid, and battery electric vehicles—what can we learn from life cycle assessment? Int. J. Life Cycle Assess. **19**, 1866–1890 (2014)

P.M.S. Nunes, F.J.G. Silva, Increasing flexibility and productivity in small assembly operations: a case study, in *Advances in Sustainable and Competitive Manufacturing Systems*, ed. by A. Azevedo. Lecture Notes in Mechanical Engineering (Springer, 2013), pp. 329–340. https://doi.org/10.1007/978-3-319-00557-7

V. Nunes, F.J.G. Silva, M.F. Andrade, R. Alexandre, A.P.M. Baptista, Increasing the lifespan of high-pressure die cast molds subjected to severe wear. Surf. Coat. Technol. **332**, 319–331 (2017)

D. O'Connor, D. Hou, J. Yeb, Y. Zhang, Y.S. Ok, Y. Song, F. Coulon, T. Peng, L. Tian, Lead-based paint remains a major public health concern: a critical review of global production, trade, use, exposure, health risk, and implications. Environ. Int. **121**, 85–101 (2018)

U. Ogbonnaya, K.T. Semple, Impact of biochar on organic contaminants in soil: a tool for mitigating risk? Agronomy **3**, 349–375 (2013)

J. Olivier, G. Janssens-Hanssens, J.A.H.W. Peters, *Trends in Global CO_2 Emissions, 2012 Report* (PBL Netherlands Environmental Assessment Agency, The Hague, 2012), https://www.pbl.nl/sites/default/files/cms/publicaties/PBL_2012_Trends_in_global_CO2_emissions_500114022.pdf. Retrieved on 20 Jan 2019

J.G.J. Olivier, G. Janssens-Maenhout, M. Muntean, J.A.H.W. Peters, *Trends in Global CO_2 Emissions, 2014 Report* (PBL Netherlands Environmental Assessment Agency, The Hague, 2014), https://www.pbl.nl/sites/default/files/cms/publicaties/PBL_2014_Trends_in_global_CO2_emisions_2014_1490.pdf. Retrieved on 20 Jan 2019

M.O. Omondi, X. Xia, A. Nahayo, X. Liu, P.K. Korai, G. Pan, Quantification of biochar effects on soil hydrological properties using meta-analysis of literature data. Geoderma **274**, 28–34 (2016)

O. Onay, Influence of pyrolysis temperature and heating rate on the production of bio-oil and char from safflower seed by pyrolysis, using a well-swept fixed-bed reactor. Fuel Process. Technol. **88**, 523–531 (2007)

O. Onay, O.M. Kockar, Slow, fast and flash pyrolysis of rapeseed. Renew. Energy **28**, 2417–2433 (2003)

R. Ortiz-Castro, M.M. Trujillo, J. López-Bucio, N-acyl-L-homoserine lactones: a class of bacterial quorum-sensing signals alter post-embryonic root development in *Arabidopsis thaliana*. Plant Cell Environ. **31**, 1497–1509 (2008)

R. Otero, M.I. López, D. Esquivel, J.M. Fernández, F.J. Romero-Salguero, Removal of S-metolachlor herbicide from aqueous solutions by meso and microporous organosilica materials. Microporous Mesoporous Mater. **278**, 35–43 (2019)

J.S. Oyepata, O. Obodeh, Cement production optimization modeling: a case study BUA plant. J. Eng. Technol. **7**(4), 53–58 (2015)

J. Palmer, *Mechanical Recycling of Automotive Composites for Use as Reinforcement in Thermoset Composites*, Ph.D. Thesis in Mechanical Engineering, University of Exeter, 2009. https://ore.exeter.ac.uk/repository/bitstream/handle/10036/72313/PalmerJ.pdf?sequence=1&isAllowed=y. Retrieved on 10 Mar 2019

T.M. Panchal, A. Patel, D.D. Chauhan, M. Thomas, J.V. Patel, A methodological review on bio-lubricants from vegetable oil based resources. Renew. Sustain. Energy Rev. **70**, 65–70 (2017)

J.H. Park, Y.S. Ok, S.H. Kim, S.W. Kang, J.S. Cho, J.S. Heo, R.D. Delaune, D.C. Seo, Characteristics of biochars derived from fruit tree pruning wastes and their effects on lead adsorption. J. Korean Soc. Appl. Biol. Chem. **58**, 751–760 (2015)

S.T. Patel, F.P. Minocheherhomji, Review: plant growth promoting rhizobacteria: blessing to agriculture. Int. J. Pure Appl. Biosci. **6**, 481–492 (2018)

R. Pavela, G. Benelli, Essential oils as ecofriendly biopesticides? Challenges and constraints. Trends Plant Sci. **21**, 1000–1007 (2016)

D.W. Pennington, J. Potting, G. Finnveden, E. Lindeijer, O. Jolliet, T. Rydberg, G. Rebitzer, Life cycle assessment. Part 2: current impact assessment practice. Environ. Int. **30**(5), 721–739 (2004)

A.E.S. Pereira, R. Grillo, N.F.S. Mello, A.H. Rosa, L.F. Fraceto, Application of poly(epsilon-caprolactone) nanoparticles containing atrazine herbicide as an alternative technique to control weeds and reduce damage to the environment. J. Hazard. Mater. **268**, 207–215 (2014)
M. Pergola, A. Persiani, A.M. Palese, V. Di Meo, V. Pastore, C. D'Adamo, G. Celano, Composting: the way for a sustainable agriculture. Appl. Soil. Ecol. **123**, 744–750 (2018)
E.D. Perry, F. Ciliberto, D.A. Hennessy, G. Moschini, Genetically engineered crops and pesticide use in U.S. maize and soybeans. Sci. Adv. **2**(8), e1600850 (2016)
B. Piechulla, J. Degenhardt, The emerging importance of microbial volatile organic compounds. Plant Cell Environ. **37**, 811–812 (2013)
H. Pinto, F.J.G. Silva, Optimisation of die casting process in Zamak alloys. Procedia Manuf. **11**, 517–525 (2017)
E.N. Pistikopoulos, S.K. Stefanis, A.G. Livingston, Pollution prevention via process and product modification. AIChE Symp. Ser. **90**(303), 139–150 (1994)
D. Pokhrel, T. Viraraghavan, Treatment of pulp and paper mill wastewater—a review. Sci. Total Environ. **333**, 37–58 (2004)
G.C. Premier, J.R. Kim, J. Massanet-Nicolau, G. Kyazze, S.R.R. Esteves, B.K.V. Penumathsa, J. Rodríguez, J. Maddy, R.M. Dinsdale, A.J. Guwy, Integration of biohydrogen, biomethane and bioelectrochemical systems. Renew. Energy **49**, 188–192 (2013)
J. Pretty, Agricultural sustainability: concepts, principles and evidence. Philos. Trans. R. Soc. Lond. B Biol. Sci. **363**, 447–465 (2008)
J. Qu, P.J. Blau, S. Dai, H. Luo, H.M. Meyer, J.J. Truhan, Tribological characteristics of aluminum alloys sliding against steel lubricated by ammonium and imidazolium ionic liquids. Wear **267**(5), 1226–1231 (2009a)
J. Qu, P.J. Blau, S. Dai, H. Luo, H.M. Meyer, Ionic liquids as novel lubricants and additives for diesel engine applications. Tribol. Lett. **35**(3), 181–189 (2009b)
L.-M. Raboin, A.H.D. Razafimahafaly, M.B. Rabenjarisoa, B. Rabary, J. Dusserre, T. Thierry Becquer, Improving the fertility of tropical acid soils: liming versus biochar application? A long term comparison in the highlands of Madagascar. Field Crops Res. **199**, 99–108 (2016)
R. Rahim, A.A.A. Raman, Cleaner production implementation in a fruit juice production plant. J. Clean. Prod. **101**, 215–221 (2015)
S.N.A. Rahman, M.S. Masdar, M.I. Rosli, E.H. Majlan, T. Husaini, S.K. Kamarudin, W.R.W. Daud, Overview biohydrogen technologies and application in fuel cell technology. Renew. Sustain. Energy Rev. **66**, 137–162 (2016)
P. Randolph, R.R. Bansode, O.A. Hassan, D. Rehrah, R. Ravella, M.R. Reddy, D.W. Watts, J.M. Novak, M. Ahmedna, Effect of biochars produced from solid organic municipal waste on soil quality parameters. J. Environ. Manage. **192**, 271–280 (2017)
M. Raugei, D. Morrey, A. Hutchinson, P. Winfield, A coherent life cycle assessment of a range of lightweighting strategies for compact vehicles. J. Clean. Prod. **108**, 1168–1176 (2015)
R. Razaghifard, Algal biofuels. Photosynth. Res. **117**, 207–219 (2013)
M.A. Reuter, A. van Schaik, *The Effect of Design on Recycling Rates for Cars* (Delft University of Technology, Delft, the Netherlands, 2005), https://www.researchgate.net/profile/Markus_Reuter3/publication/237295185_The_Effect_of_Design_on_Recycling_Rates_for_Cars/links/0deec52d540334dec1000000/The-Effect-of-Design-on-Recycling-Rates-for-Cars.pdf. Retrieved on 10 Mar 2019
M.C.S. Ribeiro, J.P. Meixedo, A. Fiúza, M.L. Dinis, A.C.M. Castro, F.J.G. Silva, C. Costa, F. Ferreira, M.R. Alvim, Mechanical behaviour analysis of polyester polymer mortars modified with recycled GFRP waste materials. World Acad. Sci. **75**, 365–371 (2011)
M.C.S. Ribeiro, A. Fiúza, A.C.M. Castro, F.G. Silva, M.L. Dinis, J.P. Meixedo, M.R. Alvim, Mix design process of polyester polymer mortars modified with recycled GFRP waste materials. Compos. Struct. **105**, 300–310 (2013)

M.C.S. Ribeiro, A.C. Meira-Castro, F.G. Silva, J. Santos, J.P. Meixedo, A. Fiúza, M.L. Dinis, M.R. Alvim, Re-use assessment of thermoset composite wastes as aggregate and filler replacement for concrete-polymer composite materials: a case study regarding GFRP pultrusion wastes. Resour. Conserv. Recycl. **104**(Part B), 417–426 (2015)

M. Ritthoff, K.O. Schallaböck, *Ökobilanzierung der Elektromobilität. Themen und Stand der Forschung. Teilbericht im Rahmen der Umweltbegleitforschung Elektromobilität im Förderschwerpunkt. Modellregionen Elektromobilität* (in German) (Wuppertal Institut für Klima, Umwelt, Energie GmbH, Wuppertal, Germany, 2012), https://wupperinst.org/uploads/tx_wupperinst/Elektromobilitaet_TB_Oekobilanzen.pdf. Retrieved on 10 Mar 2019

M.V. Rodionova, R.S. Poudyal, I. Tiwari, R.A. Voloshin, S.K. Zharmukhamedov, H.G. Nam, B.K. Zayadan, B.D. Bruce, H.J.M. Hou, S.I. Allakhverdiev, Biofuel production: challenges and opportunities. Int. J. Hydrogen Energy **42**(12), 8450–8461 (2016)

F. Ronsse, S. Van Hecke, D. Dickinson, W. Prins, Production and characterization of slow pyrolysis biochar: influence of feedstock type and pyrolysis conditions. GCB Bioenergy **5**, 104–115 (2013)

C. Rosa, F.J.G. Silva, L.P. Ferreira, Improving the quality and productivity of steel wire-rope assembly lines for the automotive industry. Procedia Manuf. **11**, 1035–1042 (2017a)

C. Rosa, F.J.G. Silva, L.P. Ferreira, R.D.S.G. Campilho, SMED methodology: the reduction of setup times for steel wire-ropes assembly lines in the automotive industry. Procedia Manuf. **13**, 1034–1042 (2017b)

C. Rosa, F.J.G. Silva, L.P. Ferreira, M.T. Pereira, R.M. Gouveia, Establishing standard methodologies to improve the production rate of assembly lines used for low added-value products. Procedia Manuf. **17**, 555–562 (2018)

A. Rosier, F.H.V. Medeiros, H.P. Bais, Defining plant growth promoting rhizobacteria molecular and biochemical networks in beneficial plant microbe interactions. Plant Soil **428**, 35–55 (2018)

R.M.A. Saboya, J.A. Cecilia, C. García-Sancho, A.V. Sales, F.M.T. de Luna, E. Rodríguez-Castellon, C.L. Cavalcante, Synthesis of biolubricants by the esterification of free fatty acids from castor oil with branched alcohols using cationic exchange resins as catalysts. Ind. Crops Prod. **104**, 52–61 (2017)

J. Shah, M.R. Jan, Tasmia, Magnetic chitosan graphene oxide composite for solid phase extraction of phenylurea herbicides. Carbohyd. Polym. **199**, 461–472 (2018)

J. Sanes, F. Carrion, M. Bermudez, G. Martinez-Nicolas, Ionic liquids as lubricants of polystyrene and polyamide 6-steel contacts. Preparation and properties of new polymer-ionic liquid dispersions. Tribol. Lett. **21**(2), 121–133 (2006)

J. Sanes, F.-J. Carrión-Vilches, M.-D. Bermúdez, New epoxy-ionic liquid dispersions. Room temperature ionic liquid as lubricant of epoxy resin-stainless steel contacts. e-Polymers **7**(1), 48–59 (2007)

J. Santos, R.M. Gouveia, F.J.G. Silva, Designing a new sustainable approach to the change for lightweight materials in structural components used in truck industry. J. Clean. Prod. **164**, 115–123 (2017)

S. Sarkar, S.C. Datta, D.R. Biswas, Synthesis and characterization of nanoclay polymer composites from soil clay with respect to their water-holding capacities and nutrient-release behavior. J. Appl. Polym. Sci. **131**(6), 39951 (2013)

F. Sarmiento, J.A. Leigh, W.B. Whitman, Genetic systems for hydrogenotrophic methanogens. Methods Enzymol. **494**, 43–73 (2011)

T. Saurabh, M. Patnaik, S.L. Bhagat, V.C. Renge, Epoxidation of vegetable oils: a review. Int. J. Adv. Eng. Technol. **2**, 491–501 (2011)

P. Schenk, S. Thomas-Hall, E. Stephens, U. Marx, J. Mussgnug, C. Posten, O. Kruse, B. Hankamer, Second generation biofuels: high-efficiency microalgae for biodiesel production. BioEnergy Res. **1**(1), 20–43 (2008)

J.-P. Schöggl, R.J. Baumgartner, D. Hofer, Improving sustainability performance in early phases of product design: a checklist for sustainable product development tested in the automotive industry. J. Clean. Prod. **140**(Part 3), 1602–1617 (2017)

U. Schuchardt, R. Sercheli, R.M. Vargas, Transesterification of vegetable oils: a review. J. Braz. Chem. Soc. **9**, 199–210 (1998)
P.T. Sekoai, C.N.M. Oumaa, S.P. du Preez, P. Modisha, N. Engelbrecht, D.G. Bessarabov, A. Ghimire, Application of nanoparticles in biofuels: an overview. Fuel **237**, 380–397 (2019)
D.K. Selvaraj, *Study on Feasibility and Viability of Applying Eco-friendly Material for the "be"-car Bonnet for a Sustainable Automotive Part*, M.Sc. Dissertation, Department of Mechanical Engineering, ISEP—School of Engineering, Polytechnic of Porto, Porto, Portugal, 2018
S.R. Senthilkumar, M. Dempsey, C. Krishnan, P. Gunasekaran, Optimization of biobleaching of paper pulp in an expanded bed bioreactor with immobilized alkali stable xylanase by using response surface methodology. Biores. Technol. **99**, 7781–7787 (2008)
H. Shapira, A. Ketchie, M. Nehe, The integration of design thinking and strategic sustainable development. J. Clean. Prod. **140**(1), 277–287 (2017)
V. Sharma, V. Krishnan, Fabrication of highly sensitive biomimetic SERS substrates for detection of herbicides in trace concentration. Sens. Actuators B Chem. **262**, 710–719 (2018)
A. Sharma, V.V. Thakur, A. Shrivastava, R.K. Jain, R.M. Mathur, R. Gupta, R.C. Kuhad, Xylanase and laccase based enzymatic kraft pulp bleaching reduces adsorbable organic halogen (AOX) in bleach effluents: a pilot scale study. Bioresour. Technol. **169**, 96–102 (2014)
P. Sharma, C. Sood, G. Singh, N. Capalash, An eco-friendly process for biobleaching of eucalyptus kraft pulp with xylanase producing *Bacillus halodurans*. J. Clean. Prod. **87**, 966–970 (2015)
Y. Shen, J.L. Linville, M. Urgun-Demirtas, R.P. Schoene, S.W. Snyder, Producing pipeline-quality biomethane via anaerobic digestion of sludge amended with corn stover biochar with in-situ CO_2 removal. Appl. Energy **158**, 300–309 (2015)
Z. Shu, J. Zhou, Y. Wang, A novel approach of preparing press-powders for cleaner production of ceramic tiles. J. Clean. Prod. **18**, 1045–1051 (2010)
E.P. Sichieri, G.C. Salado, *Materiais para construções vi (tintas para construção Civil)* (in Portuguese). Textbook of the Architectural Group Terra—arquitetura, tecnologia e materiais (São Paulo University. Department of architecture and urbanism, São Carlos, Brasil, 2007), pp. 35–92
R. Siddique, Utilization of municipal solid waste (MSW) ash in cement and mortar. Resour. Conserv. Recycl. **54**, 1037–1047 (2010)
F.J.G. Silva, A.P.M. Baptista, E. Pereira, V. Teixeira, Q.H. Fan, A.J.S. Fernandes, F.M. Costa, Microwave plasma chemical vapour deposition diamond nucleation on ferrous substrates with Ti and Cr interlayers. Diam. Relat. Mater. **11**, 1617–1622 (2002)
F.J.G. Silva, A.J.S. Fernandes, F.M. Costa, A.P.M. Baptista, E. Pereira, A new interlayer approach for CVD diamond coating of steel substrates. Diam. Relat. Mater. **13**, 828–833 (2004)
F.J.G. Silva, R.P. Martinho, R.J.D. Alexandre, A.P.M. Baptista, Increasing the wear resistance of molds for injection of glass fiber reinforced plastics. Wear **271**, 2494–2499 (2011)
F.J.G. Silva, F. Ferreira, C. Costa, M.C.S. Ribeiro, A.C. Meira Castro, Comparative study about heating systems for pultrusion process. Compos. B **43**, 1823–1829 (2012a)
F.J.G. Silva, R.P. Martinho, R.J.D. Alexandre, A.P.M. Baptista, Wear resistance of TiAlSiN thin coatings. J. Nanosci. Nanotechnol. **12**, 9094–9101 (2012b)
F.J.G. Silva, F. Ferreira, M.C.S. Ribeiro, A.C.M. Castro, M.R.A. Castro, M.L. Dinis, A. Fiúza, Optimising the energy consumption on pultrusion process. Compos. B **57**, 13–20 (2014)
F.J.G. Silva, E. Amorim, A. Baptista, G. Pinto, R.D.S.G. Campilho, M.R.A. Castro, Producing hybrid pultruded structural products based on preforms. Compos. B **116**, 325–332 (2017a)
F. Silva, R. Martinho, M. Andrade, A. Baptista, R. Alexandre, Improving the wear resistance of moulds for the injection of glass fibre-reinforced plastics using PVD coatings: a comparative study. Coatings **7**, 28–38 (2017b)
F.J.G. Silva, A. Baptista, G. Pinto, R.D.S.G. Campilho, M.C.S. Ribeiro, Characterization of hybrid pultruded structural products based on preforms. Compos. B **140**, 16–26 (2018a)
C.R.M. Silva, F.J.G. Silva, R.M. Gouveia, Investigations on the edge crack defect in dual-phase steel stamping process. Procedia Manuf. **17**, 737–745 (2018b)

C.L. Simões, R. Figueiredo de Sá, C.J. Ribeiro, P. Bernardo, A.J. Pontes, C.A. Bernardo, Environmental and economic performance of a car component: assessing new materials, processes and designs. J. Clean. Prod. **118**, 105–117 (2016)

I. Sindhu, S. Chhibber, N. Caplash, P. Sharma, Production of cellulase-free xylanase from *Bacillus megaterium* by solid state fermentation for biobleaching of pulp. Curr. Microbiol. **53**, 167–172 (2006)

G. Singh, N. Ahuja, M. Batish, N. Capalash, P. Sharma, Biobleaching of wheat straw-rich soda pulp with alkalophilic laccase from γ-proteobacterium JB: optimization of process parameters using response surface methodology. Biores. Technol. **99**, 7472–7479 (2008)

G. Singh, N. Capalash, P. Sharma, Performance of an alkalophilic and halotolerant laccase from γ-proteobacterium JB in the presence of industrial pollutants. J. Gen. Applied Microbiol. **55**, 283–289 (2010)

R. Singh, J.N. Babu, R. Kumar, P. Srivastava, P. Singh, A.S. Raghubanshi, Multifaceted application of crop residue biochar as a tool for sustainable agriculture: an ecological perspective. Ecol. Eng. **77**, 324–347 (2015a)

G. Singh, K. Kaur, S. Puri, P. Sharma, Critical factors affecting laccase-mediated biobleaching of pulp in paper industry. Appl. Microbiol. Biotechnol. **99**, 155–164 (2015b)

M. Singh, B. Sarkar, B. Biswas, N.S. Bolan, G.J. Churchman, Relationship between soil clay mineralogy and carbon protection capacity as influenced by temperature and moisture. Soil Biol. Biochem. **109**, 95–106 (2017)

S.K. Singh, S. Kulkarni, V. Kumar, P. Vashistha, Sustainable utilization of deinking paper mill sludge for the manufacture of building bricks. J. Clean. Prod. **204**, 321–333 (2018)

G. Singh, S. Kaur, M. Khatri, S.K. Arya, Biobleaching for pulp and paper industry in India: emerging enzyme technology. Biocatal. Agric. Biotechnol. **17**, 558–565 (2019)

V.R. Sinha, K. Bansal, R. Kaushik, R. Kumria, A. Trehan, Poly-ε-caprolactone microspheres and nanospheres: an overview. Int. J. Pharm. **278**, 1–23 (2004)

S. Sivasakthi, G. Usharani, P. Saranraj, Biocontrol potentiality of plant growth promoting bacteria (PGPR)—*Pseudomonas fluorescens* and *Bacillus subtilis*: a review. Afr. J. Agric. Res. **9**, 265–1277 (2014)

K. Skjånes, C. Rebours, P. Lindblad, Potential for green microalgae to produce hydrogen, pharmaceuticals and other high value products in a combined process. Crit. Rev. Biotechnol. **33**, 172–215 (2013)

P.D. Smith, M.Z. Cai, D. Gwary, H. Janzen, P. Kumar, B. McCarl, S. Ogle, F. O'Mara, C. Rice, B. Scholes, O. Sirotenko, Agriculture, in *Climate Change 2007: Mitigation*, ed. by B. Metz, O.R. Davidson, P.R. Bosch, R. Dave, L.A. Meyer (Cambridge University Press, Cambridge, UK/New York, 2007), pp. 497–540

A.E. Somers, B. Khemchandani, P.C. Howlett, J. Sun, D.R. MacFarlane, M. Forsyth, Ionic liquids as antiwear additives in base oils: influence of structure on miscibility and antiwear performance for steel on aluminum. ACS Appl. Mater. Interfaces **5**(22), 11544–11553 (2013)

S. Soni, M. Agarwal, Lubricants from renewable energy sources—a review. Green Chem. Lett. Rev. **7**, 359–382 (2014)

R. Souza, A. Ambrosini, L.M.P. Passaglia, Plant growth-promoting bacteria as inoculants in agricultural soils. Genet. Mol. Biol. **38**, 401–419 (2015)

K.A. Spokas, K.B. Cantrell, J.M. Novak, D.W. Archer, J.A. Ippolito, H.P. Collins, A.A. Boateng, A.A. Lima, M.C. Lamb, A.J. McAloon, R.D. Lentz, K.A. Nichols, Biochar: a synthesis of its agronomic impact beyond carbon sequestration. J. Environ. Qual. **41**, 973–989 (2012)

V.S. Sreenivasan, S. Somasundaram, D. Ravindran, Microstructural, physicochemical and mechanical characterization of *Sansevieria cylindrica* fibers—an exploratory investigation. Mater. Des. **32**(1), 453–461 (2011)

I. Staffell, D. Scamman, A.V. Abad, P. Balcombe, P. E. Dodds, P. Ekins, N. Shah, K.R. Ward, The role of hydrogen and fuel cells in the global energy system. Energy Environ. Sci. (2018). https://pubs.rsc.org/en/content/articlelanding/2019/ee/c8ee01157e#!divAbstract. Retrieved on 19 Jan 2019

M. Stefan, N. Munteanu, V. Stoleru, M. Mihasan, L. Hritcu, Seed inoculation with plant growth promoting rhizobacteria enhances photosynthesis and yield of runner bean (*Phaseolus coccineus* L.). Sci. Hortic. **151**, 22–29 (2013)

C.J. Stokes, M. Howden, *Adapting Agriculture to Climate Change: Preparing Australian Agriculture, Forestry and Fisheries for the Future* (CSIRO Publishing, Melbourne, Australia, 2010). ISBN 978-0643095953

S. Stoycheva, D. Marchese, C. Paul, S. Padoan, A.-S. Juhmani, I. Linkov, Multi-criteria decision analysis framework for sustainable manufacturing in automotive industry. J. Clean. Prod. **187**, 257–272 (2018)

M. Suhr, G. Klein, I. Kourti, M.R. Gonzalo, G.G. Santonja, S. Roudier, L.D. Sancho, *Integrated Pollution Prevention and Control—Reference Document on Best Available Techniques in the Pulp and Paper Industry, 2015* (European Commission, 2015), http://eippcb.jrc.ec.europa.eu/reference/BREF/PP_revised_BREF_2015.pdf. Retrieved on 30 Jan 2019

S. Sun, S. Sun, X. Cao, R. Sun, The role of pretreatment in improving the enzymatic hydrolysis of lignocellulosic materials. Biores. Technol. **199**, 49–58 (2016)

S. Sun, G. Wang, P. Wang, A cleaner approach for biodegradable lubricants production by enzymatic glycerolysis of castor oil and kinetic analysis. J. Clean. Prod. **188**, 530–535 (2018a)

M. Sun, Y. Wang, L. Shi, J.J. Klemeš, Uncovering energy use, carbon emissions and environmental burdens of pulp and paper industry: a systematic review and meta-analysis. Renew. Sustain. Energy Rev. **92**, 823–833 (2018b)

E. Sundin, *Product and Process Design for Successful Remanufacturing*. Linköping studies in science and technology. Dissertation No. 906, Department of Mechanical Engineering, Linköping University, SE-581 83 Linköping, Sweden, 2004

A.Z. Syahir, N.W.M. Zulkifli, H.H. Masjuki, M.A. Kalam, A. Alabdulkarem, M. Gulzar, L.S. Khuong, M.H. Harith, A review on bio-based lubricants and their applications. J. Clean. Prod. **168**, 997–1016 (2017)

M.C. Symington, W. Banks, O.D. West, R. Pethrick, Tensile testing of cellulose-based natural fibers for structural composite applications. J. Compos. Mater. **43**(9), 1083–1108 (2009)

H. Tabani, K. Khodaei, Y. Bide, F.D. Zare, S. Mirzaei, A.R. Fakhar, Application of pH-sensitive magnetic nanoparticles microgel as a sorbent for the preconcentration of phenoxy acid herbicides in water samples. J. Chromatogr. A **1407**, 21–29 (2015)

X.-F. Tan, Y.-G. Liu, Y.-L. Gu, Y. Xu, G.-M. Zeng, X.-J. Hu, S.-B. Liu, X. Wang, S.-M. Liu, J. Li, Biochar-based nano-composites for the decontamination of wastewater: a review. Biores. Technol. **212**, 318–333 (2016)

G. Thompson, J. Swain, M. Kay, C.F. Forster, The treatment of pulp and paper mill effluent: a review. Biores. Technol. **77**, 275–286 (2001)

D.K. Tiku, A. Kumar, S. Sawhney, V.P. Singh, R. Kumar, Effectiveness of treatment technologies for wastewater pollution generated by Indian pulp mills. Environ. Monit. Assess. **132**, 453–466 (2007)

E. Tilman, K.G. Cassman, P.A. Matson, R. Naylor, S. Polasky, Agricultural sustainability and intensive production practices. Nature **418**, 617–677 (2002)

M. Tokyay, Effect of chemical composition of clinker on grinding energy requirement. Cem. Concr. Res. **29**(4), 531–535 (1999)

A.B.H. Trabelsi, K. Zaafouri, W. Baghdadi, S. Naoui, A. Ouerghi, Second generation biofuels production from waste cooking oil via pyrolysis process. Renew. Energy **126**, 888–896 (2018)

A. Trebst, The mode of action of triazine herbicides in plants, in *Triazine Herbicides*, ed. by H.M. LeBaron, J.E. McFarland, O.C. Burnside (Elsevier, San Diego, USA, 2008), pp. 101–110. ISBN 978-0444511676

J. Trivedi, M. Aila, C.D. Sharma, P. Gupta, S. Kaul, Clean synthesis of biolubricant range esters using novel liquid lipase enzyme in solvent free medium. SpringerPlus **4**, 165–173 (2015)

K.G. Tsakalakis, G.A. Stamboltzis, Correlation of the Blaine value and the size of the cement particle size distribution. ZKG Int. **61**(3), 60–68 (2008)

N. Turpin, H. ten Berge, C. Grignani, G. Guzmán, K. Vanderlinden, H.-H. Steinmann, G. Siebielec, A. Spiegel, E. Perret, G. Ruysschaert, A. Laguna, J.V. Giráldez, M. Werner, I. Raschke, L. Zavattaro, C. Costamagna, N. Schlatter, H. Berthold, T. Sandàn, A. Baumgarten, An assessment of policies affecting sustainable soil management in Europe and selected member states. Land Use Policy **66**, 241–249 (2017)

S. Tyagi, D. Rawtani, N. Khatri, M. Tharmavaram, Strategies for nitrate removal from aqueous environment using nanotechnology: a review. J. Water Process Eng. **21**, 84–95 (2018)

UC Davis Biochar Database, *UC Davis Biochar Database,* 2015, http://biochar.ucdavis.edu/download/. Retrieved on 08 Jan 2019

UNECE, *Cleaner Electricity Production from Fossil Fuels* (2015), https://www.unece.org/fileadmin/DAM/energy/se/pdfs/Booklet_Dec2015/Booklet_Cleaner.Electricity_Dec.2015.pdf. Retrieved on 16 Jan 2019

UNEP, *Update on the Global Status of Legal Limits on Lead in Paint* (Sept 2018), https://www.unenvironment.org/resources/report/2018-update-global-status-legal-limits-lead-paint. Retrieved on 1 Feb 2019

C. Ungureanu, *Design for Sustainability: Product Life-Cycle Analysis in Aluminum Auto Body Applications*, M.Sc. Thesis, University of Kentucky, 2007. https://uknowledge.uky.edu/cgi/viewcontent.cgi?article=1452&context=gradschool_theses. Retrieved on 12 Mar 2019

United States Department of Agriculture (USDA), *Part 637 Environmental Engineering, National Engineering Handbook*, chap. 2 (2000). https://www.nrcs.usda.gov/Internet/FSE_DOCUMENTS/nrcs143_022229.pdf. Retrieved on 10 Jan 2019

E. Uosukainen, Y.Y. Linko, M. Lamsa, T. Tervakangas, P. Linko, Transesterification of trimethylolpropane and rapeseed oil methyl ester to environmentally acceptable lubricants. J. Am. Oil Chem. Soc. **75**, 1557–1563 (1998)

USDA, *World Agricultural Production, Circular Series* (United States Department of Agriculture, USA, 2016)

M.A. Vázquez, M. Soto, The efficiency of home composting programmes and compost quality. Waste Manage. **64**, 39–50 (2017)

J. Wang, Y. Yin, Principle and application of different pretreatment methods for enriching hydrogen-producing bacteria from mixed cultures. Int. J. Hydrogen Energy **42**, 4804–4823 (2017)

N. Wang, M. Liu, L. Guo, X. Yang, D. Qiu, A novel protein elicitor (PeBA1) from *Bacillus amyloliquefaciens* NC6 induces systemic resistance in tobacco. Int. J. Biol. Sci. **12**, 757–767 (2016)

H. Wanyika, E. Gatebe, P. Kioni, Z. Tang, Y. Gao, Mesoporous silica nanoparticles carrier for urea: potential applications in agrochemical delivery systems. J. Nanosci. Nanotechnol. **12**, 2221–2228 (2012)

R. Weber, S. Yoshida, K. Miwa, PCB destruction in subcritical and supercritical water evaluation of PCDF formation and initial steps of degradation mechanisms. Environ. Sci. Technol. **36**, 1839–1844 (2002)

R.G. Wetzel, *Limnology: Lake and River Ecosystems*, 3rd edn. (Academic Press, London, Cambridge, MA, USA, 2001). ISBN 978-0-127-44760-5

G. William, M. Shoukry, J. Prucz, *Analysis of Lightweighting Design Alternatives for Automotive Components*, SAE Technical Paper 2011-01-2287 (2011)

K. Wilson, T.G. Benton, R.-I. Graham, D. Grzywack, Pest control: biopesticides' potential. Science **342**(1660), 799 (2013)

World Nuclear Association, *Nuclear Power in the World Today* (2019), http://www.world-nuclear.org/information-library/current-and-future-generation/nuclear-power-in-the-world-today.aspx. Retrieved on 20 Jan 2019

S.-Y. Wu, J. Ye, Y. Tian, Y.-J. Wang, X.-D. Jian, The current status of lead paint control in China and suggestions. Mod. Chem. Ind. **57**, 1–57 (2017)

H. Xiao, Ionic liquid lubricants: basics and applications. Tribol. Trans. **60**, 20–30 (2017)

G. Yang, J. Wang, Improving mechanisms of biohydrogen production from grass using zero valent iron nanoparticles. Biores. Technol. **266**, 413–420 (2018)

M. Yates, D. Huerta, V.T.Y. Martin, M.A. Martin-Luengo, Three pathways to cleaner platform chemicals: conventional, microwave and solar transformation of a by-product from the orange juice manufacturing process. J. Clean. Prod. **168**, 746–754 (2017)

C. Ye, W. Liu, Y. Chen, L. Yu, Room-temperature ionic liquids: a novel versatile lubricant. Chem. Commun. **21**, 2244–2245 (2001)

F. Yilmaz, M.T. Balta, R. Selbaş, A review of solar based hydrogen production methods. Renew. Sustain. Energy Rev. **56**, 171–178 (2016)

J.Y. Yong, J.J. Klemeš, P.S. Varbanov, D. Huisingh, Cleaner energy for cleaner production: modelling, simulation, optimisation and waste management. J. Clean. Prod. **111**(A), 1–16 (2016)

Z.Q. You, Y. Zhu, C. Jang, S.X. Wang, J. Gao, Response surface modeling-based source contribution analysis and VOCs emission control policy assessment in a typical ozone-polluted urban Shunde, China. J. Environ. Sci. **51**, 294–304 (2017)

H. Yüksel, Design of automobile engines for remanufacture with quality function deployment. Int. J. Sustain. Eng. **3**(3), 170–180 (2010)

Y. Yusof, M.R. Ahmad, M.S. Wahab, M.S. Mustapa, M.S. Tahar, Producing paper using pineapple leaf fiber. Adv. Mater. Res. **383–390**, 3382–3386 (2012)

D.B. Žarković, V.N. Rajaković-Ognjanović, L.V. Rajaković, Conservation of resources in the pulp and paper industry derived from cleaner production approach. Resour. Conserv. Recycl. **55**, 1139–1145 (2011)

Z. Zeng, Y. Chen, D. Wang, J. Zhang, Tribological behaviors of amorphous Cr coatings electrodeposited from Cr (III) baths under ionic liquid lubrication. Electrochem. Solid-State Lett. **10**(8), D85–D87 (2007)

R. Zhang, Y. Zhang, L. Song, X. Song, H. Hänninen, J. Wu, Biochar enhances nut quality of *Torreya grandis* and soil fertility under simulated nitrogen deposition. For. Ecol. Manage. **391**, 321–329 (2017)

W. Zhao, Y. Zhang, B. Du, Q. Wei, Y. Zhao, Enhancement effect of silver nanoparticles on fermentative biohydrogen production using mixed bacteria. Biores. Technol. **142**, 240–245 (2013)

J. Zheng, J. Han, Z. Liu, W. Xia, X. Zhang, L. Li, X. Liu, R. Bian, K. Cheng, J. Zheng, G. Pan, Biochar compound fertilizer increases nitrogen productivity and economic benefits but decreases carbon emission of maize production. Agric. Ecosyst. Environ. **241**, 70–78 (2017)

Y. Zhou, J. Qu, Ionic liquids as lubricant additives: a review. ACS Appl. Mater. Interfaces **9**, 3209–3222 (2017)

Chapter 9
Drivers and Barriers to Cleaner Production

Abstract There are many barriers that get in the way of implementing sustainable practices and business models, holding back the potential for innovation and environmental improvement, and slowing it down and something even reversing it. Some examples of these barriers are related to economic issues, motivational issues, technological issues, educational issues, lobbying issues, among many others. Opposing these barriers are drivers that are beginning to gain momentum and are becoming impossible to ignore. Factors, like governmental policies, consumer trends and demands, opportunities in foreign markets, improvement of local and global market images, possibilities to work alongside well-established larger companies, etc., are drivers that are pushing companies and even industries to change their current methods, improving their overall ecological footprint, while in many cases simultaneously promoting cost savings due to a global optimization of their entire process chain. In this chapter, the drivers and barriers of Cleaner Production, sustainable systems, green manufacturing, sustainable energy sources, greener product packaging, reverse logistics, and the building construction industry will be dissected.

Keywords Cleaner Production drivers · Cleaner Production barriers · Green manufacturing · Sustainable energy sources · Reverse logistics · Building construction · Packaging recycling

Regarding what has been mentioned previously, it can be noticed that there is massive research in Cleaner Production and Sustainable Behavior, with the development of very successful methodologies and experiences both in industrial settings as in communities. It has also been mentioned that culturalism strongly influences the behavior of society and of companies in terms of sustainability. Some communities are more available to adapt to new needs, while other communities essentially need legislation and punitive measures to adapt their behavior to the current requirements of our ecosystem, in order to minimize or hopefully reverse the harm that humankind has caused to the planet.

It has not been for lack of imagination and research in this area that the results are still far from expected. After considerable research over the last 50 years, the concept that has recently received the most attention from researchers is called Circular Economy (CE), where numerous tools previously developed in a more or less

F. J. Gomes da Silva and R. M. Gouveia, *Cleaner Production*,
https://doi.org/10.1007/978-3-030-23165-1_9

dispersed way are integrated. CE is based on an economic model focused on the coordination of closed-circuit production and consumption systems, which requires adequate organization and management. CE can be defined as a system in which value is created, with special care in minimizing waste, energy use, and natural resources (Geissdoerfer et al. 2017). This objective can be achieved by using models to slow or close cycles of material and energy flows in a regenerative manner (Rizos et al. 2016), also taking into account the principles of reduction, reuse, and recycling (Jawahir and Bradley 2016).

One of the tools that have contributed to an effective increase in competitiveness has been eco-innovation (EI), which has also had a very positive influence on the well-being of society and the environment. EI can leverage the necessary departure from the initial situation of lack of concern for environmental issues, creating the conditions of awareness necessary for the progressive implementation of EI principles. EI can play a very important role in the imbalance that must be created between keeping everything as it was or starting a process of progressive change. EI is understood as a strategic enabler of transformations of the whole value chain, not just as a green technology capable of developing devices that provide beneficial effects to the environment, but only by acting as end-of-pipe technology, such as avoiding discharge of disastrous gaseous effluents into the atmosphere (Kemp 2010). Being a tool that is applied singularly to a product or company, it created increased competitiveness factors that will drag the competing products or companies to similar principles, generating an innovation chain focused on competitiveness and environmental respect, developing in a progressive way, with advantages for all stakeholders. In a study recently developed by Clark et al. (2016), it is clearly stated that innovation subject to sustainability constraints requires recirculation of resources in reuse cycles, using reconditioning and remanufacturing recycling, reformulating system inputs, and developing new output and renewal solutions through the use of clean energy and waste disposal.

9.1 Barriers and Motivations for the Implementation of Sustainable Systems

One of the main barriers to the adoption of CE practices in companies is the existence of economic cycles, which lead to some uncertainty in the recovery of investments in technology, organization, and management. On the other hand, companies like to be governed by clear and linear rules. The complexity and diversity of existing legislation on this subject and its volatility does not help companies to make clear decisions on this matter. Indeed, the legislation and policies often depend on local or national factors, as well as on political factors. Companies are able to assume risks, but in a controlled way. Companies that have their own economic potential, or obtain capital through financing, are often available to invest, and however, it is essential to know the return on the investment time frame. In the tangible case of environmental sustainability, companies realize that they are contributing to a better society, so they want to feel that governments are available to promote these policies, asking for a reduction of rates and taxes related to these investments. This may act as a motivating factor, which will break down some of the usual barriers to investment in this field.

The concept of CE depends heavily on a number of factors, including the reverse logistics system which is able to fit the circumstances. Here, also motivation and impediment factors have emerged, which, because of their specificity, will be dealt with in more detail below. All logistics operations involve energy consumption. Thus, it is also necessary for reverse logistics operations to be efficient. Zhu and Geng (2013) conducted a study to investigate the drivers and barriers described by Chinese manufacturers regarding energy savings and emission reductions in general logistics operations in China. Due to the severe problem of pollution, China has launched programs aimed at reducing emissions in general (Energy Saving and Emission Reduction), which also entails specific care by logistical operators. However, the study showed that the regulation was not effective enough to motivate operators to change their practices. Divided between motivations and doubts, Chinese manufacturers have very modestly adhered to sustainable practices in the logistics sector, since membership is voluntary and the benefits have been relatively ambiguous. The poor adherence to the Energy Saving and Emission Reduction plan is essentially due to the voluntary nature of membership, and the environmental and economic benefits to be gained by joining the program are not evident. It therefore seems clear that the nature of voluntary membership is not the most appropriate, and it is necessary to legislate more assertively in this respect, as in Germany.

The wave of awareness regarding the change to a CE will tend to increase, due to the influence of society itself (Schot and Kanger 2016). However, this path has proven to be long and difficult to navigate. It should be noted that, although there is an already greater concern of the population toward the negative effects of technological development, restrictive legislation is should continue to increase, in order to avoid certain situations of continued disrespect toward the environment to remain. According to a recent study developed by Jesus and Mendonça (2018), technical and economic factors are those that present the greatest motivations and the greatest constraints in the development of a CE. On the other hand, institutional and social factors are those that present fewer motivating factors, but also lower restrictions on the development of the same CE (Schot and Kanger 2016). Given that both motivations and limitations are factors that need to be studied in depth to understand the dynamics and inertia that affect the implementation of the concepts of sustainability and respect for the environment, the constraints are those that most concern all researchers and stakeholders interested in improving the environmental conditions on our planet. According to a collection of information performed by Jesus and Mendonça (2018) and by Tura et al. (2019) in works published by Preston (2012), Vanner et al. (2014) and Rizos et al. (2015), and many other authors, the major barriers to CE and sustainable system implementation can be listed as follows:

- Considerable initial investment costs;
- Insufficient investment in technology;

- Complex international supply chains;
- Resource-intensive infrastructure lock-in;
- Uncertainty about market demand;
- Gaps in company cooperation;
- Lack of consumer eagerness;
- Limited innovation dissemination;
- Economic factors do not encourage the efficient use of resources, pollution mitigation, and/or the demand for innovative products.
- Fair-minded consumers and businesses in this type of process;
- Lack of awareness and information;
- Public incentives for the development of sustainable policies are very limited.
- Lack of environmental culture;
- Lack of financial capacity;
- Absence of means to measure the long-term benefits brought by the adoption of these practices and technologies;
- Fear of risk and closed-loop thinking;
- Limited government support;
- Legislative gaps;
- Lack of information;
- Administrative burdens too high;
- Relatively low technical skills.

The last six factors are more directly related to the environment experienced in SMEs, but also affect the larger organizations, although not as intensely. In addition to the fact that the number of barriers identified is very high, and it is also worth noting that when one or more of these barriers are broken, there are still many others to overcome, which makes the process of evolution toward a more sustainable society very difficult. In fact, if only one of these barriers becomes insurmountable, the whole process may collapse. Indeed, even if there is a predisposition to adopt sustainable practices, there are difficult to overcome aspects such as the need of technical skills, stakeholder interest in implementing these practices, capital funding for initial investments, and financial support as well as acceptable payback time frames. On the other hand, legislation imposed as a policy, which is often seen as a motivation, may in certain cases be a fatal factor, because if companies lack financial means and there are no adequate financing programs to support the needed investments, certain companies may not be able to survive newly imposed rules.

Governmental agencies linked to the environment play an extremely important role in the institutional framework, having the necessary infrastructure to support decision making and the implementation of these practices by companies, as well as information systems that enable the legal framework. Their support regarding R&D activities and increased social awareness is also crucial. Another factor that can perfectly be considered as motivation is the governmental frameworks. These specialized bodies allow direct support to companies that want to engage in sustainability practices, through learning and training programs which can help gain access to accurate information regarding the available measures, as well as through programs promoted directly by governments or through government agencies for this purpose.

The work recently developed by Tura et al. (2019) makes an exhaustive analysis of drivers that tend to encourage CE practices. These drivers are listed below, which are more developed in the above-mentioned work:

- Limitations on resources and potential to prevent negative environmental impacts;
- Ability to act in cost reduction, find new sources of revenue, and increase income;
- Potential for the creation of new business opportunities, innovation, and synergies with other businesses;
- Increased internationalization and awareness raising in order to improve sustainability;
- Potential to increase the number of jobs and the liveliness of companies;
- Ability to comply with regulations and standard requirements;
- Government financial support, tax, and subsidies policy;
- Ability to improve existing operations;
- Existence of new technologies and willingness to adopt them;
- Access and interest in the sharing of technological information to plan new investments;
- Ability to minimize reliance on energy supply and avoid unfavorable price fluctuations;
- Collaboration and open communication with stakeholders;
- Versatility and multi-disciplinary in existing skills in the company;
- Creation and management of networks;
- Ability to strengthen the brand through differentiation;
- Adequate understanding of the demands of sustainable management;
- Integration of CE principles into the objectives and overall strategy of the company;
- Skills development and willingness to CE.

Technological development boosts companies in the direction of CE. New technologies provide cleaner solutions for the future, helping to avoid and overcome problems induced by current technologies. Information sharing platforms support cooperation between stakeholders and allow for greater clarity of information, facilitating the adoption of CE principles. Supply chain drivers can help reduce supply dependency and avoid paying higher and more volatile prices in the future. In organizational terms, companies can achieve brand benefits, protect and strengthen their image, and allow differentiation when they choose to adopt CE principles. Trade internationalization and the most demanding environmental regulations are examples of driving forces that put pressure on companies to find alternatives to traditional businesses. CE can still be seen as an opportunity to increase jobs and the vivacity of the companies. Institutional support, through legislation and regulation focused on sustainability-related factors, helps fine-tuning the strategies for adopting and improving CE practices. The adequacy of product taxation and the adoption of policies aimed at recycling reduce the business risk around a CE and create the necessary means for companies to flourish. The adoption of specific standards for environmental certification, such as ISO 14001, and the milestones imposed by legislation, end up being an extra incentive for the development of CE practices, minimizing the economic risks related to the involvement in environmental sustainability programs (Tura et al. 2019).

The following subsections are intended to provide a better understanding of the motivations and barriers identified so far for the practice of sustainable measures, focusing mainly on the highly researched sectors. The cases reported below are based on studies that were carried out by researchers based on data collected on the ground, or reports of case studies describing concrete situations. These cases are not based on isolated factors but on a broad set of factors that condition the success or failure of sustainable practices initiatives. In these cases, there are implicit economic and social conditioning factors.

9.2 Drivers and Barriers for Green Manufacturing

The manufacturing industry is one of the most important economic sectors, generating virgin raw material needs that move part of the primary sector. Together, they are also the sectors that generate the greatest environmental concern. Hence, special attention is needed to the way products are produced, while minimizing this negative impact as much as possible. Indeed, this new millennium has brought a new trilogy to the manufacturing world: There is a pressing need to combine productivity with profitability and respect for the environment. These concerns have been translated into what is known as green manufacturing. However, green manufacturing can take on different meanings or definitions and can be achieved through different approaches, depending on the size of a company, its nature and even the reality lived in each country or region. Fundamentally, green manufacturing is a new concept of manufacturing, which begins with a different strategy, much more focused on environmental sustainability, driven by motivations that, while aiming at the profit and economic health of companies, emphasizes in a much more respectful the respect for the environment, seeking to maximize efficiency. This efficiency means reducing the consumption of virgin materials and minimizing consumed energy while taking extra care of the product's life cycle. Particular attention is also given to the non-use of toxic materials which are harmful to the environment and to public health. These concerns will lead to the generation of smaller quantities of effluents, solid waste, and emissions, as desired. To this end, new technologies, designed to meet these requirements, should be used. Ultimately, it is intended to have products that have as much or even better quality than the current ones, are economically viable, and produce the least possible negative impact on the environment.

Some older classifications of drivers/motivations divide these into two groups, external and internal ones. The external group consisted essentially of financial and legislative incentives, education/market awareness, and demand, while the internal ones were based essentially on knowledge, skills, commitment, and pursuit of improved environmental performance. However, other authors (Van Hemel and Cramer 2002) refer to internal motivations as extremely important, mentioning some extremely significant factors, especially in developed countries, such as increased quality, ambition to increase market share, the ability to innovate, and the opportunity to improve company market image. These are factors that, properly managed by the top management, can leverage any company, regardless of its size, easily taking it to a green manufacturing strategy. Reputation for adopting green manufacturing is much more affected when it comes to large companies than in the case of SMEs. Thus, it may be considered that large firms have great imaging advantages when adopting green manufacturing strategies. On the other hand, SMEs are much more driven into a green manufacturing strategy through financial and legislative incentives. In fact, small businesses are not so concerned with their market image, but in compliance with regulations to avoid heavy fines that may affect their viability. Due to their weak economic power, they are dependent on government subsidies to overcome investment difficulties and gain competitiveness while in compliance with the environmental requirements that are being established. Given that they typically lack appropriate skills to deal with the new technologies, all government aid is extremely welcome, in order to overcome their limitations and aim to move to a new stage of evolution in the manufacturing landscape (Studer et al. 2008). A recent study carried out by Gandhi et al. (2018) allowed to conclude that the most important drivers regarding Indian SMEs toward green manufacturing are the commitment of top management, technological updating capacity, current legislation, the establishment of a sustainable and environmentally friendly brand image, and upcoming restrictive legislation.

There are factors that are predominant for the success of a particular strategy or set of strategies. These factors are usually referred to as drivers and are a motivation so that the objectives to which a strategy proposes can be successfully achieved. According to Ghazilla et al. (2015) and Pacheco et al. (2017), the key drivers associated with pursuing green manufacturing strategies are the need to comply with legislation imposed by governments in terms of respect for the environment, growing concerns about the environment and climate change, and increased public awareness of environmental issues.

Thus, based on an extensive review of the literature, Seth et al. (2018) pointed as the main drivers of green manufacturing in SMEs the following:

- Rigorous environmental regulation and careful enforcement of legislation;
- Penalties due to non-compliance with regulations and standards;
- Financial incentives and organizational support;
- Use of Cleaner Production technology and its permanent updating;
- Strategic and organizational changes to increase competitiveness;
- Commitment and leadership of top management;
- Involvement of all stakeholders in environmental issues;
- Reduce emissions and consumption of natural resources.
- Induce in clients the perception of quality and respect for the environment.
- Use technology for the pursuit of 3Rs policy and eco-design.

- Reduce energy consumption and preferentially use renewable energy sources.
- Update and innovate with regard to control techniques.

Larger companies live a very different reality, with the economic means to take care of environmental issues in a more decisive way. These types of companies need to take care of their market image, so they normally adopt environmental management systems and adhere to social responsibility programs as a way of being able to show and announce the newly adopted stance regarding environmental issues. Topics such as energy saving and clean water are essential, as well as compliance with all legislation that seeks to parameterize company behavior regarding the environment. Due to their size, these companies do not present difficulties in investing or obtaining credit for the adoption of newer cleaner technologies, and they also do not present considerable gaps in their organization, usually possessing their own skills to deal with the requirements imposed on them. Thus, the drivers that motivate these companies to adopt green manufacturing strategies are different from those presented previously for SMEs. These drivers are then listed based on a literature review by Seth et al. (2018):

- Use of technological means to design environmentally friendly products, as well an improved supply chain;
- Planning and using reverse logistics;
- Promote higher resource use effectiveness;
- Reduction of gaseous emissions, generation of hazardous wastes and effluents;
- Compliance with existing legislation and preparation for even more demanding legislation in the future;
- Take advantage of the support provided by the government, both in terms of training and incentives;
- Avoid payment of fees related to emissions and discharge of effluents or solid waste;
- Existence of commitment and leadership by top management;
- Involvement and commitment of employees with company strategy, benefiting from training, and reinforcement of competences;
- Interest and involvement of stakeholders in ecological concerns, meeting customer concerns;
- Maintain competitiveness and market leadership by providing more environmentally friendly products and at a lower cost;
- Have environmental awareness, education, and technical capacity;
- Knowing how to respond positively to competition, supply chain needs, and pressures exerted by public bodies;
- Know how to use energy effectively and seek to consume renewable energy;
- Preserve a company identity that is closely linked to the environment and to sustainability.

Given that the concerns and market posture are different, it turns out that the main drivers also differ. Typically, the motivation of large companies turns out to be greater than that presented by small companies due to financial difficulties and the lack of technical skills to devise an adequate company strategy.

9.3 Barriers and Motivations in the Adoption of Sustainable Energy Sources

In 1997, a pioneering experiment was carried out on the Danish island of Samsø. In fact, up to that time, this island had been completely energy dependent on imported petroleum products (Brandt and Svendsen 2016). In that year, the island won a contest to be a "Renewable Energy Island." With this aim in mind, a 10-year plan was created to make the island energy independent by using renewable energy sources. The goal was achieved in 2005 by installing wind turbines, solar panels, and biomass power plants, making it the first "Green Community" (Nielsen and Jørgensen 2015). This initiative was successful, and in 2015, a new challenge was launched, which aims to provide this community with a Circular Economy system, making this community the first community to integrate the principles of sustainability advocated by that system. The success of the creation of the first "Green Community" was due to a set of motivations based on a common goal, shared with equal enthusiasm by the majority of the citizens of that island. It is only through a convergence of motivations that an initiative of this nature can be successfully carried out. In a first phase, it can be mentioned that the motivation was heavily based on technological factors, which was based on the acquisition of wind turbines, solar panels, and heating plants, as well as all the necessary infrastructures. This investment was funded by government funds, which is a huge motivation, unlocking numerous barriers that usually arise in this type of initiatives. Moreover, incentives of various kinds have been created to encourage the production and consumption of renewable energy within the island. However, the project was not based solely on technological factors as motivating factors for success. In fact, a number of initiatives have been developed in institutional terms aimed at creating the necessary convergence of interests and ideas, as well as the necessary guidelines for the project to progress in a sustainable and coherent way. However, other motivations are also implicit, namely the cultural level of the community, the spirit of community, and the way in which the communication about the project is carried out.

A study published by Reddy (2013) states peremptorily that the most effective use of energy leads to an increase in energy consumption. On the other hand, Saunder (1992) already stated more than 20 years ago that energy efficiency led to an increase in energy consumption in two different ways: (a) The affordable price of energy creates the perception that it is preferable to consume electrical energy in detriment of other sources of energy, and (b) the economic growth leads inevitably to a higher energy consumption due to the growth of industrial and commercial activity. Concerning the electricity sector in particular, in some countries there is the existence of governmental funding for supporting electricity tariffs, which distort the reality of the market, leading people to feel that the consumption of electricity is not sufficiently painful for their budget and consuming electricity in a wasteful manner. This factor, coupled with the lack of knowledge regarding the environmental problems behind the generation of electricity, could lead to very unsustainable consumption of electricity. This leads to the realization that if the implementation of improvements in technological terms can be relatively easy, it needs to be properly accompanied by due legislative action and by a very focused environmental education, improving the way of thinking of the population. Only in this way can technological improvements in terms of energy efficiency be successful. Otherwise, technological gains may be completely nullified by a lack of public awareness or undue support for energy tariffs. In this regard, Gunn (1997) argues that the first factor to be taken into account in the assessment of energy efficiency is the legislative framework that supports the management of electricity in a given country. Without a thorough understanding of this matter and possible adjustments, devising a strategy for energy efficiency can be a waste of time. In most cases, government intervention should be limited to information on how to make energy consumption more efficient, and in certain cases regulate the market, when players are not able to do so by themselves. In fact, the market has an extremely important role to play in informing, promoting sustainable energy consumption practices and regulating consumption. The adoption of technology capable of improving energy efficiency must be deeply evaluated since its application should only be considered when technology is already at a perfectly mature stage. Investments are also discouraged in a given technology when it is expected to be replaced by a more effective one, or at significantly lower costs (Reddy 2013).

Studies carried out by Cagno et al. (2013) and by Trianni et al. (2013), specifically in the area of industrial energy efficiency, have listed a significant number of factors that act as motivations and barriers to induce or restrict efficient energy management in industrial companies. From these studies, the following factors have been considered as drivers:

- Technologies not available or not adequate;
- Lack of information on costs and paybacks;
- Unintelligible information from technology providers;
- Distrust in the available information and in its source;
- Lack of information on energy contracts;
- Elevated investment costs and poor financial preparedness;
- Unknown costs;
- External risks;
- Payback period uncertain or too long due to lower profitability;
- Different priorities and objectives;
- Apathy and lack of time;
- Disinterest in matters related to energy efficiency;
- Poor assessment of requirements and benefits;
- Poor organization and lack of internal control;
- Organization too complex to make assertive decisions;
- Archaic status of the energy management system;
- Difficulty in the implementation and control of more advanced technological systems;
- Difficulty in identifying inefficiencies;
- Difficulty in identifying opportunities;

- Difficulty in recruiting properly trained workers and provided with adequate skills;
- Lack of willingness to embark on new paths.

However, a list has also been identified of several factors which may be reasons for the adoption of more energy-efficient systems:

- Legal restrictions in terms of energy efficiency;
- Clarity of information in terms of regulations;
- Customers' requirement and/or more demanding external audits;
- Need to convey an image of "green company;"
- Rise in energy tariffs;
- Improvement of the long-term energy strategy;
- Appeal for technological updating;
- Confidence in the information received through government entities;
- Predisposition to remain extremely competitive and cut costs;
- Management support;
- Government financial incentives;
- Private financing;
- Clear and reliable information;
- Environmental and social awareness;
- External cooperation;
- Ambitious top management and staff;
- Educational training and external technical support.

A study later published by Cagno et al. (2015) on the Dutch SMEs in the metal-working sector allowed to draw some conclusions about how motivations and barriers are faced by the various actors. Some of the main motivations are induced by technology providers, as well as by installers and State agencies, who have arguments to counter all the barriers that arise before and during the changing process. However, there is also a strong divergence between companies and State bodies in terms of points of view about the existing barriers. A consistent understanding of the barriers seems to be difficult to achieve between companies and State bodies. Considering the responses obtained from State bodies and companies, there is a disagreement between the mechanisms that can positively and negatively influence decision making on the adoption of industrial energy efficiency. This disagreement could lead to a lack of efficiency in the measures stipulated by State bodies, with no impact on the programs that may have been established in the meantime with a view to increasing the efficiency of energy consumption. Although a program was implemented in the Netherlands called LTA3 to increase the Voluntary Agreements between government and industrial companies to increase sustainable energy consumption practices, its implementation fell short relatively to the planned. While companies agree that it is imperative to have an energy strategy, the same companies are unable to associate the impact of this energy strategy on the motivations and barriers to decision making in relation to the Voluntary Agreements, although an energy efficiency plan in the medium-long term has been included in these agreements. Interestingly, economic barriers do not seem to be relevant to most of the companies surveyed, and the Voluntary Agreements appear to have no impact on this type of barrier, as companies do not view such agreements as an encouraging factor for the adoption of more sustainable energy practices. This shows that there are significant differences between the way that incentives are designed by State agencies and how they are perceived by the companies for which they are designed. Thus, it is clear that there is a need to better understand how companies decode the messages issued by State agencies, improving communication efficiency and achieving greater success in policies to encourage companies to adopt more sustainable energy practices.

Both the motivations and the barriers to the application of sustainable energy consumption practices have been studied by several authors, who have promoted the division of these factors into groups, taking into account who is responsible for these factors, or how they can be faced by those who study the subject. According to Reddy (1991), barriers that impede further progress in energy efficiency can be classified into five main groups: (a) imposed by the consumer, (b) brought by the equipment manufacturer, (c) utility-related, (d) imposed by restrictions on financing, and (e) induced by the government. However, this classification has known other developments. According to Schleich and Gruber (2008), classification should safeguard the following items: (a) institutional barriers, caused by national or local political institutions; (b) market barriers, i.e., market barriers; (c) organizational barriers within the organizations themselves, with a special focus on the companies themselves; and (c) behavioral barriers due to way of being and thinking of individuals. In turn, Sorrell et al. (2011) simplified the classification, dividing the different barriers into three main factors that act against the development of sustainable practices of energy efficiency. The three categories considered are as follows: (a) neoclassical, (b) behavioral, and (c) organizational. This study describes how six different barriers to energy efficiency work and how they vary depending on the context. This author also attempts to assign relative importance to each of the different barriers. It was also clear that there are multiple barriers to energy efficiency, which can coexist, reinforcing each other, even though these barriers are interdependent. In fact, the behavior has a decisive influence on how energy efficiency progresses. This can be observed in studies focused on the energy management of residential houses, where decisions are centered only on a family, not on top management made up of people with possibly different sensitivities, and with the focus essentially on the business. A study by Vine et al. (2003) investigated market barriers in the residential sector, classifying them in three different groups: (a) lack of information on energy use, (b) lack of access to information on investment financing in general and energy efficiency technologies in particular, and (c) low importance given to energy efficiency in decision making. Another study showed that many of the market barriers can be understood as examples of deficiencies in their normal functioning, providing excellent opportunities for more incisive government intervention (Golove and Eto 1996).

9.4 Drivers and Barriers in Packaging Recycling

The collection and treatment of municipal solid waste is a problem that seems to have no end in sight and has been exacerbated by the increased consumerism, especially in developed areas and in large urban concentrations. If organic waste can be treated with some costs to the environment (but also with advantages as previously mentioned), other types of waste are much more difficult to manage. A large part of the waste generated in urban environments originates in packaging. In fact, packaging has a life cycle that may be relatively short, but goes through the extraction of virgin materials, manufacturing, distribution, and after-consumption disposal, just like other products (Gustavo et al. 2018). These packages can be made of various materials, some of which are easily recyclable, while others are the opposite. Sometimes even the separation process shows a difficulty, followed by even more serious problems such as the non-possibility of recycling and the need of about a century for them to degrade. Due to the ease of manufacturing, almost any plastic shape can be made through injection molding, leading this material to become extremely used in the packaging industry. In addition, it is lightweight, hygienic, cheap to produce and pleasing to the eye and touch. It is also lightweight and easy to drop anywhere, posing a problem in many places, including rivers and oceans, where it can be found in vast quantities. For some decades now, humanity perceived the problem that was being created, promoting less harmful alternatives to the environment such as paper and the glass. However, plastic continues to play a major role in the packaging industry, being far from sustainable. Effectively, in the work performed by Gustavo et al. (2018) sustainable packaging is defined as the one that is beneficial, healthy, and safe for citizens during its life cycle, presents a price and performance as expected by customers, essentially uses renewable energy in its manufacturing and transportation, is produced using environmentally friendly processes, and can be reused or recycled using renewable energy. Obviously, the market will hardly have conditions that allow all these conditions to be fulfilled. Packaging is a very specific product, which requires a great deal of work between the needs of the producer, the interests of the distributor, and the impact the customer feels when observe the packed product. In addition to this combination of interests, there are a large number of requirements to be met by the package, which is quite specific to each type of product.

The great majority of the polymers which are used in the packaging sector are thermoplastics, and as such are capable of being recycled. Accordingly, Hopewell et al. (2009) indicate recycling as the main solution for the plastics used and discarded by modern societies. However, the energy contained in the plastics cannot be neglected, so processes are also used that essentially aim to reuse the energy that these polymers contain (Al-Salem et al. 2009). Within this section, it is important to analyze which drivers drive sustainable recycling of the polymers used in the packaging sector. Regarding the drivers related to the management of municipal urban solids, there have been several studies that attempt to classify and divide these drivers by groups. In fact, Wilson (2007) considered as relevant six groups of drivers: public health care, environmental protection, waste value, institutional responsibility, population awareness, and product cycle closure. Regarding the drivers related to the management of municipal urban solids, there have been several studies that attempt to classify and divide these drivers by groups. In fact, Wilson (2007) considered as relevant six groups of drivers: public health care, environmental protection, waste value, institutional responsibility, population awareness, and product cycle closure. However, in a study later developed by Agamuthu et al. (2009) these groups were condensed, reducing them to four: human, institutional, economic, and environmental drivers. Concurrently, Troschinetz and Mihelcic (2009) studied in more detail the drivers related to the sustainable recycling of the urban solid waste in developed countries. This study allowed the identification of twelve drivers capable of promoting a sustainable policy for the recycling of solid urban waste, namely environmental government policies and legislation issued, government subsidies policy, waste characterization, waste collection and classification, parental education, family economic situation, solid waste management, personal sensitivity to household waste management, solid waste management planning, local materials recycling capacity, land, human resources, and appropriate technology availability. Another study developed by Contreras et al. (2010) based on the evolution in time of the strategy for two well-known cities, namely Boston and Yokohama, used as base only four groups of drivers to analyze the performance in solid urban waste management, specifically: regulation, institutional and technological development, territorial scope, and economic and social conditions. These groups of drivers are partially coincident with those found by Zaman (2013) when analyzing the development of urban solid waste management in Sweden, as well as the possibility of applying new treatment technologies to improve this management. The groups identified were social, economic, and environmental. As can be seen from the studies previously presented, most of them have a generic scope, and very few studies actually deal with the problem of plastic waste from packaging. It should also be mentioned that the drivers that may be behind the recycling of these plastics should be seen in a local perspective since the perception and legislation in each country induces a different reality, which translates into different motivations and behaviors. In addition, it should also be considered that it is very likely that there is an interaction between the different drivers, so it will not be correct to think that each driver will reach the degree of motivation that allows to unblock situations of change of habits in the treatment process of solid plastic waste. Generally, more recently developed studies tend to conclude that economic incentives, parental education, and the form and intensity of communication in terms of awareness of sustainable behavior are the main factors that to influence the behavior of people with regard to recycling (Bezzina and Dimech 2011).

Mwanza and Mbohwa (2017) conducted a literature collection study, compiling the drivers that condition citizens' behavior regarding the need to recycle municipal solid waste, with a particular focus on plastic waste. Based on previous studies carried out with several objectives, but having in common the sustainability treatment of urban solid waste, they identified three large groups of drivers, in line with some conclusions already taken from other works: economic, environmental, and social. Emphasizing sustainability from the economic point of view, waste is considered as a resource that still has a value that deserves to be used. Since the collection of waste is normally carried out by municipal entities, it is extremely positive that local political authorities perceive the value that waste still possesses, namely plastic waste, as it acts as a strong stimulus for its collection, segregation, and recycling, thus avoiding its disposal in landfills. Given that the first step to be taken is the presegregation, carried out at the home of each family, it is necessary to educate citizens so that there is a greater environmental awareness concerning the value that these residues represent. In fact, the value can be considered in two different ways: The waste can generate energy or give rise to new products that have commercial value and avoid the consumption of virgin material. This education depends first of all on the basic culture of the population, but it must be strengthened as much as is necessary, as well as economically possible by the budget of each government, in order to make citizens aware of sustainable practices. Despite representing a relatively recent trend in most developing countries, and even in some developed countries, environmental protection, climate change, and global warming are beginning to be factors that influence the way of thinking and behavior of communities and families. Thus, it is becoming commonplace to note that citizens are beginning to mobilize in defense of the environment and sustainable practices and are an extremely important driver for the political decisions of local communities and for companies involved in the production of packaging material and the selective collection of waste for recycling. This driver is directly related to the social group due to the level of behavioral education that citizens need to possess. In fact, it is necessary to bear in mind that the world population is constantly growing, and the improvement of the economic situation of a populations leads to greater economic power and consumption, which is reflected in a larger generation of solid urban waste, in which a large percentage is a plastic.

Despite all that has been previously focused on, the problem of recycling solid urban waste can also be summarized as three factors that preponderantly condition how the recycling process can evolve: It is necessary that companies have high social responsibility, realizing what is necessary to be done to ensure that the environment is preserved and that a sustainable system can be created, it is necessary to induce a strong responsibility on the consumer, from the purchase attitude to the destination to be given to the waste or end-of-life products, and it is necessary that authorities establish laws, regulations, and incentives that impel consumers and companies to realize the most responsible attitude to take, in order to reverse the situation of environmental degradation that we have all seen in recent decades. Proper management of municipal solid waste involves all these aspects. Indeed, producers, consumers and political entities can be considered the main drivers for an appropriate recycling system of plastics and other urban solid wastes.

In a recent study by Gustavo et al. (2018), based on a retail supermarket, where the interests of manufacturers, retailers, and customers are combined, the following groups of factors have been identified that influence the use of packaging:

- The main drivers are related to the need to comply with the principles of sustainability while maintaining economic profitability, meeting all requirements demanded by stakeholders, while managing to create the necessary profits.

- There are numerous opportunities to adjust the packaging to the needs and taste of consumers since it can and should be designed according to the customer's sensitivity, consumption habits, flexibility to adapt to new formats, increase or reduction in size according to the interests of consumption and distribution, possibility of refueling, and reusability of the packaging.
- The barriers to packaging design are vast, which can be described as uncertainties around any changes in the original design, problems related to the conditions offered by retailers, impediments to the dissemination of new projects due to patents and registrations, creation of barriers by competitors, consumer purchasing power, supplier participation, rule enforcement, among many others.

According to the opinion of many retailers, the customers' choice lies essentially in products with a more favorable price, within the range of quality that they consider acceptable, with packaging not being influential in a significant way. Therefore, a driver can be obtained from this fact because the package design can reduce an item's cost, becoming a motivating factor for the customer and producer.

Supermarket chains periodically analyze their products taking into account essentially two aspects: opportunities to reduce costs and opportunities to increase product sustainability. However, this analysis usually focuses only on a specific phase of the product's life cycle: the marketing phase. This analysis also takes into account customers' consumption habits, as well as the ease of storing and transporting products. The size, shape, and content of a package are extremely important factors. Size and content condition the amount consumed each time and can bring strong benefits in terms of manufacturing, filling process, customer supply and material that need to be recycled at the end-of-life. The shape can have repercussions on the amount transported, which can bring significant benefits to the environment, due to the fact that more containers can be transported in the same route, leading to fewer emissions. However, there may be a need to reduce the size of some packages when it is noted that a significant majority of consumers do not consume the entire contents of the packaging, leading to systematic waste.

The change of any packaging also entails some barriers, which are essentially linked to the retailer's uncertainty about the consumer reaction. Changing the contents of a package may also influence the number of times the customer needs to go to the grocery store, as well as the amount that he is available to spend on each trip to the supermarket. Thus, retailers may offer some resistance to changing the packaging of a given product, due to the innumerable uncertainties that this change can generate in consumer behavior.

9.5 Drivers and Barriers in Reverse Logistics

Logistics is an extremely important sector in terms of sustainability because: (a) It represents a significant portion of the gaseous emissions of pollutants discharged around the world due to the need to transport people and goods by land, sea, or air,

(b) plays an extremely important role in closing the economic circuit, transporting new products and their packaging in one direction, and transporting end-of-life products and materials which need to be reused, recycled, incinerated, or landfilled in the reverse direction. Recently, Govindan and Bouzon (2018) carried out a review of the literature devoted to reverse logistics (RL), where they identified the main stakeholders, and then studied in detail the drivers and barriers to this type of activity. Within RL, as in any other sector, it is always extremely difficult for a company to be able to respond satisfactorily to all stakeholders. Therefore, it is vitally important that companies know how to select the stakeholders to whom it is imperative to respond positively. The following is a list of eight stakeholders who were identified by Govindan and Bouzon (2018) as those who are truly important in the RL universe: government, customers, society, market, suppliers, organization, employees, and media. These stakeholders include a series of actors that are intensely related to the companies that operate in the area of logistics in their current activity. Company involvement includes relations with the government and public bodies, clients, society, local community, non-governmental sectoral associations, competitors, manufacturers or producers upstream of the logistics system, investors with economic interests in the company, workers, and means of social communication.

The competitiveness of companies is increased when they are motivated by the main drivers that lead to the differentiation of their mode of operation compared to their closest competitors. Operating in RL is something that requires motivation. In view of the aforementioned stakeholders, the factors that may be the motivation basis of companies operating in RL were listed, being 37 in total. This group must also be divided into internal and external motivational factors, as internal ones depend exclusively on the company and the way it is organized, while external motivational factors may be influenced or not in the way the company acts. A cross-reference between stakeholders and the motivations already identified in the literature allowed to find a set of eight groups that bring together these motivations, namely

- Legal matters include the need to comply with regulations and laws governing the return of products and RL.
- Issues related to supply chain management relate to supply chain drivers, cooperation, and business partners.
- Internal management issues include concerns such as employee satisfaction, human resources policy, and management of a department for the practice of RL.
- Issues related to the market and competitors include the analysis and demand of customer satisfaction, competitive pressure exerted by competitors, competitive advantage seeking, and knowledge of the green market.
- Technology and infrastructure issues include drivers for information technology, recovery technologies, and design-related issues focused on sustainability.
- Economic–financial issues include drivers strictly related to the economic–financial management of RL.

- Questions related to awareness and internal skills include the awareness and predisposition of the internal human resources for RL, knowledge on the matter, and necessary information flows.
- Social issues include RL drivers related to existing social pressures, increased public awareness of the need for RL, and increased environmental awareness of the surrounding community and associative pressure.

The classification of external and internal drivers previously referred allowed to verify that 23 drivers depend only on the company, so they are considered internal, while 14 depend on third parties, which is why they are considered external. The internal drivers depend only on the motivation of human resources to practice RL, while the external ones involve motivations that arise from stakeholders that surround the company. The work carried out so far that investigated the motivations that impel most companies to adopt RL practices, point to a crucial motivation: the need to comply with legislation, that is, an external motivation. The second most mentioned driver is the need to address the pressure of consumers with the highest environmental concerns, demonstrating that there is an increasing number of citizens with these concerns and that their influence is already felt in the management of companies. This influence is already beginning to be understood as important by companies, thus conditioning the decision making in relation to the adoption of RL practices. As the third most important motivating factor arises economic viability, which gives rise to a two-way analysis: Companies feel that there are more consumers attentive to environmental issues, acting as a stimulus to offer an RL service, but, on the other hand, companies do not see this as major stimulating factor for the adoption of this practice. In other words, companies are looking forward to market developments, hoping consumers and the supply chain give positive signs that RL is a practice to be embraced by everyone, thereby making the business more attractive.

As in drivers, barriers can also be classified as internal or external, conditioned by the dependence of their factors. External barriers are imposed essentially by actors external to the company, namely its stakeholders, which in any way prevent the pursuit of RL sustainability policies. On the other hand, internal barriers are essentially associated with factors intrinsic to the organization and resources of the company itself and which tend to prevent the normal adoption of RL practices. The different studies carried out on this subject point to the existence of about 36 barriers, which can be classified into seven different groups, taking into account the stakeholders that are involved, or the resources internal to the company to which they refer. These groups are listed below:

- Technology and infrastructures include technical barriers related to information technologies, technical skills issues, and difficulties detected regarding the lack of infrastructure for the development of RL.
- Supply chain management refers to existing barriers in the supply chain, problems of cooperation between stakeholders, and capacity to measure performance.
- Economic–financial issues include economic obstacles to the adoption and maintenance of RL services.

- Knowledge refers to difficulties felt in information flows, as well as awareness of RL in companies.
- Political issues include legal barriers, through regulations that hamper or not collaborate in RL activities, as well as legislation that makes it difficult to return products and RL.
- Market-related issues involve issues of a competitive nature, as well as market dynamics of reuse, recycling, or recovery of materials and products.
- Management and internal organization include behavioral issues of top management and intermediate management in relation to RL, and the importance that is attributed to RL in relation to other activities.

Following the same principle used in the case of drivers, barriers were classified as internal and external. Thirty-six barriers were identified, of which 13 were classified as external, 22 were internal, and one was classified as belonging to both categories. In order to determine the most influential barriers, an analysis of the number of times each barrier had been mentioned in the set of papers based on the work performed by Govindan and Bouzon (2018) was made. After this analysis, it was verified that the most frequently mentioned barrier was the lack of collaborators with the necessary skills to develop and maintain RL systems. The lack of skills was also dissected and was noted that there was a lack of training and technical knowledge. The major barriers mentioned as very important (second and third places, respectively) were the difficulty in investing in RL and the lack of involvement of top management in the development and maintenance of RL systems. This problem may be based on an even more serious problem: the absence of legislation establishing an effective waste management system in local terms. However, legislative complications may also arise from the lack of communication between various ministries of a given government, or from the existence of conflicts or inconsistencies between different laws or regulations. Official authorities may still be held liable for the establishment of certain barriers, such as the lack of policies to encourage social and environmental responsibility of producers, lack of incentives to establish an effective RL chain, or lack of compliance verification of companies regarding environmental regulations. On the other hand, there are barriers that are closely linked to the customers of these companies, remarkably through a lack of coordination between the supply chain elements, as well as the lack of information on how the costs are borne by the various players along that chain. There is also a lack of support from retailers, distributors, and traders for RL activities, which is also a serious barrier to their establishment and development. It should also be noted that when it comes to international trade, the problem arises of adapting the legislation of each country to the reality of the other participating in international trade. In fact, while some countries put immense pressure on the development of RL chains on producers, other countries distribute the responsibilities throughout the different supply chain players. This situation creates difficulties in adopting an international RL chain since stakeholders are not all equally pressured throughout the different countries involved in these commercial exchanges. Also, with regard to customers, the perception that the products are of poor quality or are made up of low-level materials creates disinterest in their reuse or recycling, which is also a problem to the proper functioning of an RL chain. However, the greatest barriers to RL are in the organizational aspect, where the lack of skills to deal with this type of issues is the problem that stands out. This problem can also be aggravated by the lack of commitment to the main objectives of RL, as well as the lack of technical capacity or lack of training of employees. Beyond the organizational barriers, there may still be economic problems that prevent the development of an RL chain, since there are insufficient economic resources to invest in this system. To exacerbate all these situations, there may still be a lack of willingness or interest of top management, which hinders, and even prevents, in most cases, the progress toward the development and implementation of a RL system. Indeed, the lack of organizational culture will certainly induce some resistance to the implementation of a system that they do not know, and which is not yet widely known in the business world. The uncertainty in the way the system will work, as well as the uncertainty as to the costs and benefits that can be extracted through the implementation of RL, leads to a natural resistance of entrepreneurs, which severely hinders the development process of this type of business. In fact, the implementation of RL implies the adaptation of information systems and workflow, which represents a new challenge. However, this challenge can be viewed positively or negatively and, when viewed negatively, significantly affects the progress of system implementation. Indeed, RL implies that there is strong coordination between companies, and even greater within each organization, which infers a great involvement of all the human resources of the company, and this coordination is usually aided by information systems. It should also be mentioned that, for a large majority of top managers or investors, investing in sustainable systems is seen more as an expense than something that adds value to your business. The compatibility of information systems (IT) is also seen as a critical factor in the development of RL networks, due to the lack of standards to better articulate the information among the various stakeholders. The existence of customs procedures and fees for returned goods may be a hidden cost to companies, which they need to know about. The ignorance of most companies about the taxes that are applied is a strong barrier for them to feel more inclined to adhere to RL practices, particularly those involving international transactions.

Unfortunately, the main motivations are essentially focused on compliance with the legislation issued by each country. Thus, the development of RL is largely based on the obligation of some countries to collect end-of-life products, leading to reuse, recycling or landfilling them in a more controlled manner. This avoids additional problems for the environment and stimulates RL. Another incentive is also promoted by the government through the granting of licenses. Indeed, some governments or entities related to environmental protection and sustainability, make it necessary for a company to operate the obligation to integrate sustainable systems that are implicit in RL principles. Thus, companies are forced to adopt these principles, which, more than stimulus, are a necessity, as they cannot operate in any other way. The cooperation between Logistics companies and customers also encourages these Logistics companies to solve the problems inherent to their customers' activities by cooperating in the closing of the product life cycle. Customer loyalty to a particular brand or product type may also be associated with how the consumer perceives the company to act in regard to the end-of-life of its products. Then, the company ends up feeling pressured and incentivized to proceed with RL, in order to improve its market image. Given the growing awareness of environmental concerns in many countries, companies are also being forced and encouraged to show their environmental concerns to consumers in order to satisfy the concerns expressed by their clients. Thus, there is a growing concern to meet the requirements of the most informed and environmentally demanding customers. In fact, green marketing is one of the motivations generated by society regarding the companies that can act in RL. This can be done in two ways: Businesses can charm their clients by acting in a more sustainable way, showing that they are proactive in encouraging the practices of recycling and reusing. On the other hand, by taking advantage of both trajectories, companies can deeply improve their efficiency, promoting more sustainable logistic practices. Thus, there is a strong motivation to participate in RL activities dynamically. Indeed, depending on the legislation imposed in each country, companies may find themselves forced to participate actively in RL, otherwise they would be engaging in illegal activities of undue deposit of products in landfills. As landfills tend to be a less available option for the end-of-life products, companies see in RL a way to solve this problem. In addition, corporate social responsibility tends to push them toward more sustainable practices, in line with ethical, legal, and economic standards. Companies also begin to realize that sustainable practices are the only way for their business to succeed in the future, both because of the legislation imposed by environmental agencies and government and because other market players tend to use RL practices more and more frequently, improving the market image. Thus, this way is the only one able to allow the companies to survive in the long term. These practices are also largely driven by other sustainable practices, such as the adoption of eco-design principles and Design for X, as previously refereed. Economically, RL practices lead to savings in raw materials and non-payment of fees by companies to deposit end-of-life products into landfills. The balance between the gains made and the costs inherent to the RL will have to be conveniently weighted so that there are clear advantages in the adoption of RL practices. Depending on the type of products and materials, RL can further encourage the market for the utilization of used components, which can easily be used to repair similar equipment in need of repair. This may be a good market opportunity as model volatility is increasing and finding replacement parts becomes a difficult task, which can be made easier by using RL and convenient separation, classification, eventual remanufacture, and reuse of products.

9.6 Drivers and Barriers in Building Construction

The construction industry has a strong impact on communities, both economically and socially, as well as environmental. Buildings account for about 40% of the energy consumed globally and account for about 40% of greenhouse gas emissions (GHG) (IEA 2013). Hence, it is extremely important to address buildings so that they are efficient during their life cycle (where the highest energy consumption resides). It is worth noting that the construction of buildings complies with a large number of conditions, which vary the typology of buildings from region to region, depending on situations as diverse as the usual climatic conditions, the economic and social priorities of families, the traditions, and cultures of each region, among many other factors. The construction of environmentally sustainable buildings has proliferated, making buildings more efficient in terms of heat and lighting during the period of use. For this, it is common to use solar technology and more energy-efficient roof tops. It has been found that technologies that make buildings more sustainable have been applied more intensively in developed countries than those still developing (Darko and Chan 2018). In a study recently done by Chan et al. (2018), twenty barriers to the application of sustainable technologies in building construction in Ghana were found. Ghana is a developing country, where some interesting sustainable technologies have been applied regarding the typology of the country, namely heating using solar panels, natural ventilation, and rainwater harvesting technologies. These barriers can be classified into five main groups, as follows:

- Barriers created by the national and local authorities and corresponding legislation;
- Barriers related to human behavior;
- Barriers imposed by lack of knowledge and/or access to information;
- Market-related barriers;
- Barriers to the cost of these technologies and risks that builders do not want to assume.

On the other hand, Darko et al. (2017) recently developed a work where he identified 16 drivers capable of driving more sustainable thinking connected to the construction project of buildings, again based on the reality lived in Ghana. These drivers were grouped into five sets, which are identified below:

- Drivers related to the environment;
- Drivers related to health and economic issues;
- Internal drivers related to companies and their organization;
- Drivers related to energy and costs;
- Drivers related to industry.

These drivers, if well understood by construction companies, can represent significant gains that can be obtained by adopting sustainable construction strategies. The adoption of these strategies provides a competitive market advantage through new sales arguments since energy-related expenses over the life of the building will be much smaller, resulting in economic benefits for owners and significant gains for the environment. In a work later developed by Darko and Chan (2018), five strategies were also recognized that could also induce the adoption of sustainable construction principles, which were categorized into five groups:

- Legislation and regulations issued by the government and standards to be complied with;
- Support provided through R&D programs as well as development incentives;
- Adequate dissemination by stakeholders of information related to the topic;

- Information and awareness programs;
- Recognition of the efforts made to achieve the proposed objectives of sustainable construction strategies.

In a study evaluating the developments achieved with previous investigations, Darko et al. (2018) concluded that for sustainable construction strategies to be implemented in a country such as Ghana, it is necessary for the government to create the conditions for the entrepreneurs in the sector to feel the necessary motivation for the pursuit and implementation of this type of construction. In addition to the incentives, it is still necessary to issue legislation and regulations that force entrepreneurs to adopt more sustainable strategies in the construction of new buildings. Thus, it can be seen that the main motivations in the current framework will have to come from within the companies themselves, which will have to identify what competitive gains they can obtain from the market through the adoption of these strategies. The differentiating factor in terms of the market offer will be the greatest motivation for the construction industry to take an environmental step forward.

References

P. Agamuthu, M.K. Khidzir, S.F. Hamid, Drivers of sustainable waste management in Asia. Waste Manage. Res. **27**(7), 625–633 (2009)

S.M. Al-Salem, P. Lettieri, J. Baeyens, Recycling and recovery routes of plastic solid waste (PSW): A review. Waste Manage **29**(10), 2625–2643 (2009)

H.F. Bezzina, S. Dimech, Investigating the determinants of recycling behaviour in Malta. Manage. Environ. Qual. Int. J. **22**(4), 463–485 (2011)

U.S. Brandt, G.T. Svendsen, When can a green entrepreneur manage the local environment? J. Environ. Manage. **183**(3), 622–629 (2016)

E. Cagno, E. Worrell, A. Trianni, G. Pugliese, A novel approach for barriers to industrial energy efficiency. Renew. Sustain. Energy Rev. **19**, 290–308 (2013)

E. Cagno, A. Trianni, C. Abeelen, E. Worrell, F. Miggiano, Barriers and drivers for energy efficiency: Different perspectives from an exploratory study in the Netherlands. Energy Convers. Manage. **102**, 26–38 (2015)

A.P.C. Chan, A. Darko, A.O. Olanipekun, E.E. Ameyaw, Critical barriers to green building technologies adoption in developing countries: The case of Ghana. J. Clean. Prod. **172**, 1067–1079 (2018)

J.H. Clark, T.J. Farmer, L. Herrero-Davila, J. Sherwood, Circular economy design considerations for research and process development in the chemical sciences. Green Chem. **18**, 3914–3934 (2016)

F. Contreras, S. Ishii, T. Aramaki, Drivers in current and future municipal solid waste management systems: Cases in Yokohama and Boston. Waste Manage. Res. **28**, 76–93 (2010)

A. Darko, A.P.C. Chan, Strategies to promote green building technologies adoption in developing countries: The case of Ghana. Build. Environ. **130**, 74–84 (2018)

A. Darko, A.P.C. Chan, S. Gyamfi, A.O. Olanipekun, B.J. He, Y. Yu, Driving forces for green building technologies adoption in the construction industry: Ghanaian perspective. Build. Environ. **125**, 206–215 (2017)

A. Darko, A.P.C. Chan, Y. Yang, M. Shan, B.-J. He, Z. Goud, Influences of barriers, drivers, and promotion strategies on green building technologies adoption in developing countries: The Ghanaian case. J. Clean. Prod. **200**, 687–703 (2018)

N.S. Gandhi, S.J. Thanki, J.J. Thakkar, Ranking of drivers for integrated lean-green manufacturing for Indian Manufacturing SMEs. J. Clean. Prod. **171**, 675–689 (2018)

M. Geissdoerfer, P. Savaget, N.M.P. Bocken, E.J. Hultink, The circular economy—A new sustainability paradigm? J. Clean. Prod. **143**, 757–768 (2017)

R.A.R. Ghazilla, N. Sakundarini, S.H. Abdul-Rashid, N.S. Ayub, E.U. Olugu, S.N. Musa, Drivers and barriers analysis for green manufacturing practices in Malaysian SMEs: A Preliminary Findings. Procedia CIRP **26**, 658–663 (2015)

W.H. Golove, J.H. Eto: *Market barriers to energy efficiency: a critical reappraisal of the rationale for public policies to promote energy efficiency,* Berkeley: Lawrence Berkeley National Laboratory. (University of California, Berkeley; 1996). http://citeseerx.ist.psu.edu/viewdoc/download?doi=10.1.1.165.3714&rep=rep1&type=pdf [ONLINE], Retrieved on the 16th Mar 2019

K. Govindan, M. Bouzon, From a literature review to a multi-perspective framework for reverse logistics barriers and drivers. J. Clean. Prod. **187**, 318–337 (2018)

C. Gunn, Energy efficiency vs. economic efficiency? New Zealand electric sector reform in the context of national energy policy objectives. Energy Policy **25**, 241–257 (1997)

J.U. Gustavo, G.M. Pereira, A.J. Bond, C.V. Viegas, M. Borchardt, Drivers, opportunities and barriers for a retailer in the pursuit of more sustainable packaging redesign. J. Clean. Prod. **187**, 18–28 (2018)

J. Hopewell, R. Dvorak, E. Kosior, Plastics Recycling: Challenges and Opportunities. Philos. Trans. Roy. Soc. B Biol. Sci. **364**, 2115–2126 (2009)

IEA 2013: *Modernising Building Energy Codes.* https://www.iea.org/publications/freepubl981ications/publication/PolicyPathwaysModernisingBuildingEnergyCodes.pdf [ONLINE], Retrieved on the 24th of March, 2019

I.S. Jawahir, R. Bradley, Technological elements of circular economy and the principles of 6R-based closed-loop material flow in sustainable manufacturing. Procedia CIRP **40**, 103–108 (2016)

A. de Jesus, S. Mendonça, Lost in transition? drivers and barriers in the eco-innovation road to the circular economy. Ecol. Econ. **145**, 75–89 (2018)

R. Kemp, Eco-innovation: Definition, measurement and open research issues. Econ. Política **3**, 397–420 (2010)

B.G. Mwanzaa, C. Mbohwa, Drivers to sustainable plastic solid waste recycling: A Review. Procedia Manuf. **8**, 649–656 (2017)

S.N. Nielsen, S.E. Jørgensen, Sustainability analysis of a society based on exergy studies—a case study of the island of Samsø (Denmark). J. Clean. Prod. **96**, 12–29 (2015)

D.A.J. Pacheco, C.S. ten Caten, C.F. Jung, J.L.D. Ribeiro, H.V.G. Navas, V.A. Cruz-Machado, Eco-innovation determinants in manufacturing SMEs: Systematic review and research directions. J. Clean. Prod. **142**, 2277–2287 (2017)

F. Preston: *Briefing Paper—A Global Redesign? Shaping the Circular Economy. Energy, Environment and Resource Governance.* https://www.chathamhouse.org/publications/papers/view/182376 [ONLINE], Retrieved on 15 Mar 2019

A.K.N. Reddy, Barriers to improvements in energy efficiency. Energy Policy **19**, 953–961 (1991)

B.S. Reddy, Barriers and drivers to energy efficiency—A new taxonomical approach. Energy Convers. Manage. **74**, 403–416 (2013)

V. Rizos, A. Behrens, T. Kafyeke, M. Hirschnitz-Garbers, A. Ioannou: *The Circular Economy: Barriers and Opportunities for SMEs* (No. 412). CEPS Working Document. (Centre for European Policy Studies (CEPS), 2015)

V. Rizos, A. Behrens, W. Van Der Gaast, E. Hofman, A. Ioannou, T. Kafyeke, A. Flamos, R. Rinaldi, S. Papadelis, M. Hirschnitz-Garbers, C. Topi, Implementation of circular economy business models by small and medium-sized enterprises (SMEs): Barriers and enablers. Sustainability **8**(11), 1212–1229 (2016)

H. Saunders, The Khazzoom-Brookes Postulate and neoclassical growth. Energy J. **13**, 131–148 (1992)

J. Schleich, E. Gruber, Beyond case studies: barriers to energy efficiency in commerce and the services sector. Energy Econ. **30**, 449–464 (2008)

J. Schot, L. Kanger: *Deep Transitions: Emergence, Acceleration, Stabilization and Directionality (SPRU Working Paper Series No. 2016–15)*, SPRU—Science and Technology Policy Research (University of Sussex, 2016)

D. Seth, M.A.A. Rehman, R.L. Shrivastava, Green manufacturing drivers and their relationships for small and medium (SME) and large Industries. J. Clean. Prod. **198**, 1381–1405 (2018)

S. Sorrell, A. Mallett, S. Nye: *Barriers to industrial energy efficiency: a literature review, working paper 10*. (Vienna: United Nations Industrial Development Organization, 2011)

S. Studer, S. Tsang, R. Welford, P. Hills, SMEs and voluntary environmental initiatives: a study of stakeholders' perspectives in Hong Kong. J. Environ. Planning Manage. **51**(2), 285–301 (2008)

A. Trianni, E. Cagno, F. Marchesani, G. Spallina: *Drivers for industrial energy efficiency: an innovative approach*. In: Proceedings of the 5th International Conference on Applied Energy (ICAE), (Pretoria, South Africa, 2013), 1–4 July 2013

M.A. Troschinetz, R.J. Mihelcic, Sustainable recycling of municipal solid waste in developing countries. Waste Manage **29**, 915–923 (2009)

N. Tura, J. Hanski, T. Ahola, M. Ståhle, S. Piiparinen, P. Valkokari, Unlocking circular business: A framework of barriers and drivers. J. Clean. Prod. **212**, 90–98 (2019)

C. Van Hemel, J. Cramer, Barriers and stimuli for eco-design in SMEs. J. Clean. Prod. **10**(5), 439–453 (2002)

R. Vanner, M. Bicket, S. Withana, P. ten Brink, P. Razzini, E. van Dijl, E. Watkins, M. Hestin, A. Tan, S. Guilcher, C. Hudson: *Scoping Study to Identify Potential Circular Economy Actions, Priority Sectors, Material Flows & Value Chains (DG Environment's Framework Contract for Economic Analysis ENV.F.1/FRA/2010/0044 Final Report)*. Policy Studies Institute (PSI), (Institute for European Environmental Policy (IEEP), BIO and Ecologic Institute, 2014)

E. Vine, J. Hamrin, N. Eyre, D. Crossely, M. Maloney, G. Watt, Public policy analysis of energy efficiency and load management in changing electricity businesses. Energy Policy **31**, 405–430 (2003)

D.C. Wilson, Development drivers for waste management. Waste Manage. Res. **25**, 198–207 (2007)

U.A. Zaman, Identification of waste management development drivers and potential emerging waste treatment technologies. Int. J. Environ. Sci. Technol. **10**, 455–464 (2013)

Q. Zhu, Y. Geng, Drivers and barriers of extended supply chain practices for energy saving and emission reduction among Chinese manufacturers. J. Clean. Prod. **40**, 6–12 (2013)

Chapter 10
Tracking Environmental Performance

Abstract Tracking the environmental performance of a giving company or sector is not a straightforward process. There are several barriers that can lead to a skewed view of the overall impact of a company. Many times, these barriers are purposefully positioned in an effort to project a market image that is better than the actual reality of a situation. Although market image perception is a very important factor for many companies and industries, tracking environmental performance has benefits that go well beyond this domain, allowing many times to cut overall costs and improve profits while reducing environmental impacts. Reduction in energy, water, raw materials, production, and labor hours, logistic optimization, among many others, are just a few examples that can be achieved by implementing and monitoring cleaner and more sustainable strategies. These accomplishments are usually the results of the application of tools and creation of indexes that allow an accurate qualitative and quantitative monitoring of a situation/process as well as its future trend. In this chapter, different environmental assessment methodologies and how these assessments are achieved in different countries and companies will be discussed. Also, the application of these assessments in sectors such as the chemical, construction, services and wastewater treatment as well as the advantages and drawbacks of publicly disclosing environmental assessments are debated.

Keywords Tracking environmental performance · Environmental performance assessment methodologies · Environmental assessment disclosure

Making the right decisions to minimize the environmental impact of industrial activities is the most important factor in achieving the goals of having a cleaner and healthier planet, without the threat of extreme weather events. In order to achieve these objectives, many companies have already taken a very significant step by adhering to environmental management and certification programs. Subsequently, it is necessary to evaluate the environmental performance of companies and institutions that intend to comply with existing environmental legislation or even take it one step further. The emergence of environmental performance indicators is closely linked with the implementation of environmental management systems, allowing through concise periodic reports, compiled with the necessary information through the use of indicators,

F. J. Gomes da Silva and R. M. Gouveia, *Cleaner Production*,
https://doi.org/10.1007/978-3-030-23165-1_10

that managers assess the company's present situation, analyze its evolution, and make decisions that will put it on the path mapped out by top management.

As the process of legislation is dynamic and evolutionary, environmental management processes in companies will have to be equally dynamic, often following a PDCA (Plan–Do–Check–Act) methodology to promote continuous improvement of their environmental management system. Thus, numerous tools have emerged to measure the environmental performance of companies and institutions, which aim to standardize the evaluation, making the results easy to obtain and at the same time comparable. Although there is some standardization in this type of evaluation, namely developed by ISO 14031, it does not show the necessary efficiency, because it allows a quantitative evaluation of only a few factors that intervene in the process, namely the evaluation of energy use and evaluation of gaseous emissions, among others. Moreover, the International Organization for Standardization presents a rather concise but also very general definition of the concept of environmental performance, referring to the measurable results achieved by the management of a given organization with respect to its environmental aspects. In this sense, many researchers have devoted their efforts to find the most appropriate tools for each situation. However, the way to evaluate environmental performance is far from reaching consensus, since some authors have used environmental costs as a determining factor in the environmental performance evaluation of industrial companies (Schmidt and Nakajima 2013; Chompu-Inwai et al. 2015), but others have used the environmental impact assessment as the main indicator for performance evaluation (Herva and Roca 2013; Elduque et al. 2015), and others the environmental risk analysis (Herva and Roca 2013; Nucci et al. 2014). These tools are essentially of voluntary application by the companies themselves to self-assess and realize whether the efforts they are making to become more environmentally friendly and more sustainable are effectively producing the desired results. The models of environmental performance analysis aim essentially to produce results, preferably quantitative, that allow top managers to make the most accurate decisions regarding these indicators, and later allow an adequate monitoring of the evolution of the situation. Obviously, the analysis can be performed in a qualitative way, but it does not have the same impact and rigor as a quantitative analysis, which allows to calculate certain indices track their evolution over time and decided upon the currently implemented situation.

Some authors tend to classify these tools essentially in two ways: procedural and analytical (Finnveden et al. 2009; Finnveden and Moberg 2005). While procedural tools deal mainly with decision procedures related to environmental management systems, analytical tools essentially focus on the analysis of technical aspects and how to quantify them (Wrisberg et al. 2002). However, analytical tools can be integrated in procedural tools, thus allowing a more informed and quantitative analysis in decision processes that aim to improve the environmental performance of a company or institution. These analyses allow a much more accurate view of the problems that may affect environmental performance, leading management to make the most appropriate decisions in each process or department of a company or institution. Therefore, if properly applied, these tools will save money and human resources, allowing to reach the desired level of performance faster, as these tools allow an easier identification of critical factors that may be affecting the overall performance of the company. Several analytical tools have been developed for this purpose. Schmidt and Nakajima (2013) and Chompu-Inwai et al. (2015) have recently used material flow cost accounting, while Hermann et al. (2007), Jawjit et al. (2015) and Elduque et al. (2015) used the Life Cycle Impact Assessment, and Herva et al. (2011) and Butnariu and Avasilcai (2014) used the ecological footprint, among many other studies in this area.

10.1 Comparing Environmental Assessment Methodologies Among Different Countries

In a more comprehensive study, García-Álvarez and Moreno (2018) developed an indicator capable of evaluating the environmental strategies and policies conducted by the different European countries, based on the 28 countries of the European Union. This work aimed to quantify the performance of different countries in complying with the existing regulations in terms of policies to reduce carbon emissions, with a view of improving the well-being of citizens. The indicator created through that work aimed to select the best policies for each country and type of society, as well as to monitor the policies already defined and in practice, letting them know if they were producing the desired results or needed adjustments. As a result of an indicator based on the same variables, the values obtained will also make it possible to compare the effectiveness of legislation and policies. This system also allows to identify the strengths and weaknesses of policies taking into account the context offered by each country and society. The time analysis of the evolution of the index will allow actions to promote a more effective reduction of environmental impacts, as well as to establish control limits, so that alerts can be issued and other types of decisions taken. In studies previously developed by Svensson et al. (2006) and Arnell et al. (2013), several categories have been proposed to be included in this type of indices, which include waste, climate change, water, agriculture, among many other factors. The number of indicators previously developed by several authors regarding the environment in general is very wide, ranging from the water and energy use index (Carrasquer et al. 2017) applied to the agro-food industry, among many others. Based on the goals set by the European Union for 2020, 2030 and 2050, Garcia-Álvarez and Moreno (2018) developed the Environmental Performance Assessment Composite Index (EPACI), supported in 24 different variables. This indicator was intended to bring together various aspects of environmental policies, mainly focused on natural resources, efficient use of resources and use of another resources, allowing for a more reliable comparison among member states, as well as the identification of the most successful policies in terms of environmental performance, and the identification of possible recommendations to be made to each member state. Of this study, five countries with a performance clearly above the European average stood out, namely Sweden, Austria, Denmark, Italy, and Germany being noticeable that they have policies that emphasize their own natural resources, are more efficient in the consumption of resources and, for example, promote the use of alternative energetic resources. On the other hand, the five countries that stood out in the negative were Lithuania, Poland, the Czech Republic, Romania, and Slovakia, with performance clearly below the European average. The study has highlighted the need to strengthen Cleaner Production policies, especially in the case of Romania and Poland, while in Slovakia the problems are essentially related to the need to significantly increase energy efficiency. In the case of these three countries, it was also clear that they will have to work hard to achieve the goals set in the 2050 Europa Vision. Thus, it has been proven through the application of the model that it allows to obtain quite interesting results, highlighting the most positive and negative points in each country, allowing to later act in a focused way, improving and carefully monitoring the areas that need the most improvement.

10.2 Analyzing Environmental Performance Assessment Methodologies in Companies

The evaluation of the environmental performance of companies, in general, is an essential tool used to facilitate top management to realize if the measures already taken are achieving the desired effect and to understand what actions are necessary to be done in order to promote a continuous and effective performance improvement. However, the evaluation of environmental performance is not easy, given the different aspects to which it must attend to and how detailed it should go. It should also be noted that it is necessary to verify whether the knowledge taken as a starting point is adequately solidified and perfectly reliable. Therefore, there are uncertainties at the outset, which need to be duly known and clarified, so that the results obtained in the end can be properly interpreted. Given the different aspects in which the analysis can be carried out, it is also necessary to select or define the indicators that can best illustrate the situation of a company at any given time. This will depend heavily on the business sector in which the company is operating, what factors are considered the most important by the company and its stakeholders, the possibility and ability to obtain certain data corresponding to performance, loyalty of the information which can be currently obtained, and also in the interest of the company in which certain data may or may not be disclosed. The information gathering issue is also of the utmost importance, as only reliable information will yield results that are truly useful. In addition, the data must be collected following the same criteria, thus eliminating another possible source of uncertainty. The chain of people related to data generation, data collection and treatment must be fully aware of the importance of the information they are dealing with, ensuring process suitability and fidelity in the obtained results. Obviously, it is of the utmost convenience the inclusion of evaluation standards, which allow comparing the performance between different companies of the same sector, in order to also measure the asymmetries in their evolution. This may help legislators to push forward certain sectors or strands within the same sector in order to stimulate the improvement of certain aspects. In most cases, the evaluation is voluntary, becoming also necessary to determine which aspects companies want to analyze for internal self-assessment, and which aspects they want to publicly display, in order to generate economic advantages as well.

Considering the ISO 14011 (2015) standard that regulates the environmental management system of companies, Trumpp et al. (2015) carried out a bibliographical review work in which the authors grouped the different perspectives understood by the researchers as important for the evaluation of the environmental performance and established a parallelism with the recommendations of the said norm. This study allowed to conclude that there are essentially two dimensions for the evaluation of environmental performance: a more qualitative one, called environmental management performance, and a more quantitative one, called environmental operational performance. These main strands are then divided into sub-strands, which can be analyzed in Fig. 10.1. It may be said, however, that the diagram in Fig. 10.1 is conditioned to the dimensions stipulated by the standard, but this does not imply that these dimensions are completely watertight. A very concrete example of this is the one that is presented by Dragomir (2018), where the case of CO_2 is pointed out, which is considered as an emission in one process, but can be partially sequestered through another process, and give rise to financial savings, which generate added value and can be included in an economic indicator different from the other indicators used to evaluate emissions and CO_2—operation indicator –, and the sequestration—management indicator.

The previous exercise serves to illustrate that the use of only one dimension in the evaluation, or the overvaluation of one dimension over others, can distort the perception that is taken away from the evaluation thus, it is extremely useful to calculate and maintain the information pattern for each indicator, illustrating more clearly what is actually happening in environmental terms. However, Dragomir (2018) also argues that the concept of environmental performance evaluation, if divided into many aspects, removes the depth of analysis and consistency in the perception of the information that the evaluation provides. This question raises some uncertainties that deserve to be properly considered. In fact, the value obtained for a given assessment must be unique, although it consists of different dimensions, to which different weights can be given, in order to adapt the general indicator to each particular case. The values for each dimension should remain accessible, so that the analysis can be carried out in a deeper way, identifying the areas where favorable developments have occurred, which areas will be necessary to invest more immediate effort into, and areas of which the maintenance of the current situation is perfectly sufficient to lead the company to the desired results. In other words, the general indicator will allow a global view of the company's behavior in environmental terms, but the information for each dimension is one that will allow selective action to be applied were needed. Obviously, the selection of the factors making up each component is essential and should be as standardized as possible in order to allow a comparative analysis between firms in the same sector. Otherwise, performance would be measured in an absolute manner and could be distorted by the company's marketing interests. In the

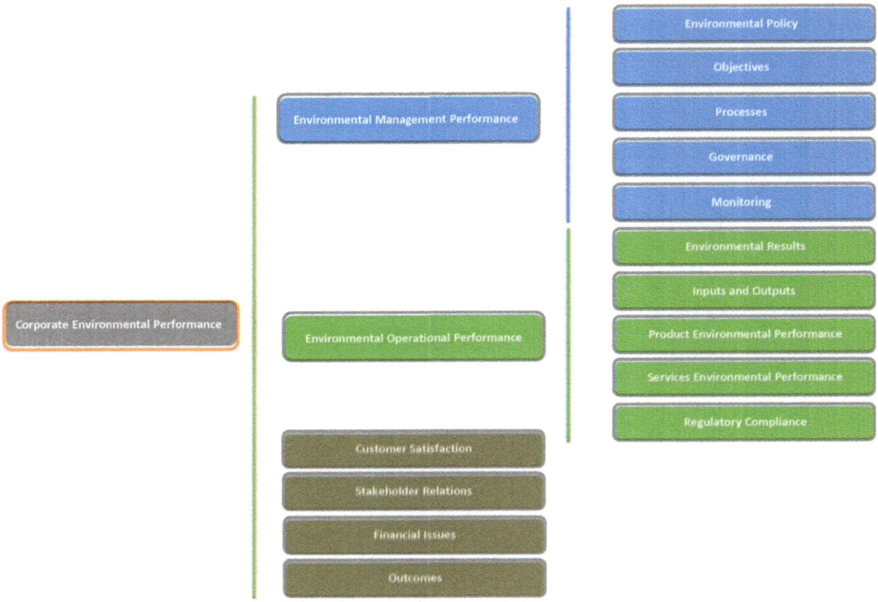

Fig. 10.1 Overview of the different dimensions and sub-strands that can be considered in an environmental performance assessment of companies

review of the literature by Dragomir (2018), it is verified that the same variable was found with different definitions from work to work, which indicates that the same variable may be translating different situations, removing credibility to the analyses that are done and making it impossible to compare performances of companies in the same sector. This situation is of some seriousness, showing the pertinence of a more evolved normalization, adaptable to different situations, but avoiding a proliferation of factors, and avoiding that the same dimension could be configured differently by researchers.

10.3 Environmental Performance Assessment in the Chemical Industry

Given that the chemical industry, in most cases, causes environmental concerns, it becomes a sector where the environmental performance assessment is extremely useful. Indeed, the evolution of the processes themselves may jeopardize the competitiveness of companies, so it is vital to monitor the behavior of each company vis-à-vis its competitors. In view of the constraints imposed by legislation and an increasingly environmentally demanding market, the design or renewal of any chemical industry requires special care from the early stages of design, making the best options from the beginning, so that problems do not worsen throughout the production chain. In order to quantify environmental performance in the chemical industry, Jia et al. (2004) developed a tool called Process Environmental Performance Assessment (PEPA) which essentially aimed to integrate the process design stage with environmental performance assessment, providing a quantitative indicator that would allow top management to perceive if the process or set of processes analyzed by the indicator is one that entails less harm to the environment. However, due to the complexity of processes usually applied in the chemical industry, reducing the environmental impact to an index did not prove to be an easy task. The authors used a multi-criteria decision-making (MCDM) analysis to develop the desired quantitative indicator. The calculation of the indicator followed a set of three steps, starting with the classification of the environmental impact data. In the second phase, these already classified data sets were characterized and quantified to finally calculate an indicator using the above-mentioned MCDM analysis. In order to validate the developed model, two distinct ethanol production processes, namely the straw cellulose-derived feedstock process and ethylene-derived feedstock process, were analyzed, concluding that the first process is more environmentally friendly than the second, with a clear difference of about 15% in the calculated index. As noted elsewhere, the authors almost always pinpoint that the indicator can still be improved by the addition of more information, better supporting the decisions that can be made on the basis of this analysis.

Maceno et al. (2018) developed a tool for environmental performance evaluation, which he called the Environmental Performance of Industrial Processes (EPIP), based on different principles regarding other analytical tools previously developed. The main objective of this development was to create a tool capable of supporting decision making by the top management regarding environmental policies, combining economic and environmental factors in a perspective focused on industrial activities. The model created encompasses different aspects of the analysis, namely the materials involved, energy consumed, as well as the environmental impact and respective costs, resulting in a quantitative indicator that allows decision makers to delineate their strategies and actions in a well-founded manner. With this analysis, it is possible to see if the previously developed actions had the desired effect, and to correct or reinforce these actions, in order to approximate the indexes of the desired values. The model was applied in a company that manufactures plastic yogurts cups and has been shown to be effective in analyzing the environmental performance of this industry, allowing the delineation of decisions to be taken in order to improve this performance. However, according to the authors, the model still needs to be improved because there are factors that are integrated in the index and which are not standardized. Moreover, it is still necessary to expand the boundaries of the system under analysis, as well as evaluate the implications and costs of external factors and take into account also some social aspects, which are not included in the current model. Thus, there is still a strong margin of progression in the analysis of environmental performance in this industry.

10.4 Environmental Performance Assessment in the Construction Sector

The construction sector also entails environmental problems, both during the construction phase and during the life stage. In fact, considering the construction phase and the useful life of buildings, they account for about 36% of the energy consumed globally and about 40% of the overall CO_2 generated, according to the International Energy Agency Report of 2017 (IEA 2017). Considering only the area corresponding to the European Union, it is worth noting that construction and completed buildings account for about 33% of global water consumption and 33% of the global waste generated. As far as construction is concerned, buildings account for 30–50% of global material consumption (CRI 2014; EC 2017). Although a number of concepts have emerged that aim to give buildings greater sustainability, which have assumed such solid names as smart buildings, sustainable buildings, intelligent buildings and green buildings, there is still a lot to be done, since energy consumption and the generation of CO_2 related to buildings has shown sustained growth of around 1% per year (IEA 2017). However, it is a fact that thermal insulation, solar exposure of buildings, use of natural light, among many other efforts, have shown a very positive development in the construction sector. There are many factors that contribute decisively to making buildings a heavy burden on the environment. Indeed, older buildings are those that are less prepared for the new environmental requirements and, given the figures typically reported by statistics, only about 1–2% of buildings are renovated with each passing year. In buildings, it is common that responsibility is shared between the owner and many tenants or between owners with different views and priorities, which conditions the rate of renewal and the quality of the improvements produced.

A study recently developed by Maslesa et al. (2018) identified eight categories of parameters commonly used to quantify the environmental performance of buildings: energy consumed, emissions generated, water consumed, waste generated, area of land/building used, quantity and type of construction materials utilized, internal environmental quality, and potential for materials reuse/recycling. It is common for environmental performance indicators to be particularly concentrated on energy consumption and the pollutant burden released in the form of emissions, effluents, or solid waste. However, there are other concerns that must also be considered, namely the proper selection of materials and products with a view to their possible reuse, or even the indoor environmental quality of a building. These indicators should also take into account the expectations of the life cycle of the building, as well as the refurbishing actions that are expected to take place during its useful life. The refurbishing operations aim essentially at increasing comfort and safety, but should be programmed to increase environmental performance, improving thermal insulation, natural ventilation conditions, daylight utilization, and among other situations. In addition, depending on the type of building and degree of refurbishing involved, it should be accounted for that these refurbishing operations usually generate waste, which often needs to be landfilled. This is also a negative factor for the environment, which must be weighted, i.e., balanced with actions that effectively improve its environmental performance after refurbishment. This is obviously valid mostly for residential buildings. However, these are the ones that cause less environmental impact, considering other buildings such as the ones for commercial and industrial purposes. It is also worthy to note that the environmental impact is usually dependent on building age. Newer buildings show a higher environmental impact during their construction phase, mainly due to the use of a greater quantities and sophistication of materials, as well as associated construction techniques, while older buildings have a greater environmental burden during their use phase, due to lesser care and poorer material selection during construction. This is particularly true for commercial and industrial buildings, where energy consumption in older buildings is much higher than it should be due to antiquated designs and materials. In terms of environmental performance assessment, there is a clear division between models that are essentially based on the Life Cycle Assessment (LCA), while others are based on the Building Research Establishment Environmental Assessment Methodology (BREEAM) and Leadership in Energy and Environmental Design (LEED) techniques. The approaches to these tools are different, with a clear understanding that the LCA is more geared toward a complete analysis of the building's life cycle, with preponderance in the use phase, while tools such as BREEAM and LEED are more geared toward the design and construction of buildings. This is emphasized below.

In the construction phase, different strategies can be adopted, which will have different environmental impacts. In fact, construction consumes a significant amount of natural resources, also affecting the environment in several ways, namely through the creation of waste and effluents from washes, dust, among others. Quantitatively assessing the impacts of each type of strategy is something that all construction companies should be concerned about. Moreover, the impact of the construction phase should be accounted for in the LCA of the constructed building. However, to do so, it is necessary to have available easy, reliable and credible tools for this purpose. Analyses carried out using the LCA methodology are not the most suitable for the construction phase of buildings, but rather for an overall analysis of their environmental performance during their overall life span. This is even clear in the essence of ISO 14040, where ISO states that this standard is only intended as a guiding tool for decisions, but not a decision-making tool. Wang et al. (2017) argue that there are far more appropriate tools for the environmental impact assessment during the construction phase of buildings than the LCA analysis, indicating as more effective alternatives the genetic algorithms, particle swarm optimization and colony optimization algorithms. This view is due to the fact that LCA does not have the most appropriate characteristics to optimize problems based on multi-objectives. Thus, a new tool was developed in which the particle swarm optimization technique was integrated into the Life Cycle Assessment, allowing for the newly developed model to become a multi-objective tool for the decision-making of constructors, with a view to the selection of the best construction process taking into account environmental concerns. The proposed model, called environmental assessment and optimization method, starts by using the LCA to determine the environmental impact of a given construction system, providing the data indispensable for the subsequent application of the particle swarm optimization method, which is better structured to manage optimization problems based on multi-objectives, as is the case of the construction phase of a building. The selection of the best solution can be obtained later using a tool such as Pareto analysis. Given that the results obtained for each different construction system are quantitative, Pareto analysis allows a quick ordering of results, highlighting the most environmentally friendly options. The model was intended to provide builders with a tool capable of enabling environmentally more informed decision-making, overcoming the usual problems of making the right choice in a wide variety of building systems, and subject to sometimes contradictory environmental metrics. In this way, builders can make decisions during a previous planning stage and can choose a more sustainable construction model, with less harmful repercussions for the environment.

On the other hand, buildings can also have their environmental performance assessed. For this, there are different tools that have evolved over time, as described by Giarma et al. (2017). The pioneering tool for this purpose was BREEAM and dates back to 1990. With the evolution of this tool, it is now possible to evaluate the environmental performance of buildings for various purposes, from residential to hospitals or schools and at any life stage. The tool presents a high versatility, allowing the adaptation to each real case scenarios, including the proper conditions for location where the building is to be implanted. The assessment is based on a series of criteria taking into account various environmental aspects. Minimum targets are set for each criterion, and in key areas the building will have to reach this level in order to be considered as environmentally friendly. The sum of the score achieved also gives rise to a qualitative classification corresponding to six different levels: unclassified, pass, good, very good, excellent, and outstanding. LEED was developed in the USA and, like BREEAM, can be adjusted to different types of buildings and to different regions of the globe, and can be used at any stage of a building's life. It has a classification by requirement, being classified according to the way the building fulfills this requirement. Adjustments made over the lifetime of this tool have allowed its power of analysis to be extended even to land phase projects or even for neighboring buildings. Following a similar philosophy to BREEAM, the score summary gives rise to a final qualitative classification that can take on one of four different levels: certified, silver, gold, and platinum. LEED tool is widely used, even in Europe. The Comprehensive Assessment System for Built Environment Efficiency (CASBEE) was developed in Japan and is aimed at the environmental assessment of buildings in three different phases of their life: new buildings, existing buildings and refurbished buildings. In terms of the purpose of use of the building and the environment where it is inserted, the classification differs clearly from previous models, with a different version for each situation (CASBEE for Cities, CASBEE for Urban Development, CASBEE for New Construction, and so on). The model is strongly geared toward Asian reality, being used more intensely in Japan. The model is structured to analyze essentially four distinct aspects: resource efficiency, energy efficiency, local environment, and interior environment, which are subdivided into more than 90 different items. The concept of evaluation is also quite different from the models described above, essentially considering the building as two distinct parts: the private part (interior) and the public part (surrounding). For the private component, it considers essentially how the quality of life of the building occupants is preserved, while in for the surrounding space it considers the environmental impact caused by the building. Each of these factors is further divided into three classes, which also comprise several categories. Classifications are assigned to the quality of the building, as well as to the reduction of the building's environmental load. The index is calculated through a ratio between the quality of the building and the reduction of the environmental load achieved. The final classification is qualitative, based on five levels, ranging from a poor performance (C level) to an excellent performance (S level). The Sustainable Building Tool (SBTool) is a computing tool that has evolved steadily thanks to the proactive collaboration of various institutions, organizations, and researchers. Like BREEAM and LEED, this model can be applied in the different phases of a building's life cycle, from the project to the service phase. In addition to the base version, specific applications have also been developed for some countries, taking into account the geographic, climatic, and social particularities of each country. Either version has a high flexibility of use, allowing easy adaptation to the conditions imposed by each analysis, as well as the evaluation needs specific to each case. These adjustments are made by fine-tuning the weight assigned to each analysis criterion. The model is based on seven different criteria, which are further subdivided into several sub-criteria. It should be noted that, during the preproject phase, only one criterion can be assessed: location, site characteristics, and available services. Each factor to be evaluated has different performance categories, which will subsequently be affected by a certain weight, and the results are then added together in an index that can assume a qualitative classification between −1 and +5.

Although all of the building rating systems described above are available on the market, there are researchers who believe that these systems are not the most appropriate for assessing the environmental performance of buildings. In fact, Yudelson (2016) recently launched a book in which he expresses high pessimism about the applicability of these classification models of buildings. In the same book, a decrease in adherence to the environmental performance evaluation tools of buildings is pointed out, also indicating several probably reasons responsible for this decrease, such as (a) being a model that takes into account too many factors, making it difficult to use, time-consuming, expensive, and excessively bureaucratic, (b) leading customers to solutions that are too costly for the city suburbs, with the aim of obtaining a higher classification, (c) in the classification model, certain critical aspects such as climate changes seem not to be taken into account, (d) high competition between different factions that struggle differently for more sustainable construction, and (e) the existence of too many assessment systems on the market, making selection and classification more confusing.

Howard (2017) is also extremely critical regarding the classification of buildings. Referring specifically to the LEED model, the same author states that this model has failed successively in performance forecasts regarding energy consumption and produced emissions. Moreover, it states that even after 20 years of improvement, the model is still not able to reliably predict the performance of buildings, nor the environmental impact they cause. It is mentioned that the dispersion of results from the LEED evaluation is too high, which means that the results obtained through the application of the model are significantly different from the real ones, discrediting the model. Some of the more aberrant faults are pointed to the fact that the model considers performance in relative terms, not in absolute terms. In this way, it is easier to capitalize on a better classification in a typically cold area compared to a moderate one. Under these conditions, any building in an area affected by typically cold climates requires greater care and innovation in its project with a view to reducing its environmental impact over its lifetime, while a building typically under moderate climates does not need such demanding levels of innovation. It should also be noted that the BREEAM model is not affected by this problem, since the evaluation is carried out in absolute rather than relative terms.

10.5 Environmental Performance Assessment in the Services Sector

Although the agricultural, mining, and industrial sectors are particularly prominent in environmental concerns related to the business world, the services sector also needs to be assessed in terms of environmental performance, as it can also contribute to overall environmental concerns. A service can be defined as a product generated by human activity, which satisfies a particular need, without assuming a physical or material form. In an abstract way, services can assume very different positions within the range of environmental concerns, as it may contain actions that aim to increase the sustainability of the most diverse activities, to others that are much more harmful to the environment, such as hospital healthcare services, car after sales repair and maintenance, industrial and commercial building cleaning services, among many others. In spite of the environmental problems that may be added to service activities, it is a fact that this activity sector is also linked to the improvement of the environmental conditions, being considered a sector that encourages and supports a sustained change to the Circular Economy (Kjaer et al. 2018a). Given the generalized application of the LCA methodology to many situations, this could also be a solution for assessing the environmental impact of activities related to services. However, Kjaer et al. (2016) is very critical regarding the use of LCA methodology in this sector due to the following difficulties: (a) difficulty in defining the boundaries of the system to be evaluated; (b) difficulties in identifying and defining the system to be evaluated; (c) difficulties in defining the functional unit. However, later a similar team of authors (Kjaer et al. 2018b) came to develop a model composed of a set of guidelines, supported by a set of research methodologies, namely structured interviews with experts, feedback from users and analysis and cross-checking of experiences obtained through several case studies. The developed model consists essentially of a preparatory phase, where needs are identified, challenges are analyzed and a list of requirements is drawn up, followed by three further stages, each of which is divided into research, methods used to develop research and obtained results. In these cycles, the process evolves through the formulation of theories, analysis of similar guidelines, experimentation of implementation, evaluation by experts, and analysis for improvement opportunities. These improvement opportunities give way to another similar cycle, which will generate new opportunities for improvement, refining the previously studied ideas, reaching a phase of refinement, and stabilization of the methodology. The last step is the validation of the model through case studies. Only after this phase is the model stabilized, fine-tuned and can have the final guidelines elaborated. Essentially, the model begins with the analysis of the environmental consequences of the service activities carried out by a given company or entity, with comparisons being made with similar activities and respective elaboration of alternatives. An internal frame of reference is then established for which improvement opportunities are planned and later analyzed to determine the effectiveness of these improvements. This will be considered as an optimization phase, being no longer based on external factors, but based on previously implemented/achieved factors. In fact, the model was designed in such a way that the evaluation is perceptible by those who use the services, thus allowing the user to opt for environmentally friendly services in a perfectly conscious way. For this, the initially enumerated difficulties related to the application of the LCA were overcome, significantly improving the definition of what was being evaluated, and how it was being evaluated. In order to validate the developed model, it was applied to three different case studies, testing the versatility of the model in truly different service cases, with the aim to identify if the evaluation was actually performed with the desired effectiveness. The first application was made to a bicycle rental system. In this case, the service is only truly sustainable if rented bicycles are to replace other polluting means of transport, since replacing owned bicycles with rented bicycles does not produce any environmental benefits. This need for verification in the evaluation system proved the usefulness of the second step of the evaluation process, avoiding that the indicator could show environmental benefits which did not really exist. In fact, the model allows a better definition of the functional unit and also the extent of the system boundaries in question, allowing a more effective analysis of the environmental benefits generated by the service. Obviously, the models always present some gaps that prevent 100% coverage of all the cases to be evaluated, but the developed model has high potential to be successfully applied in other studies evaluating the environmental performance of services.

Hospital care is also included in the services sector, being known to use a large number of environmentally harmful and non-harmful products as well as dealing with and generating dangerous pollutants such as mercury. The evaluation of its environmental performance is an essential tool in the improvement of the provided service and is therefore, more useful for an internal evaluation than an external evaluation, with a view to continuously reducing the environmental impact of its activities. Monitoring and improving healthcare activities can lead to significant savings in waste disposal costs of several toxic products and to a more systematic analysis of the procedures used in these services. The use of appropriate metrics to assess the environmental performance of these services will help intermediate managers and top management to have a more accurate view of the quality of the services being provided. Moreover, the management can also evaluate how costs are structured, which can lead to decisions aiming to improve environmental and economic performances, and can also be turned into an overall better service for patients and their families.

Blass et al. (2017) reported that hospitals in the USA produce about 6700 tons of waste daily. On the other hand, water and energy consumption are also situations that define the environmental sustainability of services. In this respect, Karlsson and Öhman (2005) report consumption of 242,000 m^3 of water and 37 GWh of electricity per year, regarding a hospital in Sweden, serving around 150,000 patients in that period. In addition, this same hospital generates about 1330 tons of biodegradable waste, 127 tons of industrial waste, 123 tons of hazardous wastes, 164 tons of paper waste, and 14 tons of glass waste per year. If we take into account the numbers revealed for the British National Healthcare System (Campion et al. 2015), it is verified that the admission of each hospitalized patient involves the equivalent of 380 kg of CO_2, the hospitalization of patients involves the equivalent of about 80 kg of CO_2 per day, and that even outpatient treatment of patients involves the generation of the equivalent of 50 kg of CO_2 per day, it will be easy to see why it is necessary to take into account the environmental performance of this type of service delivery units. These are numbers that illustrate a reality usually unknown to the general population but which must be constantly monitored, evaluated, and improved so that these services can also contribute to an improvement in overall environmental performance. The need to soften the costs and regulation that has been produced by governments and public entities related to the environment has required a much tighter control of processes, while trying to improve the service quality provided to users. There are several tools to act in this optimization of means; firstly, it is necessary to consider options to reduce the consumption of the means used in the processes, as well as trying to increase the value of the inevitable generated waste resulting from activities carried out in a hospital. However, any reduction of means cannot jeopardize the quality perceived by the user, avoiding as well any risk in the health care provided. In this way, the actions to be planned and implemented need to be based on common objectives outlined by all stakeholders. Essentially, it is a question of increasing the quality of the service provided while increasing environmental performance, a situation that will not be difficult to achieve if the solutions found to meet the global demands. Creating indicators that can cover all the above-mentioned variables is a truly challenging task. In fact, indicators capable of monitoring the situation in environmental terms cannot lose sight of patient satisfaction or the economic viability of each hospital unit. Thus, in this particular case, the objectives to be achieved must be intrinsically linked to the values that must be preserved by the hospital.

In order to establish an environmental impact assessment methodology for services provided in healthcare units, Blass et al. (2017) developed a model, which is divided into three stages, in a structure not too dissimilar from a traditional PDCA cycle: conception, implementation, and analysis. In the conception phase, the authors consider an evaluation of the initial state of things, the definition of the main objectives in environmental terms and the transformation of the objectives into strategies to be implemented in the healthcare unit. Regarding the implementation, the indicators that will allow an adequate monitoring of the environmental performance and pursuit of the previously outlined strategy are defined. In the analysis stage, the results are collected and the verification of the initially outlined objectives is done. In the end, a final report is done. The study was applied in six hospitals in Brazil, ranging each one from 17 to 100 beds. Among the hospitals were the model was tested, none had a properly defined and formalized strategy. In general, the receptivity to the application of the environmental performance evaluation model in healthcare units was well accepted, having been classified by the people involved as "Good" or "Very Good" in criteria such as utility, ease of use, and feasibility of application to this type of units. The study carried out in the field allowed researchers to identify some of the problems also identified in many industrial companies, such as (a) lack of worker training, (b) inadequate collection of waste, (c) inefficient separation of waste, (d) improper handling of infectious waste (e) disposal of hospital waste in conjunction with household waste, (f) lack of communication between people of the same hospital, and (g) insufficient and inefficient legislation. Based on the work done, these researchers elaborated 67 objectives. In pursuit of these objectives, 192 actions were planned and 81 indicators were developed, which were classified into three areas: strategic, tactical, and operational. In order for these indicators to adequately represent the intended environmental assessment, details such as data accessibility, measurability, reliability, relevancy, clarity, opportunity, and long-term vision were taken into account. This work allowed the development of a model that emphasizes the evaluation of environmental performance as a vital means to achieve the objectives normally sought by the top management of hospital units, where it is common to identify lack of focus problems in the institutions' strategy. The indicators were simple to calculate and very useful for managers in developing a strategy for continuous improvement and monitoring of this improvement. In addition, the establishment of objectives and their compliance allows hospital units to comply with what is stipulated by law. The need to fulfill the stipulated objectives allows to reinforce team spirit, which develops in each element a greater commitment to the objectives defined for the team, also reinforcing communication within the group, thus overcoming one of the gaps previously identified. Moreover, the model has become an effective means of transforming the environmental concerns of each institution into the necessary actions to achieve the defined objectives. This allows for these actions to be perfectly defined according to each organizational level of the institution, as well as permitting a facilitated implementation and minimization of the time required for the achievement of the environmental objectives. With the necessary adjustments, this model can be adapted to other types of services, allowing the evaluation of the environmental performance to become an increasingly present reality in this sector of activity.

10.6 Environmental Performance Assessment in Wastewater Treatment Plants

In the last decades, the mandatory inclusion of water treatment plants (WTPs) for municipalities, industrial parks and even medium and large industries has been implemented in many developed and developing countries. These WTPs are essentially aimed at minimizing the effect of polluted waters, forcing them to be treated in their place of origin, converting them back into ready-to-use water without risk for the environment or human health. In general, the treatment process begins (a) on the basis of a mechanical treatment wherein the thicker, suspended or floatable solids are removed by screening and retaining or settling, (b) the dissolved organic substances resulting from the treatments previously performed are biologically decomposed under aerobic conditions, and (c) finally, a treatment is performed to confirm the quality of the water that will result from this treatment process. Although this process is contributing significantly to a better environment, it also has its environmental impact, since it consumes resources in terms of materials and energy, also promoting the generation of emissions. Thus, it is vital to determine the gains obtained through each WTP, as well as to optimize its operation, to generate water of the highest quality while inducing the least possible environmental impact. As reported in the study by Teodosiu et al. (2016), the most commonly used tools for assessing the environmental performance of WTPs LCA, lifecycle costs analysis (LCCA), water footprint (WF), and environment impact quantification (EIQ).

Mustapha et al. (2017) developed a specific model for the environmental evaluation of WTPs. To do so, the collection and data analysis corresponding to factors considered as green elements must be done. This data treatment is done via the factor analysis methods. This first approach aims essentially to determine the weight to be attributed to each green factor. Then, the sustainability index is defined based on the initially calculated weight, taking into account the stock market composite index. While in the stock market trading activity, the benefit is generated when the composite index presents a positive variation, in this case, it is desirable that the value of the index be negative, since it indicates a decrease in the environmental degradation induced by the treatment activity on wastewaters. According to the authors, through this methodology, it becomes possible to calculate a quantitative value for an index that aims to represent the environmental performance of WTPs, called green index. The model was applied to a case study developed around a design of a WTP elaborated with the aid of the SuperPro Designer 8.5 software. This software allows the calculation of the weighting factors for the sustainability factors involved. For this study, three different WTP designs were considered and compared in terms of environmental performance. The following sustainability factors were considered: CO_2 and N emissions, air consumption, amount of energy and water consumed in the process, the concentration of Biochemical Oxygen Demand, the concentration of Chemical Oxygen Demand, and the concentration of Nitrite generated. The calculation of the weighting between these factors was calculated through the factor analysis method, assigning higher weights to the sustainability factors that have greater impact on the environmental performance of the process. Thus, the methodology makes it possible to perceive which factors contributed in a more relevant way to the selection of a given process. In this way, the green index is a unique indicator that allows to have an overview of the process performance and that can be applied in the design phase to select the best process to apply in a new WTP, as it can be used to monitor WTPs already in service to enable decisions to be taken regarding their management.

10.7 Advantages and Drawbacks of Environmental Assessment Disclosure

In the industrial sector, due to the nature of its activities, some companies are much more environmentally sensitive than others, particularly those acting in the petrochemical industry, chemical industry, and ore processing industry. These companies usually deal with toxic and polluting products and are therefore under much stricter rules and subject to greater vigilance by political authorities, higher pressure from surrounding communities and even under pressure of most stakeholders. These companies are viewed by populations as undesirable, being heavily scrutinized and often stigmatized (Grougiou et al. 2016). Such companies can reverse the hostility usually demonstrated populations by showing care for the environment and demonstrating progress in terms of their sustainability, displaying the assessments made regarding their environmental performance (Braam et al. 2016; Śmiechowski and Lament 2017). Opinions, however, are divided on this point. Some researchers have found a positive relationship between the disclosure of environmental performance of companies and a favorable evolution of public opinion regarding these companies (Plumlee et al. 2015; Cormier and Magnan 2015). However, another study conducted even more recently by Qiu et al. (2016) revealed that no truly positive relationship was found between the disclosure of environmental performance and an improvement in the opinion of the population regarding environmentally sensitive companies. However, there is also a strong opinion among the population that the information provided by companies is not credible or reliable, as it is not based on independent entities that ensure the adequacy of the transmitted information. Braam et al. (2016) even refer that the information disclosed by companies dealing with environmentally harmful products tend to soften the information provided to stakeholders or to emit ambiguous information, which are not compatible with the perception that the market or surrounding populations have about the activity of these companies. However, when properly used, disclosure of the company's environmental performance can be a powerful marketing tool in societies particularly sensitive to environmental issues, bringing real benefits to companies through a broadening of the customer base and greater recognition of their products. Moreover, this effect may counteract some inhibition that consumers may have on the consumption of products from a given company that acts in an environmentally non-transparent manner, bringing clear benefits for environmentally transparent companies.

On the other hand, there is a completely different perception by shareholders, who essentially want to know the extent to which the disclosure of the environmental assessment brings benefits to the company, quantifying the costs that this entails and the benefits that the company can obtain in return. Shareholders tend to view all efforts related to the environment as a cost, paying particular attention in this regard to ratio cost-effectiveness. It is also worth noting that environmental performance reports have a much higher cost for environmentally sensitive companies than for other companies that do not routinely deal with products that are environmentally hazardous. Moreover, shareholders also tend not to directly correlate an increase in sales with the disclosure of positive information concerning environmental aspects, as they are quite skeptical in this regard. Indeed, the positive effects arising from the disclosure of a good environmental performance may not be related in the short term to a better economic performance of the company, but are strongly related to a significant appreciation of the company's image by the various stakeholders, which also contribute positively for a more sustainable future of the company. A study carried out by Radhouane et al. (2018), based on the analysis of a group of French companies over a period of 11 years, aimed to understand the value attributed by shareholders to the voluntary disclosure of the environmental performance assessment of these companies. This study was elaborated taking into account factors external to corporate social responsibility, such as the impact of this disclosure on sales growth and business yield improvement. In this study, it was also possible to observe that the disclosure of environmentally negative facts by the companies was practically marginal, which confirms that the companies essentially use the disclosure of environmental performance as a means of selective information, essentially disclosing what suits them commercially, i.e., acting as a marketing tool.

References

N.W. Arnell, J.A. Lowe, S. Brown, S.N. Gosling, P. Gottschalk, J. Hinkel, B. Lloyd-Hughes, R.J. Nicholls, T.J. Osborn, T.M. Osborne, G.A. Rose, P. Smith, R.F. Warren, A global assessment of the effects of climate policy on the impacts of climate change. Nat. Climate Change **3**, 512–519 (2013)

A.P. Blass, S.E.G. da Costa, E.P. de Lima, L.A. Borges, Measuring environmental performance in hospitals: a practical approach. J. Clean. Prod. **142**, 279–289 (2017)

G. Braam, L. de Weerd, M. Hauck, M. Huijbregts, Determinants of corporate environmental reporting: the importance of environmental performance and assurance. J. Clean. Prod. **129**, 724–734 (2016)

A. Butnariu, S. Avasilcai, Research on the possibility to apply ecological footprint as environmental performance indicator for the textile industry. Proc. Soc. Behav. Sci. **124**, 344–350 (2014)

N. Campion, C.L. Thiel, N.C. Woods, L. Swanzy, A.E. Landis, M.M. Bilec, Sustainable healthcare and environmental life-cycle impacts of disposable supplies: a focus on disposable custom packs. J. Clean. Prod. **94**, 46–55 (2015)

B. Carrasquer, J. Uche, A. Martínez-Gracia, A new indicator to estimate the efficiency of water and energy use in agro-industries. J. Clean. Prod. **143**, 462–473 (2017)

R. Chompu-Inwai, B. Jaimjit, P. Premsuriyanunt, A combination of material flow cost accounting and design of experiments techniques in an SME: the case of a wood products manufacturing company in northern Thailand. J. Clean. Prod. **108**, 1352–1364 (2015)

D. Cormier, M. Magnan, The economic relevance of environmental disclosure and its impact on corporate legitimacy: an empirical investigation. Bus. Strategy Environ. **24**(6), 431–450 (2015)

Copenhagen Research Institute, M. Herczeg, D. McKinnon, L. Milios, I. Bakas, E. Klaassens, K. Svatikova, O. Widerberg, *Resource Efficiency in the Building Sector* (2014) (ECORYS Nederland BV), http://ec.europa.eu/environment/eussd/pdf/Resource%20efficiency%20in%20the%20building%20sector.pdf. Retrieved on 06 Apr 2019

V.D. Dragomir, How do we measure corporate environmental performance? a critical review. J. Clean. Prod. **196**, 1124–1157 (2018)

European Commission, *Building Sustainability Performance* (2017). http://ec.europa.eu/environment/eussd/buildings.htm. Retrieved on 06 Apr 2019

A. Elduque, D. Elduque, C. Javierre, Á. Fernández, J. Santolaria, Environmental impact analysis of the injection molding process: analysis of the processing of high-density polyethylene parts. J. Clean. Prod. **108**, 80–89 (2015)

G. Finnveden, Å. Moberg, Environmental systems analysis tools - an overview. J. Clean. Prod. **13**, 1165–1173 (2005)

G. Finnveden, M.Z. Hauschild, T. Ekvall, J. Guinée, R. Heijungs, S. Hellweg, A. Koehler, D. Pennington, S. Suh, Recent developments in life cycle assessment. J. Environ. Manag. **91**, 1–21 (2009)

M.T. García-Álvarez, B. Moreno, Environmental performance assessment in the Eu: a Challenge for the sustainability. J. Clean. Prod. **205**, 266–280 (2018)

C. Giarma, K. Tsikaloudakia, D. Aravantinos, Daylighting and visual comfort in buildings' environmental performance assessment tools: a critical review. Proc. Environ. Sci. **38**, 522–529 (2017)

V. Grougiou, E. Dedoulis, S. Leventis, Corporate social responsibility reporting and organizational stigma: the case of "Sin" industries. J. Bus. Res. **69**(2), 905–914 (2016)

B.G. Hermann, C. Kroeze, W. Jawjit, Assessing environmental performance by combining life cycle assessment, multi-criteria analysis and environmental performance indicators. J. Clean. Prod. **15**, 1787–1796 (2007)

M. Herva, E. Roca, Review of combined approaches and multi-criteria analysis for corporate environmental evaluation. J. Clean. Prod. **39**, 355–371 (2013)

M. Herva, A. Franco, E.F. Carrasco, E. Roca, Review of corporate environmental indicators. J. Clean. Prod. **19**, 1687–1699 (2011)

N. Howard, Environmental assessment and rating – have we lost the plot? Proc. Eng. **180**, 640–650 (2017)

International Energy Agency (IEA), *Energy Technology Perspectives*, (Paris, France, 2017). ISBN 978-92-64-27050-3

ISO 14001:2015. Environmental management systems - Requirements with guidance for use (International Organization for Standardization, Geneva, Switzerland, 2015)

W. Jawjit, P. Pavasant, C. Kroeze, Evaluating environmental performance of concentrated latex production in Thailand. J. Clean. Prod. **98**, 84–91 (2015)

X.-P. Jia, F.-Y. Han, X.-S. Tan, Integrated environmental performance assessment of chemical processes. Comput. Chem. Eng. **29**, 243–247 (2004)

M. Karlsson, D.P. Öhman, Material consumption in the healthcare sector: strategies to reduce its impact on climate changed: the case of Region Scania in South Sweden. J. Clean. Prod. **13**(10), 1071–1081 (2005)

L.L. Kjaer, A. Pagoropoulos, J.H. Schmidt, T.C. McAloone, *Challenges when evaluating product/service-systems through life cycle assessment*. J. Clean. Prod. **20**, 95–104 (2016)

L.L. Kjaer, D.C.A. Pigosso, M. Niero, N.M. Bech, T.C. McAloone, Product/service-systems for a circular economy: the route to decoupling economic growth from resource consumption? J. Ind. Ecol. **23**(1), 22–35 (2018a)

L.L. Kjaer, D.C.A. Pigosso, T.C. McAloone, M. Birkved, Guidelines for evaluating the environmental performance of product/service-systems through life cycle assessment. J. Clean. Prod. **190**, 666–678 (2018b)

M.M.C. Maceno, U. Pawlowsky, K.S. Machado, R. Seleme, Environmental performance evaluation – a proposed analytical tool for an industrial process application. J. Clean. Prod. **172**, 1452–1464 (2018)

E. Maslesa, P.A. Jensen, M. Birkved, Indicators for quantifying environmental building performance: a systematic literature review. J. Build. Eng. **19**, 552–560 (2018)

M.A. Mustapha, Z.A. Manan, S.R.W. Alwi, A new quantitative overall environmental performance indicator for a wastewater treatment plant. J. Clean. Prod. **167**, 815–823 (2017)

B. Nucci, M. Puccini, L. Pelagagge, S. Vitolo, C. Nicolella, Improving the environmental performance of vegetable oil processing through LCA. J. Clean. Prod. **64**, 310–322 (2014)

M. Plumlee, D. Brown, R. Hayes, R. Marshall, Voluntary environmental disclosure quality and firm value: further evidence. J. Acc. Public Policy **34**(4), 336–361 (2015)

Y. Qiu, A. Shaukat, R. Tharyan, Environmental and social disclosures: link with corporate financial performance. Br. Acc Rev. **48**(1), 102–116 (2016)

I. Radhouane, M. Nekhili, H. Nagati, G. Paché, Customer-related performance and the relevance of environmental reporting. J. Clean. Prod. **190**, 315–329 (2018)

M. Schmidt, M. Nakajima, Material flow cost accounting as an approach to improve resource efficiency in manufacturing companies. Resources **2**, 358–369 (2013)

K. Śmiechowski, M. Lament, Impact of corporate social responsibility (CSR) reporting on pro-ecological actions of tanneries. J. Clean. Prod. **161**, 991–999 (2017)

N. Svensson, L. Roth, M. Eklund, A. Mårtensson, Environmental relevance and use of energy indicators in environmental management and research. J. Clean. Prod. **14**, 134–145 (2006)

C. Teodosiu, G. Barjoveanu, B.R. Sluser, S.A.E. Popa, O. Trofin, Environmental assessment of municipal wastewater discharges: a comparative study of evaluation methods. Int. J. Life Cycle Assess. **21**, 395–411 (2016)

C. Trumpp, J. Endrikat, C. Zopf, E. Guenther, Definition, conceptualization, and measurement of corporate environmental performance: a critical examination of a multidimensional construct. J. Bus. Ethics **126**, 185–204 (2015)

Y. Wang, K. Feng, W. Lu, An environmental assessment and optimization method for contractors. J. Clean. Prod. **142**, 1877–1891 (2017)

N. Wrisberg, H.A. Udo de Haes, U. Triebswetter, P. Eder, R. Clift, *Analytical Tools for Environmental Design and Management in a Systems Perspective* (Springer, The Netherlands, 2002). ISBN 978-1-4020-0453-7

J. Yudelson, *Reinventing Green Building* (New Society Publishers, Gabriola Island, BC, Canada, 2016). ISBN 978-0865718159

Printed by Printforce, the Netherlands